Soil: Science and Technology

Soil: Science and Technology

Edited by **Brian Bechdal**

Callisto Reference
New York

Published by Callisto Reference,
106 Park Avenue, Suite 200,
New York, NY 10016, USA
www.callistoreference.com

Soil: Science and Technology
Edited by Brian Bechdal

International Standard Book Number: 978-1-63239-712-6 (Hardback)

Printed in China.

Contents

Preface

Every book is a source of knowledge and this one is no exception. The idea that led to the conceptualization of this book was the fact that the world is advancing rapidly; which makes it crucial to document the progress in every field. I am aware that a lot of data is already available, yet, there is a lot more to learn. Hence, I accepted the responsibility of editing this book and contributing my knowledge to the community.

This book outlines the processes and applications of soil science in detail. It further elucidates the concepts studied and practiced in this field. Soil science refers to the study of soil as a natural resource. It talks about the categorization and composition of soil and about the various properties related with it. This book unravels the recent studies in the field. It also provides interesting topics for research which interested readers can take up. With state-of-the-art inputs by experts from across the globe, this book aims to provide an in-depth understanding of the subject. This book is an essential guide for both academicians and those who wish to pursue soil science further.

While editing this book, I had multiple visions for it. Then I finally narrowed down to make every chapter a sole standing text explaining a particular topic, so that they can be used independently. However, the umbrella subject sinews them into a common theme. This makes the book a unique platform of knowledge.

I would like to give the major credit of this book to the experts from every corner of the world, who took the time to share their expertise with us. Also, I owe the completion of this book to the never-ending support of my family, who supported me throughout the project.

Editor

1

Biochar Source and Application Rate Effects on Soil Water Retention Determined Using Wetting Curves

Katy E. Brantley, Kristofor R. Brye, Mary C. Savin, David E. Longer

Department of Crop, Soil, and Environmental Sciences, University of Arkansas, Fayetteville, USA
Email: kbrye@uark.edu

Abstract

Biochar is a stable carbon substance produced by the pyrolysis of biomass and used as a soil amendment. Biochar application to soil has resulted in agronomic benefits, including improved water-holding capacity. However, limited studies exist quantifying different biochars' roles in soil water retention, especially when the soil is drier than field capacity. The objective of this study was to investigate the effects of poultry litter and woodchip biochars, applied at various rates (*i.e.*, 0, 5, and 10 Mg·ha^{-1} incorporated to a depth of 10 cm) to a loam soil, on the relationship between soil water potential and water content across a wide range of moisture conditions. Based on analysis of variance, the relationship between water potential and water content differed ($P < 0.05$) between biochar sources based on differing ($P < 0.02$) power-function coefficients, with poultry litter biochar having greater water retention capacity across a wide range of water potentials, but was unaffected by application rate. Based on regression analyses, application rate affected the relationship between water potential and water content for poultry litter biochar. Results indicate that biochar may not generally improve water retention at all water contents with one-time application rates. Additional experiments will be necessary to understand the impact on water retention of biochars produced from different feedstocks under varying pyrolysis conditions.

Keywords

Biochar, Poultry Litter, Water Retention

1. Introduction

Biochar is a charcoal product produced in high-heat and low-oxygen conditions that can be used as a soil amendment with the potential for enhancing soil properties and plant growth. Biochar can have large surface area and high porosity, tending to increase with increasing pyrolysis temperature until around 850˚C [1] [2]. Brown *et al.* [1] reported pitch pine (*Pinus rigida*) biochar surface area peaking near 400 $m^2 \cdot g^{-1}$, while Lua *et al.* [2] reported a surface area over 700 $m^2 \cdot g^{-1}$ and pore volume exceeding 0.45 $cm^3 \cdot g^{-1}$ with pistachio (*Pistacia vera*)-nut biochar. Like ashes, many biochar products have alkaline pH [3] [4], which can decrease soil acidity, creating a more favorable habitat for many plants and microbes. Soils with biochar addition have been shown to increase nutrient contents and improve nutrient-holding capacity, potentially through direct nutrient addition [3] [5]-[7], as well as increase soil water-holding capacity [8] [9].

Research is ongoing to confirm these previous reports since the assumption that benefits will occur may not hold true with all biochar products (*i.e.*, feedstocks and processing characteristics), field and soil conditions, management practices, and crops grown. In a greenhouse experiment, Rajkovich *et al.* [10] investigated the effects of over 30 different biochars produced from corn (*Zea mays*) stover, nut shells, woodchips, manures, and sawdust bedding at four different pyrolysis temperatures on corn growth and reported differences between feedstock source, pyrolysis temperature, application rate on corn biomass, plant growth, nitrogen concentration, and total nitrogen uptake. Novak *et al.* [11] reported that greater biochar production temperatures could produce more alkaline pH, greater ash contents, and greater surface areas in the biochar and could result in increased soil pH.

Biochar has also been observed to increase soil-water retention [8] [12]-[14]. Laird *et al.* [8] reported greater water retention at gravity-drained equilibrium and −100 and −500 kPa water potentials in a Hapludoll from Iowa with mixed hardwood biochar. Utilizing an incubation and leaching experiment to determine the percentage of water retained in pots, Novak *et al.* [11] reported varied results depending on the biochar feedstock used; switchgrass (*Panicum virgatum*) biochar maximized soil-water retention when applied at a 40 $Mg \cdot ha^{-1}$ application rate incorporated to a depth of 15 cm to loamy sand compared to poultry litter, pecan (*Carya illinoinensis*) shell, and peanut (*Arachis hypogaea*) hull biochars. A similar experimental design was used to test soil-water retention at −5 and −60 kPa with switchgrass biochar produced at 500˚C resulting in the greatest soil-water retention [9]. Birch (*Betula* spp.) biochar added to a silt loam in Finland increased soil water-holding capacity by 11% compared to the unamended control [13]. Based on moisture release curve data from a laboratory experiment with loamy sand, only the addition of peanut hull biochar applied at 88 $Mg \cdot ha^{-1}$ increased soil water-holding capacity compared to lower biochar rates and various woodchip biochar types [12]. In a Brazilian sandy clay loam, sugarcane (*Saccharum* spp.) filtercake biochar addition led to greater water retention within the plant available water range, even though drier soils were also evaluated [15].

Many of the preexisting studies related to the effects of biochar on water-holding capacity involved the use of techniques that evaluate maximum water-holding capacity and plant-available water (*i.e.*, field capacity at ~−0.03 MPa to permanent wilting point at ~−1.5 MPa) in soil [8] [16]-[18]. The relationship between soil water potential and soil water content can be determined from data generated by drying saturated soil cores with the application of pressure using a pressure-plate apparatus [8] [16] [17] [19] [20]. Another method quantified the volume of water in samples centrifuged at speeds corresponding to matric potentials between saturation and the permanent wilting point [21]. Pot water-holding capacities were used by Novak *et al.* [9] [11] to represent gravimetric soil moisture contents in incubation leaching experiments, while Major *et al.* [22] used tensiometers in the field. Investigating a wider range of moisture conditions, Sun *et al.* [23] investigated water retention to matric potentials close to −7 MPa, and sugarcane water retention effects were examined in Brazil using a Dewpoint Potentiameter [15]. Systematically, rewetting dry soil samples and measuring corresponding water-potential responses (*i.e.*, determining the soil wetting curve) facilitates a thorough and orderly comparison among soil amendment effects of water retention characteristics.

Though most previous studies have concentrated on moist-soil conditions, it is not always possible to keep managed fields at field capacity. Dry-land agriculture in particular is prone to drought conditions and may become more common as changes in climate continue. Consequently, additional research is necessary to shed light on potential water-holding capacity improvements from biochar amendment over a wider range of soil moisture conditions. Therefore, the objective of this study was to investigate the effect of poultry litter and woodchip biochars, applied at various rates to a loam soil, on the relationship between a wide range of soil-water potentials and water contents using soil wetting curves. It was predicted that the type (*i.e.*, feedstock source) of biochar added to soil at varying rates would differentially affect water-retention characteristics determined using soil wetting curves. Specifically, it was hypothesized that poultry litter biochar applied at the greatest rate would alter water-retention characteristics more than at lower rates or with the woodchip biochar due to the intrinsic differences in the biochar products.

2. Materials and Methods

2.1. Soil Sample Collection and Initial Characterization

Soil was collected from approximately the top 10 cm of a floodplain at the University of Arkansas Agricultural Research and Extension Center in Fayetteville, Arkansas (36.09390°N, 94.16494°W) in an area that was classified as a Razort loam (fine-loamy, mixed, active, mesic Mollic Hapludalf) [24], which was where soil was collected for asimilar greenhouse experiment. The soil was homogenized, air-dried for 72 hours, ground, and sieved to pass a 2-mm mesh screen. Four sub-samples were dried for 24 hours at 105°C to quantify the initial moisture content of the air-dried soil. Soil particle-size analysis was conducted on oven-dry sub-samples of the soil sieved through a 2-mm mesh screen using an adaptation of the 12-hr hydrometer method [25]. Mehlich-3 extractable nutrients (*i.e.*, phosphorus (P), potassium (K), calcium (Ca), magnesium (Mg), sulfur (S), iron (Fe), manganese (Mn) and copper (Cu)) were determined by inductively coupled plasma (ICP) spectrometry (SPECTRO ARCOS, SPECTRO Analytical Instruments GmbH, Kleve, Germany) [26]. Electrical conductivity (EC) and pH were determined potentiometrically on a 1:2 soil mass-to-water volume paste, and organic matter concentration was determined by loss-on-ignition in a muffle furnace.

2.2. Biochar Characteristics

Granular pine (*Pinus* spp.) woodchip biochar (Waste To Energy Solutions Inc., Destin, FL), produced through pyrolysis at 500°C, was obtained for use in this study. Pelletized poultry litter biochar (Whitfield Biochar, Burlington, WA), produced through pyrolysis at 500°C to 520°C, was also obtained for use in this study. These two biochar types were used as soil amendments to represent a cultivated-soil scenario.

Total recoverable minerals (*i.e.*, P, K, Ca, Mg, S, sodium (Na), Fe, Mn, zinc (Zn), Cu and boron (B)) were determined from acid digests [27] by ICP spectrometry. Both biochars were dried for at least 48 hours at 70°C then ground to pass a 40-mesh screen. Biochar pH and EC were determined potentiometrically on a 1:2 sample mass-to-water volume paste. Sub-samples of each biochar were subsequently dried for 24 hours at 105°C to obtain moisture contents of both biochar materials. The water droplet penetration time (WDPT) test was measured using a modification of the procedure described by Herath *et al.* [28]. Three drops of water from a Pasteur pipet were added to 2 g of each biochar and the time of penetration was recorded [29] [30].

2.3. Wetting Curve Determinations

Following the procedures of Brye [31], the relationship between water potential and water content was determined for each of five replicate treatment combinations by re-wetting air-dry soil from randomly selected biochar source-rate treatment combinations. Brye [31] used this technique to determine the effect of years of cultivation and land use on soil water-retention characteristics. Eight sub-samples of air-dried sieved soil from each treatment combination were prepared by mixing 5 ± 0.01 g of soil with 0, 28 mg, or 57 mg of biochar (*i.e.*, equivalent to 0, 5, and 10 $Mg{\cdot}ha^{-1}$ rates incorporated to a depth of 10 cm at the field scale). Soil and biochar samples were placed into small cups to which 1, 2, 4, 6, 10, 12, 15, or 20 drops of distilled water were added to each sample cup, respectively, using a Pasteur pipet and mixed thoroughly. The moist soil and biochar mixtures were transferred to small plastic measurement vessels, 3.8 cm in diameter and 1 cm tall, and lightly tamped to fit into the bottom 0.5 cm of the cup, resulting in a dry bulk density of each sample < 1 $g{\cdot}cm^{-3}$. Sample vessels were covered and allowed to equilibrate overnight at room temperature. The following day, a WP4 Dewpoint Potentiameter (Decagon Devices, Inc., Pullman, WA) was used to measure the resulting water potential for each sample. The potentiameter was calibrated using 0.5 molal KCl standard solution. After the water potential was measured for each sample, the gravimetric water content was determined by weighing, oven-drying at 105°C for at least 12 hours, and reweighing.

2.4. Data Analyses

Initial biochar properties were compared between biochar sources by analysis of variance (ANOVA) using SAS (version 9.2, SAS Institute, Inc., Cary, NC). When appropriate, means were separated by least significant difference at $P = 0.05$.

The relationship between soil water potential (ψ) and gravimetric water content (θg) was determined using nonlinear regression as described in Brye [31] using the power function:

$$\Psi = a(\theta g)^{-b} \quad (1)$$

The power function was fit to the soil wetting curve data for each replicate treatment combination, with ψ representing soil water potential (in MPa), θg representing the gravimetric water content of a given sample (in %), and a and b were empirically derived coefficients. Based on a completely random design, a two-factor ANOVA was conducted using SAS to determine the effects of biochar source, rate, and their interaction on modeled water-retention characteristics (*i.e.*, *a* and *b* coefficients). When appropriate, means were separated by least significant difference at $P = 0.05$. A multiple regression analysis was also conducted on the modeled water-retention characteristics using Minitab (version 13.31, Minitab, Inc., State College, PA) to further investigate the effect of biochar rate.

3. Results and Discussion

3.1. Initial Soil and Biochar Properties

Initial soil and biochar properties were similar to expectations. Particle-size analyses confirmed the texture of the soil to be loam and the soil pH was determined to be 6.7 (**Table 1**). The poultry litter and woodchip biochars possessed alkaline pH of 8.0 and 8.9, respectively (**Table 2**). The majority of the nutrient concentrations and EC differed ($P < 0.05$) between the two biochars. Except for iron, the poultry litter biochar possessed greater ($P < 0.05$) concentrations of all other total recoverable elements measured (*i.e.*, P, K, Ca, Mg, S, Na, Mn, Zn, Cu; **Table 2**). The water repellency differed between the two biochars ($P < 0.05$), with greater water repellency in the poultry litter (>60 seconds for water to penetrate into biochar) compared to the woodchip biochar (2.5 seconds).

Biochars can have large surface areas (200 to >500 $m^2 \cdot g^{-1}$) and are highly porous when produced at intermediate temperatures around 450°C to 750°C [1] [2] [32], a range which includes the temperatures that produced the biochars used in this study. These physical biochar attributes have affected water relationships in soil by increasing soil porosity and changing the pore-size distribution [17] [33]. In this study, the short duration of the experiment and the pre-treatment of the biochars, including air-drying and sieving, eliminated some possible mechanisms of biochar effect on soil water-retention (e.g., improved soil structure, aggregate stability, and the increase of macropores). Grinding the biochar created a more uniform comparison between biochars so that their intrinsic property differences rather than their structural differences (*i.e.*, pellet vs. granular) were likely responsible for subsequent measured differences in the relationship between water potential and water content.

3.2. Soil Wetting Curve Differences

The power function (Equation (1)) fit the soil wetting curve data well, with R^2 values of 0.94 and 0.95 (**Figure 1**). Water contents measured in this study included those in the plant available water range as well as those drier than permanent wilting point (*i.e.*, water potentials ranging from ~−0.2 to ~−80 MPa; **Figure 1**). The *a* coefficient of the wetting-curve model, an experimentally derived coefficient, differed ($P = 0.018$) between biochar sources (**Table 3**). Averaged across biochar rate, the *a* coefficient from the wetting-curve model averaged 277.1 for the poultry litter and 392.8 for the woodchip biochars (**Figure 1**). However, despite the differences in the rate of biochar addition, biochar rate did not affect ($P > 0.05$) the *a* coefficient of the soil wetting-curve model (**Table 3**). These results suggest that soil response was similar across biochar rates, but that differences existed based on the biochar source, potentially different pore structures, requiring significant adjustments in the model's *a* coefficient [31].

The *b* coefficient of Equation (1) characterizes the rate of water potential decrease as the water content decreased [31]. Similar to the *a* coefficient, the *b* coefficient also differed between biochar sources ($P = 0.001$), but was unaffected ($P > 0.05$) by biochar rate (**Table 3**). Combined across biochar rates, the *b* coefficient from the wetting-curve model averaged −2.36 for the poultry litter and −2.62 for the woodchip biochars (**Figure 1**). The more negative *b* coefficient for the woodchip biochar treatments indicated that the water content decrease was greater for the woodchip than for the poultry litter biochar for the same decrease in water potential. Conse-

Table 1. Initial mean (±standard error (SE)) particle-size distribution, pH, electrical conductivity (EC), organic matter, and Mehlich-3 extractable nutrients for the Razort soil (n = 4).

Soil property	Mean (±SE)
Particle-size distribution ($g \cdot g^{-1}$)	
Sand	0.40 (0.01)
Silt	0.47 (0.01)
Clay	0.13 (< 0.01)
pH	6.7 (0.1)
EC ($dS \cdot m^{-1}$)	0.3 (0.2)
Organic matter (%)	3.5 (0.1)
Phosphorus ($\mu g \cdot g^{-1}$)	5.2 (0.5)
Potassium ($\mu g \cdot g^{-1}$)	84.9 (15)
Calcium ($\mu g \cdot g^{-1}$)	1191 (26)
Magnesium ($\mu g \cdot g^{-1}$)	47.4 (0.2)
Sulfur ($\mu g \cdot g^{-1}$)	14.0 (5.9)
Iron ($\mu g \cdot g^{-1}$)	48.9 (1.8)
Manganese ($\mu g \cdot g^{-1}$)	93.1 (3.7)
Copper ($\mu g \cdot g^{-1}$)	1.1 (0.02)

Table 2. Initial mean (±standard error (SE)) pH, electrical conductivity (EC), total carbon (C), total nitrogen (N), C:N ratio, and total recoverable minerals for poultry litter and woodchip biochar sources (n = 2).

Biochar property	Biochar source	
	Poultry litter	Woodchip
pH	8.0 (0.04)* *a*	8.9 (0.03) *b*
EC ($dS \cdot m^{-1}$)	42.0 (0.01) *a*	4.6 (0.02) *b*
Total Carbon ($g \cdot kg^{-1}$)	337.2 (2.5) *a*	244.5 (21) *b*
Total Nitrogen ($g \cdot kg^{-1}$)	34.9 (0.06) *a*	0.7(0.2) *b*
C:N ratio	9.7:1 (0.05) *a*	366:1 (64) *b*
Phosphorus ($g \cdot kg^{-1}$)	25.7 (0.2) *a*	0.8 (0.02) *b*
Potassium ($g \cdot kg^{-1}$)	52.4 (0.2) *a*	2.1 (0.1) *b*
Calcium ($g \cdot kg^{-1}$)	45.4 (5.8) *a*	10.1 (0.5) *b*
Magnesium ($g \cdot kg^{-1}$)	12.6 (0.9) *a*	2.7 (0.2) *b*
Sulfur ($g \cdot kg^{-1}$)	13.6 (0.3) *a*	0.1 (<0.01) *b*
Sodium ($g \cdot kg^{-1}$)	15.3 (0.6) *a*	0.3 (0.01) *b*
Iron ($g \cdot kg^{-1}$)	1.4 (0.1) *a*	0.9 (0.1) *a*
Manganese ($mg \cdot kg^{-1}$)	715.0 (18) *a*	420.5 (30) *b*
Zinc ($mg \cdot kg^{-1}$)	829.5(26) *a*	0.01** (0) *b*
Copper ($mg \cdot kg^{-1}$)	583.0 (40) *a*	6.5 (0.04) *b*
Boron ($mg \cdot kg^{-1}$)	78.0 (4.0) *a*	10.4 (0.7) *b*

*Means in the same row followed by different letters are different ($P < 0.05$). **Zinc in the woodchip biochar was below the detection limit of the method. Therefore, the detection limit of 0.01 was used for statistical analysis.

Table 3. Analysis of variance summary of the effects of biochar (BC) source, rate, and their interaction on water-retention characteristics from soil wetting curve data fit to the model $\Psi = a(\theta_g)^{-b}$.

Source of variation	Degrees of freedom	*a* coefficient	*b* coefficient
		P	
Biochar source	20	0.018	0.001
Biochar rate	20	0.293	0.085
Biochar source × biochar rate	20	0.365	0.291

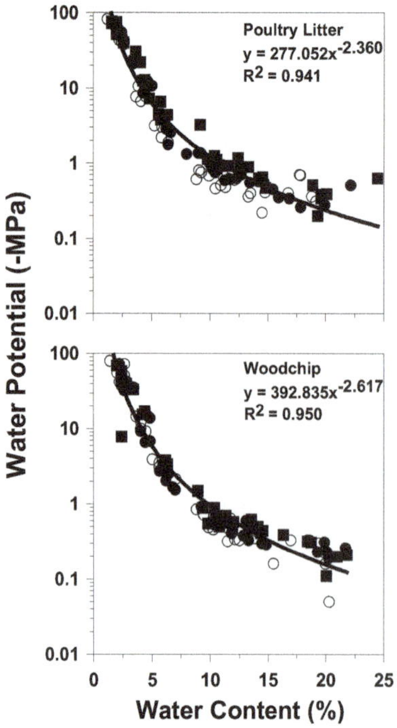

Figure 1. Relationship between water potential and gravimetric water content for poultry litter and woodchip biochars averaged over rate. The resulting fitted power function and associated coefficient of determination (R^2) are reported for each biochar source. The open circles are data points with 0 Mg·ha^{-1}, the filled circles with 5 Mg·ha^{-1}, and the filled squares with 10 Mg·ha^{-1} biochar application.

quently, more water was retained at a given water potential with the poultry litter biochar than with the woodchip biochar.

Though ANOVA did not demonstrate a biochar rate effect, regression analyses were conducted to investigate a potential relationship between biochar rate and water retention assuming that biochar rate represented a continuous variable. The *a* coefficient for the poultry litter biochar varied significantly ($P = 0.037$) as biochar rate increased, while the *a* coefficient for the woodchip biochar and neither *b* coefficients were affected by biochar rate (**Table 4**). The *a* coefficient for the poultry litter biochar increased from the 0 to the 5 Mg·ha^{-1} rate, then decreased from the 5 to the 10 Mg·ha^{-1} rate (**Figure 2** and **Table 4**), suggesting a change in biochar effect at the 5 Mg·ha^{-1} application rate compared to the control that was no longer observed at the greatest rate.

Considering results for both the *a* and *b* coefficients, the poultry-litter-biochar-amended soil possessed different overall water-retention characteristics than the woodchip-biochar-amended soil. Since the soil was uniformly treated by air-drying, grinding, and sieving and the biochars were oven-dried and ground, these results suggest that there were factors intrinsic to the biochars that potentially led to water-retention differences between biochar sources. The poultry litter biochar, with a significantly smaller C:N ratio, significantly greater nutrient contents, and almost ten-fold numerically greater EC (**Table 2**), possessed a strikingly different composition that could have affected water-retention characteristics differently than those of the woodchip biochar. Water could have reacted with the inorganic nutrients and compounds in the poultry litter biochar, creating a physical adhesion effect to retain the water in the soil [9].

The greater C:N ratio of the woodchip biochar, which suggests a greater lignin and cellulose concentration than in the poultry litter biochar due to its origin as plant material, could lead to greater hydrophobicity and water repellency. However, the poultry litter biochar possessed greater water repellency than the woodchip biochar. Fresh biochar tends to be water repellent [34], and Briggs *et al.* [35] speculated that the water repellency of their pine biochar samples had to do with the aliphatic compounds that make up biochar as well as the mineral-rich composition clogging outer pores. Tars in biochars can also repel water, and they remain in charcoal produced at temperatures below 600˚C [32] [36]-[38], such as with the biochars used in this study. The original poultry litter biochar was a hydrophobic pellet, while the woodchip was a coarse, less compacted granular-like form, which could have resulted in differences in the biochars based on the processing and production conditions. Based on

Table 4. Summary of multiple regression results for *a* and *b* coefficients from poultry litter and woodchip biochar and the effect of biochar rate and rate².

Variable	Biochar source	Regression coefficients: Rate	Rate²	Intercept	Overall model	R^2
					P	
a coefficient	Poultry litter	78.2*	−7.16*	244*	0.037	0.42
	Woodchip	4.05	−0.34	479*	0.998	0.00
b coefficient	Poultry litter	−0.075	0.0086	−2.40*	0.100	0.32
	Woodchip	0.0057	0.00052	−2.70*	0.790	0.04

*Asterisks indicate values significantly different from 0 ($P < 0.05$).

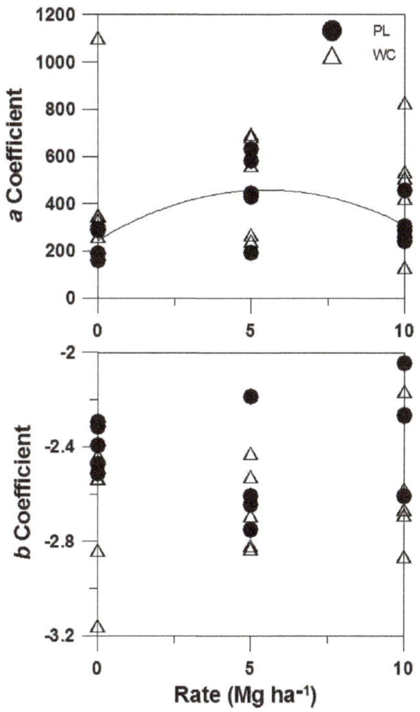

Figure 2. Effect of biochar rate on *a* and *b* coefficients from fitting the power function to soil wetting-curve data for poultry litter (PL) and woodchip (WC) biochars. The plotted line represents a significant regression model, where model parameters are summarized in **Table 4**.

visual observation, following grinding to facilitate comparison among intrinsic properties between biochars, the poultry litter biochar was composed of finer particles than the woodchip biochar. Grinding of the biochars for use in this study may have produced finer particles in the poultry litter biochar that filled soil pores more than the coarser woodchip biochar material to help retain water once soil applied.

Different biochar feedstocks, production temperatures, and environmental conditions (*i.e.*, a greenhouse compared to a field setting) could also produce drastically different results than those observed in this study in terms of their effects on soil water-retention characteristics [1]. Thirty months after field application of 47 $Mg{\cdot}ha^{-1}$ *Acacia* spp. Green waste biochar and incorporated to a depth of 10 cm in a silt loam, Hardie *et al.* [39] reported no significant effects on soil-water retention or soil moisture content near field capacity, but reported numerically greater water contents near saturation. Soil water-holding capacity was unaffected by 20 $Mg{\cdot}ha^{-1}$ of mixed wood species biochar incorporated to a depth of 5 cm in an Oxisol [22] or by 11 or 22 $Mg{\cdot}ha^{-1}$ of peanut (*Arachis hypogaea*) hull, hardwood, or various pine biochars added to loamy sand [12]. In a laboratory experiment, plant-available water was shown to increase in sandy soil, remain unaffected in loamy soil, and decrease in clayey soil with the additions of hardwood and pine biochars from 0% to 15%, 30%, and 45% (w/w) application rates [40]. After investigating 60 combinations of soils and biochars, Strebel *et al.* [41] reported 25 combinations resulting in significantly greater soil water-holding capacity. Furthermore, Abel *et al.* [42] reported biochar addition to a sandy soil decrease bulk density and increased available water-holding capacity and the water content at permanent wilting point, but biochar did not impact soil wettability six months after incorporation.

Though only a one-time biochar application was used in this study, there is potential that differing results could be observed if greater rates were used or if multiple biochar additions were made over time in a longer study. However, greater rates could prove uneconomical for field application [20]. The Na content in poultry litter biochar was roughly 50 times greater than that in the woodchip biochar; thus, repeated biochar additions, especially with the poultry litter biochar, could negatively impact soil water-holding capacity by increasing soil dispersion and off-setting the potential for improved soil structure. Future studies should also investigate the effects of other biochar products produced from different feedstocks and under various production and environmental conditions on soil water-retention characteristics.

4. Conclusion

Water-retention characteristics of a loam soil differed as a result of a single addition of poultry litter and woodchip biochar. Poultry litter biochar had different chemical characteristics and affected water-retention characteristics differently compared with woodchip biochar, although both biochars were produced from pyrolysis at approximately 500˚C. More water was retained at given water potential with the poultry litter biochar than with the woodchip biochar, but rate effects were minimal. The results of this study demonstrate that all biochars may not improve soil water retention over a wide range of soil moisture conditions at all application rates. Additional research with various biochar products applied at a range of application rates and with multiple applications would enhance knowledge of the effects of biochar amendment on soil water-retention characteristics.

Acknowledgements

Laboratory assistance provided by Melissa Fulenwider is gratefully acknowledged.

References

[1] Brown, R.A., Kercher, A.K., Nguyen, T.H., Nagle, D.C. and Ball, W.P. (2006) Production and Characterization of Synthetic Wood Chars for Use as Surrogates for Natural Sorbents. *Organic Geochemistry*, **37**, 321-333. http://dx.doi.org/10.1016/j.orggeochem.2005.10.008

[2] Lua, A.C, Yang, T. and Guo, J. (2004) Effects of Pyrolysis Conditions on the Properties of Activated Carbons Prepared from Pistachio-Nut Shells. *Journal of Analytical and Applied Pyrolysis*, **72**, 279-287. http://dx.doi.org/10.1016/j.jaap.2004.08.001

[3] Gaskin, J.W., Steiner, C., Harris, K., Das, K.C. and Bibens, B. (2008) Effect of Low-Temperature Pyrolysis Conditions on Biochar for Agricultural Use. *Transactions of the American Society of Agricultural and Biological Engineers*, **51**, 2061-2069.

[4] Spokas, K.A., Cantrell, K.B., Novak, J.M., Archer, D.W., Ippolito, J.A., Collins, H.P., Boateng, A.A., Lima, I.M., Lamb, M.C., McAloon, A.J., Lentz, R.D. and Nichols, K.A. (2012) Biochar: A Synthesis of Its Agronomic Impact beyond Carbon Sequestration. *Journal of Environmental Quality*, **41**, 973-989. http://dx.doi.org/10.2134/jeq2011.0069

[5] Glaser, B., Haumaier, L., Guggenberger, G. and Zech, W. (2001) The "Terra Preta" Phenomenon: A Model for Sustainable Agriculture in the Humid Tropics. *Naturwissenschaften*, **88**, 37-41. http://dx.doi.org/10.1007/s001140000193

[6] Lehmann, J., da Silva Jr., J.P., Rondon, M., da Silva, C.M., Greenwood, J., Nehls, T., Steiner, C. and Glaser, B. (2002) Slash-and-char: A Feasible Alternative for Soil Fertility Management in the Central Amazon? *17th World Congress of Soil Science*, Bangkok, 14-21 August 2002, 1-12.

[7] Lehmann, J., da Silva Jr., J.P., Steiner, C., Nehls, T., Zech, W. and Glaser, B. (2003) Nutrient Availability and Leaching in an Archaeological Anthrosol and a Ferralsol of the Central Amazon Basin: Fertilizer, Manure and Charcoal Amendments. *Plant and Soil*, **249**, 343-357. http://dx.doi.org/10.1023/A:1022833116184

[8] Laird, D.A., Fleming, P., Davis, D.D., Horton, R., Wang, B. and Karlen, D.L. (2010) Impact of Biochar Amendments on the Quality of a Typical Midwestern Agricultural Soil. *Geoderma*, **158**, 443-449. http://dx.doi.org/10.1016/j.geoderma.2010.05.013

[9] Novak, J.M., Busscher, W.J., Watts, D.W., Amonette, J.E., Ippolito, J.A., Lima, I.M., Gaskin, J., Das, K.C., Steiner, C., Ahmedna, M., Rehrah, D. and Schomberg, H. (2012) Biochars Impact on Soil-Moisture Storage in an Ultisol and Two Aridisols. *Soil Science*, **177**, 310-320. http://dx.doi.org/10.1097/SS.0b013e31824e5593

[10] Rajkovich, S., Enders, A., Hanley, K., Hyland, C., Zimmerman, A.R. and Lehmann, J. (2012) Corn Growth and Nitro-

gen Nutrition after Additions of Biochars with Varying Properties to a Temperate Soil. *Biology and Fertility of Soils*, **48**, 271-284. http://dx.doi.org/10.1007/s00374-011-0624-7

[11] Novak, J.M., Lima, I., Xing, B., Gaskin, J.W., Steiner, C., Das, K.C., Ahmedna, M., Rehrah, D., Watts, D.W., Busscher, W.J. and Schomberg, H. (2009) Characterization of Designer Biochar Produced at Different Temperatures and Their Effects on a Loamy Sand. *Annals of Environmental Science*, **3**, 195-206.

[12] Gaskin, J.W., Speir, A., Morris, L.M., Ogden, K., Harris, K., Lee, D. and Das, K.C. (2007) Potential for Pyrolysis Char to Affect Soil Moisture and Nutrient Retention Status of a Loamy Sand Soil. *Proceedings of the* 2007 *Georgia Water Resources Conference*, Athens, GA, 27-29 March 2007, 1-3.

[13] Karhu, K., Mattila, T., Bergström, I. and Regina, K. (2011) Biochar Addition to Agricultural Soil Increased CH_4 Uptake and Water Holding Capacity—Results from a Short-Term Pilot Field Study. *Agriculture, Ecosystems, and Environment*, **140**, 309-313. http://dx.doi.org/10.1016/j.agee.2010.12.005

[14] Major, J., Steiner, C., Downie, A. and Lehmann, J. (2009) Biochar Effects on Nutrient Leaching. In: Lehmann, J. and Joseph, S., Eds., *Biochar for Environmental Management: Science and Technology*, Earthscan Publishers Ltd., London, 271-287.

[15] Eykelbosh, A.J., Johnson, M.S., de Queiroz, E.S., Dalmagro, H.J. and Couto, E.G. (2014) Biochar from Sugarcane Filtercake Reduces Soil CO_2 Emissions Relative to Raw Residue and Improves Water Retention and Nutrient Availability in a Highly-weathered Tropical Soil. *PLoS ONE*, **9**, 1-9. http://dx.doi.org/10.1371/journal.pone.0098523

[16] Fellet, G., Marchiol, L., Delle Vedove, G. and Peressotti, A. (2011) Application of Biochar on Mine Tailings: Effects and Perspectives for Land Reclamation. *Chemosphere*, **83**, 1262-1267. http://dx.doi.org/10.1016/j.chemosphere.2011.03.053

[17] Jones, B.E.H., Haynes, R.J. and Phillips, I.R. (2010) Effect of Amendment of Bauxite Processing Sand with Organic Materials on its Chemical, Physical and Microbial Properties. *Journal of Environmental Management*, **91**, 2281-2288. http://dx.doi.org/10.1016/j.jenvman.2010.06.013

[18] Mukherjee, A. and Lal, R. (2013) Biochar Impacts on Soil Physical Properties and Greenhouse Gas Emissions. *Agronomy Journal*, **3**, 313-339. http://dx.doi.org/10.3390/agronomy3020313

[19] Basso, A.S. (2012) Effect of Fast Pyrolysis Biochar on Physical and Chemical Properties of a Sandy Soil. Master's Thesis, Iowa State University, Ames.

[20] Piccolo, A., Pietramellara, G. and Mbagwu, J.S.C. (1997) Use of Humic Substances as Soil Conditioners to Increase Aggregate Stability. *Geoderma*, **75**, 267-277.

[21] de Melo Carvalho, M.T., de Holanda Nunes Maia, A., Madari, B.E., Bastiaans, L., van Oort, P.A.J., Heinemann, A.B., Soler da Silva, M.A., Petter, F.A., Marimon Jr., B.H. and Meinke, H. (2014) Biochar Increases Plant-Available Water in a Sandy Loam Soil under an Aerobic Rice Crop System. *Solid Earth*, **5**, 939-952. http://dx.doi.org/10.5194/se-5-939-2014

[22] Major, J., Rondon, M., Molina, D., Riha, S.J. and Lehmann, J. (2012) Nutrient Leaching in a Colombian Savanna Oxisol Amended with Biochar. *Journal of Environmental Quality*, **41**, 1076-1086. http://dx.doi.org/10.2134/jeq2011.0128

[23] Sun, Z., Moldrup, P., Elsgaard, L., Arthur, E., Bruun, E.W., Hauggaard-Nielsen, H. and de Jonge, L.W. (2013) Direct and Indirect Short-Term Effects of Biochar on Physical Characteristics of an Arable Sandy Loam. *Soil Science*, **178**, 465-473. http://dx.doi.org/10.1097/SS.0000000000000010

[24] United States Department of Agriculture, Natural Resources Conservation Service. Web Soil Survey. http://websoilsurvey.sc.egov.usda.gov/App/WebSoilSurvey.aspx

[25] Gee, G.W. and Bauder, J.W. (1986) Particle-Size Analysis. In: Klute, A., Ed., *Methods of Soil Analysis*, *Part* 1, *Agronomy Monographs*, ASA and SSSA, Madison, 383-411.

[26] Tucker, M.R. (1992) Determination of Phosphorus by Mehlich 3 Extractant. In: Donohue, S.J., Ed., *Reference Soil and Media Diagnostic Procedure for the Southern Region of the United States*, *Southern Cooperative Series Bulletin* 374, Virginia Agricultural Experiment Station, Blacksburg, 9-12.

[27] United States Environmental Protection Agency (1996) Method 3050B: Acid Digestion of Sediments, Sludges, and Soils. http://www.epa.gov/wastes/hazard/testmethods/sw846/pdfs/3050b.pdf

[28] Herath, H., Camps-Arbestain, M. and Hedley, M. (2013) Effect of Biochar on Soil Physical Properties in Two Contrasting Soils: An Alfisol and an Andisol. *Geoderma*, **209-210**, 188-197. http://dx.doi.org/10.1016/j.geoderma.2013.06.016

[29] Bisdom, E.B.A., Dekker, L.W. and Schoute, J.F.T. (1993) Water Repellency of Sieve Fractions from Sandy Soils and Relationships with Organic Material and Soil Structure. *Geoderma*, **56**, 105-118. http://dx.doi.org/10.1016/0016-7061(93)90103-R

[30] Dekker, L.W. and Jungerius, P.D. (1990) Water Repellency in the Dunes with Special Reference to the Netherlands.

Catena Supplement, **18**, 173-183.

[31] Brye, K.R. (2003) Long-Term Effects of Cultivation on Particle Size and Water-Retention Characteristics Determined Using Wetting Curves. *Soil Science*, **168**, 459-468. http://dx.doi.org/10.1097/01.ss.0000080331.10341.36

[32] Downie, A., Crosky, A. and Munroe, P. (2009) Physical Properties of Biochar. In: Lehmann, J. and Joseph, S., Eds., *Biochar for Environmental Management: Science and Technology*, Earthscan, London, 13-32.

[33] Ouyang, L., Wang, F., Tang, J., Yu, L. and Zhang, R. (2013) Effects of Biochar Amendment on Soil Aggregates and Hydraulic Properties. *Journal of Soil Science and Plant Nutrition*, **13**, 991-1002.

[34] Lehmann, J., Czimezik, C., Laird, D. and Sohi, S. (2009) Stability of Biochar in the Soil. In: Lehmann, J. and Joseph, S., Eds., *Biochar for Environmental Management: Science and Technology*, Earthscan, London, 184-205.

[35] Briggs, C., Breiner, J.M. and Graham, R.C. (2012) Physical and Chemical Properties of *Pinus ponderosa* Charcoal: Implications for Soil Modification. *Soil Science*, **177**, 263-268. http://dx.doi.org/10.1097/SS.0b013e3182482784

[36] Amonette, J.E. and Joseph, S. (2009) Characteristics of Biochar: Microchemical Properties. In: Lehmann, J. and Joseph, S., Eds., *Biochar for Environmental Management: Science and Technology*, Earthscan, London, 33-52.

[37] Antal, M.J. and Grønli, M. (2003) The Art, Science, and Technology of Charcoal Production. *Industrial and Engineering Chemistry Research*, **42**, 1619-1640. http://dx.doi.org/10.1021/ie0207919

[38] Kameyama, K., Miyamoto, T. and Shinogi, Y. (2010) Increase in Available Water Content of Soils by Applying Bagasse-Charcoals. *Proceedings of the* 19*th World Congress of Soil Science, Soil Solutions for Changing World*, Brisbane, 1-6 August 2010, P-0155.

[39] Hardie, M., Clothier, B., Bound, S., Oliver, G. and Close, D. (2014) Does Biochar Influence Soil Physical Properties and Soil Water Availability? *Plant and Soil*, **376**, 347-361. http://dx.doi.org/10.1007/s11104-013-1980-x

[40] Tryon, E.H. (1948) Effect of Charcoal on Certain Physical, Chemical, and Biological Properties of Forest Soils. *Ecological Monographs*, **18**, 81-115. http://dx.doi.org/10.2307/1948629

[41] Streubel, J.D., Collins, H.P., Garcia-Perez, M., Tarara, J., Granatstein, D. and Kruger, C.E. (2011) Influence of Contrasting Biochar Types on Five Soils at Increasing Rates of Application. *Soil Science Society of America Journal*, **75**, 1402-1413. http://dx.doi.org/10.2136/sssaj2010.0325

[42] Abel, S., Peters, A., Trinks, S., Schonsky, H., Facklam, M. and Wessolek, G. (2013) Impact of Biochar and Hydrochar Addition on Water Retention and Water Repellency of Sandy Soil. *Geoderma*, **202-203**, 183-191.

2

Physical and Biochemical Properties of Apple Orchard Soils of Different Productivities

Takamitsu Kai, Masaki Mukai, Kiwako S. Araki, Dinesh Adhikari, Motoki Kubo*

Department of Biotechnology, Faculty of Life Sciences, Ritsumeikan University, Biwako-Kusatsu Campus, Kusatsu, Shiga 525-8577, Japan

Email: *kubo@sk.ritsumei.ac.jp

Abstract

Physical and biochemical properties of apple orchard soils of different productivities (orchard A: 30 t·ha^{-1}; orchard B: 20 t·ha^{-1}) were analyzed. Most of the physical properties were similar in both orchards. In orchard A, the horizontal saturated hydraulic conductivity at 10-cm depth was 16.42 cm·day^{-1}, and was about four times higher than that in orchard B (4.41 cm·day^{-1}). Total carbon, total nitrogen, and total phosphorus were about two times higher in orchard B soil, whereas total potassium, bacterial biomass, nitrification, and phosphorus mineralization activities were similar in both orchards. Excess nutrients accumulated on the top 15 cm layer of orchard B soil because the topographical and physical conditions were associated with reduced apple productivity. Appropriate management of fallen leaves and reduction of chemical fertilizer seem necessary for a high level of apple productivity.

Keywords

Apple Orchard Soil, Physical Properties, Biochemical Properties, Fruit Yield

1. Introduction

Apple (*Malus domestica* Borkh.) is cultivated in most of the world's temperate regions, including Japan, where it is one of the country's oldest crops [1] [2]. Apple cultivation became widespread following the importation of 75 cultivars from America in 1871 [3]. It is cultivated mainly in the northern part of Japan, viz. Aomori, Nagano, Iwate, Yamagata, Akita, and Fukushima prefectures. Aomori and Nagano prefectures are the two major apple-

*Corresponding author.

producing areas, contributing about 77% of the national production [4]. Average productivity of apple in Japan during 2012 was 21.2 t·ha^{-1}, which was lower than that in other developed countries, such as 77.9 t·ha^{-1} in Austria, 47.4 t·ha^{-1} in New Zealand, and 31.9 t·ha^{-1} in the USA [5].

Apple requires deep and well-drained soil with a pH of 5.5 to 6.5 [6]. Soils rich in organic matter and high in available nitrogen, phosphorus, and potassium are suitable for its cultivation. Apple is cultivated conventionally in Japan, relying on use of chemical fertilizers and pesticides [4]. Recommendations for chemical fertilizers are generally based on the soil nutrient status and rate of plant uptake [7]. Some fertilizer recommendations for apple are based on leaf nutrient analyses, because orchard soils receive additional nutrients from fallen leaves [8]-[10]. However, most producers in Japan apply fertilizers based on national recommendations as follows: 144 kg N, 72 kg P_2O_5, 48 kg K_2O, 4.8 kg MnO, and 2.4 kg·B·ha^{-1}·year^{-1} [4].

Low apple productivity has been reported from several orchards in Japan. Although poor soil fertility is one of the major problems causing lower apple production in several prefectures [11], there is limited information on the interrelationship between soil properties and apple productivity [12] [13]. Assessment of soil properties could help to improve soil management and enhance productivity on marginal sites.

In this study, physical and biochemical properties of soils from two orchards with different productivities were analyzed with the goal being to improve apple production on marginal sites. In addition, nutrient status of leaves and fruit was analyzed in relation to soil properties and orchard productivity.

2. Materials and Methods

2.1. Study Area

This study was conducted in two adjacent apple orchards located in Iizuna Town in Nagano Prefecture (36°26'44.3"N and 138°9'11.6"E). The local climate is humid temperate, where July is the warmest month and January is the coolest month. The main orchard in the study site is distributed along a north-facing slope with 32-year-old Fuji/Marubakaido (*Malus prunifolia* Borkh. var. *ringo* Asami). The two orchard blocks (0.3 haeach) selected for this study are situated along a south-facing slope. Each orchard block consists of 15 trees (spacing: 16.6 m × 12 m).The orchard blocks on uphill (orchard A) and downhill (orchard B) positions are higher and lower in productivity, respectively (**Figure 1**). The slope between the orchards was 12.2% with horizontal and vertical distances of 90 m and 11 m, respectively. Fertilizers were applied at the rate of 144 kg N, 72 kg P_2O_5, 48 kg K_2O, 4.8 kg MnO, and 2.4 kg B·ha^{-1}. Total fertilizer is split into two halves and applied in December (after harvesting) and in March (before flowering). Pesticides, irrigation, and cultural operations are applied as per the recommendations for Nagano Prefecture by the Japanese Government [14].

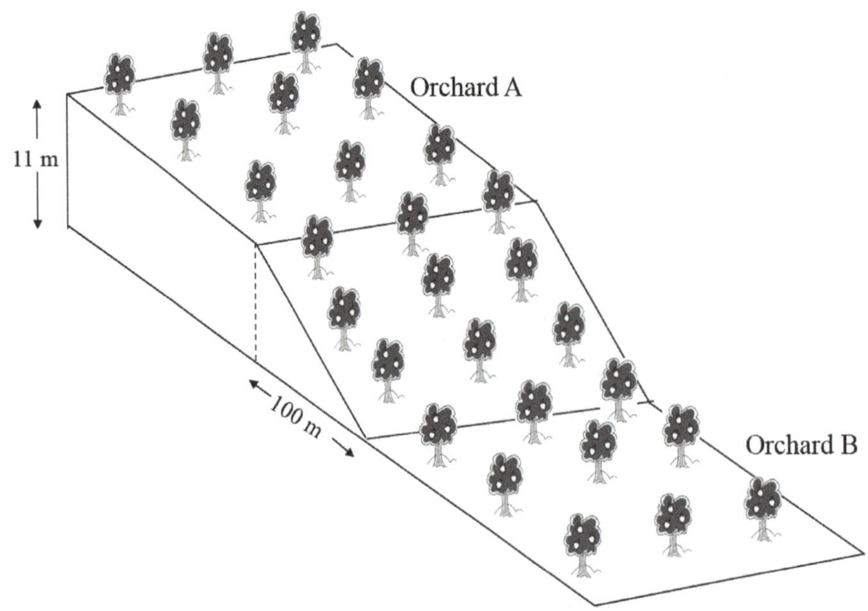

Figure 1. Schematic diagram of the two apple orchards used in this study.

2.2. Determination of Physical Properties

The following physical properties of soils at 15 cm and 35 cm depths were analyzed: color, texture, bulk density, porosity, soil solid, soil air, soil water, and hydraulic conductivity. Soil color was determined using Munsell Soil Color Charts [15]. The three soil phases were measured by a soil three-phase meter (Daiki Rika Kogyo Co., Ltd., Saitama, Japan), which is a type of gas pycnometer. Gaseous phase, total porosity and water saturation percentage were then calculated. Wet and dry bulk densities were estimated from the oven-dried mass of the solid phase in core samples. Vertical and horizontal saturated hydraulic conductivity were analyzed by the falling head permeability test using a falling head permeameter (Model: DIK-4050; Daiki Rika Kogyo Co., Ltd., Saitama, Japan).

2.3. Biochemical Properties of Soils

Composite soils samples (top 15 cm layer excluding the 2 - 3 cm surface crust) were takennearthe bases of five randomly selected trees in April, July, and December. The following biochemical properties of the composite soil samples were analyzed: pH, total carbon (TC), ammonium-nitrogen (NH_4^+-N), nitrate-nitrogen (NO_3^--N), total nitrogen (TN), water soluble phosphorus (SP), total phosphorus (TP), water soluble potassium (SK), total potassium (TK), total bacterial biomass, nitrification activity, and phosphorus mineralization activity. TC was analyzed with a TOC analyzer (Model: SSM-5000A, Shimadzu, Kyoto, Japan). NH_4^+-N was analyzed by extracting the soil sample with 1 N KCl. Soil-water suspension (1:20, w/v) was shaken reciprocally at 100 rpm for 1 h and the extracts were analyzed for NO_3^--N , SP, and SK by brucine, indophenol blue method and atomic absorption spectrophotometry, respectively. The same methods were used to detect TN, TP, and TK by digesting soils in a Kjeldahltherm (Gerhardt, Königswinter, Germany) with H_2SO_4 and H_2O_2. The pH of the soil-water suspension (1:2.5, w/v) was analyzed.

Total bacterial biomass was estimated by quantification of environmental DNA (eDNA) using the slow-stirring method following the procedures of Aoshima *et al.* [16]. Nitrification activity was analyzed by pooling the values of ammonium oxidizing activity, nitrite oxidizing activity, and total bacterial number. Ammonium and nitrite oxidizing activities were estimated by analyzing the percent reduction in N during an incubation period of 3 days at 25˚C in the soil samples with added ammonium sulfate and sodium nitrate, respectively. Similarly, phosphorus mineralization activity was estimated following the procedures of Horii *et al.* [17] by analyzing percent increment in SP from the substrate (phytic acid) under similar incubation conditions.

2.4. Nutrient Contents in Leaves and Fruits, Soluble Sugar Content, and Yield

Leaf and fruit samples were collected at different growth stages for chemical analysis. TN, TP, and TK in leaves and fruits were analyzed by digestion following the same procedures described above for soil samples. Soluble sugar content in the ripened fruit was analyzed with a palette digital refractometer (Model: DBX-55; ATAGO, Tokyo, Japan). Fruit were harvested during December, and yields were recorded separately for each orchard block.

3. Results

3.1. Weather

Weather data averaged over a 31-year period (1980 to 2010) shows that the site annually receives sunshine and precipitation of 1931 hours and 930 mm, respectively (**Figure 2**). The sunshine duration, precipitation, and temperature around the site during the study period (2011) were similar to normal, except for higher than normal rainfall during May.

3.2. Soil Physical Properties

Physical properties of soils at different depths in two orchards were analyzed during April 2011. The colors of the surface (0 cm to 15 cm) and sub-surface (15 cm to 40 cm) in orchard A were dark brown (7.5YR 4/4) and brown (7.5YR 3/3), respectively (**Figure 3**). Those of the surface and sub-surface in orchard B were dark brown (7.5YR 4/4). Likewise, textures of the surface and sub-surface soils in orchard A were clay loam and light clay, respectively, and those of the surface and sub-surface in orchard B were clay loam.

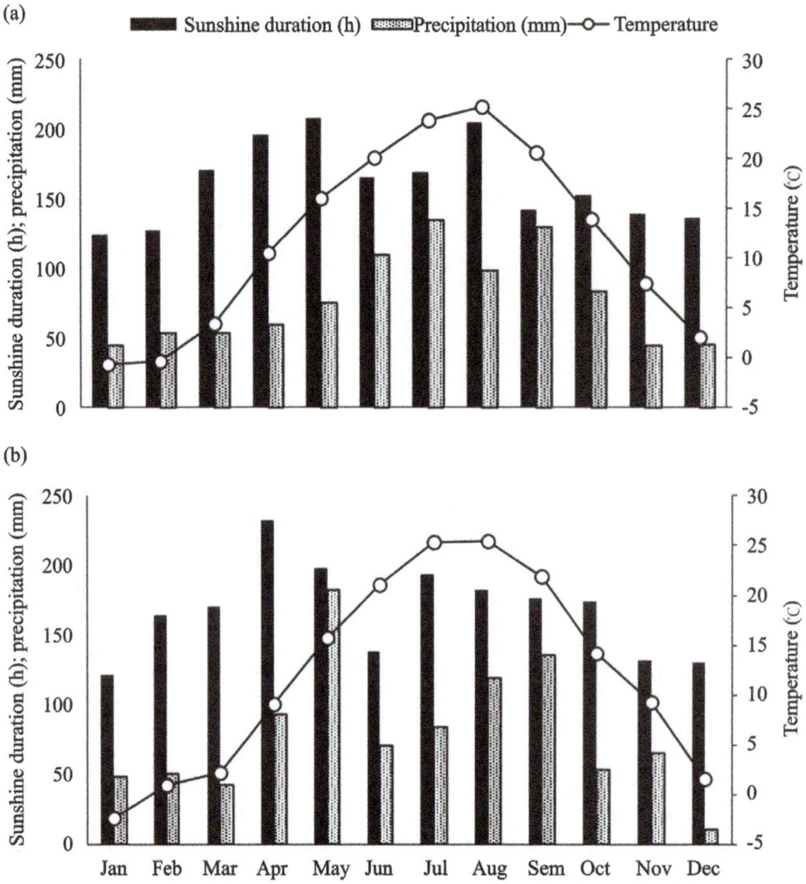

Figure 2. Monthly averages of sunshine duration, precipitation, and temperature around of the study site during 1980 and 2010 (a) and in 2011 (b).

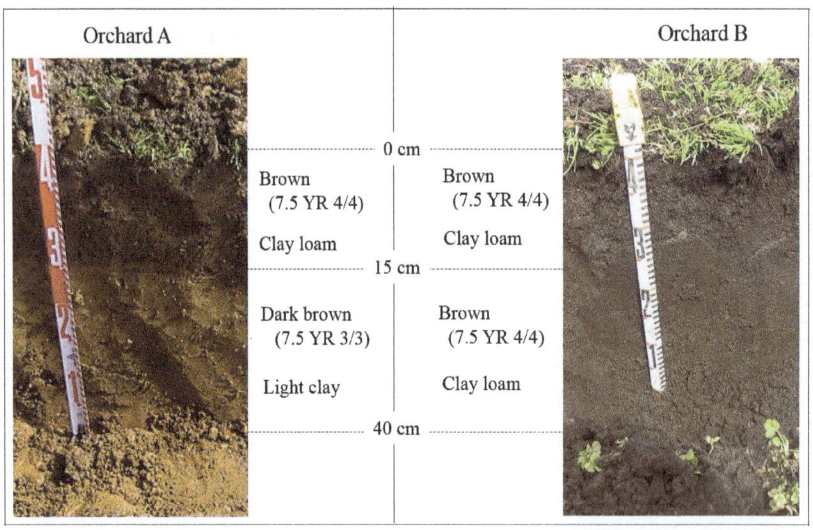

Figure 3. Soil profiles of orchard A and B showing color and texture at different depths. Photographs were taken on April, 2011.

Clear differences in dry and wet bulk densities of soils between the two orchards were not observed, except for dry bulk density at 10 cm (1.420 g·cm^{-3} and 1.296 g·cm^{-3} in orchards A and B, respectively; **Table 1**). Relative proportions of soil solids, water, and air also did not differ between the two orchards, except for the porosity at 10 cm (46.70% in orchard A and 50.25% in orchard B). Similarly, no clear differences in vertical saturated hydraulic conductivity were observed between orchards A and B (**Table 1**). In contrast, the horizontal hydraulic conductivity at 10-cm depth of orchard A soil was 3.7 times greater than that in orchard B (16.42 cm·day^{-1} in orchard A and 4.41 cm·day^{-1} in orchard B), while the horizontal hydraulic conductivity at 35-cm depth in orchard A was 1.7 times higher than that in orchard B.

3.3. Soil Biochemical Properties

Bacterial biomass, nitrification activity, and phosphorus mineralization activity were almost equal between orchards A and B when examined in April, July, and December (**Table 2**). Bacterial biomass in soils of both orchards gradually increased from April to December. In contrast, nitrification activity gradually decreased during this period. Phosphorus mineralization activity was highest in July.

Concentrations of TC, TN, and TP were almost two times higher in orchard B than in orchard A (**Table 2**). Similarly, concentrations of NO_3^--N, SP, and SK were also higher in orchard B. The large differences in nutrients show that organic materials and inorganic ions are accumulated in orchard B.

Table 1. Physical properties of soils in two apple orchards.

Property	Orchard A		Orchard B	
	10 cm	35 cm	10 cm	35 cm
Gravimetric water (w/w) (%)	24.40	35.10	27.70	25.60
Specific gravity	2.66	2.73	2.61	2.66
Wet bulk density (g/cm^3)	1.78	1.64	1.67	1.60
Dry bulk density (g/cm^3)	1.42	1.21	1.30	1.28
Void ratio	0.88	1.25	1.01	1.08
Porosity (%)	46.70	55.53	50.25	52.03
Degree of saturation (%)	74.05	76.77	71.44	62.65
Soil solid (v/v) (%)	53.30	44.47	49.75	47.97
Soil water (v/v) (%)	34.58	42.63	35.90	32.60
Soil air (v/v) (%)	12.12	12.90	14.35	19.43
Vertical saturated hydraulic conductivity (cm/day)	40.61	1.03	57.89	1.04
Horizontal saturated hydraulic conductivity (cm/day)	16.42	3.54	4.41	6.05

Table 2. Biochemical properties of soils in two apple orchards.

Property	Orchard A			Orchard B		
	Apr	Jul	Dec	Apr	Jul	Dec
Nitrification activity (%)	96	59	33	92	77	33
Bacterial biomass (×10^9 cells·g^{-1} soil)	1.4	1.7	1.8	1.3	1.6	1.7
NH_4^+ oxidation activity (%)	93	38	0	96	66	0
NO_2^- oxidation activity (%)	100	100	100	91	100	100
P mineralization activity (%)	23	96	47	39	100	53
pH	7.0	7.4	7.5	7.2	7.6	6.8
Total carbon (TC) (mg/kg)	35,700	45,900	46,900	59,800	85,600	67,400
Total nitrogen (TN) (mg/kg)	4100	4500	5100	10,500	8500	7500
NH_4^+-N (mg/kg)	2	3	8	3	3	2
NO_3^--N (mg/kg)	0	4	0	8	20	2
C/N ratio	9	10	9	6	10	9
Total phosphorus (TP) (mg/kg)	1280	1140	900	2720	2900	2720
Soluble phosphorus (SP) (mg/kg)	20	10	30	200	190	55
Total potassium (TK) (mg/kg)	5300	4700	4300	5300	4700	5000
Soluble potassium (SK) (mg/kg)	151	140	350	840	420	340

3.4. Nutrient Concentrations in Plant Tissues, Fruit Soluble Sugar, and Yield

Concentrations of TN, TP, and TK in the leaves were determined (**Table 3**). TP and TK were higher in orchard A in July and December. Leaf TN in orchard A was higher than in orchard B in July, but not in December. Nutrient concentrations in fruit also differed between orchards A and B (**Table 4**). Concentrations of TN, TP, and TK in fruit were higher in orchard A at both stages compared with those in orchard B. Soluble solids were also higher in orchard A. The results indicate that soil properties can affect the levels of nutrients in apple leaves and fruit.

4. Discussion

To improve the productivity of apple, physical and biochemical properties of soils and nutrient concentrations in leaves and fruit were investigated in two apple orchards with different productivities. Several characteristic differences between the two orchards indicate that apple productivity can be improved with greater attention to soil properties.

Apple productivity and soil nutrient status in two orchards were oppositely related in this study. Lower apple productivity in orchard B may be due to the toxicity of TN and TP. The slope of the land from orchard A to orchard B may have favored leaching of nutrients towards orchard B. Apple trees can grow vigorously and produce good yield with optimum supply of nitrogen and phosphorus. Excessive nitrogen application increases tree growth and leaf nitrogen content, but negatively affects fruit yield and quality [18]. In a similar study, a very high nitrogen application caused for the reduction in trunk size, tree height, and floral buds [19].

Leaf litter contributes significant amounts of biomass and nutrients to orchard soil [10] [20]. It is reported that approximately 30 $kg \cdot N \cdot ha^{-1}$ is contained in leaves that fall each year [21]. In this study, nutrients and biomass from the fallen leaves may contribute to the higher accumulation of TC, TN, and TP in orchard B soil.

Hydraulic conductivity is an important factor for productivity of orchards. In this study, lower apple productivity in orchard B was associated with the lower hydraulic conductivity. Previous studies have shown reduction

Table 3. Nutrient concentrations in leaves of apple grown in two orchards.

Property	Orchard A		Orchard B	
	Jul	Dec	Jul	Dec
TC ($mg \cdot kg^{-1}$)	422,500	491,500	433,800	485,300
TN ($mg \cdot kg^{-1}$)	18,900	22,000	25,200	21,200
NO_3^--N ($mg \cdot kg^{-1}$)	80	700	60	690
TP ($mg \cdot kg^{-1}$)	1800	2410	1520	2030
SP ($mg \cdot kg^{-1}$)	870	260	620	190
TK ($mg \cdot kg^{-1}$)	14,900	20,100	12,700	16,600
SK ($mg \cdot kg^{-1}$)	8650	20,550	6910	16,400

Table 4. Nutrient concentrations in fruit and yield of apple grown in two orchards.

Property	Orchard A		Orchard B	
	Jul	Dec	Jul	Dec
TC ($mg \cdot kg^{-1}$)	390,400	415,000	380,700	413,000
TN ($mg \cdot kg^{-1}$)	10,100	2400	9200	2600
NO_3^--N ($mg \cdot kg^{-1}$)	50	600	80	880
TP ($mg \cdot kg^{-1}$)	1410	850	1100	830
SP ($mg \cdot kg^{-1}$)	840	680	500	500
TK ($mg \cdot kg^{-1}$)	10,900	11,900	8700	10,100
SK ($mg \cdot kg^{-1}$)	7540	11,500	5750	9380
Soluble sugar (%)	-	16.6	-	15.4
Yield ($t \cdot ha^{-1}$)	-	30.0	-	20.0

in hydraulic conductivity in the soils rich in organic matter [22]-[24]. Accumulation of organic matter in orchard B may have reduced the hydraulic conductivity in this study. Soils with a high amount of litters, pore spaces are partially clogged by degraded organic materials [25].

Large amounts of nutrients in soils often result in over-uptake and imbalanced uptake of nutrients, and quality and quantity of apple can be adversely affected [26]-[28]. A negative relationship between soil nutrient content and fruit yield in this study indicates that chemical fertilizers should be carefully applied. Moreover, judging from the accumulation of nutrients, appropriate management of fallen leaves from the orchard is also an important factor for continuing a high level of apple productivity. Better management of biological and chemical properties of orchard soils is necessary for improving apple productivity.

References

[1] Forsline, P.L., Aldwinckle, H.S., Dickson, E.E., Luby, J.J. and Hokanson, S.C. (2003) Collection, Maintenance, Characterization, and Utilization of Wild Apples of Central Asia. *Horticultural Reviews*: *Wild Apple and Fruit Trees of Central Asia*, **29**, 1-62.

[2] Cornille, A., Gladieux, P., Smulders, M.J., Roldán-Ruiz, I., Laurens, F., Le Cam, B., Nersesyan, A., Clavel, J., Olonova, M., Feugey, L., Gabrielyan, I., Zhang, X.G. and Giraud, T. (2012) New Insight into the History of Domesticated Apple: Secondary Contribution of the European Wild Apple to the Genome of Cultivated Varieties. *PLoS Genetics*, **8**, e1002703. http://dx.doi.org/10.1371/journal.pgen.1002703

[3] Soejima, J., Bessho, H., Tsuchiya, S., Komori, S., Abe, K. and Kotoda, N. (1998) Breeding of Fuji Apples and Performance on JM Root Stocks. *Compact Fruit Tree*, **31**, 22-24.

[4] MAFF (2014) Monthly Statistics of Agriculture, Forestry and Fisheries. Ministry of Agriculture, Forestry and Fisheries, Japan.

[5] FAO (2014) FAOSTAT. faostat3.fao.org/home/E

[6] Krishna, K.R. (2014) Agroecosystems: Soils, Climate, Crops, Nutrient Dynamics and Productivity. Apple Academic Press, Inc., Oakville. http://dx.doi.org/10.1201/b16300

[7] Neilsen, D. and Neilsen, G.H. (2002) Efficient Use of Nitrogen and Water in High-Density Apple Orchards. *HorTechnology*, **12**, 19-25.

[8] Han, M.Y., Zhang, L.X., Fan, C.H., Liu, L.H., Zhang, L.S., Li, B.Z. and Alva, A.K. (2011) Release of Nitrogen, Phosphorus, and Potassium during the Decomposition of Apple (*Malus domestica*) Leaf Litter under Different Fertilization Regimes in Loess Plateau, China. *Soil Science and Plant Nutrition*, **57**, 549-557. http://dx.doi.org/10.1080/00380768.2011.593481

[9] Tagliavini, M., Tonon, G., Scandellari, F., Quinones, A., Palmieri, S., Menarbin, G., Gioacchini, P. and Masia, A. (2007) Nutrient Recycling during the Decomposition of Apple Leaves (*Malus domestica*) and Mowed Grasses in an Orchard. *Agriculture, Ecosystems and Environment*, **118**, 191-200. http://dx.doi.org/10.1016/j.agee.2006.05.018

[10] Tonon, G., Ciavatta, C., Solimando, D., Gioacchini, P. and Tagliavini, M. (2007) Fate of ^{15}N Derived from Soil Decomposition of Abscised Leaves and Pruning Wood from Apple (*Malus domestica*) Trees. *Soil Science and Plant Nutrition*, **53**, 78-85. http://dx.doi.org/10.1111/j.1747-0765.2007.00112.x

[11] Fujisawa, M. and Kobayashi, K. (2013) Shifting from Apple to Peach Farming in Kazuno, Northern Japan: Perceptions of and Responses to Climatic and Non-Climatic Impacts. *Regional Environmental Change*, **13**, 1211-1222. http://dx.doi.org/10.1007/s10113-013-0434-6

[12] Dris, R. (2002) Influence of Orchard Management on Growth and Production of Fruits. In: Dris, T., Jain S.M. and Khan I.A., Eds., *Environment and Crop Production*, Science Publishers Inc., Enfield, NH, 1-3.

[13] Aggelopoulou, K.D., Wulfsohn, D., Fountas, S., Gemtos, T.A., Nanos, G.D. and Blackmore, S. (2010) Spatial Variation in Yield and Quality in a Small Apple Orchard. *Precision Agriculture*, **11**, 538-556. http://dx.doi.org/10.1007/s11119-009-9146-9

[14] Nagano Agricultural Administration Department (2000) The Guide of Environment-Friendly Agricultural Technology.

[15] Munsell Color Company (1975) Munsell Soil Color Charts. Munsell Color Company, Baltimore.

[16] Aoshima, H., Kimura, A., Shibutani, A., Okada, C., Matsumiya, Y. and Kubo, M. (2006) Evaluation of Soil Bacterial Biomass Using Environmental DNA Extracted by Slow-Stirring Method. *Applied Microbiology and Biotechnology*, **71**, 875-880. http://dx.doi.org/10.1007/s00253-005-0245-x

[17] Horii, S., Matsuno, T., Tagomori, J., Mukai, M., Adhikari, D. and Kubo, M. (2013) Isolation and Identification of Phytate-Degrading Bacteria and Their Contribution to Phytate Mineralization in Soil. *The Journal of General and Applied Microbiology*, **59**, 353-360. http://dx.doi.org/10.2323/jgam.59.353

[18] Komamura, K., Suzuki, A., Fukumoto, M., Kato, K. and Sato, Y. (2000) Effects of Long-Term Nitrogen Application on Tree Growth, Yield, and Fruit Qualities in a "Jonathan" Apple Orchard. *Journal of the Japanese Society for Horticultural Science*, **69**, 617-623. http://dx.doi.org/10.2503/jjshs.69.617

[19] Ro, H.M. and Park, J.M. (2000) Nitrogen Requirements and Vegetative Growth of Pot-Lysimeter-Grown "Fuji" Apple Trees Fertilized by Drip Irrigation with Three Nitrogen Rates. *Journal of Horticultural Science and Biotechnology*, **75**, 237-242.

[20] Kumar, K. and Goh, K.M. (1999) Crop Residues and Management Practices: Effects on Soil Quality, Soil Nitrogen Dynamics, Crop Yield, and Nitrogen Recovery. *Advances in Agronomy*, **68**, 197-319. http://dx.doi.org/10.1016/S0065-2113(08)60846-9

[21] Haynes, R.J. and Goh, K.M. (1980) Some Effects of Orchard Soil Management on Sward Composition, Levels of Available Nutrients in the Soil, and Leaf Nutrient Content of Mature "Golden Delicious" Apple Trees. *Scientia Horticulturae*, **13**, 15-25. http://dx.doi.org/10.1016/0304-4238(80)90017-5

[22] Marshall, M.R., Francis, O.J., Frogbrook, Z.L., Jackson, B.M., McIntyre, N., Reynolds, B., Solloway, I., Wheater, H.S. and Chell, J. (2009) The Impact of Upland Land Management on Flooding: Results from an Improved Pasture Hillslope. *Hydrological Processes*, **23**, 464-475. http://dx.doi.org/10.1002/hyp.7157

[23] Gonzalez-Sosa, E., Braud, I., Dehotin, J., Lassabatère, L., Angulo-Jaramillo, R., Lagouy, M., Branger, M., Jacqueminet, C., Kermadi, S. and Michel, K. (2010) Impact of Land Use on the Hydraulic Properties of the Topsoil in a Small French Catchment. *Hydrological Processes*, **24**, 2382-2399. http://dx.doi.org/10.1002/hyp.7640

[24] Archer, N.A.L., Bonell, M., Coles, N., MacDonald, A.M., Auton, C.A. and Stevenson, R. (2013) Soil Characteristics and Landcover Relationships on Soil Hydraulic Conductivity at a Hillslope Scale: A View towards Local Flood Management. *Journal of Hydrology*, **497**, 208-222. http://dx.doi.org/10.1016/j.jhydrol.2013.05.043

[25] Oliveira, M.T. and Merwin, I.A. (2001) Soil Physical Conditions in a New York Orchard after Eight Years under Different Groundcover Management Systems. *Plant and Soil*, **234**, 233-237. http://dx.doi.org/10.1023/A:1017992810626

[26] Drake S.R., Raese, J.T. and Smith, T.J. (2002) Time of Nitrogen Application and Its Influence on "Golden Delicious" Apple Yield and Fruit Quality. *Journal of Plant Nutrition*, **25**, 143-157. http://dx.doi.org/10.1081/PLN-100108786

[27] Nava, G. and Dechen, A.R. (2009) Long-Term Annual Fertilization with Nitrogen and Potassium Affect Yield and Mineral Composition of "Fuji" Apple. *Scientia Agricola*, **66**, 377-385. http://dx.doi.org/10.1590/S0103-90162009000300013

[28] Teravest, D., Smith, J.L., Carpenter-Boggs, L., Hoagland, L., Granatstein, D. and Reganold, J.P. (2010) Influence of Orchard Floor Management and Compost Application Timing on Nitrogen Partitioning in Apple Trees. *HortScience*, **45**, 637-642.

3

Analysis Methods for the Determination of Anthropogenic Additions of P to Agricultural Soils

Richard L. Haney[1], Virginia L. Jin[2], Mari-Vaughn V. Johnson[3], Elizabeth B. Haney[4], R. Daren Harmel[1], Jeffrey G. Arnold[1], Michael J. White[1]

[1]United States Department of Agriculture-Agriculture Research Service, Grassland, Soil and Water Research Laboratory, Temple, TX, USA
[2]United States Department of Agriculture-Agriculture Research Service, Agroecosystem Management Research Laboratory, University of Nebraska, East Campus, Lincoln, NE, USA
[3]United States Department of Agriculture-Natural Resource Conservation Service, Temple, TX, USA
[4]Texas A&M University, Texas AgriLife Research & Extension Center, Temple, TX, USA
Email: rick.haney@ars.usda.gov

Abstract

Phosphorus loading and measurement is of concern on lands where biosolids have been applied. Traditional soil testing for plant-available P may be inadequate for the accurate assessment of P loadings in a regulatory environment as the reported levels may not correlate well with environmental risk. In order to accurately assess potential P runoff and leaching, as well as plant uptake, we must be able to measure organic P mineralized by the biotic community in the soil. Soils with varying rates of biosolid application were evaluated for mineralized organic P during a 112-day incubation using the difference between P measured using a rapid-flow analyzer (RFA) and an axial flow Varian ICP-OES. An increase in the P mineralized from the treated soils was observed from analysis with the Varian ICP-OES, but not with the RFA. These results confirm that even though organic P concentrations have increased due to increasing biosolid application, traditional soil testing using an RFA for detection, would not accurately portray P concentration and potential P loading from treated soils.

Keywords

Phosphorus, Anthropogenic Additions, Biosolids, Rapid-Flow Analyzer (RFA), Inductively Coupled Plasma (ICP), Texas Commission of Environmental Quality (TCEQ), Soil Organic C (SOC), Total N (TN), Water-Soluble Organic C (WSOC), Water Soluble Organic N (WSON)

1. Introduction

Phosphorus (P) is an essential nutrient for plant growth and is applied to agricultural and urban lands in inorganic (fertilizer) and organic forms (manure, biosolids). While the application of inorganic P is relatively uniform on urban and most agricultural lands, intensive P application may occur with the application of biosolids from municipal waste processing centers on agricultural fields. The land application of biosolids is a means of waste management that may provide many benefits to soils including soil conditioning [1] [2] and increased aggregate stability [3] [4] soil water content [5] [6], soil organic C (SOC) [7]-[9], soil nitrogen (N) and phosphorus (P) [6] [10] and potential C mineralization and net N mineralization and immobilization [11]. When biosolids are applied at normal agronomic rates, agricultural productivity may be increased; however, the benefits are limited when crop requirements are exceeded [12]-[14].

Phosphorus concentration in surface water can be tied directly to water quality degradation and can accelerate freshwater eutrophication [15]. Eutrophication is the accumulation of high concentration of inorganic chemicals such as N or P in bodies of water, which is often associated with algal blooms. The negative effects of eutrophication range from decreased species diversity to decreased aesthetics [16]. When large populations of algae die, they decompose on the bottom of the lake or reservoir, resulting in a depletion of oxygen in the ecosystem, which can cause fish and aquatic plant to kill. The algal blooms are not only visually unappealing, but also potentially hazardous to human and animal health.

Total P concentrations of 50 $\mu g \cdot L^{-1}$ are sufficient to sustain chlorophyll concentrations of 10 $\mu g \cdot L^{-1}$, the lower boundary for a trophic state in temperate streams [17] and eutrophy in lakes [18]. Using the Redfield ratio [19], P concentrations exceeding 65 $\mu g \cdot L^{-1}$ are sufficient to support algal growth in many water bodies when nitrate-N concentrations exceed 10 $mg \cdot L^{-1}$. It has been shown that these concentrations of total P can be from ground water in intensive agricultural regions [20]. These non-point sources of P, including both surface runoff and groundwater, can contribute to eutrophication of streams, lakes and rivers. USEPA national lakes assessment survey indicates that based chlorophyll-*a* concentrations, 20% of lakes are hypereutrophic [21].

Because of the negative environmental effects of the overloading of P in surface waters, P regulations for agricultural lands are being considered, as well as for the land application of biosolids. Regulations for agricultural lands may be centered on a soil-test P value [15]. Soil-test P is most commonly related to plant-available P and may be assessed using many different extractions and analysis methods depending upon which part of the country the soil is analyzed in and the laboratory performing the analysis. As a result, the soil-test P numbers obtained may vary widely.

The most common laboratory analysis for solution P is by colorimetric analyzers (rapid flow analyzer (RFA), which use chemistries to develop color intensity) or inductively coupled plasma (ICP) which uses heat generated by producing plasma to excite atoms to a higher state of energy and measure the light given off as they fall back to a normal state. Differences between P values obtained from colorimetric and ICP analyses have been noted by some studies, while others indicate there is no significant difference between the analysis methods [22]. The difference between the two instruments is often attributed to the fact that the colorimetric method measures H_2PO_4 and HPO_4 (inorganic P) whereas ICP measures orthophosphate P plus other forms of organic and inorganic P [23]. In other words, the organic P may not be available to form the phosphomolybdate complex needed for color formation and detection in colorimetric analysis [22] [24]. The instrumentation used to analyze P can have a significant influence on P values, even though values obtained using the RFA and ICP are generally correlated [25]. Therefore, depending on instrumentation and soil inputs, phosphate values in soils treated with biosolids, where a significant portion of P may be attributed to the organic additions, may be underestimated if using colorimetric analyses versus ICP.

In addition, soil-test P recommendations may not be relevant to the environmental risk associated with soil P levels [15] because they do not reflect organic P concentrations associated with soil organic matter or P mineralized over time due to natural processes in the soil. Flushes of P are associated with the natural drying and rewetting of soil with natural rainfall events or irrigation and are partially due to the death of the microbes and decomposition of their cells [26]. Even though P associated with organic matter that is converted to dissolved inorganic P contributes to P loading, mineralized P is not routinely tested for regulatory or fertility assessment.

It is important that the methods of analysis for P be as sound and comprehensive as possible in order to accurately assess the P loading associated with runoff from agricultural lands with biosolid application. The objective of this study is to: 1) determine if there is a significant difference between analyses of P from a RFA as compared to an ICP in soils treated with varying levels of biosolids; 2) determine the fraction of organic P mineralized during 112-day incubation.

2. Materials and Methods

The study site where samples were obtained is a municipally operated 485-ha, zero-discharge facility in Travis County, TX. Approximately 220 ha of the facility are used in a year round, continuous field rotation for hay production using coastal Bermuda grass (*Cynodon dactylon* L.). Currently, one-third of the total biosolids processed by the facility is land-applied as anaerobically digested, Class B biosolids. Following anaerobic digestion, biosolids are belt-pressed to reduce water content then loaded into manure-spreaders for surface application to a perennial no-tillage forage crop system. Annual mean characteristics of dry biosolids based on monthly analysis (November 2006-October 2007) were pH 8.46 (SM 4500-H B); 17.5% total solids (SM 2540 G); 10,745 mg NH_3-N kg^{-1} (SM 4500-NH3 D); 14.7 mg NO_2-N + NO_3-N kg^{-1} (EPA 353.2); 52,127 mg TKN kg^{-1} (EOA 351.2). Regulated biosolids elemental constituents for dry biosolids at all samplings fall well below the EPA 40 CRF Part 503 limits [27] [28].

A subset of fields within the 220-ha area was sampled for soils used in the present study. Since 1985, a 14-ha field has been treated with 22 dry Mg biosolids $ha^{-1}\cdot y^{-1}$ (25-yr treatment), an agronomic rate below the maximum annual metal loading rate regulated by the Texas Commission of Environmental Quality (TCEQ); 30 TAC §312.43 (b) (4); TAC (2005). Annual applications in this field, however, were changed to a single application every 2 years (*i.e.* bi-annual applications) starting in 2006 to comply with nutrient management plan recommendations to address high soil N levels. In 2001, 105 ha of adjacent row-crop land was acquired and approved by the former Texas Natural Resource Conservation Commission (now TCEQ) as an experimental exemption area to evaluate the potential soil impacts of applying higher rates of biosolids amendments. The exemption area was split into 20 ha, 36 ha, and 49 ha fields for experimental applications of 22, 45, and 67 Mg $ha^{-1}\cdot y^{-1}$, respectively, starting in 2002 (8-yr treatment). This newer area was converted to coastal Bermuda grass hay production in 2003. A 2-ha unamended control field is located next to the lowest application rate field. The control field has not received applications of biosolids or commercial fertilizer of any kind since 2000. Soils across the site are Bergstrom silt-loams (fine-silty, mixed, superactive, and thermic Cumulic Haplustolls) and are very deep, well-drained, and moderately permeable, with slopes <1%.

Surface soils (0 - 10 cm; 20 cm × 20 cm sampling area) were collected in January 2009 from each field (unamended control; 22 dry Mg biosolids $ha^{-1}\cdot y^{-1}$ for 25 years; 22, 45, and 67 Mg $ha^{-1}\cdot y^{-1}$ for 8 years) using a hand spade. No biosolids had been applied to any of the treatment fields for at least 3 months prior to soil sampling. Visible biosolids and plant material were removed from the soil surface prior to sampling. Soils were sampled from three random locations per treatment area and composited by treatment. Soils were sieved through a 2 mm mesh, large roots removed, and air-dried prior to laboratory incubations.

Initial total soil organic C (SOC) and total N (TN) in soils were determined by dry combustion at 680˚C (Vario Max CHN, Elementar, Hanau, Germany). Soils were extracted with water (10:1 deionized water: dry soil ratio) to simulate available soil nutrients following a natural rain event, as would occur in a field setting. Water-extracts were measured for water-soluble organic C (WSOC) and organic N (WSON) to evaluate the potential in field availability of labile C and N for microbial consumption (Apollo 9000, Teledyne Tekmar, Mason, OH). Inorganic N concentrations in water extracts were measured using continuous flow colorimetry (Flow Solution IV, OI Analytical, College Station, TX). Soil C and N concentrations in the 0 - 10 cm soil layer were converted to an area basis using soil bulk densities (Mg ha^{-1} for SOC, TN; kg·ha^{-1} for water-soluble C and N). Net N mineralization (mg N $kg^{-1}\cdot d^{-1}$) over the incubation period was determined as the difference between water-extractable soil inorganic N concentrations $\left(NH_4^+\text{-N} + NO_2^-\text{-N} + NO_3^-\text{-N}\right)$ between two consecutive destructive sampling days. Initial P concentrations were colorimetrically determined using a water extract and H3A extract [29]. CO_2-C mineralized after 1-day was determined for each of the surface soil samples using the Haney-Brinton Method [30].

Triplicate soil samples from each treatment were moistened to 50% water-filled pore space and incubated at 25˚C ± 1˚C in the dark, with destructive sampling at days 0, 7, 14, 28, 56, and 112. Soil samples (40 g each) were incubated in 1-quart canning jars containing vials with 10 mL water to maintain humidity. Soils that were destructively harvested were measured for net P mineralization over time. On the appropriate sampling day, each sample was dried to 40˚C, ground and passed through a 2 mm sieve. These sub-samples were weighed into

50-ml plastic centrifuge tubes and received 40 ml of the soil extractant H3A [29], shaken for 10 minutes and centrifuged for 5 min at 3500 rpm. The extracts were filtered through Whatman 2V filter paper.

Orthophosphate P concentrations were measured using an OI Analytical segmented rapid flow analyzer (RFA) (Flow Solution IV, OI Analytical, College Station, Texas) and elemental P concentrations were measured with an axial flow Varian ICP-OES from the same extractant. Net P mineralization (mg P kg^{-1} soil) over the incubation period was determined as the difference between soil extractable P concentrations (PO_4-P and elemental P) from the initial P concentrations (day 0) and consecutive destructive sampling days (days 7, 14, 28, 56, and 112). P mineralization values were compared between the results obtained from the rapid flow analyzer and the axial flow Varian ICP-OES.

Data were tested for normality using the Shapiro-Wilk statistic. One-way ANOVA was used to test the effects of biosolids treatment on net P mineralization rates over time using pair-wise multiple comparison procedures (Holm-Sidak method). Linear regressions were used to examine correlations between net P mineralization from both instruments. All statistical tests were performed using SigmaPlot ver. 11 (Systat, Inc.).

3. Results and Discussion

Surface soils ranged in pH from 7.2 to 7.7, with the control (unamended soil) having the highest pH value. The amended soils had significantly higher total soil organic carbon (SOC) than the unamended soil (**Table 1**). The amended soils had correspondingly increased C mineralization after 1 day, except the 22 Mg dry biosolids ha^{-1} 8-yr treatment. Total N, total C:N, WSON, soluble organic C:N, NH_4^+-N, NO_3^--N, inorganic N, total soluble N, and total soluble C:N were greatest in the soil amended with the two highest application rates. There was a statistically significant difference in H3A extractable P among each treatment ($P < 0.001$), with the 22 Mg dry biosolids ha^{-1} 25-yr treatment having the greatest H3A extractable P. The 22 Mg dry biosolids ha^{-1} 25-yr treatment also had significantly more water soluble P than the other treatments ($P < 0.001$). All treatments had significantly greater water soluble P than the control ($P < 0.001$).

Table 1. Mean initial soil pH, bulk densities, and total and water-soluble C and N in surface soils (0 - 10 cm) collected from control and biosolids-applied fields.

Constituent	Unit	Biosolids application rate (Mg dry biosolids ha^{-1})				
		0	22	22	45	67
		Control	25-year	8-year	8-year	8-year
pH (1:10)		7.7	7.4	7.4	7.4	7.2
Bulk density	$g\cdot cm^{-3}$	1.5	1.5	1.5	1.5	1.5
Total SOC	$mg\cdot kg^{-1}$	31.4	45.7	41.8	48.1	59.8
H3A extractable P	$mg\cdot kg^{-1}$	5.7	73.3	30.4	39.0	46.3
Water soluble P	$mg\cdot kg^{-1}$	2.0	12.1	7.7	6.7	6.2
Total N	$mg\cdot kg^{-1}$	3.0	4.6	3.8	5.6	9.1
Total C:N		10.4	9.9	11.6	8.7	6.5
Water soluble OC	$kg\cdot ha^{-1}$	545.0	521.0	534.0	483.0	555.0
Water soluble ON	$kg\cdot ha^{-1}$	37.7	45.9	55.0	143.5	165.4
Soluble organic C:N		14.5	11.6	9.8	3.4	4.8
NH_4^+-N	$kg\cdot ha^{-1}$	7.5	7.9	84.0	10.0	12.6
NO_3^--N	$kg\cdot ha^{-1}$	11.1	55.9	71.4	374.7	573.6
Inorganic N	$kg\cdot ha^{-1}$	18.6	63.8	79.8	384.7	586.2
Total soluble N	$kg\cdot ha^{-1}$	56.3	109.7	134.8	528.2	751.7
Total soluble C:N		9.7	4.7	4.1	0.9	0.7
1-day CO_2-C production	mg CO_2-C kg^{-1}	47.7	64.5	47.1	53.4	61.5

P mineralization at 7, 28, 56, and 112 days was highly correlated between the ICP and RFA, while the 14-day mineralization numbers were not as strongly correlated, but still related (**Figure 1**). When P mineralization values obtained from the RFA and ICP were compared for each treatment, we found that values for the control were not correlated (**Figure 2**). These results were to be expected since very little P was mineralized from the control when measured both by the ICP and the RFA. P mineralization values for the remaining treatments were highly correlated. Furthermore, H3A and water extractable P, as well as P mineralization values, follow the increased additions of biosolids in the treatments. We speculate that as organic P from the biosolids is mineralized and converted to inorganic P, the inorganic P accumulates in the soil.

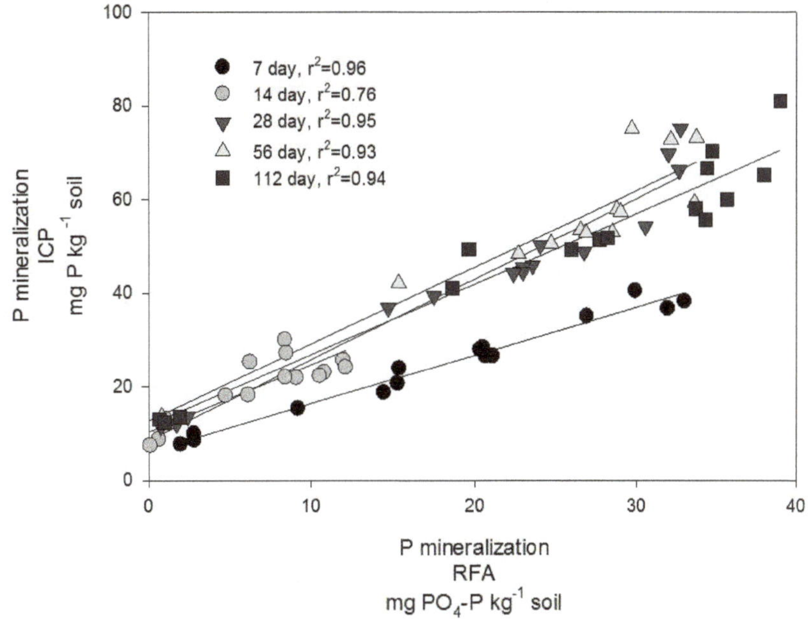

Figure 1. Regression analysis among P mineralized from rapid flow analyzer vs. inductively coupled plasma analyzer across days of mineralization.

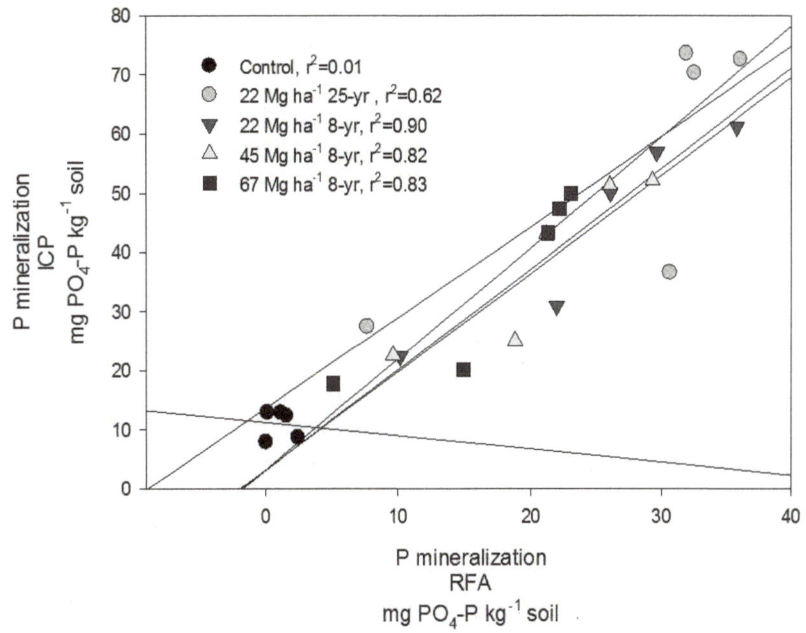

Figure 2. Regression analysis among P mineralized from rapid flow analyzer vs. inductively coupled plasma analyzer across treatments.

Even though the relationship among the amounts of P mineralized as measured by the ICP were highly significant ($P < 0.001$), the values measured by the ICP were greater than those measured by the RFA for all treatments (**Figure 3**). The difference between the ICP and RFA results are clearly depicted in **Figure 4**, which illustrates the P mineralized from both methods at 56 days (the point at which organic P mineralization peaked).

There was not a significant difference in P mineralization among the biosolid treatments when measured using the RFA, excepting the 22 Mg ha^{-1} 25-yr treatment at 28 days incubation (**Figure 5**). When measured using the ICP, a significant difference was seen in P mineralization between the 22 Mg ha^{-1} 25-yr treatment and the remaining 8 year treatments after 14 days incubation. If the RFA method was used exclusively, the observer

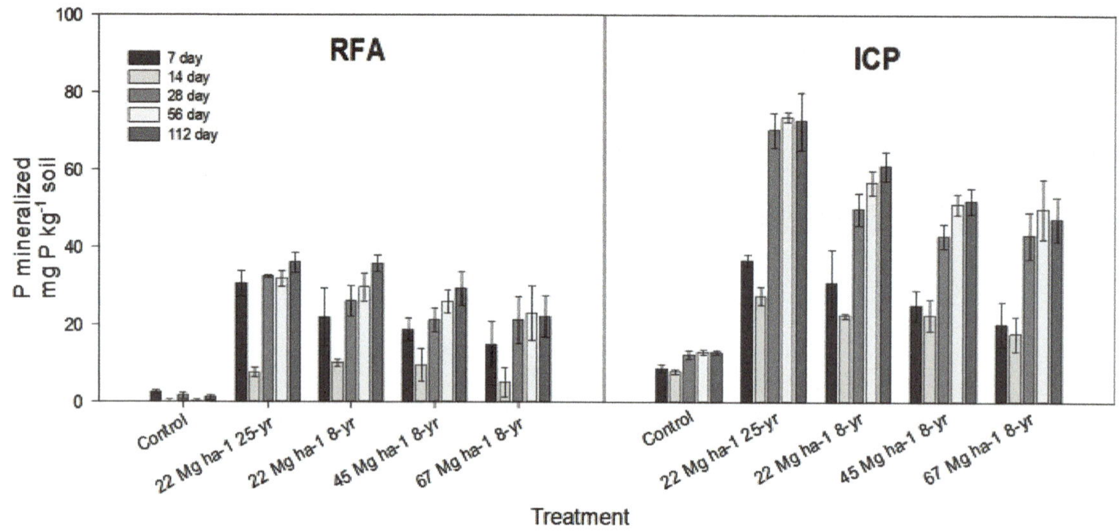

Figure 3. P mineralized from rapid flow analyzer and inductively coupled plasma analyzer across treatments. Error bars represent one standard deviation.

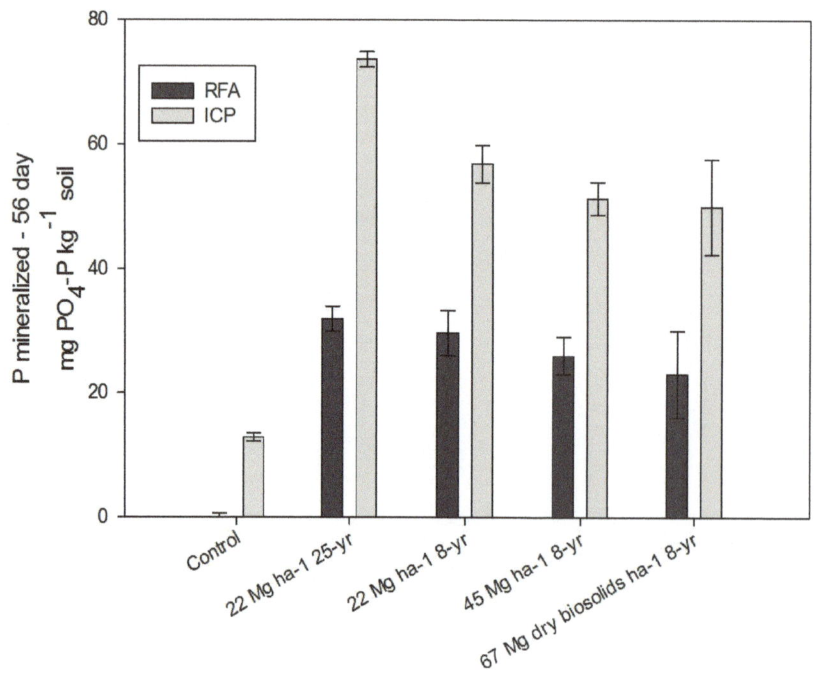

Figure 4. P mineralized from rapid flow analyzer and inductively coupled plasma analyzer across treatments after 56 days. At 56 days organic P mineralized had peaked. Error bars represent one standard deviation.

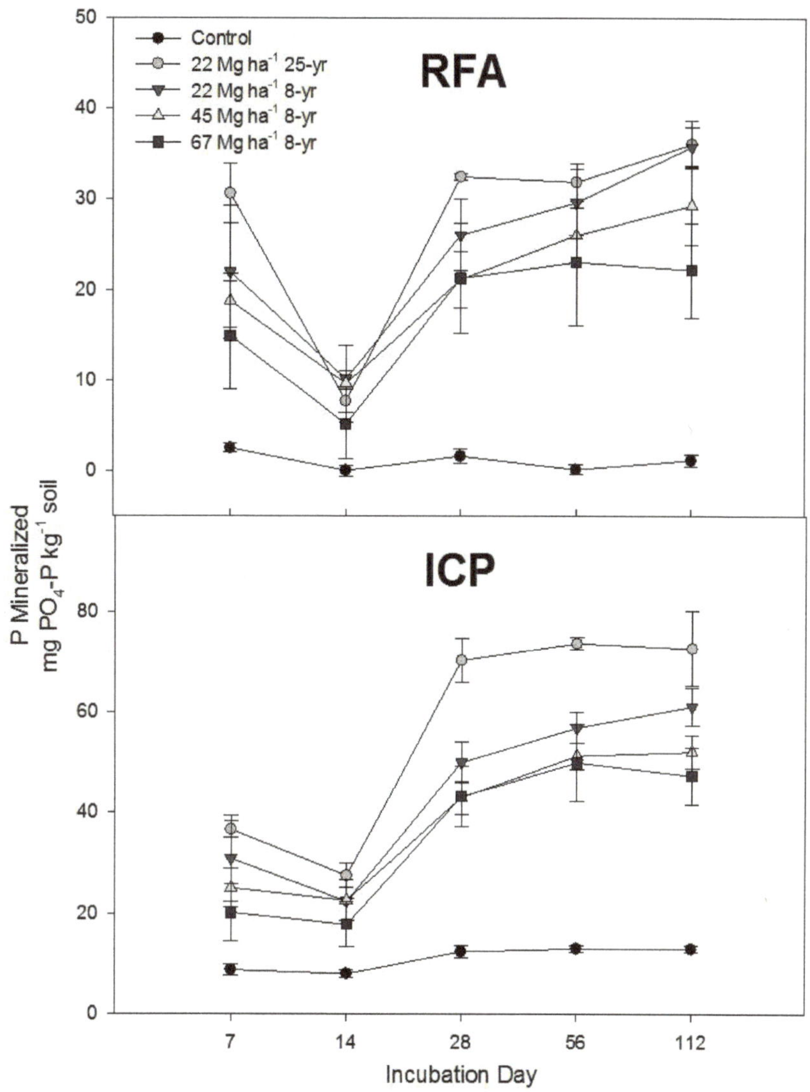

Figure 5. P mineralization over time as determined using the rapid flow analyzer and inductively coupled plasma analyzer methods.

would not see a significant difference among biosolid treatments, which may lead them to believe that there would be no difference in the amount of P mineralized in each soil.

As shown in **Figure 6**, differences in the mean organic P mineralization values between the control and all treatments and the 22 Mg dry biosolids ha^{-1} 25-yr treatment and all of the 8-yr treatments are greater than would be expected by chance; meaning there is a statistically significant difference (P = <0.001). The 22 Mg dry biosolids ha^{-1} 25-yr treatment had the greatest amount of organic P mineralization, followed by the three 8-yr treatments, then the control. As expected, as organic matter increases organic P mineralization increases, which was easily discernable using the difference in P measured from the ICP and RFA.

There was no correlation between the P mineralized obtained using the RFA method and 1-day CO_2 production ($r^2 = 0.19$, **Figure 7**). The correlation increased, but was still weak, when using the ICP method, with 35% of the variability in P mineralized being related to 1-day CO_2 production. One-day CO_2 results are a direct reflection of the activity of the soil microbial biomass. Fierer and Schimel [31] demonstrated that the flush of CO_2 after drying/rewetting soil largely originated from the microbial biomass and Franzluebbers *et al.* [32] and Haney *et al.* [33] observed a strong correlation observed between short-term C mineralization and microbial biomass C. Release of potentially mineralizable nutrients and decomposition are generally functions of the soil microbial biomass and its activity. The flush of CO_2 after 1 day following rewetting is closely related to potential C

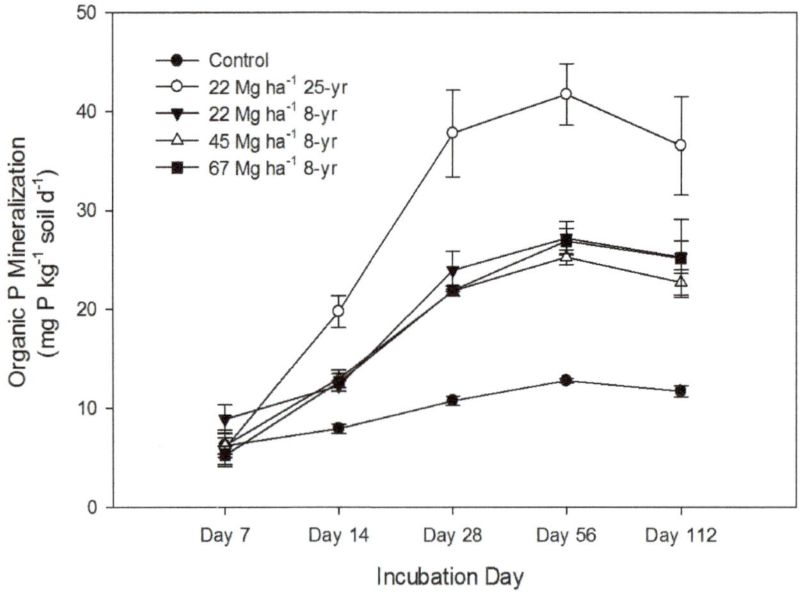

Figure 6. The difference between P mineralized from inductively coupled plasma analyzer minus rapid flow analyzer from biosolids application over time.

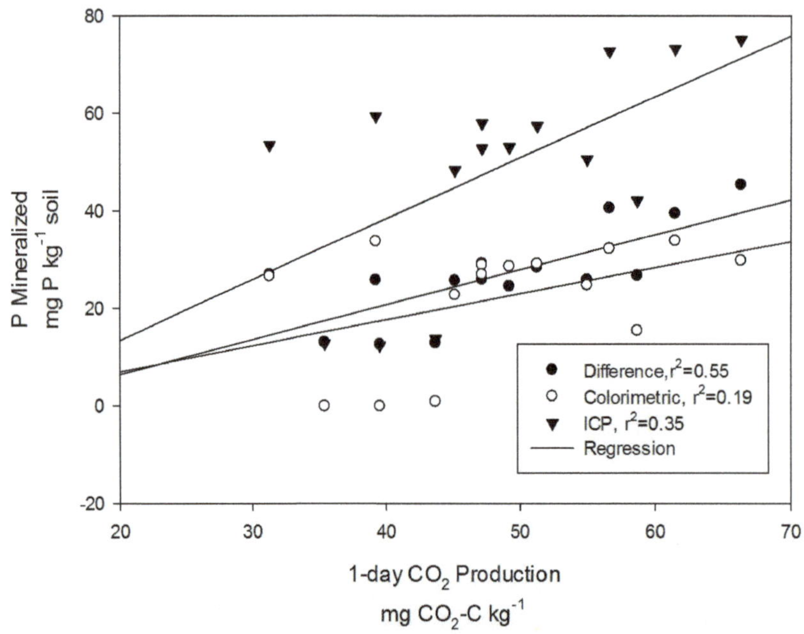

Figure 7. Correlation between P mineralized as measured by the colorimetric method, inductively coupled plasma analyzer, and organic P mineralized and 1-day CO_2 production.

and N mineralization [32]. It follows that this flush should also be correlated to P mineralization; however, 1-day CO_2 is only correlated with difference between the P measured using the ICP and the colorimetric method, or the organic P mineralized. When considering only the difference in P mineralized by the ICP and the RFA, the correlation further increased as 55% of the variability in P mineralized can be explained by 1-day CO_2 values. The additional P that is detectable by the ICP only may be from the organic P mineralized from the organic fraction (biosolids) of the soil. Little is actually known about the factors that drive microbial P mineralization, but researchers have indicated that microbes mineralize P as a byproduct of C acquisition, without incorporating the P mineralized [34].

The results of this study indicate that as organic P concentrations increase due to increasing biosolid application, traditional soil testing using an RFA for detection would not accurately portray P concentration and potential P loading from treated soils. We recommend using the ICP to measure P mineralized from soils treated with organic P amendments.

References

[1] Hargreaves, J.C., Adl, M.S. and Warman, P.R. (2008) A Review of the Use of Composted Municipal Solid Waste in Agriculture. *Agriculture, Ecosystems & Environment*, **123**, 1-14. http://dx.doi.org/10.1016/j.agee.2007.07.004

[2] Singh, R.P. and Agrawal, M. (2008) Potential Benefits and Risks of Land Application of Sewage Sludge. *Waste Manage*, **28**, 347-358. http://dx.doi.org/10.1016/j.wasman.2006.12.010

[3] Ojeda, G., Alcañiz, J.M. and Ortiz, O. (2003) Runoff and Loses by Erosion in Soils Amended with Sewage Sludge. *Land Degradation & Development*, **14**, 563-573. http://dx.doi.org/10.1002/ldr.580

[4] Tejada, M. and Gonzalez, J.L. (2007) Application of Different Organic Wastes on Soil Properties and Wheat Yield. *Agronomy Journal*, **99**, 1597-1606. http://dx.doi.org/10.2134/agronj2007.0019

[5] Nielsen, G.H., Hogue, E.J., Forge, T. and Nielsen, D. (2003) Surface Application of Mulches and Biosolids Affect Orchard Soil Properties After 7 Years. *Canadian Journal of Soil Science*, **83**, 131-137. http://dx.doi.org/10.4141/S02-034

[6] Kelly, J.J., Favila, E., Hundal, L.S. and Marlin, J.C. (2007) Assessment of Soil Microbial Communities in Surface Applied Mixtures of Illinois River Sediments and Biosolids. *Applied Soil Ecology*, **36**, 176-183. http://dx.doi.org/10.1016/j.apsoil.2007.01.006

[7] Klavidko, E.J. and Nelson, D.W. (1979) Changes in Soil Properties from Application of Anaerobic Sludge. *Journal (Water Pollution Control Federation)*, **51**, 325-332.

[8] Crecchio, C., Curci, M., Mininni, R., Ricciuti, P. and Ruggiero, P. (2001) Short-Term Effects of Municipal Solid Waste Compost Amendments on Soil Carbon and Nitrogen Content, Some Enzyme Activities and Genetic Diversity. *Biology and Fertility of Soils*, **34**, 311-318. http://dx.doi.org/10.1007/s003740100413

[9] Parat, C., Chaussod, R., Lévêque, J. and Andreux, F. (2005) Long-Term Effects of Metal Containing Farmyard Manure and sewage Sludge on Soil Organic Matter in a Fluvisol. *Soil Biology and Biochemistry*, **37**, 673-679. http://dx.doi.org/10.1016/j.soilbio.2004.08.025

[10] Bayley, R.M., Ippolito, J.A., Stromberger, M.E., Barbarick, K.A. and Paschke, M.W. (2008) Water Treatment Residuals and Biosolids Coapplications Affect Semiarid Rangeland Phosphorus Cycling. *Soil Science Society of America Journal*, **72**, 711-719. http://dx.doi.org/10.2136/sssaj2007.0109

[11] Jin, V.L., Johnson, M.V.V., Haney, R.L. and Arnold, J.G. (2011) Potential Carbon and Nitrogen Mineralization in Soils from a Perennial Forage Production System Amended with Class B Biosolids. *Agriculture, Ecosystems Environment*, **141**, 3-4. http://dx.doi.org/10.1016/j.agee.2011.03.016

[12] Nielson, G.H., Hogue, E.J., Nielson, D. and Zebarth, B.J. (1998) Evaluation of Organic Wastes as Soil Amendments for Cultivation of Carrot and Chard on Irrigated Sandy Soils. *Canadian Journal of Soil Science*, **78**, 217-225. http://dx.doi.org/10.4141/S97-037

[13] Speir, T.W., Horswell, J., van Schaik, A.P., McLaren, R.G. and Fietje, G. (2004) Composted Biosolids Enhance Fertility of a Sandy Loam Soil under Dairy Pasture. *Biology and Fertility of Soils*, **40**, 349-358. http://dx.doi.org/10.1007/s00374-004-0787-6

[14] Sigua, G.C. and Adjei, M.B. (2005) Cumulative and Residual Effects of Repeated Sewage Sludge Applications: Forage Productivity and Soil Quality Implications in South Florida, USA. *Environmental Science and Pollution Research*, **12**, 80-88. http://dx.doi.org/10.1065/espr2004.10.220

[15] Sharpley, A.N., McDowell, R.W. and Kleinman, P.J.A. (2001) Phosphorus Loss from Land to Water: Integrating Agricultural and Environmental Management. *Plant and Soil*, **237**, 287-307.http://dx.doi.org/10.1023/A:1013335814593

[16] Steffen, W., Crutzen, P.J. and McNeill, J.R. (2007) The Anthropocene: Are Humans Now Overwhelming the Great Forces of Nature. *AMBIO: A Journal of the Human Environment*, **36**, 614-621. http://dx.doi.org/10.1579/0044-7447(2007)36[614:TAAHNO]2.0.CO;2

[17] Dodds, W.K., Jones, J.R. and Welch, E.B. (1998) Suggested Classification of Stream Trophic State: Distributions of Temperate Stream Types by Chlorophyll, Total Nitrogen, and Phosphorus. *Water Research*, **32**, 1455-1462. http://dx.doi.org/10.1016/S0043-1354(97)00370-9

[18] Smith, V.H. (2003) Eutrophication of Freshwater and Coastal Marine Ecosystems a Global Problem. *Environmental Science and Pollution Research*, **10**, 126-139. http://dx.doi.org/10.1065/espr2002.12.142

[19] Redfield, A.C. (1958) The Biological Control of Chemical Factors in the Environment. *American Scientist*, **46**, 205-221.

[20] Burkart, M.R., Simpkins, W.W., Morrow, A.J. and Gannon, J.M. (2004) Occurrence of Total Dissolved Phosphorus in Unconsolidated Aquifers and Aquitards in Iowa. *Journal of the American Water Resources Association*, **40**, 827-834.

[21] USEPA (2009) National Lakes Assessment: A Collaborative Survey of the Nation's Lakes. US Environmental Protection Agency, Office of Water and Office of Research and Development, Washington DC, EPA 841-R-09-001.

[22] Pittman, J.J., Zhang, H. and Schroder, J.L. (2005) Differences of Phosphorus in Mehlich-3 Extracts Determined by Colorimetric and Spectroscopic Methods. *Communications in Soil Science and Plant Analysis*, **36**, 1641-1659. http://dx.doi.org/10.1081/CSS-200059112

[23] Kuo, S. (1996) Phosphorus. In: Sparks, D.L., Ed., *Methods of Soil Analysis*: *Part* 3, SSSA Book Series No. 5, SSSA and ASA, Madison, 869-919.

[24] Sobeck, S.A. and Ebeling, D.D. (2007) Mass Spectrometric Analysis for Phosphate in Soil Extracts; Comparison of Mass Spectrometry, Colorimetry, and Inductively Coupled Plasma. *The Journal of Analytical Sciences Digital Library.*

[25] Wolf, A.M., Kleinman, P.J.A., Sharpley, A.N. and Beegle, D.B. (2005) Development of a Water-Extractable Phosphorus Test for Manure: An Interlaboratory Study. *Soil Science Society of America Journal*, **69**, 695-700. http://dx.doi.org/10.2136/sssaj2004.0096

[26] Oehl, F., Frossard, E., Fliessbach, A., Dubois, D. and Oberson, A. (2004) Basal Organic Phosphorus Mineralization in Soils under Different Farming Systems. *Soil Biology and Biochemistry*, **36**, 667-675. http://dx.doi.org/10.1016/j.soilbio.2003.12.010

[27] USEPA (1992) Technical Support Document for Land Application of Sewage Sludge, Volume I. Office of Water, United States Environmental Protection Agency, Washington DC, EPA 822/R-93-001a.

[28] USEPA (1993) Standards for the Use or Disposal of Sewage Sludge; Final Rules. United States Environmental Protection Agency, Washington DC, 40 CFR Part 257 and 503.

[29] Haney, R.L., Haney, E.B., Hossner, L.R. and Arnold, J.G. (2006) A New Soil Extractant for Simultaneous Phosphorus, Ammonium, and Nitrate Analysis. *Communications in Soil Science and Plant Analysis*, **37**, 1511-1523. http://dx.doi.org/10.1080/00103620600709977

[30] Haney, R.L., Brinton, W.F. and Evans, E. (2008) Soil CO_2 Respiration: Comparison of Chemical Titration, CO_2 IRGA Analysis and the Solvita Gel System. *Renewable Agriculture and Food Systems*, **23**, 171-176. http://dx.doi.org/10.1017/S174217050800224X

[31] Fierer, N. and Schimel, J.P. (2003) A Proposed Mechanism for the Pulse in Carbon Dioxide Production Commonly Observed Following the Rapid Rewetting of a Dry Soil. *Soil Science Society of America Journal*, **67**, 798-805. http://dx.doi.org/10.2136/sssaj2003.0798

[32] Franzluebbers, A.J., Haney, R.L., Honeycutt, C.W., Schomberg, H.H. and Hons, F.M. (2000) Flush of CO_2 Following Rewetting of Dried Soil Relates to Active Organic Pools. *Soil Science Society of America Journal*, **64**, 613-623. http://dx.doi.org/10.2136/sssaj2000.642613x

[33] Haney, R.L., Franzluebbers, A.J. and Hons, F.M. (2001) A Rapid Procedure for Prediction of N Mineralization. *Biology and Fertility of Soils*, **33**, 100-104. http://dx.doi.org/10.1007/s003740000294

[34] Spohn, M. and Kuzyakov, Y. (2013) Phosphorus Mineralization Can Be Driven by Microbial Need for Carbon. *Soil Biology Biochemistry*, **61**, 69-75. http://dx.doi.org/10.1016/j.soilbio.2013.02.013

4

Phosphorus Adsorption of Some Brazilian Soils in Relations to Selected Soil Properties

Valdinar Ferreira Melo[1*], Sandra Cátia Pereira Uchôa[1], Zachary N. Senwo[2*], Ronilson José Pedroso Amorim[3]

[1]Department of Soil and Agricultural Engineering, Federal University of Roraima, Boa Vista, Brazil
[2]Department of Biological & Environmental Sciences, Alabama A&M University, Huntsville, USA
[3]Agronomy, Federal University of Roraima, Boa Vista, Brazil
Email: *valdinar@yahoo.com.br, *zachary.senwo@aamu.edu

Abstract

A major nutritional problem to crops grown in highly weathered Brazilian soils is phosphorus (P) deficiencies linked to their low availability and the capacity of the soils to fix P in insoluble forms. Our studies examined factors that might influence P behavior in soils of the Amazon region. This study was conducted to evaluate the maximum phosphate adsorption capacity (MPAC) of the soils developed from mafic rocks (diabase), their parent materials and other factors resulting in the formation of eutrophic soils having A chernozemic horizon associated with Red Nitosols (Alfisol) and Red Latosols (Oxisol) of the Amazonian environment. The MPAC was determined in triplicates as a function of the remnant P values. The different concentrations used to determine the MPAC allowed maximum adsorption values to be reached for all soils. The Latosol (Oxisol) and Nitosol (Alfisol) soils presented higher phosphate adsorption values that were attributed to the oxidic mineralogy and high clay texture while the Chernosol (Mollisol) soils presented the lowest phosphate adsorption values.

Keywords

Amazon Soil, Eutrophic Soils, Oxidic Soils, Phosphate, Alfisol, Oxisol

1. Introduction

Due to the climatic variations impacting the Amazon region of Brazil, the Northeast area of Roraima has transitioned from a semi-arid climatic condition to one with high precipitation [1]. The increase in precipitation has

*Corresponding authors.

greatly affected the soil's biogeochemical weathering attributes, erosion and deposition processes, thus contributing to the formation of deep, highly weathered soils dominant in kaolinite and oxidic mineralogy [2]. Some areas of the Amazon have soils originated from mafic rocks (diabase), resulting in the formation of Eurtophic Red Nitosol (Haplustalf), Dystrophic Yellow Red Latosol (Xanthic Haplustox), Eutropic Haplic Tb Cambisols, with A chernozem, Orthic black Chernosol (Mollisol) and vertic Orthic Ebanic Cheronosol (Mollisol) soil series [3].

Studies have shown that the soils consist typically of kaolinitic mineralogy and low fertility, with the predominant classes being Yellow Latosols (Oxisol) and Yellow Argisols (Ultisol) [4]-[6]. The indigenous areas of Roraima in general are located where these soils are found and reflect certain aspects of the sustainability and quality of life of the indigenes cultivating the soils for agricultural production. In the Raposa Serra do Sol Indigenous Reserve (approximately 1,747,464 ha); the proportion of soils (approximately 19,900 ha) considered as fertile for sustainable agricultural production is situated on gentle undulating relief and includes diabase soils of the "Pedra Preta" Sill [7].

A major nutritional problem to crops grown in these soils is P deficiency linked to low available P content and capacity to fix fertilizer P in insoluble form. Evaluating the P availability in these soils is important because soil P adsorption capacity directly influences plant response to phosphate fertilization of soils. Phosphorus fixation in soils, widely designated as adsorption, is a slow process that can take years to reach a balance and may lead to decrease in P availability. In most Brazilian soils, studies have shown that the major factors influencing soil P adsorption are the clay fractions, mineralogy, amorphous colloid content, pH, exchangeable aluminum and organic matter [8]-[10].

Phosphorus adsorption depends on the nature and quality of sites available on the mineral surfaces and is therefore affected by high clay contents within the same mineralogy [11]. Studies by Ker *et al.* [9] indicate that the mineralogy, crystal size, and specific surfaces of soils will be more important than the quantity of clay in determining soil's P adsorption capacities [12]-[14]. Bedin *et al.* [14] state that the presence of large proportions of sesquioxides on clays influences phosphate adsorption and precipitation with iron and aluminum. The objective of this study was to evaluate the maximum P adsorption capacity (MPAC) of the soils developed from mafic rocks (diabase) in northeastern Roraima and its relationships with certain physical, chemical and mineralogical attributes.

2. Materials and Methods

The soils used in this study were sampled from a toposequence and consisted of five soil profiles, developed from mafic rocks that form one of the largest mafic rock bodies in North Amazon's "Pedra Preta" Sill. The Maloca do Flechal study areas (**Figure 1**) is inhabited by the Macuxi Indians and situated in the northeast of the state of Roraima, Brazil. The climate average about 1200 mm in rainfall, with sunshine duration of 12 hours/day. The region's relief is characterized as undulated with step savannah vegetation and patches of seasonal forest on the East-West transect line.

The fieldwork consisted of opening trenches in a toposequence, collecting the soil samples in each profile soil and describing them according to Santos *et al.* [15]. Soils were classified based on the Brazilian Soil Classification System manual published by Embrapa [16] and analyzed for chemical properties (**Table 1**) as described in a soil analysis manual published by Embrapa [17]. The profiles were classified as Eutrophic Red Nitosol (Alfisol)-NVe, Tb Eutrophic Haplic Cambisol (Inceptisol)-CXbe; Orthic Ebanic Chernosol (Mollisol)-MEo; vertic Orthic Ebanic Chernosol (Mollisol)-MEov and Distgrophic Yellow Red Latosol (Oxisol)-LVAdf.

The Fe associated with the low crystal minerals in the surface horizons was extracted with ammonium oxalate [18], and the free Fe extracted using dithionite-citrate bicarbonate (DCB) [19]. The Fe content was determined using induced plasma emission spectrometry while the Fe associated with the crystalline minerals was calculated by the difference between the Fe extracted by DCB (Fe_d) and the Fe extracted with oxalate (Fe_o). The mineralogy was determined by X-ray diffraction of the clay and silt fraction. From the natural, Fe-free clay fraction, oriented slides were prepared and irradiated in an X-ray diffractometer at sweeping angle (2θ) intervals between 2° and 40° and a goniometer speed of 2° in 2θ/min, using CuKα radiation with Ni filter [20]. Powdered silt slides were analyzed under the same conditions while the diffraction patterns were interpreted according to Chen [21].

Phosphorus (P) was evaluated as described by Alvarez and Fonseca [22] using 5 g of fine ground soils mixed with a solution of $CaCl_2$ (0.01 mol·L^{-1}) containing 50 mg·L^{-1} P and shaken for an hour then filtered and the P

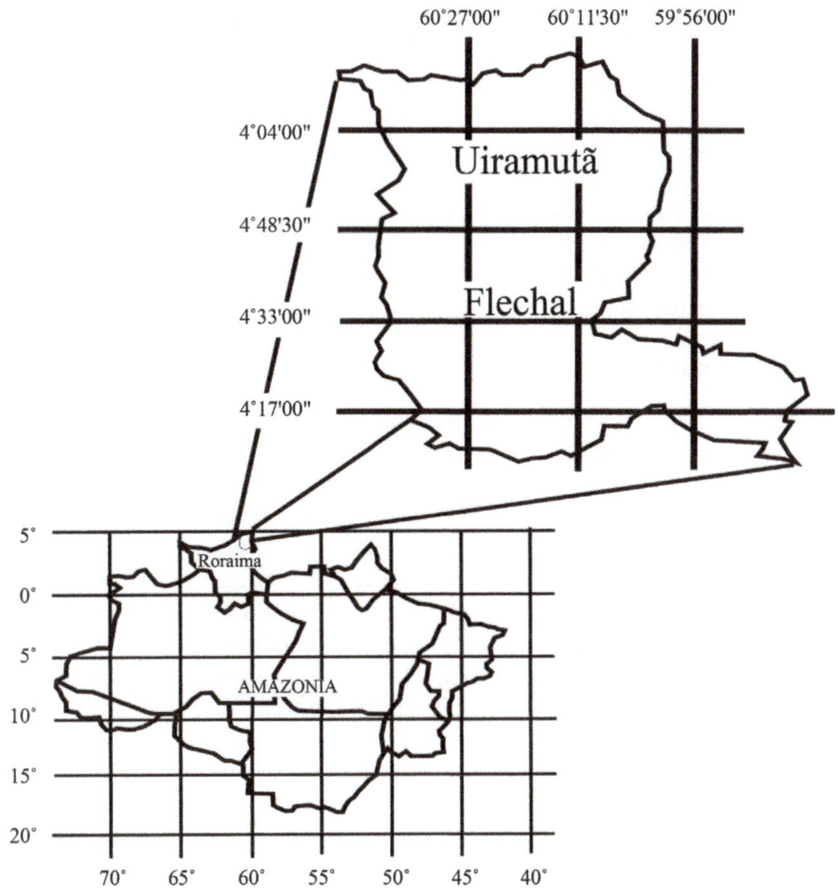

Figure 1. The maloca do flechal study areas.

Table 1. Chemical characteristics of soils studied.

Soil	Hor.	pH	Ca^{2+}	Mg^{2+}	Na^{+}	K^{+}	Al^{3+}	H + Al	S	CEC	V	SOM	P
			----------------------------$cmol_c \cdot kg^{-1}$							------------------------------------	-(%)-	$g \cdot kg^{-1}$	$mg \cdot kg^{-1}$
NVe	Ap	5.7	2.63	2.48	0.13	0.10	0.05	6.87	5.34	12.21	43	24.6	0.96
	Bnit	5.9	1.03	0.91	0.03	0.01	-	2.20	1.98	4.18	65	9.3	0.72
MEo	Ap	6.1	6.87	3.16	0.32	0.15	-	4.34	10.50	14.84	71	23.8	2.50
	Bnit	6.2	5.80	2.62	0.04	0.05	0.05	3.57	8.51	12.08	71	14.2	1.01
CXbe	Ap	6.8	8.69	2.79	0.11	0.05	0.05	2.75	11.64	14.39	81	15.3	1.33
	Bi_2	7.3	9.21	0.00	0.12	0.05	-	1.54	9.38	10.92	86	9.5	1.44
MEov	Ap	7.0	9.31	0.04	0.09	0.05	-	2.53	9.35	11.88	79	20.6	1.16
	Bi_2	7.4	12.7	0.06	0.11	0.05	-	1.43	12.92	14.35	90	3.7	1.39
CXbe	Ap	6.5	6.75	0.37	0.07	0.04	-	3.24	7.23	10.63	69	19.3	1.22
	Bi_2	7.8	7.31	0.00	0.09	0.03	-	0.71	7.43	8.14	91	3.7	1.08
LVAd	A	5.4	0.00	0.49	0.14	0.11	0.50	6.60	0.74	7.34	10	24.6	1.48
	Bw	4,3	0.00	0.50	0.04	0.04	0.50	8.52	0.52	9.04	06	22.5	0.53

Hor. = Horizon. 1) pH in water. 2) S = sum of bases. CEC = Cation exchange capacity at pH 7.0. V = percentage of saturation by bases; SOM = Soil organic matter. P = Phosphorus extractable by Mehlich-1. NVe-Red Nitosol. MEo-Orthic Ebanic Chernosol. Cxbe-Tb Eutrophic Haplic Cambisol. MEov-Vertic Orthic Ebanic Chernosol, LVAd-Distrophic Red Yellow Latosol.

concentration using Murphy and Riley [23] procedure. The Langmuir isotherm were obtained by adding 25 mL $CaCl_2$ (10 $mol \cdot L^{-1}$) solution containing P in doses ranging from 0 to 60 $mg \cdot L^{-1}$. After 24 hours agitation, the samples were centrifuged and the P in the equilibrium solution was quantified colorimetrically [23].

The data were fitted to the non-linear Langmuir model, relating the concentration of the adsorbed element per adsorption unit (soil) and the concentration of the element in the equilibrium solution (supernatant), and the maximum adsorption capacity calculated. The sorption values of each soil were plotted according to the Langmuir isotherm: $C/(x/m) = (1/Kb) + (C/b)$. Where C, is equilibrium P concentration ($mg \cdot L^{-1}$), x/m is the amount of sorbed P ($mg \cdot kg^{-1}$), m and b, are the constants related to P sorption maximum ($mg \cdot kg^{-1}$) and K is the bonding energy ($L \cdot mg^{-1}$), respectively. The plot of C/(x/m) versus C should give a straight line from which 1/b (slope) and K (slope or intercept) can be calculated [24]. The Pearson analysis of simple linear correlation using the Statistic program 7.0 between MPAC and some soil characteristics was carried out, and the correlations greater than 95% were considered significant ($\alpha < 0.05$).

3. Results and Discussion

The Fe contents determined from successive extractions using DCB (2), ranged from 29.4 (MEov) to 258.7 $g \cdot kg^{-1}$ (LVAdf). The values seemed high and atypical of Roraima [5] [6] and Amazonia soils [25] [26], given that the predominant soils in these environments have iron contents averaging less than 73 $g \cdot kg^{-1}$. The LVAdf stood out among the soils studied with lower Fe_o/Fe_d ratio indicating the predominance of high crystalline iron oxide.

The soils mineralogical compositions (**Table 2**) showed the presence of high mineral activities. Illite was common in all the soil horizons but smectite was restricted to soils with the A Chernozem horizon (CXbe and MEov). Kaolinite also occurred generally in most of the profiles, indicating intense weathering with strong lixiviation of bases, low fertility and lower (CEC) especially for LVAdf. The mineralogy of the 2:1 group might have resulted from the geology and climatic variations in the Amazon over geological time. The northeast of Roraima has passed through a recent sub-period of semi-arid climate to the current conditions of greater precipitations [27] [28]. However, this mineralogy contrasts with the data obtained by Vale Júnior [5] and Melo *et al.* [6] for other Roraima soils.

The P values varied greatly among the soil profiles and horizons. The highest values were observed in the surface horizons (**Table 3**), ranging from 19.8 (LVAdf) to 50.2 $mg \cdot L^{-1}$ (MEo). For the surface horizons, the values ranged from 6.6 $mg \cdot L^{-1}$ (LVAdf) to 46.9 (MEov) and this could be attributed to the type of clay mineralogy present in the soils. According to Alvarez *et al.* [29], the soils and horizons analyzed presented low P retention, except for LVAdf, which could be attributed to the highly negative charge on the smectites surfaces of the clays. The high P values in most of the soils could be explained by the striking presence of 2:1 minerals. Alves and Lavorenti [30] worked with Latosols in the state of São Paulo, Brazil and reported the lack of correla-

Table 2. Mineralogical composition of the clay, silt and sand fractions of soils studied.

Soil	Hor.	Clay	Silt	Sand
NVe	Ap	Ct, Gt	Qz, Ct, Bt, Fs, Ilm	Qz, Mi
	Bnit	Ct, Il, Gt	Qz, Ct, Bt, Fs, Ilm, Pg	Qz, Mi,
MEo	Ap	Ct, Il, Gt	Qz, Bt, Fs, Px	Qz, Mi
	Bnit	Ct, Il, Gt	Qz, Ct, Mi	Qz, Mi
CXbe	Ap	Ct, Il/Mt, Es, Gt	Qz, Fs, Ct, Px, Af	Qz, Fs, Pg, Mi
	Bi_2	Ct, Il, Es	Pg, Fs, Qs, Ct, Mi, Af	Pg, Qz, Fs, Mi, Mg, Cn
MEov	Ap	Ct, Il, Es	Fs, Pg, Qz, Ct, Mi, Af	Qz, Pg, Af, Fs
	Bi_2	Es, Ct, Il	Ac, Cl, Mi, Fs	Qz, Pg, Fs, Cn
CXbe	Ap	Ct, Il, Gt	Ct, Qz, Zl, Pg	Qz, Fs, Ru
	Bi_2	Ct, Il, Gt	Ct, Qz, Hb, Bt	Qz
LVAd	Ap	Gb, Ct, Il, Gt	Qz, Ct, Gb, Ru, Hm	Qz
	Bw_1	Ct, Gb, Il	Qz, Ct, Gb	Qz

Qz-Quartz, Ct-Kaolinite, Es-Smectite, Fs-Feldspar, Mi-Mica, Pg-Plagioclase, Af-Amphiboles, Px-Piroxene, Il-Illite, Bt-Biotite, Ac-Actinolite, Cl-Chlorite, Zl-Zeolite, Hb-Hornblendite, Ru-Rutile, Gb-Gibbsite, Gt-Goethite; Hm-Hematite, Mg-Manganite, Cn-Corundum.

Table 3. Maximum phosphate adsorption capacity for the soils studied.

Soil	Horizon	Clay $g \cdot kg^{-1}$	P-remaining $mg \cdot L^{-1}$	MAPC $mg \cdot kg^{-1}$
NVe	Ap	520	39.8	439.0
	Bnit.	640	4.2	1280.0
MEo	Ap	280	50.2	353.0
	Bnit.	370	36.2	475.0
CXbe	Ap	320	43.3	409.0
	Bi_2	470	29.5	551.0
MEov	Ap	290	47.9	370.0
	Bi_2	290	46.9	378.0
CXbe	Ap	350	40.7	431.0
	Bi_2	380	37.7	460.0
LVAdf	Ap	650	19.8	701.0
	Bw_1	700	6.6	1111.0

Nve-Red nitosol, MEo-Orthic ebanic chernosol, Cxbe-Tb eutrophic haplic cambisol, MEov-vertic orthic ebanic chernosol LVAdf-distrophic red latosol.

tions between P and goethite and P and kaolinite. High iron-sesquioxides contents are often considered as the determining factor of P adsorption in soils [31]; however, the low crystalline forms should fix more P than the crystalline forms [30]. The differences in behavior observed for LVAdf was more related to the presence of gibbsite than to the presence of the low crystalline forms, considering the low Fe_o/Fe_d ratio presented by this soil (**Table 4**).

The MPAC values (**Table 3**) ranged from 353 to 701 $mg \cdot kg^{-1}$ in the surface horizons and from 378 to 1280 $mg \cdot kg^{-1}$ for the sub-surface horizons. Except for the values obtained in the MEov profile and the MEo surface layer, the rest of the soils were in the range of MPAC values as indicated by Novais and Smyth [32] for Cerrado Latosols. In Amazon soils, Lima *et al.* [26] reported MPAC values ranging from 210 to 2170 $mg \cdot kg^{-1}$ in Oxisol, while Singh *et al.* [33] obtained values between 160 and 980 $mg \cdot kg^{-1}$ P. For central Amazon, Falcão and Silva [34] reported MPAC ranging from 298 to 888 $mg \cdot kg^{-1}$ for Ultisols. The MPAC values were lower than those normally found in highly weathered soils. In basalt-derived soils, Oxisol and Alfisol, Bognola [35] found MPAC values between 1620 and 2740 $mg \cdot kg^{-1}$, while Valladares *et al.* [36] obtained values between 526 and 1.667 $mg \cdot kg^{-1}$ for Ultisols. Soils originating from this material have higher MPAC because of the higher clay content and occurrence of oxides such as hematite and goethite [37] [38]. Despite the parental materials, climate and relief contributed to the soil formation with high smectite contents in undulated relief, highly weathered rich Fe and Al oxides contents. The structure, chemical compositions, exchangeable ion type and small crystal size of smectite clays are responsible for the unique properties, including a large chemically active surface area, a high cation exchange capacity and low anion exchange capacity.

In the high clay soils (**Table 3**), phosphate adsorption was more pronounced. The Yellow Red Latosol (Oxisol) presented the greatest MPAC in the two horizons, which can be explained by the high clay content, high Fe_d and low Fe_c contents (**Table 4**). The Nitosol also presented high MPAC, greater than LVAdf, in the subsurface horizon, which can be due to the high clay content. Vertic Orthic Ebanic Chernosol had the least adsorbed P in the two horizons, which can be explained by the low clay contents and low Fe_d and Fe_c values. Generally the soils showed increase in P adsorption with increase in soil depth, which can be attributed to increase in the clay contents from horizon A to horizon B. The high phosphate adsorption in the Latosol and Nitosol in the Amazon soils can be attributed to the low Fe contents predominate in this environment [6].

The clay, Fe_d and Fe_c contents correlated positively and significantly with the MPAC, suggesting that these attributes contributed to an increase in the MPAC in the horizons and profiles studied (**Table 5**). Given the strict correlations between MPAC and clay, Fe_d and Fe_c contents, it can be postulated that the presence of goethite in all the soils was a major determinant for the MPAC and not kaolinite soils. Bahia Filho *et al.* [39] tested the effects of the mineralogical components on MPAC and observed that goethite (Gt) was mainly responsible, con-

Table 4. Soil iron and silicon dioxide contents of soils.

Soil	Hor.	Fe_d	Fe_o	Fe_o/Fe_d	Fe_c	SiO_2
				$g \cdot kg^{-1}$		
NVe	Ap	58.3	3.5	0.06	54.8	1.41
	Bnit	75.5	3.2	0.04	72.3	1.13
MEo	Ap	106.0	7.7	0.07	98.3	1.26
	Bnit	95.1	5.6	0.06	86.9	1.90
CXbe	Ap	52.0	6.4	0.12	45.6	2.01
	Bi_2	51.1	4.9	0.10	46.2	2.65
MEov	Ap	34.9	7.7	0.22	27.2	2.01
	Bi_2	29.4	1.6	0.05	27.8	1.37
CXbe	Ap	77.1	7.2	0.09	69.9	1.61
	Bi_2	64.8	2.5	0.04	62.3	2.03
LVAdf	A	180.3	1.8	0.01	178.9	0.73
	Bw	258.7	1.0	0.004	257.7	0.88

Nve-Red nitosol, MEo-Orthic ebanic chernosol, Cxbe-Tb eutrophic haplic cambisol, MEov-vertico orthic ebanic chernosol, LVAdf-distrophic red yellow latosol. Fe_d-Dithionite-extracted Fe. Fe_o-Oxalate-extracted iron. Fe_c-crystalline Fe.

Table 5. Simple correlations between maximum phosphate adsorption capacity of soils and some soils properties.

	A Horizon	B Horizon
Clay	0.79^{*}	0.94^{**}
T	-0.87^{*}	-0.79^{*}
C	$0.38^{n.s.}$	$0.54^{n.s.}$
Fe_d	0.83^{*}	0.93^{**}
Fe_o	-0.87^{*}	-0.62^{*}
Fe_c	0.84^{*}	0.94^{**}

T-Clay. Fe_d-Dithionite-extracted Fe. Fe_o-Oxalate-extracted iron. Fe_c-crystalline Fe. n.s.-not significant. *mean of 0.01 < probability < 0.05; **significant at probability < 0.01.

tributing 6% of the total MPAC. The high affinity of geothite soils for P is likely due to the easy access of surface phosphate anions. The $H_2PO_4^-$ occupies site of the hydroxides (OH^-) previously coordinated to Fe^{3+}, forming much more stable surface complexes.

The correlation between the soil organic C and MPAC was not significant for the soils studied. This might be because organic C can influence P adsorption positively in numerous ways [40], negatively [41] or no effect [42]. The latter has been observed in Cerrado ecosystem (Savanna) soils where the C stock increases linearly with the soil clay and silt contents, demonstrating that total C does not increase the P adsorption, but enhances the richness of the minerals that retain P.

The CEC values and Fe_o correlated significantly and negatively, respectively, with MPAC. Indeed P adsorption tends to diminish with increase in CEC in the soils, because the negative charges repel the phosphate ion. On the other hand, negative correlation of MPAC with Fe_o is in line with reported studies. Hernández and Meurer [43] studied three forms of soils in Uruguay, an Argiudoll, a Hapludert, and a Natraqualf, and observed positive correlations between P adsorption and low Fe crystalline forms (Fe_o).

4. Conclusion

The Oxisol and the Alfisol soils presented the highest phosphate adsorption values, due to Fe and Al oxides rich minerals. The Molissol soils showed the lowest adsorption values due to the high presence of smectite clay. The soils that showed positive correlations with MPAC were in the order: A Horizon-Fe_c > Fe_d > clay and B Horizon-clay = Fe_c > Fe_d. The soils of the northern border of the Amazon exhibit P adsorption behavior different from other areas of the Amazon. Sustaining agricultural production in these soils will require management approaches that will enhance P availability. An adequate pool of labile P in these soils could be enhanced with inorganic P fertilizations and other organic residues.

References

[1] Schaefer, C.E.R. and Dalrymple, J. (1996) Pedogenesis and Relict Properties of Soils with Columnar Structure from Roraima, North Amazonia. *Geoderma*, **71**, 1-17. http://dx.doi.org/10.1016/0016-7061(95)00073-9

[2] Benedetti, U.G., Vale Jr., J.F., Schaefer, C.E.G.R., Melo, V.F. and Uchôa, S.C.P. (2011) Genesis, Chemistry and Mineralogy of Soils Derived from Plio-Plestocene Sediments and from Volcanic Rocks in Roraima—North Amazonia. *Brazilian Journal of Soil Science*, **35**, 299-312.

[3] Melo, V.F., Schaefer, C.E.G.R. and Uchôa, S.C.P. (2010) Indian Land Use in the Raposa-Serra do Sol Reserve, Roraima, Amazonia, Brazil: Physical and Chemical Attributes of a Soil Catena Developed from Mafic Rocks under Shifting Cultivation. *Catena*, **80**, 95-105. http://dx.doi.org/10.1016/j.catena.2009.09.004

[4] Schaefer, C.E.G.R., Resende, S.B., Correia, G.F. and Lani, J.L. (1993) Chemical and Minerological Characteristics of Sodium-Afected Soils from Northeeastern Roraima. *Brazilian Journal of Soil Science*, **17**, 431-438.

[5] Vale Jr., J.F. (2000) Pedogenesis and Changing of Soil under Shifting Cultivation in Areas of Acid and Basic Volcanic Rocks in Northeastern Roraima. Ph.D. Thesis, Universidade Federal de Viçoa, Viçosa.

[6] Melo, V.F., Schaefer, C.E.G.R., Fontes, L.E.F., Chagas, A.C., Lemos Jr., J.B. and Andrade, R.P. (2006) Physical Chemical and Mineralogical Characteristics of Soils from the Agricultural Colony of Apiaú (Roraima, Amazonia), under Different Land Uses and after Burning. *Brazilian Journal of Soil Science*, **30**, 1039-1050.

[7] Melo, V.F. (2002) Soil and Indicators of Agricultural Use in Roraima: Indigenous Areas of Maloca Flechal and Colonization of Apiaú. Ph.D. Thesis, Universidade Federal de Viçoa, Viçosa.

[8] Brennan, R.F., Bolland, M.D.A., Jeffery, R.C. and Allen, D.G. (1994) Phosphorus Adsorption by a Range of Western Australian Soils Related to Soil Properties. *Communications in Soil Science and Plant Analysis*, **25**, 2785-2795. http://dx.doi.org/10.1080/00103629409369225

[9] Motta, P.E.F., Curi, N., Siqueira, J.O., Van Raij, B., Furtini Neto, A.E. and Lima, J.M. (2002) Adsorption and Forms of Phosphorus in Latosols: Influence of Mineralogy and Use. *Brazilian Journal of Soil Science*, **26**, 349-359.

[10] Moreira, F.L.M., Mota, F.O.B., Clemente, C.A., Azevedo, B.M. and Bomfim, G.V. (2006) Phosphorus Adsorption in Ceará State Soils, Brazil. *Revista Ciência Agronômica*, **37**, 7-12.

[11] Bahia Filho, A.F.C., Braga, J.M., Resende, M. and Ribeiro, A.C. (1983) Relationship between Phosphorus Adsorption and Mineralogical Components of Latosols Clay Fraction of the Central Plateau. *Brazilian Journal of Soil Science*, **7**, 221-226.

[12] Torrent, J., Barron, V. and Schwertmann, U. (1990) Phosphate Adsorption and Desorption by Goethites Differing in Crystal Morphology. *Soil Science Society of American Journal*, **54**, 1007-1012. http://dx.doi.org/10.2136/sssaj1990.03615995005400040012x

[13] Ruan, H.D. and Gilkes, R.J. (1996) Kinetics of Phosphate Sorption and Desorption by Synthetic Aluminous Goethite before and after Thermal Transformation to Hematite. *Clay Minerals*, **31**, 63-74. http://dx.doi.org/10.1180/claymin.1996.031.1.06

[14] Bedin, I., Furtini Neto, A.E., Resende, A.V., Faquin, V., Tokura, A.M. and Santos, J.Z.L. (2003) Phosphate Fertilizers and Soybean Production in Soils with Different Phosphate Buffer Capacities. *Brazilian Journal of Soil Science*, **27**, 639-646.

[15] Santos, R.D., Lemos, R.C., Santos, H.G., Ker, J.C. and Anjos, L.H. (2005) Description Handbook and Collecting of Soil in Field. 5 Edition, Sociedade Brasileira de Ciência do Solo, Viçosa, MG.

[16] Embrapa (Empresa Brasileira de Pesquisa Agropecuaria) (2006) Brazilian System of Soil Classification. 2nd Edition, Embrapa Solos, Rio de Janeiro.

[17] Embrapa (Empresa Brasileira de Pesquisa Agropecuaria) (1997) Manual of Soil Analysis Methods. 2nd Edition, National Soil Research Center, Rio de Janeiro.

[18] Mckeague, J.A. and Day, J.H. (1966) Dithionite and Oxalate Extractable Fe and Al as Aids in Differentiating Various Classes of Soils. *Canadian Journal of Soil Science*, **46**, 13-22. http://dx.doi.org/10.4141/cjss66-003

[19] Mehra, O.P. and Jackson, M.L. (1960) Iron Oxide Removal from Soils and Clay by a Dithionite-Citrate System Buffered

with Sodium Bicarbonate. *Clays and Clay Minerals*, **7**, 317-327. http://dx.doi.org/10.1346/CCMN.1958.0070122

[20] Whittig, L.D. and Allardice, W.R. (1986) X-Ray Diffraction Techniques. In: Klute, A., Ed., *Methods of Soil Analysis, Part* 1: *Physical and Mineralogical Methods*, American Society of Agronomy, Madison, 331-362.

[21] Chen, P.-Y. (1977) Table of Key by Lines in X-Ray Power Diffraction Patterns of Minerals in Clays and Associated Rocks. Occasional Paper 21, Department of Natural Resources, Geological Survey, Bloomington, 67 p.

[22] Alvarez, V.V.H. and Fonseca, D.M. (1990) Definition of Doses of Phosphorus to Determine the Maximum Adsorption Capacity for Phosphate in Greenhouse Trials. *Brazilian Journal of Soil Science*, **14**, 49-55.

[23] Murphy, J. and Riley, J.P. (1962) A Modified Single Solution Method for the Determination of Phosphate in Natural Waters. *Analytica Chimica Acta*, **27**, 31-36. http://dx.doi.org/10.1016/S0003-2670(00)88444-5

[24] Bera, R., Seal, A., Bhattacharyya, P., Mukhopadhyay, K. and Giri, R. (2006) Phosphate Sorption Desorption Characteristics of Some Ferruginous Soils of Tropical Region in Eastern India. *Environmental Geology*, **51**, 399-407. http://dx.doi.org/10.1007/s00254-006-0335-9

[25] Silva, J.R.T. (1999) Soils of Acre: Physical, Chemical and Mineralogical and Phosphate Adsorption. Thesis of Ph.D., Federal Universtiy of Viçosa, MG, Brazil.

[26] Lima, H.N., Melo, J.W.V., Schaefer, C.E.G.R., Ker, J.C. and Lima, A.M.N. (2006) Mineralogy and Chemistry of Three Soils along a Toposequence from the Upper Solimões Basin, Western Amazonia. *Brazilian Journal of Soil Science*, **30**, 59-68.

[27] Schaefer, C.E.G.R. (1994) Soils and Paleosols from Northeast Roraima, North Amazonia: Geomorphology, Genesis and Landscape Evolution. Ph.D. Dissertation, University of Reading, Reading, Berkshire.

[28] Schaefer, C.E.G.R. and Vale Júnior, J.F. (1997) Climate Changing and Landscape Evolution in Roraima: A Review from Cretaceous to Recent. In: Barbosa, R.I., Ferreira, E.J.G. and Castellón, E.G., Eds., *Man, Environment and Ecology in the State of Roraima*, INPA, Manaus, 231-293.

[29] Alvarez, V.V.H., Novais, R.F., Dias, L.E. and Oliveira, J.A. (2000) Determination and Use of the Remaining Phosphorus. *Newsletter, Sociedade Brasileira de Ciência do Solo*, **25**, 27-33.

[30] Alves, M.E. and Lavorenti, A. (2004) Remaining Phosphorus and Sodium Fluoride pH in Soils with Different Clay Contents and Clay Mineralogies. *Pesquisa Agropecuária Brasileira*, **39**, 241-246. http://dx.doi.org/10.1590/S0100-204X2004000300006

[31] Uehara, G. and Gillman, G. (1981) The Mineralogy, Chemistry and Physics of Tropical Soils with Variable Charge Clays. Westview, Boulder.

[32] Novais, R.F. and Smyth, T.J. (1999) Phosphorus in Soil and Plant in Tropical Conditions. Universidade Federal de Viçosa-MG, Viçosa, 399 p.

[33] Singh, R., Moller, M.R.F. and Ferreira, W.A. (1983) Phosphorus Adsorption Kinetics in Amazonian Soils. *Revista Brasileira de Ciência do Solo*, **7**, 227-231.

[34] Falcão, N.P. and Silva, J.R.A. (2003) Phosphorus Adsorption Characteristics in Some Central Amazonian Soils. *Acta Amazônica*, **34**, 337-342. http://dx.doi.org/10.1590/S0044-59672004000300001

[35] Bognola, I.A. (1995) Chemical, Physical and Mineralogical Soil Intermediaries between Latosols Brunos and Latosols Purple. Master's Thesis, Universidade Federal de Viçosa, Viçosa.

[36] Valladares, G.S., Pereira, M.G. and Anjos, L.H.C. (2003) Phosphate Sorption in Low Activity Clay Soils. *Bragantia*, **62**, 111-118. http://dx.doi.org/10.1590/S0006-87052003000100014

[37] Curi, N., Camargo, O.A., Guedes, G.A.A. and Silveira, J.V. (1988) Phosphorus Sorption in Latosols of Southeast and South of Brazil. In: *Reunião de Classificação e Correlação de Solos e Interpretação de Aptidão Agrícola*, 3rd Edition, Empresa Brasileira de Pesquisa Agropecuária, Rio de Janeiro, 150.

[38] Fontes, M.P.F. and Weed, S.B. (1996) Phosphate Adsorption by Clays from Brazilian Oxisols: Relationships Whit Specific Surface Area and Mineralogy. *Geoderma*, **72**, 37-51. http://dx.doi.org/10.1016/0016-7061(96)00010-9

[39] Bahia Filho, A.F.C., Vasconcellos, C.A., Santos, H.L., Mendes, J.F., Pitta, G.V.E. and Oliveira, A.C. (1982) Inorganic Phosphorus and Available Phosphorus Forms in a Dark Red Latosol, Fertilized with Different Phosphates. *Brazilian Journal of Soil Science*, **6**, 99-104.

[40] Parfitt, R.L. (1978) Anion Adsorption by Soils and Soil Materials. *Advances in Agronomy*, **30**, 1-50. http://dx.doi.org/10.1016/S0065-2113(08)60702-6

[41] Borggaard, O.K., Jorgensen, S.S., Moberg, J.P. and Raben-Lage, B. (1990) Influence of Organic Matter on Phosphate Adsorption by Aluminum and Iron Oxides in Sandy Soils. *European Journal of Soil Science*, **41**, 443-449. http://dx.doi.org/10.1111/j.1365-2389.1990.tb00078.x

[42] Zinn, Y.L., Lal, R. and Resck, D.V.S. (2005) Texture and Organic Carbon Relations Described by a Profile Pedotrans-

fer Function for Brazilian Cerrado Soils. *Geoderma*, **127**, 168-173. http://dx.doi.org/10.1016/j.geoderma.2005.02.010

[43] Hernández, J. and Meurer, E.J. (1990) Phosphorus Sorption in Soils from Uruguay and Its Relationship with Iron Oxides. *Brazilian Journal of Soil Science*, **22**, 223-230.

5

The Effect of the Geomorphologic Type as Surrogate to the Time Factor on Digital Soil Mapping

Huixia Chai[1], Sheng Rao[1], Reibo Wang[2], Jing Liu[3], Qi Huang[1], Xuejie Mou[1]

[1]Chinese Academy for Environmental Planning, MEP, Beijing, China
[2]State Key Laboratory of Resources and Environmental Information System, Institute of Geographical Sciences and Natural Resources Research, CAS, Beijing, China
[3]Department of Geography, University of Wisconsin-Madison, Madison, USA
Email: chaihx@caep.org.cn

Abstract

Many environmental variables are frequently used to predict values of soil in locations where they are not measured. Digital soil mapping (DSM) has a long-standing convention to describe soils as a function of climate, organisms, topography, parent material, time and space. It is obvious that terrain, climate, parent material and organisms are used frequently in the prediction of soil properties while time and space factors are rarely used. Time is the indirect factor for the formation and development of soil. Moreover, it is very useful to explicit and implicit estimates of soil age for DSM. However, it is often difficult to obtain time factor. In the absence of explicit soil age data, geomorphologic data are commonly related to soil relative age. Consequently, this study adopts the geomorphologic types (genesis type of geomorphology) as surrogate to the time factor and analyzes its effect on DSM. To examine this idea, we selected the Ili region of northwestern China as the study area. This paper uses geomorphologic data from a new digital geomorphology map as the implicit soil age in predictive soil mapping. For this study, Soil-landscape inference model (SoLIM) was used to predict soil properties based on the individual representation of each sample. This model applies the terrain (topography), climate, parent material (geology) and time (geomorphologic type) to predict soil values in the study area where they are not measured. And the independent sample validation method was used to estimate the precision of results. The validation result shows that the use of geomorphologic data as surrogate to the time factor in the individual representation leads to a considerable and significant increase in the accuracy of results. In other words, implicit estimates of soil age by genesis type of geomorphology are very useful for DSM. This increase was due to the high purity of the geomorphologic data. This means that the geomorphologic variable, if used, can improve the quality of DSM. Predicted value through the proposed approach comes closer to the real value.

Keywords

Geomorphologic Type, Environmental Variables, Digital Soil Mapping, Soil-Landscape Inference Model (SoLIM), Individual Validation

1. Introduction

Digital soil mapping (DSM) is a technology and method to describe the spatial distribution information of soil with a grid format on the basis of field samples and/or auxiliary data, such as soil environmental factors and other soil properties [1]. DSM has a long-standing convention to describe soils as an equation of climate, organisms, topography, parent material, and time [2] [3]. McBratney *et al.* [4] put forward a so-called "scorpan model" as a spatial predicted model of soil properties based on the environmental variables of climate, organisms, topography, parent material, time and space. Those variables, if their information is available for the location, are frequently used to predict values of soil in locations where they are not measured [5]-[9]. Therefore, these factors should consider into the methods or models of predicted soil properties, at least in theory. But in practice it is hard to do because it is too difficult to get or measure some environmental factors like time.

Studies use various approaches to predict soil properties by single-variable and multi-variable statistics [10]-[12], geostatistical [13]-[15] and hybrid methods [16] [17], and process-based models that relate soils to environmental covariates consider spatial and temporal dimensions [1] [4] [18]. Recently, most studies adopt the synthetic analysis and comprehensive applications of more environmental variables to predict the soil properties or the spatial distribution of soil [19]-[22]. However, it is obvious that terrain, climate, parent material (geology) and organisms (vegetation) are used frequently in the prediction of soil properties while time and space factors are rarely used.

Actually, it is very necessary to get the time factor into DSM. Time is the indirect factor for the formation and development of soil because it is a way to embody the effect of climate, organisms, topography, and parent material on the soil formation through time [3]. Explicit and implicit estimates of soil age are very useful environmental covariants for DSM, but are often difficult to obtain [12]. The temporal information of soil formation is unavailable, or not widely available, especially the absolute time, although some studies use the 14C detection or fossil analysis to obtain the soil age of sample points [23] [24]. Moonjun *et al.* [25] tried to find a correlation between the observed data (in the sample areas) and the environmental characteristics derived from the DEM, and they distinguish three relative soil ages: old (Ultisols and Oxisols), mature (Alfisols), and young (Inceptisols and Entisols). Noller [26] reviews various approaches to add geochronology in DSM and obtains the improved prediction of soil on the landscape in addition to several implicit age covariates, including custom thematic maps of lithology and geologic age. Therefore, it is important to get the time factor into DSM. Furthermore, how to accurately get the temporal information of soil formation is always a problem in the whole study area only based on few observed data of field samples. Generally, in the absence of explicit soil age data, implicit topographic data representing landforms that are commonly related to soil relative age and remotely sensed spectral data that focus on surface spectral characteristics represent surface or soil age [12].

Consequently, the purpose of this study is to analyze the effect of the genesis types of geomorphology as surrogate to the time factor on DSM. In other words, this study adopts the geomorphologic types as the relative time in the process of DSM. To analyze the effect of geomorphologic types on DSM, this study uses the SoLIM (a professional software for DSM) to predict soil organic matter (%) of the top layer in locations where they are unknown.

2. Basic Idea

2.1. Theory

Geomorphology is one of the predominant factors determining the features of the earth's surface and an important symbol and dominant factor of regional differentiation. It (geomorphology) directly influences and even decides the spatial distribution and transformation of hydrology, climate, soil, and other ecological and environmental factors [27]. Since in 1980, Knott *et al.* [28] had been analyze and find that the geomorphologic mapping boundaries are coincident with soil material boundaries, which implies a correlation between soil material and geomorphology to some degree. Geomorphologic processes, both erosional and depositional, create distinctive landforms which have a great influence on soil types and soil distribution [29]. The soil condition is different in places with various geomorphologic types due to the forming reason of geomorphology, as well as in different parts of geomorphology. The forming reason of geomorphology means that it will appear in some dominant geomorphologic processes, and produce characteristic landforms in particular climatic conditions [30]. Bacon *et al.* [31] point that systematic relationships exist between landscapes position (terrain conditions of desert landform) and soil formation processes. In their study, the assignment of relative age classes to each landform or geomorphic surface map unit was based on crosscutting relations, surface morphology and roughness, and topographic relief observable on the multispectral imagery. Moreover, Stum *et al.* [32] use landform surface age as the implicit soil age in predictive soil mapping by spatially modeling the shoreline of Pleistocene pluvial Lake Bonneville.

Therefore, we consider that the geomorphologic type could be used to approximate the time factor at the categorical level in soil formation processes. In other words, this study wants to take the geomorphologic type as surrogate to the time factor on DSM. The geomorphologic type here specifically means the genesis type of geomorphology. The genesis type is classified by the forming reason of geomorphology. It is a nominal variable and only represents the relative time.

2.2. Experiment Design

The process of DSM in this study was based on the Soil-landscape inference model (SoLIM). SoLIM is a new technology for soil mapping based on recent developments in Geographic Information Sciences (GIS), Artificial Intelligence, and Information Representation Theory. SoLIM was designed to overcome the limitations of existing soil survey methods and to improve the efficiency and accuracy of soil surveys [3]. Case studies have shown that SoLIM is more efficient and accurate than traditional soil survey methods, in that it generates a range of products which the traditional approaches couldn't provide, and it can be employed in a production mode of soil survey.

This study adds the geomorphologic type as surrogate to the relative time of soil and use the sampled-based approach under the SoLIM framework to examine the effect of including geomorphologic type. The environmental variables in this study (**Table 1**) including topography (Terrain, derive from DEM), climate, parent material (obtain from geology map) and time (relative time, using geomorphologic type as indicator).

This study describes two experiments to test the idea mentioned above. Each experiment has different environmental variables. The first group of environmental variables includes climate, parent material and topography. The second group includes climate, parent material, topography and geomorphology. Based on two groups of environmental variables, this paper presents two soil maps by the same research methods. Then we compare and analyze the precision between the two soil maps.

3. Method and Data

This section provided an introduction to study area, data and method in this study. By the way, readers can refer to literatures [3] [33] if you are interested in a comprehensive description of prediction method. Based on the knowledge of the relationship between soil and environment, this paper selected four environmental (**Table 1**) variables to describe environmental features. Climate data and geomorphic types collected by State Key Laboratory of Resource and Environmental Information System, IGSNRR, CAS. Parent material rasterized from Chinese Geology Map (1:1000,000). Seven topographic covariates (altitude (m), slope gradient (%), aspect, contour curvature, profile curvature, Plane curvature, Surface area ratio (SAR), Topographical wetness index (TWI)) were derived from DEM, which was downloaded from SRTM website.

3.1. Study Area

The study area of this research is the Ili region located between 42°14'16"N and 44°50'30"N, 80°09'42"E and 84°56'50"E, Xinjiang Uygur Autonomous Region, China (**Figure 1**). The Ili region is about 350 km (east-west) by 280 km (north-south) covering about 56,400 km^2. Terrain conditions here are complex, containing arduous mountains as well as flat plains. Influenced by local geomorphology, climate and vegetation conditions deviate

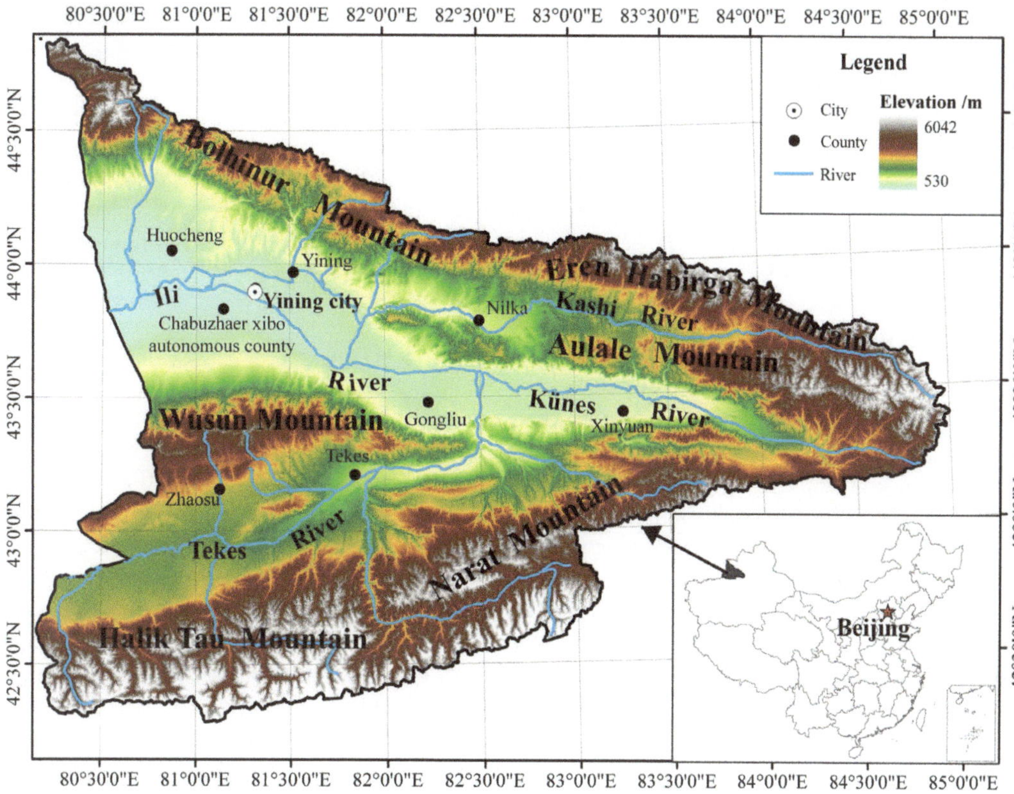

Figure 1. Study area.

Table 1. Environmental variables and their resolution.

Environmental variables		Resolution	Data sources
Climate	Precipitation Temperature Relative humility	1 km	State Key Laboratory of Resource and Environmental Information System, IGSNRR, CAS
Parent material	Parent material type	1 km	Rasterized from Chinese Geology Map (1:1000,000)
Topography	Altitude Slope gradient Aspect Profile curvature Plane curvature Surface area ratio, SAR Topographical wetness index, TWI	90 m	Original DEM was downloaded from SRTM website and derived indexes were generated based on either ArcGIS or specific algorithms
Geomorphology	Geomorphologic type (genesis type of geomorphology)	1 km	State Key Laboratory of Resources and Environmental Information System, IGSNRR, CAS

from the general distribution pattern at certain locations. The amount of annual precipitation on the flat plain is about 140 - 500 mm and, as elevation increases from west to east, annual precipitation increases at the same time. Vegetation types mainly include meadows and swamps on flat plains and forests on mountains.

The main parent materials are quartz sand, mica sand, red mudstones and alluvial loess or alluvial loess-like materials. Thanks to the variation of climate, terrain, vegetation and parent material conditions, soil types are relatively abundant on both vertical and horizontal distributions, mainly including chestnut soil, chernozem soil, forest soil, meadow soil, and Sierozem soil.

The basic geomorphology in the study area is composed of mountain landform and river-valley landforms. Geomorphologic types in the mountains at extremely high or high altitudes are mainly glacial and periglacial landforms. The lower part of the mountains is mainly fluvial landforms which formed by the erosion action of water and arid landforms which formed by a dry eroding effect. Alluvial plain, proluvial plain, and alluvial-proluvial plain are mainly distributed in the basin area. Fluvial landform is the dominant geomorphologic type in the plain landform area. In addition, the study area also has a small number of aeolian landform and loess landform.

3.2. Geomorphologic Data and Field Data

This paper collected all necessary data of study area, including climate data, SRTM-DEM, Geology Map, and the data of geomorphic types, to gain corresponding variables.

Geomorphology has a profound influence on the formation and development of soil by redistributing parent material, water, and heat. The geomorphologic type derives from the new 1:1,000,000 set of geomorphologic atlas of China. The digital geomorphologic database was created by visual interpretation from Landsat TM/ETM imageries, SRTM-DEM, and published geomorphologic maps, etc. [34]. The atlas was compiled and finished from the database, published by Science Press in 2009 [35]. The geomorphologic data include seven layers, *i.e.* basic morphology, genesis, sub-genesis, morphology, micro-morphology, slope and aspect, material composition and lithology. Here, we only use the genesis type of geomorphology to represents the relative time.

In addition, this paper also collected 107 soil sample data from the following sources: 1) National Native Soil Type Log (the Second National Soil Survey Office, 1995); 2) Field survey carried out by Dr. Ma Xingwang, Institute of Soil and Fertilizer, Xinjiang Academy of Agricultural Sciences; 3) Offered by Dr. Zhang Baiping and Dr. Zhang Hongqi, Institute of Geographical Sciences and Natural Resources Research, Chinese Academy of Sciences; 4) The book named Soil in Ili area, Xinjiang (1985).

3.3. Prediction Method and Computational Details

This paper adopts the independent sample validation method to estimate the performance of the predicted result. First, 75 of the 107 sampling were chosen to serve as sample points to predict soil properties. Then the other 32 soil samples serve as validation points to test the precision of DSM. Finally, estimate value of validation points would be extracted from the predicted result of 75 sample points through add validation points layer upon the result layer. This method can examine the precision of single point forecasting on two soil maps.

3.3.1. Environmental Variables

Climate data include annual-averaged precipitation, temperature and relative humidity, as well as the maximum and minimum of monthly-averaged precipitation, temperature and relative humidity within a year. These three variables have great influence on the decomposition and transformation of soil organic matter.

Topographical indexes include slope, aspect, profile curvature, plane curvature, surface area ratio and topographical wetness index. Topographical Wetness Index (TWI) is estimated according to a modified Multiple-Flow-Direction (MFD) algorithm [36] for calculating upslope drainage area. Surface Area Ratio (SAR), which represents the roughness within a mapping grid, is generated based on the algorithm developed by Jenness [37]. Other topographical variables are generated through ArcGIS. The content of surface soil organic matter often has a negative correlation to elevation [38] [39]. Slope, which represents the tilt degree of the local surface topography, has a direct relationship to the stability of soil, the drainage and gathering ability of surface water [39] [40]. Aspect affects the accumulation and decomposition of surface soil organic matter [41]. Curvature affects the spatial distribution of surface soil organic matter [42]. Surface Area Ratio (SAR) is used for distinguishing mountains and plains by terrain characteristics [43]. Topographical wetness index, which represents the drainage and gathering ability of soil, affects the content of surface soil organic matter [39].

Parent material types are often used to describe the different characteristics of parent material [4]. They include quartz sand, mica sand, red mudstones, alluvial loess, alluvial loess-like materials, glacier and snow cover.

The geomorphologic type which mentioned in this paper is the genesis type of geomorphology. This study adopts the genesis type of geomorphology as surrogate to the time factor. In other words, the geomorphologic type should be regard as the relative time for DSM in this study. According to the forming reason of geomorphology, there are six geomorphologic types in the study area: loess landform, fluvial landform, Aeolian landform, arid landform, glacial landform, and periglacial landform.

3.3.2. Quantifying and Estimating Similarity

as the resolution of climate, parent material and geomorphology information is approximate to the mapping grid, the single value method was adopted to represent climate and geological characteristics of each grid. As the resolution of topographical factors is refined, the probability distribution of each index was approximated to represent topographical characteristics of each grid. For estimating a smooth probability distribution curve, the Kernel Density Estimation (KDE) method has been adopted in this case study.

In this study GOWER distance was adopted to calculate climate similarity. GOWER distance measures the similarity between a given location (i, j) and a sample k on a series of environmental variables, standardized by the range of each variable [44] (Equation (1)):

$$Clim\,ate\_simi_{i,j}^{k} = \frac{1}{m}\sum_{v=1}^{m}\left(1 - \frac{\left|e_{ij}^{v} - e_{k}^{v}\right|}{range\left(e^{v}\right)}\right) \tag{1}$$

where m is the number of climate variables; $\left|e_{ij}^{v} - e_{k}^{v}\right|$ denotes the absolute difference on the v-th climate variable between location (i, j) and sample k; $\mathrm{range}\left(e^{v}\right)$ is the global value-range of the v-th climate variable. The value of overall climate similarity $Clim\,ate\_simi_{ij}^{k}$ will be between 0 and 1.

Parent material similarity and geomorphologic similarity were decided through Boolean function. If the parent material type and geomorphologic type at the location (i, j) are identical with sample point k, it means parent material similarity and geomorphologic similarity are equal to 1, conversely, parent material similarity and geomorphologic similarity equal 0.

Similarity on each topographical index is calculated through the Consistency Measure (CM) algorithm, and finally, the integrated topographical similarity for all topographical indexes was estimated by an average-operator (Equation (2)).

$$Topographic\_simi_{i,j}^{k} = \frac{1}{m}\sum_{v=1}^{m} CM_{ij,k}^{v} \tag{2}$$

where $Topographic\_simi_{i,j}^{k}$ denotes the topographical similarity between location (i, j) and sample k; m is the number of involved topographical indexes; $CM_{ij,k}^{v}$ is the Consistency index of the probability density distribution function on the v-th topographical variable between location (i, j) and sample k.

Based on the importance of each environmental variable which indicates spatial distribution of soil at different scales, this paper generates the similarity between location point and sample point through comprehensive analysis of climate similarity, parent material similarity, topographical similarity and similarity of geomorphologic type (**Figure 2** and **Figure 3**).

Please note the similarities and differences between **Figure 2** and **Figure 3**.

Bothe figures have three things in common. Firstly, C_similarity, Topo_similarity means climate similarity, topographical similarity separately. Secondly, weights of climate similarity and topographical similarity both are 0.5. Thirdly, climate similarity cut is 0.5.

Figure 2 and **Figure 3** have one difference. In **Figure 2**, environmental variables include climate, parent material and topography. **Figure 3** has one more environmental variable than **Figure 2**, environmental variables include climate, parent material, topography and geomorphologic type.

First, this paper examines the climate similarity between an unsampled location and a certain case (a climate-similarity-cut was set at 0.5 and has been adopted generally as we had no background information on deciding this threshold).

Then, we test the parent material similarity if the unsampled location was similar enough with the case on climate.

Third, this paper continue to examine the geomorphologic similarity if the parent material of the unsampled location is the same as the case (it means the parent material similarity is equal to 1).

Finally, the topographical similarity and climate similarity would be integrated to estimate the final similarity using the weighted-average method if the geomorphologic similarity is equal to 1 (it means the geomorphologic

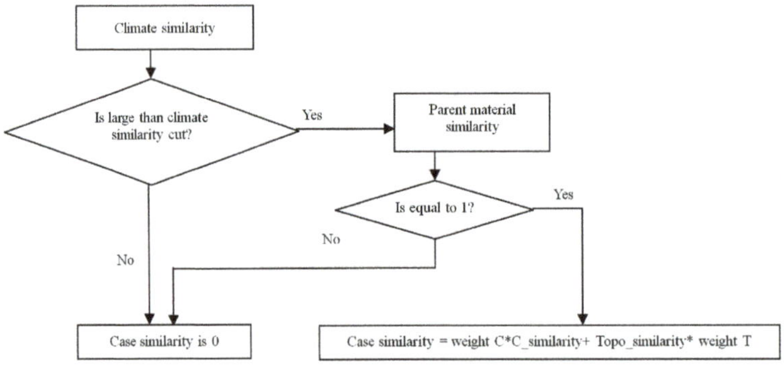

Figure 2. P-function for integrating environmental similarity into case-similarity.

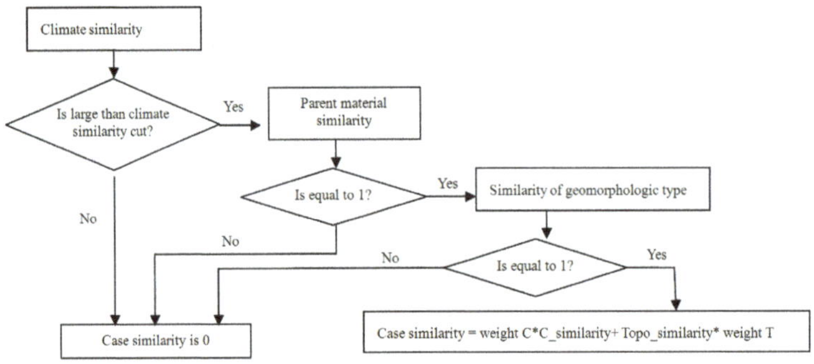

Figure 3. P-function for integrating environmental similarity into case-similarity.

type of the unsampled location is the same as the case).

Here, the weights of the topographical similarity and climate similarity should be assigned considering the relative importance between them. As we had no knowledge to decide which one is more important, we assigned 0.5 to both as the weights.

The estimation of the value of the targeted soil property is calculated with a weighted average method [3]:

$$V_{ij} = \frac{\sum_{k=1}^{n} S_{ij}^{k} \cdot V^{k}}{\sum_{k=1}^{n} S_{ij}^{k}} \quad (3)$$

in which *n*' is the number of the selected soil samples whose environmental similarity to the unvisited location (*i*, *j*) exceeds 1 minus the uncertainty threshold, S_{ij}^{k} is the environmental similarity of the unvisited location (*i*, *j*) to the soil type *k* in sample location k, and V^k is the value of the targeted soil property of soil sample location k.

4. Result and Discussion

4.1. Predicted and Discussion

Figure 4 shows soil organic matter (%) of the top layer predicted through the proposed approaches. The gray area with "No-Data" is the grid which cannot be represented by current samples. **Figure 4(b)** is the result on the basis of three environmental variables: climate, parent material and topography. **Figure 4(b)** is the result on the basis of four environmental variables: climate, parent material, topography and geomorphology.

The result shows that estimated value through the proposed approaches is closer to the real value than to the average value. In other words, the proposed approach in this paper is valid. These graphs (**Figure 4**) show that the spatial distribution of soil organic matter content has little difference between the two results. It can help avoid wasting time and energy if the geomorphologic type is added as surrogate to the time factor in the process

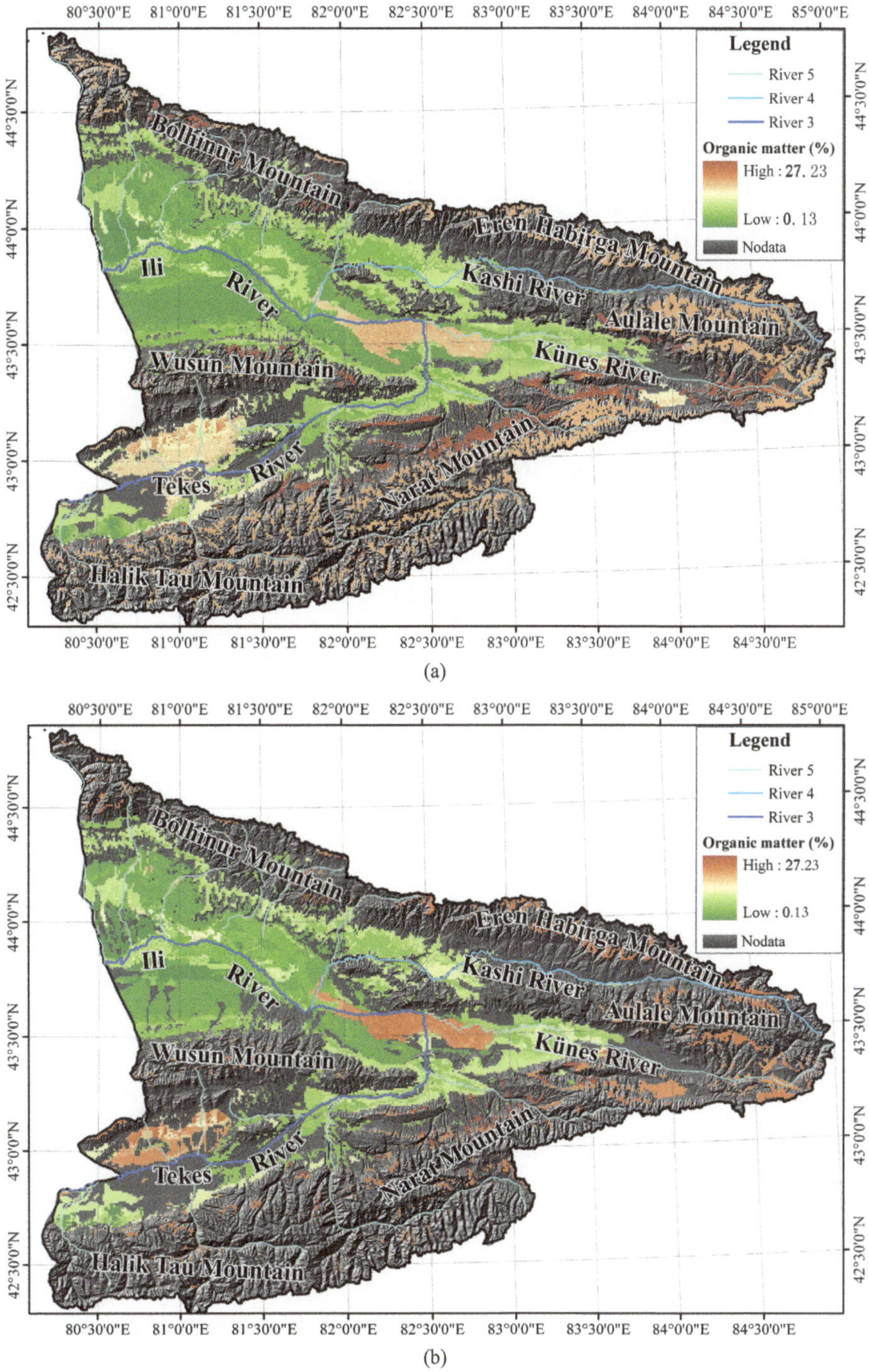

Figure 4. Soil organic matter (%) of top layer predicted though the proposed approach. (a) Without geomorphology; (b) With geomorphology.

of DSM. This means it will improve work efficiency.

4.2. Validation and Discussion

Generally, the mean/mean-bias error (ME or MBE), the mean absolute error (MAE), and the root mean squared error (RMSE) are used to test the error statistic. In this paper, the deviations of validation results are summarize by RMSE and MAE.

The root mean squared error (RMSE) provide a measure of estimation error that was more sensitive to outliers than the MAE

$$RMSE = \sqrt{\frac{1}{n}\sum_{i=1}^{n}\left[V'(s_i)-V(s_i)\right]^2} \tag{3}$$

where the generated predicted value represent by $V'(s_i)$, the true observation value represent by $V(s_i)$. So, predicted error at each sample point describe by $V'(s_i)-V(s_i)$.

The mean absolute error (MAE) reflects the absolute error of the estimates. It is expressed as follows

$$MAE = \frac{1}{n}\sum_{i=1}^{n}\left|V'(s_i)-V(s_i)\right| \tag{4}$$

For 32 validation points, only 20 points have effective values, the rest of the validation points are located in the place where without predicted value. So, this paper just adopts the predicted values of these 20 points to calculate RMSE and MAE.

Table 2 shows the validation results of two experiments within different environmental variables. **Table 2** indicates that the root mean squared error (RMSE) and the mean absolute error (MAE) are lower significantly after the data layer of geomorphologic type is added. It explains that the geomorphology variable can improve the accuracy of the predicted results in soil organic matter mapping. The values of RMSE and MAE with geomorphology are less than that without geomorphology. The value of RMSE without geomorphology is three times as big as the values of RMSE with geomorphology. This result indicating that the accuracy of predicted results with geomorphologic type is higher than one without geomorphologic type.

According to the validation results, most predicted values with geomorphology of validation samples are closer to real values than predicted values without geomorphology (**Figure 5** and **Figure 6**). It can veritably reflect the soil properties if the geomorphologic type is added as the relative time. After adding the geomorphology variable, the accuracy rate of predicted values have been increased by an average of 34%. In other words, the predicted result with geomorphology is much better than the predicted result without geomorphology.

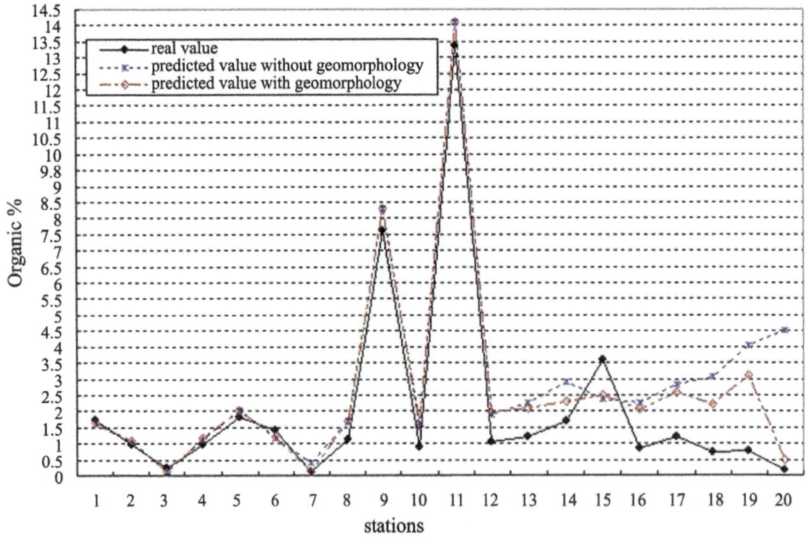

Figure 5. Comparative analysis of the real values, predicted values without geomorphology and with geomorphology.

Table 2. Validation results.

	Without geomorphology	With geomorphology
RMSE	2.33	0.81
MAE	1.06	0.69

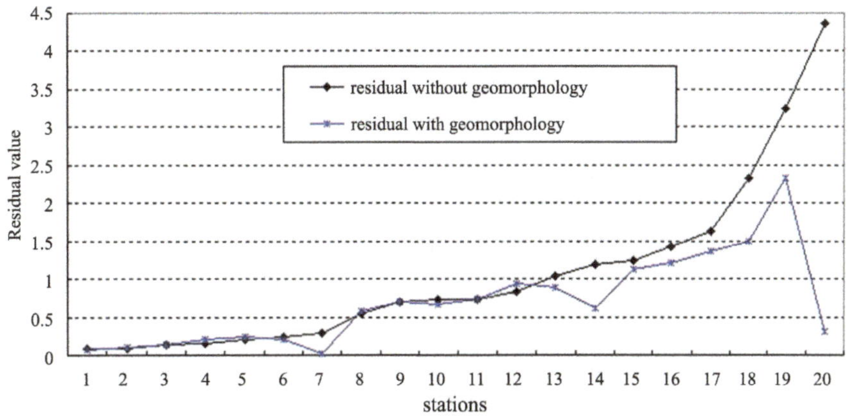

Figure 6. Comparative residual of the predicted values without geomorphology and the predicted values with geomorphology.

5. Conclusions

There is no universal soil equation or digital soil prediction model that fits all regions and purposes [11]. Hence, the soil prediction at different scales (local, national and global scales) needs to build various equations and models respectively. This study adopts the geomorphologic type to represent the relative time of soil age and estimate the effect of the geomorphologic type on the predicted precision of DSM. We consider genesis type of geomorphology as surrogate to the time factor and predict soil organic matter through the SoLIM. This study tries to analyze the effect of the geomorphologic types as surrogate to the time factor on DSM by the case study.

The result shows that using the geomorphologic type as surrogate to the time factor can improve the result quality of DSM. The genesis type of geomorphology results in the estimated value closer to the real value. Validation with a probability sample shows that use of geomorphology data as the relative time factor leads to a considerable and significant increase of accuracy. In other words, implicit estimates of soil age by genesis type of geomorphology are very useful for DSM. There are 75% predicted values having been increased. This increase in accuracy is due to the high precision of the existing geomorphology data.

In addition, with such a limited sample-set and such complex natural conditions, it is probably not reasonable to adopt any of the current interpolation method to predict soil property values over the entire study area. That's why there are many empty values in **Figure 4**. Therefore, in order to further test the degree of influence of the geomorphology variable on the precision of DSM, more samples are needed. This will be the author's next step, a quantitative description of the degree of influences of the geomorphology variable on the precision of DSM.

Acknowledgements

The authors would like to thank anonymous reviewers for very helpful suggestions that were used to improve the manuscript.

References

[1] McBratney, A.B., Odeh, I.O.A., Biship, T., Dunbar, M.S. and Shatar, T.M. (2000) An Overview of Pedometric Techniques for Use in Soil Survey. *Geoderma*, **97**, 293-327. http://dx.doi.org/10.1016/S0016-7061(00)00043-4

[2] Jenny, H. (1941) Factors of Soil Formation: A System of Quantitative Pedology. Dover Publications, New York, 281 p. http://books.google.com/books?id=p94TjEnzCTcC&dq=Factors+of+Soil+Formation:+A+System+of+Quantitative+Pedology&printsec=frontcover&source=bn&hl=en&ei=XfXZS_udG5OINtr74Ug&sa=X&oi=book_result&ct=result&resnum=4&ved=0CBkQ6AEwAw#v=onepage&q&f=false

[3] Zhu, A.X., Li, B.L. and Pei, T. (2008) Model and Method of Detail Digital Soil Survey. Science Press, Beijing. (In Chinese)

[4] McBratney, A.B., Mendonça Santos, M. and Minasny, B. (2003) On Digital Soil Mapping. *Geoderma*, **117**, 3-52. http://dx.doi.org/10.1016/S0016-7061(03)00223-4

[5] Zhu, H.J. and He, Y.G. (1991) Soil Geography. Higher Education Press, Beijing. (In Chinese)

[6] Wu, G.H., Tian, L.S., Hu, S.X. and Wang, N.A. (2000) Natural Geography. Higher Education Press, Beijing. (In Chinese)

[7] Florinsky, I.V., Eilers, R.G., Manning, G.R. and Fuller, L.G. (2002) Prediction of Soil Properties by Digital Terrain Modelling. *Environmental Modelling & Software*, **17**, 295-311. http://dx.doi.org/10.1016/S1364-8152(01)00067-6

[8] Li, T.J., Zhao, Y., Zhang, K.L., Zheng, Y.S. and Wang, Y. (2004) Soil Geography. Higher Education Press, Beijing. (In Chinese)

[9] Mendonça-Santos, M.L., Dart, R.O., Santos, H.G., Coelho, M.R., Berbara, R.L.L. and Lumbreras, J.F. (2010) Digital Soil Mapping of Topsoil Organic Carbon Content of Rio de Janeiro State, Brazil. In: Boettinger, J.L., Howell, D.W., Moore, A.C., Hartemink, A.E. and Kienast-Brown, S., Eds., *Digital Soil Mapping*, Springer Science + Business Media B.V., New York, 255-266. http://dx.doi.org/10.1007/978-90-481-8863-5_21

[10] Moore, I.D., Gessler, P.E., Nielsen, G.A. and Peterson, G.A. (1993) Soil Attributes Prediction Using Terrain Analysis. *Soil Science Society of America Journal*, **57**, 443-452. http://dx.doi.org/10.2136/sssaj1993.03615995005700020026x

[11] Grunwald, S. (2009) Multi-Criteria Characterization of Recent Digital Soil Mapping and Modeling Approaches. *Geoderma*, **152**, 195-207. http://dx.doi.org/10.1016/j.geoderma.2009.06.003

[12] Boettinger, J.L. (2010) Environmental Covariates for Digital Soil Mapping in the Western USA. In: Boettinger, J.L., Howell, D.W., Moore, A.C., Hartemink, A.E. and Kienast-Brown, S., Eds., *Digital Soil Mapping*, Springer Science + Business Media B.V., New York, 17-27. http://dx.doi.org/10.1007/978-90-481-8863-5_2

[13] Odeh, I.O.A., McBratney, A.B. and Chittleborough, D.J. (1992) Fuzzy-c-Means and Kriging for Mapping Soil as a Continuous System. *Soil Science Society of America Journal*, **56**, 1848-1854. http://dx.doi.org/10.2136/sssaj1992.03615995005600060033x

[14] Bourgault, G., Journel, A.G., Rhoades, J.D., Corwin, D.L. and Lesch, S.M. (1997) Geostatistical Analysis of a Soil Data Set. *Advances in Agronomy*, **58**, 241-292.

[15] Hendricks Frassen, H.J.W.M., van Eijnsbergen, A.C. and Stien, A. (1997) Use of Spatial Prediction Techniques and Fuzzy Classification for Mapping Soil Pollutants. *Geoderma*, **77**, 243-262. http://dx.doi.org/10.1016/S0016-7061(97)00024-4

[16] Odeh, I.O.A., McBratney, A.B. and Chittleborough, D.J. (1995) Further Results on Prediction of Soil Properties from Terrain Attributes: Heterotopic Cokriging and Regression-Kriging. *Journal of Geoderma*, **67**, 215-225. http://dx.doi.org/10.1016/0016-7061(95)00007-B

[17] Odeh, I.O.A. and McBratney, A.B. (1998) Using NOAA Advanced Very High Resolution Radiometric Imageries for Regional Soil Inventory. *Proceedings of the* 16*th World Congress of Soil Science*, Montpellier, 20-26 August 1998. http://natres.psu.ac.th/Link/SoilCongress/bdd/symp17/498-r.pdf

[18] Grunwald, S. (2010) Current State of Digital Soil Mapping and What Is Next. In: Boettinger, J.L., Howell, D.W., Moore, A.C., Hartemink, A.E. and Kienast-Brown, S., Eds., *Digital Soil Mapping*, Springer Science + Business Media B.V., New York, 3-12. http://dx.doi.org/10.1007/978-90-481-8863-5_1

[19] Boettinger, J.L., Ramsey, R.D., Bodily, J.M., Cole, N.J., Kienast-Brown, S., Nield, S.J., Saunders, A.M. and Stum, A.K. (2008) Landsat Spectral Data for Digital Soil Mapping. In: Hartemink, A.E., McBratney, A.B. and MendonÇa-Santos, M.L., Eds., *Digital Soil Mapping with Limited Data*, Springer Science + Business Media B.V., New York, 193-202.

[20] Gonzalez, J.P., Jarvis, A., Cook, S.E., Oberthür, T., Rincon-Romero, M., Bagnell, J.A. and Dias, M.B. (2008) Digital Soil Mapping of Soil Properties in Honduras Using Readily Available Biophysical Datasets and Gaussian Processes. In: Hartemink, A.E., McBratney, A.B. and MendonÇa-Santos, M.L., Eds., *Digital Soil Mapping with Limited Data*, Springer Science + Business Media B.V., New York, 367-380.

[21] Roecker, S.M., Howell, D.W., Haydu-Houdeshell, C.A. and Blinn, C. (2010) A Qualitative Comparison of Conventional Soil Survey and Digital Soil Mapping Approaches. In: Boettinger, J.L., Howell, D.W., Moore, A.C., Hartemink, A.E. and Kienast-Brown, S., Eds., *Digital Soil Mapping*, Springer Science + Business Media B.V., New York, 369-384. http://dx.doi.org/10.1007/978-90-481-8863-5_29

[22] Wiesmeier, M., Barthold, F., Blank, B. and Kögel-Knabner, I. (2011) Digital Mapping of Soil Organic Matter Stocks Using Random Forest Modeling in a Semi-Arid Steppe Ecosystem. *Plant and Soil*, **340**, 7-24. http://dx.doi.org/10.1007/s11104-010-0425-z

[23] Wysocki, D.A., Schoeneberger, P.J. and LaGarry, H.E. (2000) Geomorphology of Soil Landscapes. In: Sumner, M.E., Ed., *Handbook of Soil Science*, CRC Press, New York, E5-E39.

[24] Yang, M.S., Zhang, H.C., Ding, H., Lei, G.L., Fan, H.F., Chang, F.Q., Li, B., Zhang, W.X., Niu, J. and Chen, Y. (2005) 14C Dating of Paleosol and Animal Remains in Loess Deposit: A Comparative Study. *Journal of China University of Geosciences*, **30**, 589-596.

[25] Moonjun, R., Farshad, A., Shrestha, D.P. and Vaiphasa, C. (2010) Artificial Neural Network and Decision Tree in Predictive Soil Mapping of Hoi Num Rin Sub-Watershed, Thailand. In: Boettinger, J.L., Howell, D.W., Moore, A.C., Hartemink, A.E. and Kienast-Brown, S., Eds., *Digital Soil Mapping*, Springer Science + Business Media B.V., New York, 151-164. http://dx.doi.org/10.1007/978-90-481-8863-5_13

[26] Stratton, N.J. (2010) Applying Geochronology in Predictive Digital Mapping of Soils. In: Boettinger, J.L., Howell, D.W., Moore, A.C., Hartemink, A.E. and Kienast-Brown, S., Eds., *Digital Soil Mapping*, Springer Science + Business Media B.V., New York, 43-53. http://dx.doi.org/10.1007/978-90-481-8863-5

[27] Chai, H.X., Zhou, C.H., Chen, X., Cheng, W.M., Ou, Y. and Yuan, Y.C. (2008) The New Methodology of Geomorphological Zonalization in Xinjiang Based on Geographical Grid. *Journal of Geophysical Research*, **27**, 481-492. (In Chinese)

[28] Knott, P.A., Doornkamp, J.C. and Jones, R.H. (1980) The Relationship between Soils and Geomorphologic Mapping Units—A Case Study from Northamptonshire. *Bulletin of the International Association of Engineering Geology*, **21**, 186-193.

[29] Gerrard, J. (1992) Soil Geomorphology: An Integration of Pedology and Geomorphology. Chapman and Hall, London.

[30] Zhou, C.H. (2006) A Dictionary of Geomorphology. China Water Power Press, Beijing. (In Chinese)

[31] Bacon, S.N., McDonald, E.V., Dalldorf, G.K., Baker, S.E., Sabol Jr., D.E., Minor, T.B., Bassett, S.D., MacCabe, S.R. and Bullard, T.F. (2010) Predictive Soil Maps Based on Geomorphic Mapping, Remote Sensing, and Soil Databases in the Desert Southwest. In: Boettinger, J.L., Howell, D.W., Moore, A.C., Hartemink, A.E. and Kienast-Brown, S., Eds., *Digital Soil Mapping*, Springer Science + Business Media B.V., New York, 411-421. http://dx.doi.org/10.1007/978-90-481-8863-5_32

[32] Stum, A.K., Boettinger, J.L., White, M.A. and Ramsey, R.D. (2010) Random Forests Applied as a Soil Spatial Predictive Model in Arid Utah. In: Boettinger, J.L., Howell, D.W., Moore, A.C., Hartemink, A.E. and Kienast-Brown, S., Eds., *Digital Soil Mapping*, Springer Science + Business Media B.V., New York, 179-190. http://dx.doi.org/10.1007/978-90-481-8863-5_15

[33] Liu, J. (2010) Mapping Soil Properties Using Individual Representativeness of Samples over Large Areas. Master's Dissertation, Beijing Normal University, Beijing.

[34] Cheng, W.M., Zhou, C.H., Chai, H.X., Zhao, S.M., Liu, H.J. and Zhou, Z.P. (2011) Research and Compilation of the Geomorphologic Atlas of the People's Republic of China (1:1,000,000). *Journal of Geographical Sciences*, **21**, 89-100. http://dx.doi.org/10.1007/s11442-011-0831-z

[35] Li, J.J., Zhou, C.H. and Cheng, W.M. (2009) Geomorphologic Atlas of the People's Republic of China. Science Press, Beijing. (In Chinese)

[36] Qin, C.Z., Zhu, A.X., Pei, T., Li, B.L., Zhou, C.H. and Yang, L. (2007) An Adaptive Approach to Selecting a Flow-Partition Exponent for a Multiple-Flow-Direction Algorithm. *International Journal of Geographical Information Science*, **21**, 443-458. http://dx.doi.org/10.1080/13658810601073240

[37] Jenness, J.S. (2004) Calculating Landscape Surface Area from Digital Elevation Models. *Wildlife Society Bulletin*, **32**, 829-839. http://dx.doi.org/10.2193/0091-7648(2004)032[0829:CLSAFD]2.0.CO;2

[38] Qiu, Y., Fu, B.J., Wang, J. and Chen, L.D. (2004) Spatio-Temporal Variation and Its Influencing Factors of Soil Nutrients in a Small Catchment of the Loess Plateau. *Journal of Progress in Natural Science*, **14**, 294-299. (In Chinese)

[39] Lian, G., Guo, X.D., Fu, B.J. and Hu, C.X. (2006) Spatial Variability and Prediction of Soil Organic Matter at County Scale on the Loess Plateau. *Progress in Geography*, **25**, 112-122. (In Chinese)

[40] Huang, P., Li, T.X., Zhang, J.B. and Liao, G.T. (2009) Effects of Slope's Gradient and Aspect on Spatial Heterogeneity of Soil Organic Matter in Low Mountainous Region. *Journal of Soils*, **41**, 264-268. (In Chinese)

[41] Zhong, Y.P., Tang, J. and Wang, L. (2006) Distribution Characteristic of Soil Organic Carbon in Three Gorges Reservoir District. *Journal of Soil and Water Conservation*, **20**, 73-76. (In Chinese)

[42] Zhou, B., Xu, H.W. and Wang, R.C. (2003) Soil Organic Matter Mapping Based on Classification Tree Modeling. *Acta Pedologica Sinica*, **40**, 801-808. (In Chinese)

[43] Jenness, J. (2009) Surface Area and Ratio for ArcGIS (surface_area.exe). *Proceedings of the ESRI International User Conference*, San Diego, 13-17 July 2009. http://www.jennessent.com/arc gis/surface_ area.htm

[44] Gower, J.C. (1971) A General Coefficient of Similarity and Some of Its Properties. *Biometrics*, **27**, 857-871. http://dx.doi.org/10.2307/2528823

6

Biochar in Nutrient Recycling—The Effect and Its Use in Wastewater Treatment

Bente Foereid

Bioforsk—Norwegian Institute for Agricultural and Environmental Research, Ås, Norway
Email: bente.foreid@bioforsk.no

Abstract

Biochar, *i.e.* charred organic material, appears to increase nutrient retention in soil. The mechanism for how this happens is not clear. Here two possible mechanisms, adsorption and microbial immobilisation, are suggested and compared. It is also suggested that we use biochar in wastewater treatment, and so potentially integrate it better into the total waste management cycle.

Keywords

Biochar, Nutrients, Wastewater Treatment

1. Introduction

Increase in population and relative wealth has led to a greater demand for food. Population growth is estimated to continue at a slowing rate until about 2100 and peak at about 10.1 billion [1]. It is projected that agriculture can supply enough food for the growing population, but at a cost to the environment [2] [3]. In the future, we will need to increase agricultural production without increasing and ideally decreasing environmental impacts. Furthermore, in the future, farmland will need to serve multiple purposes, such as carbon storage in addition to food production.

Many of the adverse environmental impacts on agriculture are caused by the use of nitrogen fertilizers [4] [5]. In turn, many waste products, including sewage and animal waste are rich in nutrients and should be used as fertilizers in agriculture to close the nitrogen cycle as much as possible. This would reduce pollution and reduce the need to use energy intensive mineral fertilizers. Mineral nitrogen occurs in soil in two forms: ammonium and nitrate [6]. In agriculture, soil microbial transformation of ammonium into more soluble and mobile nitrate (nitrification) is usually quick. Nitrogen is lost from agricultural systems principally by leaching of nitrate, but also to some extent through leaching of ammonium and gaseous losses.

Biochar is usually defined as charred organic material used as a soil amendment [7] [8]. It is promoted as a way to increase storage of carbon in the soil [9]. There are also reports of biochar applications having agronomic benefits, particularly on nutrient use efficiency, crop yield and reduced leaching losses [10]-[15] and pollutant immobilization [16]. However, no effect or even opposite effects have also been reported, so biochar remains a controversial issue [17]. Varying results may be expected as there are many types of biochar (different feedstock and production temperature and time) and many conditions under which it is applied (soil type, climate and crop type). To be able to give meaningful advice to farmers and other practitioners on whether to apply biochar and what type of biochar to use, a mechanistic understanding of how biochar works and interacts with other soil factors is necessary.

2. What Is the Mechanism for Biochar Effect?

Whilst some of the short term effects observed upon biochar application may be explained by pH changes [18], or direct release of nutrients and other substances from the biochar itself [19], the longer-term effects are almost certainly explained by its porous structure and large internal surface area, in a similar way as activated carbon that has been used as an adsorbent for water purification for a long time [20].

The most direct explanation for the effect of biochar on nutrient retention is that it acts as an adsorbent [13]. Adsorption is a phenomenon that occurs at the interface between a liquid and solid phase with the formation of bonds between the surface of solid phase and molecules present in liquid phase [20]. Organic material generally has some adsorption capacity, and charring the material usually increases the adsorption capacity [21]. Most evidence suggests that biochar adsorbs more cations than anions, and several experiments have shown removal of ammonium from aqueous solution [22]-[24]. Gaseous ammonia also adsorbs to biochar, and this ammonia is available to plants [25] [26]. Biochar has also been shown to increase cation exchange capacity in soil [27], and it is therefore likely that biochar can reduce leaching of ammonium and store it in a plant available form in the soil. Most studies suggest little or no adsorption of nitrate, although some adsorption has been observed with some high temperature biochars [22] [28]. Applied ammonium usually reacts quickly to form nitrate in agricultural soils, and nitrate is easily lost through leaching. It is possible that this process can be stopped if the applied released ammonium is quickly adsorbed to soil or biochar particles. Only if all the applied nitrogen is in the form of nitrate, it becomes difficult to explain the effect on nitrogen leaching through adsorption.

In addition to adsorption, biochar may also affect nutrient cycles in soil through microbial processes, in the sense that microbial biomass could act as a storage pool for nutrients, *i.e.* substances that microbes preferentially take up [29] [30]. Biochar may enhance microbial growth by providing pores as refuges or just larger surfaces to grow on [31] [32]. This may explain why increased plant growth is sometimes only observed a few years after the application of biochar [11]. If this mechanism is at play, both nitrate and ammonium could be immobilized, as both can be taken up by microbes. The nitrogen in microbial biomass is probably not directly available to plants, but may become available as the microbes die, and plants or their mycorrhiza may be able to stimulate this release in some way [33]-[35]. It is also possible that microbial processes and adsorption can interact; microbes may be better able to utilize nutrients adsorbed to the surfaces they grow on, but this has not been investigated.

Soil types and their suitability for cropping have been studied for more than a century, but the research has focused more on the impact which different types of soil have on cropping rather than understanding the underlying mechanisms. It is well known that nitrogen losses tend to be greater from light sandy soils than from clay soils, and that clay soils usually have more carbon and nitrogen stored in organic form [36]-[38]. Clay soils also have higher internal surface area and cation exchange capacity than sandy soils [39]. There is also evidence that the difference between soil textures is related to pore structure [40] [41]. The effect of adding biochar may therefore be similar to that of increasing clay content. If this is the case, it would be expected that biochar would have a stronger effect in sandy soils than in clay soils. This was also found by Liu *et al.* [15] in a meta-analysis and by Zheng *et al.* [42] when comparing two soils.

3. Possibilities for Biochar in Wastewater Treatment

Biochar has so far mainly been promoted as a soil amendment, but as a good adsorbent, it may also hold promise in low cost wastewater treatment. Activated carbon is already being used, but biochars may form a lower cost alternative. Several authors have found that biochar can remove various heavy metals [16] [43]-[47], phosphorous [24] [48], and as already discussed, nitrogen from solution. Biochar could potentially be used to adsorb nutrients from wastewater, and then the loaded biochar or mixture of sludge and biochar could be added to soil. Digestate from anaerobic digestion of sewage sludge, manure and residues from the food/beverage or agricultural industry usually contains high levels of nutrients which makes it suitable for utilization as an organic soil fertilizer [49]. Using biochar to extract and concentrate the nutrients could eliminate current difficulties associated with the storage of high volumes of the digestate and other waste liquids, and also eliminate environmental concerns relating to spreading the high volumes of liquid fertilizer on soil. In some wastewaters, adsorption of heavy metals and other harmful substances in addition to nutrient could be a problem, but in the case of farm waste or residues from the food beverage industry it should be safe.

4. Benefits of Pre-Loading with Nutrients

There is evidence that biochar increases nutrient retention [7] [8] [10]-[14] [17] [18] [22] in soil and that it can absorb nutrients from solution. This opens the possibility of using biochar as an adsorbent in wastewater treatment and then applying the biochar. It is possible that such "pre-loading" with nutrients could have additional benefits compared to adding biochar nutrient-rich wastewater separately, but this has not been investigated.

There are some reports that suggest that adding biochar to farm waste during storage and "pre-loading" it with nutrients may be beneficial compared to adding both separately to soil [50] [51]. Sarkhot *et al.* [52] found that added biochar pre-loaded with manure has many advantages compared to adding only biochar, but did not compare it to adding both separately. If there is an advantage in terms of nitrogen-use efficiency in pre-adsorbing nitrogen to biochar compared to adding both separately, it further strengthens the case for using biochar in wastewater treatment. Better understanding of the importance of adsorption and microbial immobilization as mechanisms for biochar effect, as well as possible interactions between the two, is necessary to predict the effect of "pre-loading" biochar with nutrients.

5. Recommendations for Future Work

The mechanism(s) for biochar effect on nutrient retention needs to be resolved. To distinguish between adsorption and microbial uptake as mechanisms, one can take advantage of the fact that most biochar do not adsorb nitrate [22] [53]. If microbial immobilization is the main mechanism, retention of nitrate and ammonium nitrogen will be similar. If there is no effect at all on retention of nitrogen added as nitrate, adsorption would be the mechanism, intermediate results would support some sorts of interaction.

Furthermore, if pre-loading of biochar with nutrients can have benefits compared to adding both separately, this would further strengthen the case for integrating biochar into the wastewater management system. This question can be investigated by applying biochar pre-adsorbed with labelled nitrogen and comparing it to the same amount of biochar and labelled nitrogen applied separately.

Acknowledgements

The author wishes to thank Dr. Alessio Cibati and Dr. Joseph Akunna for useful discussions. Three reviewers are acknowledged for useful suggestions.

References

[1] Population Division, Department of Economic and Social Affairs, United Nations (2011) World Population Prospects: The 2010 Revision, Highlights and Advance Tables. United Nations, Working Paper No. ESA/P/WP.220.

[2] Evans, L.T. (1998) Feeding the Ten Billion—Plants and Population Growth. Cambridge University Press, Cambridge.

[3] Godfray, H.C.J., Beddington, J.R., Crute, I.R., Haddad, L., Lawrence, D., Muir, J.F., *et al.* (2010) Food Security: The Challenge of Feeding 9 Billion People. *Science*, **327**, 812-818. http://dx.doi.org/10.1126/science.1185383

[4] Vitousek, P.M., Aber, J.D., Howarth, R.W., Likens, G.E., Matson, P.A., Schindler, D.W., *et al.* (1997) Human Alteration of the Global Nitrogen Cycle: Sources and Consequences. *Ecological Applications*, **7**, 737-750.

[5] Galloway, J.N., Dnetener, F.J., Capone, D.G., Boyer, E.W., Howarth, R.W., Seitzinger, S.P., *et al.* (2004) Nitrogen Cycles: Past, Present, and Future. *Biogeochemistry*, **70**, 153-226. http://dx.doi.org/10.1007/s10533-004-0370-0

[6] MacDonald, J.G. and Nielsen, D.R. (1978) Nitrogen in the Environment. Volume 1: Nitrogen Behaviour in Field Soil. Academic Press.

[7] Lehmann, J. and Rondon, M. (2006) Bio-Char Soil Management on Highly Weathered Soils in the Humid Tropics. In:

Uphoff, N., Ed., *Biological Approaches to Sustainable Soil Systems*, CRC Press, Boca Raton, 517-530. http://dx.doi.org/10.1201/9781420017113.ch36

[8] Lehmann, J. and Joseph, S. (2009) Biochar for Environmental Management: Science and Technology. Earthscan, London.

[9] Woolf, D., Amonette, J.E., Street-Perrott, F.A., Lehmann, J. and Joseph, S. (2010) Sustainable Biochar to Mitigate Global Climate Change. *Nature Communications*, **1**, Article No. 56.

[10] Lehmann, J., da Silva Jr., J.P., Steiner, C., Nehls, T., Zech, W. and Glaser, B. (2003) Nutrient Availability and Leaching in an Archaeological Anthrosol and a Ferralsol of the Central Amazon Basin: Fertilizer, Manure and Charcoal Amendments. *Plant and Soil*, **249**, 343-357. http://dx.doi.org/10.1023/A:1022833116184

[11] Major, J., Rondon, M., Molina, D., Riha, S.J. and Lehmann, J. (2010) Maize Yield and Nutrition during 4 Years after Biochar Application to a Columbian Savanna Oxisol. *Plant and Soil*, **333**, 117-128. http://dx.doi.org/10.1007/s11104-010-0327-0

[12] Knowles, O.A., Robinson, B.H., Contangelo, A. and Clucas, L. (2011) Biochar for Mitigation of Nitrate Leaching from Soil Amended with Biosolids. *Science of the Total Environment*, **409**, 3206-3210. http://dx.doi.org/10.1016/j.scitotenv.2011.05.011

[13] Clough, T.J., Condron, L.M., Kamman, C. and Muller, C. (2013) A Review of Biochar and Soil Nitrogen Dynamics. *Agronomy*, **3**, 275-293. http://dx.doi.org/10.3390/agronomy3020275

[14] Zheng, H., Wang, Z., Deng, X., Herbert, S. and Xing, B. (2013) Impacts of Adding Biochar on Nitrogen Retention and Bioavailability in Agricultural Soil. *Geoderma*, **206**, 32-39. http://dx.doi.org/10.1016/j.geoderma.2013.04.018

[15] Liu, X., Zhang, A., Joseph, C.J.S., Bian, R., Li, L., Pan, G. and Paz-Ferreiro, J. (2013) Biochar's Effect on Crop Productivity and the Dependence on Experimental Conditions—A Meta-Analysis of Literature Data. *Plant and Soil*, **373**, 583-594. http://dx.doi.org/10.1007/s11104-013-1806-x

[16] Beesley, L., Moreno-Jiménez, E., Gomez-Eyles, J.L., Harris, E., Robinson, B. and Sizmur, T. (2011) A Review of Biochars' Potential Role in the Remediation, Revegetation and Restoration of Contaminated Soils. *Environmental Pollution*, **159**, 3269-3282. http://dx.doi.org/10.1016/j.envpol.2011.07.023

[17] Spokas, K.A., Cantrell, K.B., Novak, J.M., Archer, D.W., Ippolito, J.A., Collins, H.P., *et al.* (2012) Biochar: A Synthesis of Its Agronomic Impact beyond Carbon Sequestration. *Journal of Environmental Quality*, **41**, 973-989. http://dx.doi.org/10.2134/jeq2011.0069

[18] Van Zwieten, L., Kimber, S., Morris, S., Chan, K., Downie, A., Rust, J., *et al.* (2010) Effects of Biochar from Slow Pyrolysis of Papermill Waste on Agronomic Performance and Soil Fertility. *Plant and Soil*, **327**, 235-246. http://dx.doi.org/10.1007/s11104-009-0050-x

[19] Smith, J.L., Collins, H.P. and Bailey, V.L. (2010) The Effect of Young Biochar on Soil Respiration. *Soil Biology & Biochemistry*, **42**, 2345-2347. http://dx.doi.org/10.1016/j.soilbio.2010.09.013

[20] Gray, N.F. (1999) Water Technology: An Introduction for Scientists and Engineers. Arnold, London, 548 p.

[21] Bhatnagar, A. and Sillanpaa, M. (2010) Utilization of Agro-Industrial and Municipal Waste Materials as Potential Adsorbents for Water Treatment: A Review. *Chemical Engineering Journal*, **157**, 277-296. http://dx.doi.org/10.1016/j.cej.2010.01.007

[22] Yao, Y., Gao, B., Zhang, M., Inyang, M. and Zimmermann, A.R. (2012) Effect of Biochar Amendment on Sorption and Leaching of Nitrate, Ammonium, and Phosphate in a Sandy Soil. *Chemosphere*, **89**, 1467-1471. http://dx.doi.org/10.1016/j.chemosphere.2012.06.002

[23] Dempster, D.N., Gleeson, D.B., Solaiman, Z.M., Jones, D.L. and Murphy, D.V. (2012) Decreased Soil Microbial Biomass and Nitrogen Mineralisation with Eucalyptus Biochar Addition to a Coarse Textured Soil. *Plant and Soil*, **354**, 311-324. http://dx.doi.org/10.1007/s11104-011-1067-5

[24] Hollister, C.C., Bisogni, J.J. and Lehmann, J. (2012) Ammonium, Nitrate, and Phosphate Sorption to and Solute Leaching from Biochars Prepared from Corn Stover (*Zea mays* L.) and Oak Wood (*Quercus* spp.). *Journal of Environmental Quality*, **1**, 137-144.

[25] Taghizadeh-Toosi, A., Clough, T.J., Sherlock, R.R. and Condron, L.M. (2012) A Wood Based Low Temperature Biochar Captures NH_3-N Generated from Ruminant Urine-N, Retaining Its Bioavailability. *Plant and Soil*, **353**, 73-84. http://dx.doi.org/10.1007/s11104-011-1010-9

[26] Taghizadeh-Toosi, A., Clough, T.J., Sherlock, R.R. and Condron, L.M. (2012) Biochar Adsorbed Ammonia Is Bioavailable. *Plant and Soil*, **350**, 57-69. http://dx.doi.org/10.1007/s11104-011-0870-3

[27] Liang, B., Lehmann, J., Solomon, D., Kinyangi, J., Grossman, J., O'Neill, B., *et al.* (2006) Black Carbon Increases Cation Exchange Capacity in Soils. *Soil Science Society American Journal*, **70**, 1719-1730. http://dx.doi.org/10.2136/sssaj2005.0383

[28] Mizuta, K., Matsumoto, T., Hatate, Y., Nishihara, K. and Nakanishi, T. (2004) Removal of Nitrate-Nitrogen from Drink-

ing Water Using Bamboo Powder Charcoal. *Bioresource Technology*, **95**, 255-257. http://dx.doi.org/10.1016/j.biortech.2004.02.015

[29] Lehmann, J., Rillig, M.C., Thies, J., Masiello, C.A., Hockaday, W.C. and Crowley, D. (2011) Biochar Effects on Soil Biota: A Review. *Soil Biology and Biochemistry*, **43**, 1812-1836. http://dx.doi.org/10.1016/j.soilbio.2011.04.022

[30] Güereña, D.T., Lehmann, J., Hanley, K., Enders, A., Hyland, C. and Riha, S. (2013) Nitrogen Dynamics Following Field Application of Biochar in a Temperate North American Maize-Based Production System. *Plant and Soil*, **365**, 239-254. http://dx.doi.org/10.1007/s11104-012-1383-4

[31] Steiner, C., Glaser, B., Teixaira, W.G., Lehmann, J., Blum, W.E.H. and Zech, W. (2008) Nitrogen Retention and Plant Uptake on a Highly Weathered Central Amazonian Ferralsol Amended with Compost and Charcoal. *Journal of Plant Nutrition and Soil Science*, **171**, 893-899. http://dx.doi.org/10.1002/jpln.200625199

[32] Steinbiss, S., Gleixner, G. and Antonietti, M. (2009) Effect of Biochar on Soil Carbon Balance and Soil Microbial Activity. *Soil Biology & Biochemistry*, **41**, 1301-1310. http://dx.doi.org/10.1016/j.soilbio.2009.03.016

[33] Robinson, D., Griffiths, B., Ritz, K. and Wheatley, R. (1989) Root-Induced Nitrogen Mineralisation: A Theoretical Analysis. *Plant and Soil*, **117**, 185-193. http://dx.doi.org/10.1007/BF02220711

[34] Liljeroth, E., Baath, E., Mathasson, I. and Lundborg, T. (1990) Root Exudation and Rhizoplane Bacterial Abundance of Barley (*Hordeum vulgare* L.) in Relation to Nitrogen Fertilisation and Root Growth. *Plant and Soil*, **127**, 81-89. http://dx.doi.org/10.1007/BF00010839

[35] Liljeroth, E., Kuikman, P. and Van Veen, J.A. (1994) Carbon Translocation to the Rhizospere of Maize and Wheat and Influence on the Turnover of Native Soil Organic Matter at Different Soil Nitrogen Levels. *Plant and Soil*, **161**, 233-240. http://dx.doi.org/10.1007/BF00046394

[36] Parton, W.J., Schimel, D.S., Cole, C.V. and Ojima, D.S. (1987) Analysis of Factors Controlling Soil Organic-Matter Levels in Great-Plains Grasslands. *Soil Science Society of America Journal*, **51**, 1173-1179. http://dx.doi.org/10.2136/sssaj1987.03615995005100050015x

[37] Parton, W.J., Ojima, D.S., Cole, C.V., Bryant, R.B. and Arnold, R.W. (1994) A General Model for Soil Organic Matter Dynamics: Sensitivity to Litter Chemistry, Texture and Management. In: Bryant, R.B. and Arnold, R.W., Eds., *Quantitative Modeling of Soil Forming Processes*, Soil Science Society of America, Madison, 147-167.

[38] Schmidt, M.W.I., Torn, M.S., Abiven, S., Dittmar, T., Guggenberger, G., Janssens, I.A., Kleber, M., Kögel-Knabner, I., Lehmann, J., Manning, D.A.C., Nannipieri, P., Rasse, D.P., Weiner, S. and Trumbore, S.E. (2011) Persistence of Soil Organic Matter as an Ecosystem Property. *Nature*, **478**, 49-56. http://dx.doi.org/10.1038/nature10386

[39] Osman, K.T. (2013) Soils: Principles, Properties and Management. Springer, Dordrecht. http://dx.doi.org/10.1007/978-94-007-5663-2

[40] Strong, D.T., De Wever, H., Merckx, R. and Recous, S. (2004) Spatial Location of Carbon Decomposition in the Soil Pore System. *European Journal of Soil Science*, **55**, 739-750. http://dx.doi.org/10.1111/j.1365-2389.2004.00639.x

[41] Sleutel, S., Bouckaert, L., Buchan, D., Van Loo, D., Cornelis, W.M. and Sanga, H.G. (2012) Manipulation of the Soil Pore and Microbial Community Structure in Soil Mesocosm Incubation Studies. *Soil Biology & Biochemistry*, **45**, 40- 48. http://dx.doi.org/10.1016/j.soilbio.2011.09.016

[42] Zheng, J., Stewart, C.E. and Cotrufo, F. (2012) Biochar and Nitrogen Fertilizer Alters Nitrogen Dynamics and Greenhouse Gas Fluxes from Two Temperate Soils. *Journal of Environmental Quality*, **41**, 1361-1370.

[43] Beesley, L. and Marmiroli, M. (2011) The Immobilisation and Retention of Soluble Arsenic, Cadmium and Zink by Biochar. *Environmental Pollution*, **159**, 474-480. http://dx.doi.org/10.1016/j.envpol.2010.10.016

[44] Chen, X.C., Chen, G.C., Chen, L.G., Chen, Y.X., Lehmann, J., McBride, M.B. and Hay, A.G. (2011) Adsorption of Copper and Zinc by Biochars Produced from Pyrolysis of Hardwood and Corn Straw in Aqueous Solution. *Bioresource Technology*, **102**, 8877-8884. http://dx.doi.org/10.1016/j.biortech.2011.06.078

[45] Uchimiya, M., Klasson, T.K., Wartelle, L.H. and Lima, I.M. (2011) Influence of Soil Properties on Heavy Metal Sequestration by Biochar Amendment: 1. Copper Sorption Isotherms and Release of Cations. *Chemosphere*, **82**, 1431- 1437. http://dx.doi.org/10.1016/j.chemosphere.2010.11.050

[46] Inyang, M., Gao, B., Yao, Y., Xue, Y.W., Zimmermann, A.R., Pullammanappalill, P. and Cao, X.D. (2012) Removal of Heavy Metals from Aqueous Solution by Biochars Derived from Anaerobically Digested Biomass. *Bioresource Technology*, **110**, 50-56. http://dx.doi.org/10.1016/j.biortech.2012.01.072

[47] Kolodynska, D., Wnetrzak, R., Leahy, J.J., Hayes, M.H.B., Kwapinski, W. and Hubicki, Z. (2012) Kinetic and Adsorptive Characterization of Biochar in Metal Ions Removal. *Chemical Engineering Journal*, **197**, 295-305. http://dx.doi.org/10.1016/j.cej.2012.05.025

[48] Yao, Y., Gao, B., Inyang, M., Zimmermann, A.R., Cao, X.D., Pullammanappallil, P. and Yang, L.Y. (2011) Biochar Derived from Anaerobically Digested Sugar Beet Tailings: Characterization and Phosphate Removal Potential. *Bioresource*

Technology, **102**, 6273-6278. http://dx.doi.org/10.1016/j.biortech.2011.03.006

[49] Abdullahi, Y.A., Akunna, J.C., White, N.A., Hallett, P.D. and Wheatley, R. (2008) Investigating the Effects of Anaerobic and Aerobic Post-Treatment on Quality and Stability of Organic Fraction of Municipal Solid Waste as Soil Amendment. *Bioresource Technology*, **99**, 8631-8636. http://dx.doi.org/10.1016/j.biortech.2008.04.027

[50] Schmidt, H.-P. (2012) Treating Liquid Manure with Biochar. *Ithaka Journal*, **1**, 273-276.

[51] Schmidt, H.-P. (2012) 55 Uses of Biochar. *Ithaka Journal*, **1**, 286-289.

[52] Sarkhot, D.V., Berhe, A.A. and Ghezzehei, T.A. (2012) Impact of Biochar Enriched with Dairy Manure Effluent on Carbon and Nitrogen Dynamics. *Journal of Environmental Quality*, **41**, 1107-1114. http://dx.doi.org/10.2134/jeq2011.0123

[53] Mizuta, K., Matsumoto, T., Hatate, Y., Nishihara, K. and Nakanishi, T. (2004) Removal of Nitrate-Nitrogen from Drinking Water Using Bamboo Powder Charcoal. *Bioresource Technology*, **95**, 255-257. http://dx.doi.org/10.1016/j.biortech.2004.02.015

7

Corn and Soybean Responses to Two Tillage Systems in a Cool Growing Season

Jesper K. V. Nielsen[1], Howard J. Woodard[2]

[1]Carrington Research Station, North Dakota State University, Carrington, ND, USA
[2]Plant Science Department, South Dakota State University, Brookings, SD, USA
Email: NDSU.Carrington.REC@ndsu.edu, Howard.Woodard@sdstate.edu

Abstract

A field study in 2014 documented corn and soybean biomass and nutrient responses between conventional-till and no-till tillage systems at Beresford, SD during cooler than normal weather conditions with adequate soil moisture. The overall study was established in 1992. Each treatment plot was monitored weekly from June to August for soil moisture, temperature, and plant growth stages. Biomass was harvested during and at the end of the growing season for yield and nutrient content. Soil moisture measured throughout the early and middle part of the growing season was determined to be sufficient for crop growth, since precipitation was much greater than normal in June (33.2 cm). However, air temperature was below normal early in the growing season and lowered Growing Degree Days (939˚C) compared to the 30-year average (139˚C). Soil temperatures (5 cm depth) were not significant between tillage treatments in the corn plots during the growing season for 12 observation dates (range 16.3˚C - 28.0˚C). Plant growth was not significantly different between tillage treatments, reflecting the lack of soil temperature differences (5 cm depth) between tillage treatments. The mid-season plant tissue and crop residue at harvest nutrient content (P, K, and Zn) were not significant between tillage treatments. Corn grain yields were 10.3 T·ha^{-1} and 10.1 T·ha^{-1} for conventional tillage and no-till, respectively. Soybean grain yields were 3.9 T·ha^{-1}and 3.3 T·ha^{-1} for conventional tillage and no-till, respectively. These results would more than likely have been much different in a warmer growing season, when soil temperature and moisture differences between tillage treatments would likely stimulate crop growth in the conventional-tilled soil. This would have also increased nutrient uptake and grain yield levels to greater degree than observed in this study.

Keywords

Tillage, Corn, Soybean

1. Introduction

Tillage implementation for corn and soybean production depends on agronomic, soil, and economic issues involved. A study in Southern Nebraska showed that corn grain yield was greatest with a rotation-chisel-plow system compared to disk, no-tillage, and ridge-tillage systems in years with average temperature and moisture [1]. Corn grain yield was greater for a no-till system in dry, wet and warmer years. Soybean grain yield was not different among tillage systems in years with average temperature and moisture. Soybean grain yield was greater for a chisel plow system in dry years. Soybean grain yield was greater for a no-till system in warmer and wet years. Furthermore, in extremely hot and dry years, corn and soybean grain yield was greater in the no-till system.

A study in Michigan discussed the effects of no-till corn from wheat root shoot residue (WRSR), no wheat residue (NWR), and wheat root residue treatments (WRR) [2]. The WRSR treatments were greater in soil moisture content, while soil in the NWR and WRR treatments had a faster drying rate. The WRSR treatments resulted in lower soil temperature and less fluctuation from day and night compared to the NRW and WRR due to the insulating effect of the wheat residue on the soil surface.

A study in China showed that corn grain yield was greatest when planted in a ridge-till system compared to no-till and conventional-till treatments [3]. The moisture depletion from ridge-till compared to no-till was not significantly different at two sites. Soil moisture reduction from conventional-till was significantly lower at one site compared to no-till.

A study in Central Mexico showed that increasing residue resulted in greater soil moisture content through the growing season [4]. Ponding time (of surface water) and corn yield increased with the zero-till treatment when crop residue was maintained on the field. Moisture depletion was greatest in conventional-till with or without residue and no-till treatment without residue.

A study in Washington showed that soil moisture was significantly greater with no-till compared to the sweep-till treatment which had less residue retention and less evaporation near the soil surface [5]. Lower temperature fluctuations occurred in the no-till compared to the sweep-till treatment for which more dark soil was exposed to the sun compared to no-till. The temperature of soils with greater water content was more difficult to change since it has a higher heat capacity than dry soils. A study in Iowa showed that corn grain yield had no significant difference between strip-till, chisel-plow and no-till treatment [6]. The same trend was observed for throughout the soil profile.

A study in Northeast China showed that soil moisture depletion in no-till was lower compared to reduced-till and conventional-till treatment, especially in the beginning and late season [7]. Soil moisture depletion was greater in reduced-till and conventional-till treatment. The increased soil moisture content from no-till resulted in an increase in soybean grain yield but a decrease in corn grain yield. A study in North Dakota showed that corn grain yield for no-till was significantly lower than the strip-till treatment. However, grain yields in conventional-till were not significantly different from the other two tillage treatments [8].

In summary, these research studies showed that crop residue cover not disturbed by tillage resulted in retention of greater soil moisture content and reduced temperature fluctuation compared to tilled soils. Soil moisture content decreased because of increased evaporation and soil temperature fluctuated more throughout the day in conventional-till. Corn grain yield also seemed to decrease with no-till and increase with conventional till treatment. However soybean grain yield seemed to increase with no-till and decrease with conventional-till treatment. It can also be concluded that during years of extreme heat and drought, the increased surface residue from no-till may benefit the growing environment by moderating temperature and maintaining moisture levels through the growing season. This would have implications for influencing corn and soybean grain yield in the dryer parts of the soybean production areas of the Great Plains.

The weather in South Dakota can be variable and tillage influences on crop growth and final grain yield can respond strongly to climatic conditions. The objectives for this research were to consider how corn and soybean crops respond to a cooler than normal growing season with adequate available soil water by: 1) Documenting how two tillage systems can influence topsoil soil temperature and soil moisture; and 2) Comparing corn and soybean growth, crop biomass production, and plant nutrient up take among two tillage treatments.

2. Methods and Materials

2.1. Background of Study Site

The study site was located at the SDSU Southeast Research Farm (SERF) near Beresford, SD at latitude

43.04270°N, and longitude 96.892232°W. The elevation of the site was 464 m above sea level. The soil type was an Egan-Trentsilty clay loam which is described as a fine-silty, mixed, and super active mesic Udic Haplustolls [9]. The field site had a slope of less than 2% and slopes from northeast to southwest.

2.2. Experimental Design

The field study began in 1992 when tillage treatments of conventional-till (CT) and no-till (NT) were established for corn and soybean in rotation. The plot dimensions for the soybean crop were 22 m wide by 100 m long. The treatments of tillage and crop were assembled as a randomized complete block with three replications. The CT system for corn production consisted of fall disk and chisel-plow operation followed by a spring field-cultivation operation. However, for soybean production only a disking or field cultivation was performed in the spring.

2.3. Field Preparation

In spring 2014, fertilizer potassium was surface-applied as potash (0-0-60) at the rate of 240 kg K ha^{-1} to both crops with a broadcast application before the tillage operation. Nitrogen was applied to the corn plots before planting as surface application of urea-ammonium-nitrate solution (28-0-0) at a rate of 550 liters ha^{-1} to provide 220 kg N ha^{-1} with drop nozzles as a pre-plant application for both tillage systems. Corn (hybrid PIOO193AM) was planted at the rate of 71,000 seeds ha^{-1} in mid-May. Soybean (variety AG2135) was planted at the rate of 330,000 seeds ha^{-1} in late May in 27 cm rows for all treatments.

2.4. Soil Measurements

Soil cores were removed from the 0 - 15 cm depth randomly within each plot area between rows at the beginning of the growing season from each treatment plot. The soil was dried and ground to prepare for analysis at the SDSU soil testing lab using standard procedures [10]. Soil particle analysis was determined by the hydrometer method [11]. The organic matter was determined by the Loss-On-Ignition (LOI) method based on [12]. Soil pH was determine by the pH-electrode method based on [13]. Soil nitrate analysis was determined by the Nitrate Electrode method based on [10]. Soil orthophosphate-P analysis was extracted by the Olsen-P methods based on [14] and was determined colori metrically based on [15] Soil-K analysis was extracted by the Ammonium Acetate method and determined by the Atomic absorption based on [16]. Soil-Zn was extracted by DTPA extraction and determined by the Atomic adsorption method based on [17].

2.5. Plant Measurements

The plant growth stage rating was documented weekly from June to August, as an average for each measurement in which three plants per plot were selected. Corn plant samples were collected during the growing season from at least four to five rows from the border of the plots. Within those rows, the middle three plants that were most representative of the whole plot were harvested.

2.6. Soil Measurement

Soil temperature and moisture were measured on the same day and time of day on a weekly basis. Soil moisture content was measured at a 5 cm soil depth with a model Field scout TDR 300 Soil Moisture Meter Item # 6430f3, (Spectrum Technologies, 3600 Thayer Court, Aurora, IL 60,504). Raw moisture measurements were calibrated according to actual volumetric water content measurements determined from another experiment (data not shown). Soil temperature was measured at a two in. depth with a model HH21 Microprocessor Thermometer, Type J-K-T Thermocouple, Ω-OMEGA (OMEGA Engineering, INC. One Omega Drive, P.O. Box 4047, Stamford, Connecticut 06907-0047).

Ten whole above-ground corn plants were sampled at the V6 [18] growth stage and from the corn ear leaf at tasseling (R1) stage. Ten plants were collected from each of two rows. Plant samples were oven dried and ground. Plant samples were prepared by microwave digestion [19] and analyzed for total nitrogen (N), phosphorus (P), potassium (K), and zinc (Zn).

Total above-ground soybean plant tissue was sampled at the R1 [18] growth stage along a one-meter length of row from the second row of the plot. Samples were oven dried, ground and analyzed for total N, P, K, and Zn and analyzed with the same analytical methodology used for corn sample analysis.

Corn and soybean grain was harvested by small plot combine after reaching physiological maturity along the entire length of the treatment plots from the middle of the plot. Grain was weighed and analyzed for moisture content and test- weight. Grain yields were estimated on an acre basis. Grain samples were prepared for analysis as the tissue was before. Data were entered into an Excel spreadsheet and analyzed using R software and using the "Agricolae" package [20].

Corn and crop soybean residue was collected from a 6.5 m^2 surface area after grain harvest for all treatment-replications. The crop residue was weighed in the field then a subsample was dried to determine biomass harvest weight. The biomass was prepared for analysis (P, K, and Zn) as for the plant tissue analysis. Grain yield was also analyzed for the nutrient content as previously described.

3. Results and Discussion

3.1. Climatic Parameters

The average mean air temperature was just slightly below normal for May and June compared to the 30-year average, but 2.6˚C below normal for July, and 1.1˚C below normal for August (**Table 1**). Lower mean air temperatures were reflected a lower cumulative Growing Degree Days (GDD) of 20(˚C) lower than normal for May, 62(˚C) lower than normal for June, 85(˚C) lower than normal for July, and 68(˚C) lower than normal for August. The four-month growing season cumulative GDD was 939(˚C) which was 20% below the normal GDD of 1173(˚C). Lower than normal GDD in June and July during critical corn biomass and ear development stage, probably impacted final corn grain test weight and yield. Cooler than normal air temperatures in the growing season most likely decreased soil evaporation as well. The monthly precipitation was slightly below normal for May, but 23.2 cm above normal for June compared to the 30-year average. The four-month cumulative precipitation of 48.5 cm was 38% greater than for the 30-year average of 35.0 cm. Most of this difference was attributed to higher than normal rainfall received in June. There was adequate soil moisture for crop growth at the early part of the growing season since no plant stresses were observed.

3.2. Physical and Chemical Properties

Soil organic matter content ranged from 4.80% to 4.97% and soil pH ranged from 5.40 to 5.47 among treatment replications which was very narrow (**Table 2**). The extractable or tho-P ranged from 11.7 to 14.4 ppm. These levels were determined to have a "medium" status for corn [21]. The extractable potassium was determined to be at a "very high" status for both corn and soybean across all treatments. This high level reflects the addition of fertilizer K added to all treatment plots in the previous spring before planting in order to minimize any K stress to the growing crops. Extractable soil zinc was determined to be "very high" for corn and soybean across all treatments.

3.3. Soil Temperature

The mean soil temperature for corn with conventional-till was generally greater in the early part of the growing season (June 6: 28.6˚C) compared to the late growing season with (August 21: 23.3˚C) (**Table 3**). In the no-till treatment, the mean soil temperature in corn was slightly lower than the conventional-till treatments in the early part of the growing season (June 6: 24.3˚C) and in the later part of the growing season (August 21: 25.4˚C). In each of the observation dates, there was no statistical significance (N.S.) in the soil temperature between tillage treatments. It was expected that the residue cover of no-till treatment to reflect the sun's energy and result in lower soil temperatures than average compared to the 30-year average. However similar soil temperatures between tillage treatments were observed during this growing season.

The mean soil temperature for soybean with conventional-till was generally higher in the early part of the growing season (June 6: 26.1˚C) compared to the late growing season with (August 21: 23.7˚C) (**Table 3**). In the no-till plots, the soil temperature was slightly lower than in the tilled treatments in the early part of the growing season (June 6: 23.9˚C) but similar to the tilled treatments in the later part of the growing season (August 21: 22.8˚C). Only for four of the five observations between June 14 and July 10 were the soil temperature differences statistically greater for the conventional-tilled treatments compared to the no-till treatments.

Table 1. Mean monthly temperature, growing degree days (GDD), and precipitation at Beresford, SD, in 2014 compared to a 30-year average.

Month	Mean Temp. 2014	Mean Temp. 1971-2000	Monthly GDD 2014	Average Monthly GDD 1971-2000	Monthly Precip. 2014	Average Monthly Precip. 1971-2000
	˚C	˚C	(˚C)	(˚C)	cm	cm
May	15.2	15.1	141	161	6.3	9.3
June	20.1	20.8	203	265	33.2	10.0
July	20.6	23.2	298	382	2.2	8.5
August	20.8	21.9	297	365	6.8	7.2
Total			939	1173	48.5	35.0

Table 2. Mean chemical and physical properties for soil for crop and tillage (conventional tillage—CT; No tillage—NT) treatment plots at Beresford, SD, in 2014.

Crop	Tillage	Organic Matter	pH	Extr. NO_3-N	Applied Fert. N*	Extr. Orto-P	Soil P Status†	Extr. K**	Soil K Status†	Extr. Zn	Soil Zn Status†
		%		ppm	kg·ha^{-1}	ppm		ppm		ppm	Status
Corn	CT	4.80	5.47	78.8	220.0	13.8	Medium	298.7	Very High	2.7	Very High
Corn	NT	4.83	5.43	62.8	220.0	12.4	Medium	218.0	Very High	1.7	Very High
Soybean	CT	4.83	5.40	51.8	0	11.7	Medium	261.0	Very High	2.1	Very High
Soybean	NT	4.97	5.43	42.7	0	14.4	Medium	226.7	Very High	1.6	Very High

*Corn yield goal was 11.3 Mg ha^{-1}; Soybean yield goal was 4.0 Mg ha^{-1}.Fertilizer Nitrogen was applied preplantto the corn treatment plots as U.A.N. liquid (28-0-0) at a rate of 550 liters ha^{-1} to provide 220 kg N ha^{-1}. **Potash (0-0-60) was applied at the rate to provide 240 kg K ha^{-1} to both crops. †Adapted from [21].

Table 3. Mean soil temperature (5 cm depth) in (˚C) for crop and tillage (conventional tillage—CT; No tillage—NT) treatments at Beresford, SD in 2014. Each single date was a response between tillage-treatment within a single crop.

Crop	Tillage	6-Jun	14-Jun	20-Jun	27-Jun	3-Jul	10-Jul	17-Jul	25-Jul	1-Aug	11-Aug	14-Aug	21-Aug
								˚C					
Corn	CT	28.6	23.7	28.2	24.4	22.6	23.3	24.2	20.9	20.8	20.4	16.1	23.3
Corn	NT	24.3	20.4	27.2	24.1	20.6	22.2	24.3	20.4	19.9	20.3	16.4	25.4
P-values (α = 0.05)		0.211	0.135	0.165	0.610	0.274	0.328	0.946	0.635	0.396	0.787	0.816	0.510
Significance		N.S.	N.S.	N.S.	N.S.	N.S.	N.S.	N.S.	N.S.	N.S.	N.S.	N.S.	N.S.
								˚C					
Soybean	CT	26.1	22.1	27.3	24.6	22.2	23.6	26.4	20.7	20.9	19.7	16.3	23.7
Soybean	NT	23.9	17.4	23.3	23.2	19.6	21.4	23.5	20.2	20.3	20.2	16.5	22.8
P-values (α = 0.05)		0.189	0.015	0.025	0.062	0.006	0.016	0.109	0.439	0.518	0.479	0.304	0.556
Significance		N.S.	*	*	N.S.	**	*	N.S.	N.S.	N.S.	N.S.	N.S.	N.S.

N.S. Not significant paired comparison at the alpha = 0.05 level, between tillage treatments within a sampling date. *Significant paired comparison at the alpha = 0.05 level. **Significant paired comparison at the alpha = 0.01 level.

3.4. Soil Moisture

The calculated mean soil moisture content for corn for the tillage treatments was generally adequate for the early part of the growing season (June 3: 30.2% vs. 28.9%, respectively, for CT vs. NT) as well in the later part of the growing season (August 21: 28.1% vs. 28.0%, respectively) (**Table 4**). The mean soil moisture for soybean for both conventional-till and no-till treatment was generally adequate for the early part of the growing season (June 3: 29.0% vs. 31.6% respectively, for CT vs. NT) and was slightly greater for the conventional-till compared to the no-till treatment in the late growing season (August 21: 31.7% vs. 27.5%, respectively, for CT vs. NT). However for each of the observation dates, there was no statistical significance in the soil moisture (N.S.) between till treatments.

One would expect that the residue cover in the no-tillage treatment to reduce soil evaporation compared to the conventional-till treatment since the residue acted as a barrier to evaporation and reflected the sun's energy. Cooler than normal growing season air temperatures reduced the soil evaporation potential in these treatments. The result was that there was no difference in the soil moisture content between tillage treatments. These results could have been much different in a drier or warmer growing season. Soil moisture content was usually high among tillage treatment. Perhaps higher rainfall in the early part of the growing season reduced nitrogen availability because of leaching. It is also possible that denitrification occurred in these soils (not measured) which was usually wet during the early part of the growing season. This could also have influenced crop growth since available soil N levels may have been reduced.

3.5. Plant Growth Stages

The growth stage for both crops progressed in both tillage systems from the vegetative to the reproductive growth stage from June 3 to August 21 (**Table 5**). The transition from the vegetative stage to the reproductive stage occurred from July 10 to July 17 for both crops. However, there was no statistical difference in the growth stage progression (N.S.) or between tillage treatments for most dates. Generally the soil temperatures are greater in conventional-till treatments due to greater exposure of the disturbed soil to the sun's energy and the absence of a crop residue surface barrier to reduce evaporation as in no-till. However, tillage had little influence on growth progression.

During a warmer growing season, one would expect that crops growing in the tilled treatments would advance farther compared to the no-till treatments. Slightly higher soil temperatures were observed in the middle of the growing season for the conventional-till plot (**Table 3**). The result was that the conventional-till plots generally stimulated crop growth somewhat compared to no-till plots. This resulted in a slightly greater advancement in growth stages in the middle of the growing season. But the difference was not statistically significant.

Adequate soil moisture levels together with cooler than normal air temperatures probably lowered the evaporation potential in the conventional-till treatments. This probably slowed growth in the conventional-till treatment and so the crop advanced for both tillage treatments similarly in the early part of the growing season.

3.6. Plant Tissue Nutrient Level

The total tissue P, K, and Zn-nutrient concentration for corn at the V6 stage of growth was sufficient for the conventional-till treatment by (0.34%, 3.3%, and 25.6 ppm). The total P, K, and Zn-nutrient concentration was also sufficient for the no-till treatment by (0.35%, 3.3%, and 22.6 ppm) (**Table 6**). The total tissue N concentration was (2.6%), slightly lower than the ranges of Mills and Jones (1996) for corn at ear-leaf growth stage for conventional-till treatment. However total N concentration was within the sufficiency range (2.7%) for no-till treatment at the ear-leaf growth stage. The total tissue P and K concentration was sufficient (0.26%, 1.7% respectively) for corn at ear-leaf growth stage for conventional-till treatment. The total tissue P and K concentration was sufficient (0.28%, 1.7% respectively) for corn at ear-leaf growth stage for the no-till treatment. The total tissue Zn concentration was slightly lower (15.3 ppm) than the average range for corn at ear-leaf growth stage for conventional-till treatment. The total tissue Zn concentration was slightly lower (14.3 ppm) than the sufficiency range for corn at ear-leaf growth stage for no-till treatment cross all tillage treatments. Cooler than normal soil conditions in the early part of the growing season probably slowed the diffusion rate of nutrients that move by mass flow such as P, K, and Zn and lowered tissue concentration. Overall, there were no significant differences in the plant nutrient concentration of P, K, and Zn between tillage treatments. Because tissue P, K,

Table 4. Mean calculated volumetric water content (5 cm depth) soil moisture of crop and tillage (conventional tillage—CT; No tillage—NT) treatments at Beresford, SD in 2014.

Crop	Tillage	3-Jun	6-Jun	14-Jun	20-Jun	10-Jul	17-Jul	25-Jul	1-Aug	11-Aug	14-Aug	21-Aug
							%**					
Corn	CT	30.2	31.4	33.0	48.3	30.1	26.5	23.1	22.6	26.1	21.7	28.1
Corn	NT	28.9	32.2	36.2	46.7	26.7	26.4	24.7	23.1	26.5	22.9	28.0
P-values ($\alpha = 0.05$)		0.790	0.784	0.143	0.841	0.296	0.971	0.493	0.882	0.906	0.541	0.910
Significance		N.S.	N.S.	N.S.	N.S.	N.S.	N.S.	N.S.	N.S.	N.S.	N.S.	N.S.
							%**					
Soybean	CT	29.0	36.9	32.2	43.5	29.5	28.0	25.5	21.5	29.2	27.0	31.7
Soybean	NT	31.6	37.5	38.3	43.1	27.7	26.4	24.9	21.9	26.7	22.8	27.5
P-values ($\alpha = 0.05$)		0.426	0.882	0.161	0.924	0.709	0.738	0.838	0.887	0.512	0.0271	0.817
Significance		N.S.	N.S.	N.S.	N.S.	N.S.	N.S.	N.S.	N.S.	N.S.	*	N.S.

N.S. Not significant paired comparison at the alpha = 0.05 level, between tillage treatments within a sampling date. *Significant paired comparison at the alpha = 0.05 level. **Calculated from raw TDR readings using the equation: $0.61^{*}x + 10.2$ calibrated previously from a separate field study.

Table 5. Mean growth stage for crop and tillage (conventional tillage—CT; No tillage—NT) treatments at Beresford, SD in 2014.

Crop	Tillage	3-Jun	6-Jun	14-Jun	20-Jun	27-Jun	3-Jul	10-Jul	17-Jul	1-Aug	11-Aug	14-Aug	21-Aug
Corn*	CT	V3.0	V4.0	V4.6	V5.0	V6.0	V6.6	V8.0	R1.7	R2.0	R3.0	R3.0	R4.0
Corn*	NT	V3.0	V4.0	V5.0	V5.3	V5.3	V6.0	V8.0	R1.3	R2.0	R3.3	R3.0	R4.0
P-values ($\alpha = 0.05$)		1.000	1.000	0.894	1.000	0.253	0.511	1.000	0.995	1.000	0.599	1.000	0.599
Significance		N.S.	N.S.	N.S.	N.S.	N.S.	N.S.	N.S.	N.S.	N.S.	N.S.	N.S.	N.S.
Soybean**	CT	U0.0	U0.0	V1.0	V2.3	V4.3	V4.6	V6.0	R1.0	R3.0	R4.0	R4.0	R5.0
Soybean**	NT	U0.0	U0.0	V1.0	V2.0	V3.3	V4.3	V6.0	R1.0	R3.0	R4.0	R4.0	R4.6
P-values ($\alpha = 0.05$)		1.000	1.000	1.000	0.374	0.101	0.518	1.000	1.000	1.000	1.000	1.000	0.374
Significance		N.S.	N.S.	N.S.	N.S.	N.S.	N.S.	N.S.	N.S.	N.S.	N.S.	N.S.	N.S.

N.S. Not significant paired comparison at the alpha = 0.05 level. V signifies the vegetative-growth stage and R signifies the reproductive growth stage for both corn and soybean. *V3—three leaf stage and growing point is under the soil surface;V6—six leaf stage and growing point is above soil surface; R1—silk is visible; R2—blister stage; R3—milk stage; R5—dent stage [18]. **U—Unifoliate leaf (Cotyledon leaf); V1—first trifoliate leaf unfolded; V2—two fully unfolded trifoliate leaves; V3 to V5—plantdevelops 3 - 5 trifoliate leaves, respectively; V6—six fully developed trifoliate leaves [22].

Table 6. Mean total nutrient concentration for above-ground plant tissue for crop and tillage (conventional tillage—CT; No tillage—NT) treatments at Beresford, SD in 2014.

Crop	Tillage	Tissue N	N Sufficiency-Level*	Tissue P	P Sufficiency-Level*	Tissue K	K Sufficiency-Level*	Tissue Zn	Zn Sufficiency-Level*
		%						ppm	
Corn (1)	CT	3.1	3.5 - 5.0	0.34	0.30 - 0.50	3.3	2.5 - 4.0	25.6	20 - 60
Corn (1)	NT	2.4	3.5 - 5.0	0.35	0.30 - 0.50	3.3	2.5 - 4.0	22.6	20 - 60
P-values ($\alpha = 0.05$)		0.201		0.784		0.984		0.415	
Significance		N.S.		N.S.		N.S.		N.S.	
Corn (2)	CT	2.6	2.7 - 4.0	0.26	0.25 - 0.50	1.7	1.7 - 3.0	15.3	25 - 100
Corn (2)	NT	2.7	2.7 - 4.0	0.28	0.25 - 0.50	1.7	1.7 - 3.0	14.3	25 - 100
P-values ($\alpha = 0.05$)		0.846		0.318		0.523		0.251	
Significance		N.S.		N.S.		N.S.		N.S.	
		%						ppm	
Soybean (3)	CT	3.7	3.5 - 5.0	0.34	0.3 - 0.5	2.2	2.5 - 4.0	44.6	20 - 60
Soybean (3)	NT	3.6	3.5 - 5.0	0.34	0.3 - 0.5	2.2	2.5 - 4.0	44.3	20 - 60
P-values ($\alpha = 0.05$)		0.387		0.209		0.322		0.969	
Significance		N.S.		N.S.		N.S.		N.S.	

N.S. Not significant paired comparison at the alpha = 0.05 level. (1) corn plant samples harvested at V6growth stage on June 27; (2) corn pant samples harvested at tassel R1 growth stage on July 25; (3) soybean plant samples harvested at R1-early bloom stage on July 17. *Adapted from [23].

and Zn were similar between tillage treatments, these nutrients were available at similar levels in both tillage systems in the early part of the growing season. These results could have been much different in a drier season where the average temperature could also have been greater and growth stimulated to a greater extent in the conventional-till treatment.

The total tissue N, P, and Zn-nutrient concentration was sufficient (3.7%, 0.34%, and 44.6 ppm) for soybean at R1 growth stage [22], for conventional-till treatment across all tillage treatments. The total tissue N, P, and Zn-nutrient concentration, was sufficient (3.6%, 0.34%, 44.6 ppm) for soybean at R1 growth stage for no-till treatment across all tillage treatments. The total tissue K concentration was deficient by 2.2% vs. the sufficiency level (2.5% - 4.0%) for soybean at R1 growth stage across all tillage treatments, according to [23] (**Table 7**). There was no significant difference in the plant nutrient concentration between tillage treatments. These results could have been much different in a warmer season where growth could have been stimulated to a greater extent.

3.7. Grain Yield

The mean corn grain yield for the conventional-till treatment (10,311 kg·ha^{-1}) was only slightly greater than the yield of the no-till treatment (10,123 kg·ha^{-1}) and was not significant (**Table 7**). In addition, the corn grain yield for the conventional-till treatment was less than the four year average (10,311 kg·ha^{-1} vs.11,121 kg·ha^{-1} respectively), but statistics were not applied to the comparison. The lower grain yield compared to the four-year average reflected the probable lower yield potential due to the cooler early weather compared to a longer term average. The yield of the no-till treatment was slightly greater than the four-year average (10,123 kg·ha^{-1} vs. 9966 kg·ha^{-1} respectively). The cooler than average weather conditions did not impact the yield potential, which was similar to the four-year average yield. The lack of an effect of tillage on corn grain yield somewhat agreed with the research of [6]. They reported that the mean corn grain yield differences between conventional-till and no-till was within 125 kg·ha^{-1} and was not significant. Corn grain yields between conventional-till and no-till in this study were within 200 kg·ha^{-1}.

The mean soybean grain yield (3961 kg·ha^{-1}) for the conventional-till treatment was greater than the yield of the no-till treatment (3295 kg·ha^{-1}) but was not significant (**Table 7**). The soybean grain yield for conventional-till treatment was less than the four-year average (3961 kg·ha^{-1} vs. 3550 kg·ha^{-1}, respectively). The lower yield reflected the cooler early season weather, which probably reduced the yield potential. The yield of no-till was similar to the yield of the four-year average (3295 kg·ha^{-1} vs. 3362 kg·ha^{-1}, respectively). The cooler than average weather conditions and lower than normal GDD (**Table 1**) of 2014 did not impact the yield potential of the no-till treatment compared to the four-year average of the region. Soybean grain yield differences between conventional-till and no-till in this study were 665 kg·ha^{-1}, but were not significant. Unlike the corn response, the difference in soybean grain yield between tillage treatments in this study was somewhat different than the research results of [7]. They reported that the mean soybean grain yield different between conventional-till and no-till was similar, about 200 kg·ha^{-1}. Soils and growing season conditions probably played a role in these response differences.

3.8. Grain Nutrient Removal

The mean corn grain nutrient concentration for P, K, and Zn was 0.24%, 0.32%, and 12.2 ppm, respectively, for the conventional-till treatment (**Table 8**). The corn grain nutrient concentration for P, K, and Zn was 0.22%, 0.31%, and 11.6 ppm respectively for the no-till treatment. The difference in corn grain P, K, and Zn, content was not significant different between tillage treatments.

The mean soybean grain nutrient concentration for P, K, and Zn was 0.50%, 1.66%, and 31.3 ppm, respectively, for the conventional-till treatment. The soybean grain nutrient concentration for P, K, and Zn was 0.54%, 1.70%, and 33.4 ppm, respectively for the no-till treatment. The difference in soybean grain nutrient concentration for P and Zn was not significant different between tillage treatments. However P, nutrient was significantly different within the conventional-till treatment. This indicated that nutrient uptake to the grain was not influenced by tillage.

3.9. Plant Tissue Nutrients in Crop Residue

The mean nutrient concentration of corn residue for P, K, and Zn was 0.05%, 0.85%, and 13.20 ppm respectively, for the conventional-till treatment (**Table 9**). The residue nutrient concentration for P, K, and Zn was 0.04 %, 0.91 %, 14.43 ppm, respectively, for the no-till treatment.

The mean nutrient concentration of soybean residue for P, K, and Zn was 0.08%, 1.14%, and 32.67 ppm respectively for the conventional-till treatment. The nutrient concentration of soybean residue for P, K, and Zn was

Table 7. Mean grain yield for crop and tillage (conventional tillage—CT; No tillage—NT) treatments at Beresford, SD comparing 2014 yield to the mean of 2010-2013 yield.

Filed Crop	Tillage	Grain Yield 2014	Mean Grain Yield 2010-2013
		kg·ha^{-1}	kg·ha^{-1}
Corn	CT	10,311	11,121
Corn	NT	10,123	9966
P-values ($\alpha = 0.05$)		0.803	
Significance		N.S.	
Soybean	CT	3961	3550
Soybean	NT	3295	3362
P-values ($\alpha = 0.05$)		0.167	
Significance		N.S.	

N.S. Not significant paired comparison at the alpha = 0.05 level.

Table 8. Mean grain nutrient concentration of crop and tillage (conventional tillage—CT; No tillage—NT) treatments at Beresford, SD in 2014.

Crop	Tillage	P	K	Zn
		%		ppm
Corn	CT	0.24	0.32	12.2
Corn	NT	0.22	0.31	11.6
P-values ($\alpha = 0.05$)		0.295	0.802	0.230
Significance		N.S.	N.S.	N.S.
Soybean	CT	0.50	1.66	31.3
Soybean	NT	0.54	1.70	33.4
P-values ($\alpha = 0.05$)		0.014	0.090	0.155
Significance		*	N.S.	N.S.

N.S. Not significant paired comparison at the alpha = 0.05 level. *Significant paired comparison at the alpha = 0.05 level.

Table 9. Mean crop residue nutrient concentration for crop and tillage (conventional tillage—CT; No tillage—NT) treatments at Beresford, SD in 2014.

Crop	Tillage	P	K	Zn
		%		ppm
Corn	CT	0.05	0.85	13.20
Corn	NT	0.04	0.91	14.43
P-values ($\alpha = 0.05$)		0.408	0.796	0.329
Significance		N.S.	N.S.	N.S.
Soybean	CT	0.08	1.14	32.67
Soybean	NT	0.08	0.97	44.80
P-values ($\alpha = 0.05$)		0.539	0.026	0.327
Significance		N.S.	*	N.S.

N.S. Not significant paired comparison at the alpha = 0.05 level. *Significant paired comparison at the alpha = 0.05 level.

Table 10. Mean dry weight for stover and residue of crop and tillage (conventional tillage—CT; No tillage—NT) treatments at Beresford, SD in 2014.

Crop	Tillage	Dry Weight
		$kg{\cdot}plot^{-1}$
Corn	CT	3.36
Corn	NT	2.86
P-values ($\alpha = 0.05$)		0.394
Significance		N.S.
Soybean	CT	2.36
Soybean	NT	2.13
P-values ($\alpha = 0.05$)		0.346
Significance		N.S.

N.S. Not significant paired comparison at the alpha = 0.05 level.

0.08%, 0.97%, 44.80 ppm, respectively for the no-till treatment.

When comparing the mean nutrient concentration of corn residue for conventional-till to the no-till treatment, the nutrient differences for P, K, and Zn levels were + 0.01%, −0.06%, and −1.2 ppm, respectively for the specific nutrient. Similar to the other tissue nutrient levels, the difference in corn residue nutrient concentration was not significantly different between tillage treatments.

When comparing the mean nutrient concentration of soybean residue for conventional-till to no-till treatment, the nutrient differences for P, K, and Zn levels were +0.00%, −0.17%, and −12.13 ppm, respectively for the specific nutrient. The difference in soybean residue nutrient concentration for P and Zn was not significant between tillage treatments. However, the K nutrient levels were significantly different between the tillage treatments.

3.10. Stover and Crop Residue Dry Weight

The mean corn residue dry weight for conventional-till treatment was greater (3.36 $kg{\cdot}plot^{-1}$) than that (2.86 $kg{\cdot}plot^{-1}$) of the no-till treatment (**Table 10**). However this 15% difference for the no-till treatment was not statistically different than the conventional-till treatment weight. The mean soybean crop residue dry weight for conventional treatments of 2.36 $kg{\cdot}plot^{-1}$ was greater than the 2.13 $kg{\cdot}plot^{-1}$ in the no-till treatment. However, the 10% difference was not statistically different between tillage treatments.

4. Conclusion

The precipitation in the early growing season was higher than normal (30-year average) and air temperature in the early part of the growing season was lower than normal in June 2014. This influenced the GDD's which were lower than normal in the early part of the growing season. As the result, there was no significant difference in soil temperature (5 cm depth) between tillage treatments in the early part of the growing season, removing the advantage of tillage for increasing early season soil temperatures. In the middle of the growing season, growth stages of the conventional-till treatments were slightly more advanced than the no-till treatments, but were not significantly different. The grain yields and nutrient removal were not influenced by tillage treatment. The growth and grain parameters of crops growing in the conventional-till treatment were suppressed due to the lower than normal air temperature and higher than normal soil moisture content during the early part of the growing season. This provided roughly equivalent growing environment for plants growing in the conventional and no-till treatments. As a result, no differences in growth and yield parameters of the crop were observed between tillage treatments. A warmer and a drier growing season would probably have shown a significant difference in soil temperature, soil moisture, dry matter production, nutrient concentration, and grain yield between tillage treatments.

References

[1] Wilhelm, W.W. and Wortmann, C.S. (2004) Tillage and Rotation Interaction for Corn and Soybean Grain Yield as Affected by Precipitation and Air Temperature. *Agronomy Journal*, **96**, 425-432. http://dx.doi.org/10.2134/agronj2004.0425

[2] Kravechenko, G.A. and Thelen, K.D. (2007) Effect of Winter Wheat Crop Residue on No-Till Corn Growth and Development. *Agronomy Journal*, **99**, 540-555.

[3] He, J., Li, H., Kuhn, N.J., Wang, Q. and Zhang, X. (2010) Effect of Ridge-Tillage, No-Tillage, and Conventional Tillage on Soil Temperature, Water Use, and Crop Performance in Cold and Semi-Arid Areas in Northeast China. *Australian Journal of Soil Research*, **48**, 737-744. http://dx.doi.org/10.1071/SR09155

[4] Govaerts, B., Sayre, K.D., Goudeseune, B., Corte, P.D., Lichter, K., Dendooven, L. and Deckers, J. (2009) Conservation Agriculture as a Sustainable Option for the Central Mexican Highlands. *Soil and Tillage Research*, **103**, 222-230. http://dx.doi.org/10.1016/j.still.2008.05.018

[5] Schwartz, R.C., Baumhardt, R.L. and Evett, S.R. (2010) Tillage Effects on Soil Water Redistribution and Bare Soil Evaporation Throughout a Season. *Soil and Tillage Research*, **110**, 221-229. http://dx.doi.org/10.1016/j.still.2010.07.015

[6] Licht, M.A. and Al-Kaisi, M. (2004) Strip-Tillage Effect on Seedbed Soil Temperature and Other Soil Physical Properties. *Soil and Tillage Research*, **80**, 233-249. http://dx.doi.org/10.1016/j.still.2004.03.017

[7] Chen, Y., Liu, S., Li, H., Li, X.F., Song, C.Y., Cruse, R.M. and Zhang, X.Y. (2011) Effects of Conservation Tillage on Corn and Soybean Yield in the Humid Continental Climate Region of Northeast China. *Soil and Tillage Research*, **115-116**, 56-61. http://dx.doi.org/10.1016/j.still.2011.06.007

[8] Franzen, D.W., Overstreet, L.F., and Cattanach, N. (2007) NDSU Campus Tillage Trials, Year 4. Production Practices. *Sugar Beet Res. and Ed. Board*, **38**, 119-123. N.D.S.U. http://www.sbreb.org/research/prod/prod07/Campustill07.pdf

[9] USDA-NRCS (2012) Web Soil Survey. USDA-NRCS, Washington DC. http://websoilsurvey.nrcs.usda.gov/app/WebSoilSurvey.aspx

[10] Gelderman, R. and Beegle, D. (1998) Nitrate-Nitrogen. In: Brown, J.R., Ed., *Recommended Chemical Soil Test Procedures for the North Central Region*, N.C.R. Pub. 221 (Revised), 17-20.

[11] Bouyoucos, G.J. (1962) Hydrometer Method Improved for Making Particle Size Analysis of Soils. *Agronomy Journal*, **54**, 464-465. http://dx.doi.org/10.2134/agronj1962.00021962005400050028x

[12] Combs, S.M. and Nathan, M.V. (1998) Soil Organic Matter. In: Brown, J.R., Ed., *Recommended Chemical Soil Test Procedures for the North Central Region*, NCR Publication No. 221, Missouri Agricultural Experiment Station, Columbia, MO, 53-58.

[13] Watson, M.E. and Brown, J.R. (1998) pH and Lime Requirement. In: Brown, J.R., Ed., *Recommended Chemical Soil Test Procedures for the North Central Region*, NCR Publication No. 221, Missouri Agricultural Experiment Station, Columbia, MO, 13-16.

[14] Watanabe, F.S. and Olsen, S.R. (1965) Test of an Ascorbic Acid Method for Determining Phosphorus in Water and $NaHCO_3$ Extracts from the Soil. *Soil Science Society of America Journal*, **29**, 677-678. http://dx.doi.org/10.2136/sssaj1965.03615995002900060025x

[15] Frank, K., Beegle, D. and Denning, J. (1998) Phosphorus. In: Brown, J.R., Ed., *Recommended Chemical Soil Test Procedures for the North Central Region*, NCR Publication No. 221, Missouri Agricultural Experiment Station, Columbia, MO, 21-30.

[16] Warncke, D. and Brown, J.R. (1998) Potassium and Other Basic Cations. In: Brown, J.R., Ed., *Recommended Chemical Soil Test Procedures for the North Central Region*, NCR Publication No. 221, Missouri Agricultural Experiment Station, Columbia, MO, 31-33.

[17] Whitney, D.A. (1998) Micronutrients: Zinc, Iron, Manganese and Copper. In: Brown, J.R., Ed., *Recommended Chemical Soil Test Procedures for the North Central Region*, NCR Publication No. 221, Missouri Agricultural Experiment Station, Columbia, MO, 41-44.

[18] McWilliams, D.A., Berglund, D.R. and Endres, G.J. (1999) Corn Growth and Management Quick Guide. North Dakota Cooperative Extension Service, Fargo, 1-8.

[19] Sah, R.N. and Miller, R.O. (1992) Spontaneous Reaction for Acid Dissolution of Biological Tissues in Closed Vessels. *Analytical Chemistry*, **64**, 230-233. http://dx.doi.org/10.1021/ac00026a026

[20] de Mendiburu, F. (2013) Statistical Procedures for Agricultural Research. Package "Agricolae" Version 1.4-4. Comprehensive R Archive Network, Institute for Statistics and Mathematics, Vienna. http://cran.r-project.org/web/packages/agricolae/agricolae.pdf

[21] Gerwing, J. and Gelderman, R. (2005) Fertilizer Recommendations Guide. South Dakota Cooperative Extension Service, South Dakota State University and USDA Cooperating, Brookings, SD, 8-20.

[22] Pedersen, P. (2004) Soybean Growth and Development. Iowa State University Extension Document PM 1945, 1-27. http://extension.agron.iastate.edu/soybean/documents/SoybeanGrowthandDevelopment.pdf

[23] Mills, H.A. and Benton Jones Jr., J. (1996) Plant Analysis Handbook II. Micro-Macro Publishing Inc., Jefferson, MO, 187-191.

8

Evaluation of Soil Potassium Test to Improve Fertilizer Recommendations for Corn

Manbir K. Rakkar, David W. Franzen, Amitava Chatterjee

Soil Science Department, North Dakota State University, Fargo, USA
Email: manbir.rakkar@ndsu.edu

Abstract

The soil potassium (K) test methodology is under increased evaluation due to the soil sample drying effect, temporal variations of test results and inconsistent crop response to applied K fertilizers. Ten on-farm trials were conducted in 2014 in eastern North Dakota to determine the corn response to different K-fertilizer rates and to assess the variation of soil K test levels between air-dried (KDry) and field moist (KMoist) soil samples during the corn growing season. Significant differences were observed between KDry and KMoist soil K test results. The ratio of KDry/KMoist showed high correlation with cation exchange capacity ($r = 0.63$, $p < 0.10$), Organic matter ($r = 0.61$, $p < 0.10$) and (Ca + Mg)/K ratio ($r = 0.64$, $p < 0.10$) from the 1 M ammonium acetate extractant, while pH, electrical conductivity, clay (%), and soil moisture showed non-significant correlation. On average, KDry resulted in higher soil K test levels than KMoist and pattern of deviation was different for surface and sub-surface soil samples. Soil K analysis of samples collected during the fall and spring showed large enough variations to affect the soil test interpretation category which was used to make fertilizer recommendations. Corn yield increased significantly with applied K fertilizer at only three out of 8 sites with beginning K levels below the current critical level of 150 ppm, and one response was at a site with K level above the critical level. Therefore, use of either the KDry or KMoist method alone may not be adequate to predict K response in some North Dakota soils.

Keywords

Potassium, Soil Test Methodology, Fertilizer Recommendations, Grain Yield

1. Introduction

The corn (*Zea mays*) growing belt of the United States is shifting north and west of the traditional Corn Belt due to changing climate patterns and improved corn hybrid varieties with short-season yield potential. Corn yields have increased more than two folds in North Dakota in past three decades [1]. The increase in corn yield in North Dakota is the net result of improved corn genetics and higher rainfall during the growing season [2]. Since higher yields are often accompanied with high nutrient removal from the soil [3], maintaining an adequate supply of nutrients is the next major challenge for the corn growers of North Dakota.

Providing an adequate supply of nutrients to corn is important for gaining yield benefits from other management practices. Corn is known to take up substantial amounts of K during the growing season. For instance, corn yielding 10.11 Mt/ha can accumulate about 165 $kg \cdot ha^{-1}$ of potassium [4]. Crop response to K is not as great as that of N, but K plays a vital role in every facet of crop growth. Positive correlation has been reported among K content of crops and photosynthesis, carbohydrate metabolism, lodging and disease resistance [5]. Potassium plays an important role in water uptake and helps in maintenance of yields in adverse climatic conditions such as drought [6]-[8]. Therefore, maintaining an adequate level of K is important in the rain-fed agricultural system of North Dakota.

Soil testing is an important diagnostic tool for estimating nutrient supplying capacity of soils for growing crops. The most widely used procedure for estimating plant-available potassium is extraction of K from air-dried soil samples using 1 M ammonium acetate [9]. However, air-drying of soil samples is known to collapse or scroll up the clay lattice structure leading to release or entrapment of K depending upon soil solution K concentration and clay mineralogy [10], which can lead to over- or under-estimation of soil-K levels [11]. To overcome this issue, Iowa State University has reintroduced the procedure of using field-moist soil samples for plant-available K analysis. Analysis of field-moist soil samples from Iowa for available K has resulted in improved correlation with corn yields compared with air-dried soil K analysis [12]. Therefore, performance of this new methodology needs to be reviewed with the soils of North Dakota.

Soil K results not only are subject to change due to the air-drying of soil samples, but also may vary depending on the date of sampling [13]. The seasonality effect is likely due to seasonal variability in moisture (high moisture in winters and comparatively low moisture towards the end of growing season when soils are driest), K leaching from crop residues, freezing and thawing, and microbial activity [14]. Switching from fall to spring sampling can lead to significant changes in soil K values, affecting the rate of K-fertilizer application [15]. Therefore, a better understanding of fluctuations of soil K level during the growing season will be helpful in improving K-fertilizer recommendations.

In North Dakota, fertilizer recommendations for corn were formulated in the late 1970s and early 1980s when yields were much lower than they are today. The new corn varieties for the region are much more productive and generally soil tests K levels are much lower today.

To address the increase in corn acres in North Dakota, the relevance of the current soil K test and response of modern corn hybrids to K fertilizer, a study was conducted with three main objectives:

1) To compare soil K test values based on air-dried and field moist samples;
2) To determine the effect of sampling time on soil K test levels during the corn growing season;
3) To determine the corn response to applied K-fertilizer based on the predictability of the soil K test.

2. Materials and Methods

2.1. Site Descriptions

During 2014, trials were conducted at ten locations in the eastern part of North Dakota including the Cass, Barnes, Richland and Sargent counties (**Figure 1**). All of these sites are involved in agricultural production with corn and soybean as the main crops. These areas have a humid-continental climate with mean precipitation about 55 cm and mean temperature varying about 5˚C (Mean of temperature and precipitation from 1981 to 2010).

Soil series descriptions are listed in **Table 1**. Most of these soils are developed from glacial lacustrine sediments, glacial outwash or till/moraines with somewhat poorly drained to well drained characteristics.

2.2. Experimental Design

Each experimental location was established with a minimum distance of 30 m from the field edge. The experimental design of the trials was a randomized complete block design with six K-fertilizer treatments and four

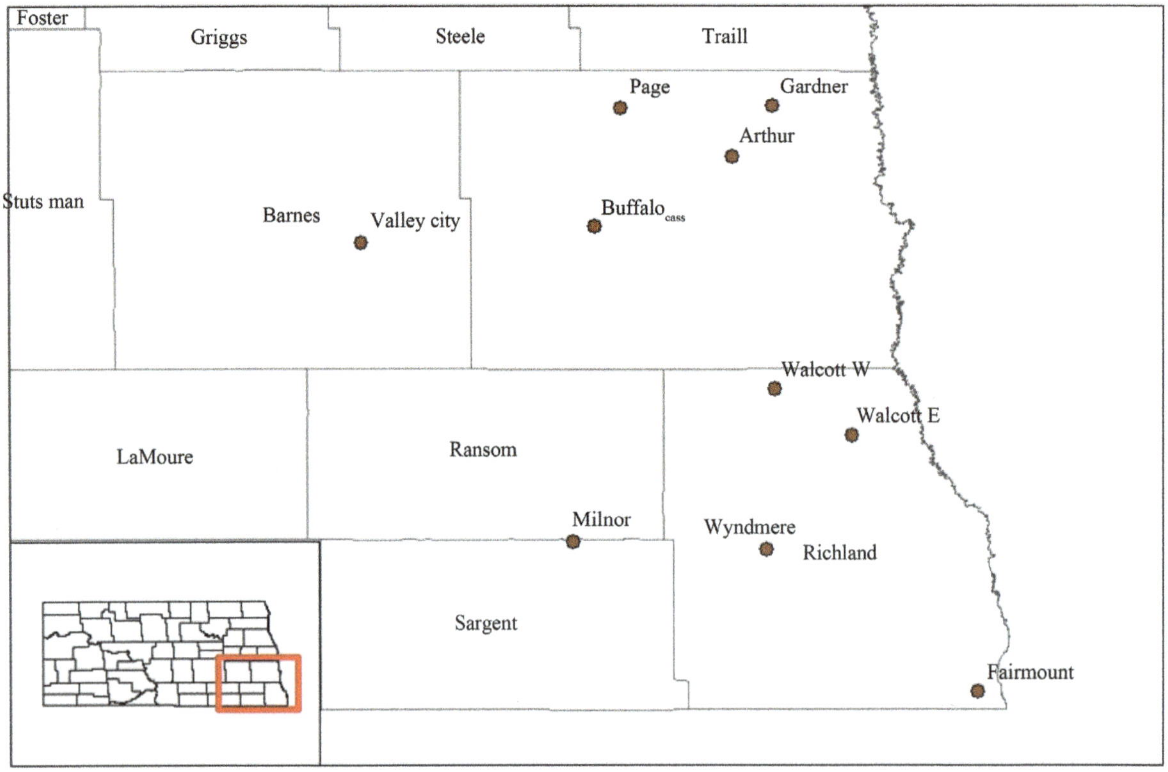

Figure 1. North Dakota map showing experimental sites of 2014.

Table 1. Location and soil characterization information of K-experimental sites.

Location	Latitude and longitude	Soil series	Taxonomic classification
Buffalo	46°55'12.582"N 97°25'18.338"W	Lankin-Gilby	Fine-loamy, mixed, superactive, frigid Pachic Hapludolls
Gardner	47°09'57.830"N 97°03'04.561"W	Galchutt	Fine, smectitic, frigid Vertic Argialbolls
Walcott E	46°29'43.090"N 96°53'05.196"W	Wheatville-Mantador-Delamere	Coarse-silty over clayey, mixed over smectitic, superactive, frigid Aeric Calciaquolls
Wyndmere	46°15'38.809"N 97°03'50.155"W	Glyndon	Coarse-silty, mixed, superactive, frigid Aeric Calciaquolls
Fairmount	45°58'18.719"N 96°37'08.665"W	Gardena	Coarse-silty, mixed, superactive, frigid Pachic Hapludolls
Milnor	46°16'33.843"N 97°28'01.110"W	Embden-Wyndmere	Coarse-loamy, mixed, superactive, frigid Pachic Hapludolls
Walcott W	46°35'16.546"N 97°02'50.090"W	Hecla-Garborg	Sandy, mixed, frigid Oxyaquic Hapludolls
Arthur	47°03'46.590"N 97°08'03.730"W	Glyndon-Tiffany	Coarse-silty, mixed, superactive, frigid Aeric Calciaquolls
Valley city	46°53'17.843"N 97°54'54.062"W	Barnes-Svea	Fine-loamy, mixed, superactive, frigid Calcic Hapludolls
Page	47°09'38.226"N 97°22'02.788"W	Swenoda	Coarse-loamy, mixed, superactive, frigid Pachic Hapludolls

replications. Nine of the total sites received a fertilizer application of potassium chloride-KCl (0-0-60) at the rate of 0, 33.6, 67.2, 100.9, 134.5, 168.1 K_2O kg·ha^{-1} while the Milnor site received K application of 0, 67.2, 134.5, 201.7, 269.0, 336.2 K_2O kg·ha^{-1}. Dimensions of all plots were 9.14 m long by 3.05 m wide, with a 1.52 m of alley between each replication. The alleyways were cut out when the corn had 8 - 12 leaves. Corn planting and all agronomic and cultural operations were carried out by the farmers and were uniform for all plots within a location (**Table 2**). The farmer did not apply K fertilizer within the boundaries of experimental plots. When the grower applied K with N or P fertilizer, the plot area was excluded from his field application and N, P and any other nutrients determined necessary by the pre-plant soil test were broadcast applied by the researchers.

2.3. Soil Sampling

Initial composite soil samples were collected from 0 - 15 cm depth from each site before planting and were analyzed for plant available nutrients and other basic soil properties. During the growing season, soil samples were collected from the control plots (plots with no K-fertilizer application) twice each month with an interval of about 15 days. A 2.5 cm diameter Hofer soil tube was used to take the samples from the 0 - 15 cm and 15 - 30 cm depth throughout the growing season. Soil samples were not taken from 15 - 30 cm on the second August sampling at Page and Valley City due to soil hardness. Soil samples were collected by taking four to five cores at each depth from the interior inter-row area within each plot. Samples from each depth were then composited and stored in zip-lock polythene bags to maintain the moisture level comparable to the field conditions. Samples were transported in a cooler to the laboratory and stored in laboratory refrigerator at 7˚C for one to three weeks.

2.4. Laboratory Analysis

Initial soil samples-Initial composite soil samples were analyzed for pH, N, P, K, EC and organic matter by the NDSU Soil and Water Testing Laboratory using approved methods for the North Central Region of the USA (**Table 3**). Soil texture was determined by a hydrometer method [16]. Cation exchange capacity of the soil was determined by saturating the soil with 1 M sodium acetate solution and then washing the soil with 90% ethanol solution and replacing the sodium ions from exchange complex using 1 M ammonium acetate [17].

Methodology for KDry (plant-available-K test of air-dried soil samples) and KMoist (plant-available-K test of field-moist soil samples)-Each soil sample was thoroughly mixed and subdivided into two sub-samples. One of them was analyzed with standard procedure of soil K test which involves air-drying of soil, grinding and passing through 2 mm sieve. Two grams of air-dried sample was extracted with 20 ml of 1M NH_40Ac, shaken for 5 min and filtered through Whatman No. 2 filter paper. Gravimetric water content of air-dried and field-moist soil was determined by oven drying a sub-sample at 105˚C for at least 24 hours [18]. For KMoist, sub-sample was not air-dried but was sieved through a 2 mm sieve. Two grams of sieved field-moist soil was treated with 20 ml of NH_4O Ac by adjusting the molarity of extracting solution to 1 M according to the moisture content of the sample.

Table 2. Corn production details for all experimental sites.

Site	Corn variety	Planting density --seeds ha^{-1}--	Sowing date	Harvesting date
Buffalo	Dekalb DKC 36-30 RIB	80,000	5/15/2014	10/8/2014
Gardner	NuTech 5B782	-	5/18/2014	9/24/2014
Walcott E	Dekalb DKC 36-30RIB	85,000	5/30/2014	10/15/2014
Wyndmere	Dekalb DKC 43-10	87,250	5/27/2014	10/14/2014
Fairmont	GC 95-33 VT3P	87,340	5/23/2014	10/16/2014
Milnor	Pioneer 9917	81,250	5/17/2014	10/14/2014
Walcott W	Dekalb 39-07	85,000	5/23/2014	10/15/2014
Arthur	ProSeed 11-91 VT2P RIB	90,000	5/18/2014	10/3/2014
Page	REA 2A550	-	5/25/2014	10/17/2014
Valley City	Crop Plan 2417 VT2	75,000	5/5/2014	10/13/2014

Table 3. Soil test results of initial soil samples collected from 0 - 15 cm depth.

Location	NO_3-N[†]	P[§]	K[¶]	pH[#]	EC[††]	OM[‡‡]	Clay[§§]	CEC[¶¶]
	kg·ha^{-1}	---ppm--			dS·m^{-1}	-----%-----		cmol·kg^{-1}
Buffalo	18	12	115	7.6	0.19	2.1	10.8	12.9
Gardner	10	13	110	5.9	0.09	2.2	11.3	12.5
Walcott E	6	3	105	7.4	0.45	2.3	11.5	12.1
Wyndmere	20	8	100	7.9	0.27	2.3	11.5	15.6
Fairmount	23	10	140	7.6	0.30	2.7	15.5	19.9
Milnor	9	18	110	6.2	0.43	2.2	7.30	14.1
Walcott W	10	16	80	5.8	0.10	1.5	4.50	10.6
Arthur	15	10	170	8.2	0.26	3.1	14.5	23.1
Page	20	12	200	7.5	0.48	2.4	10.0	14.9
Valley City	10	27	485	6.5	0.30	3.1	17.5	19.7

[†]NO_3-N extracted with water, [§]P extracted with Olsen procedure, [¶]K extracted with 1M ammonium acetate, [#]pH in water, [††]EC using 1:1 (soil:water) ratio, [‡‡]Organic matter-Loss on Ignition method, [§§]Clay (%)-Hydrometer method, [¶¶]Cation Exchange capacity estimated by 1M sodium acetate method.

The resulting slurry was then shaken for 5 min and filtered through Whatman No-2 filter paper. Soil K concentration of filtrate was determined with necessary dilutions using a Buck Scientific Atomic Absorption Spectrometer-Model 200A (Norwalk, CT, USA) using 766.5 nm wavelength.

2.5. Yield Analysis

For yield analysis, corn ears were harvested from one of the middle two rows leaving first and last plant in each row. Ears were shelled and grain weight was measured in grams. Grain moisture and test weight were measured using Dickey-John Grain Moisture tester (GAC 500 XT). Grain yield was calculated in kg·ha^{-1} adjusted to 15.5% grain moisture content.

2.6. Statistical Analysis

Statistical software-SAS 9.3 and SAS Enterprise Guide 4.3 were used for data analyses [19] [20]. Linear regression was imposed on KDry and KMoist collectively over all sites as well as separately at very low, low, medium, high and very high K soil test K-levels. Pearson correlation coefficients were used to evaluate the relationship of KDry/KMoist ratio with clay content, soil moisture, cation exchange capacity, organic matter, and (Ca + Mg)/K at $p < 0.10$. Analysis of variance for yield response was calculated by SAS PROC GLM procedure using Randomized Complete Block Design with K-fertilizer rates as the main factor. Means of main effects were compared using Fisher's least significant difference (LSD) at 90% confidence level.

3. Results and Discussion

3.1. Basic Soil Properties

Initial soil test results of all experimental sites are presented in **Table 3**. The pH of soils ranged from moderately acidic to moderately alkaline [21]. Based upon the EC levels, all sites had non-saline soils [22]. Seven of the total sites had sandy loam texture, while two of them had loam and one of the sites was categorized as loamy sand. Organic matter determined by loss of weight on Ignition method [23] ranged from 1.5% to 3.1%. The CEC level of soils varied from 10.6 to 23.1 cmol·kg^{-1}.

3.2. Comparison of Soil Potassium Test Based upon Air-Dried and Field Moist Samples

Soil test-K values of surface soil samples (0 - 15 cm depth) determined by KDry ranged from 21 ppm to 824 ppm across all sites with an average of 93 ppm. The KMoist test values had an average of 99 ppm with K values ranging from 14 ppm to 837 ppm. On average, KDry test of surface soils (0 - 15 cm) were 1.07 times higher in K compared to KMoist values but the change of Soil K test varied between soils. Out of 366 soil samples, 47% showed a decrease in K content upon drying while 53% of samples showed an increase in K content. The ratio of KDry/KMoist varied from 0.32 to 2.66 across all sites for surface soil samples. The KDry of sub-surface soil samples (15 - 30 cm) was 1.52 times greater in K content compared with KMoist. Only 20% of the total samples showed a decrease in K content upon drying while 80% samples showed an increase in K values. The linear trend line deviated from the 1:1 line, with the greatest difference in the high and very high K range (**Figure 2**). Such variation in soil K levels of moist and dried soil samples had been observed in various earlier studies in Iowa [12] [24]

Since the variation between KDry and KMoist was different for different sites throughout the growing season, probable factors that might contribute to the difference in drying response were correlated to the KDry/KMoist ratio and summarized in **Table 4**.

Soil moisture content was poorly correlated ($r = -0.02$) with KDry/KMoist ratio. Similar conclusions were found in Iowa where they determined $r^2 = 0.03$ between KDry and KMoist ratio and soil moisture [12]. Clay percentage of initial soil samples was not significantly correlated with ratio of KDry/KMoist ($r = 0.45$, $p = 0.19$). Texture has previously been reported as the main factor for influencing of the degree of K release or fixation [25]. However, clay type may have influenced the KDry/KMoist ratio [26]. Presence of illite is usually responsible for release while montmorillonite (a smectitic clay) is known to fix K [10]. Analysis of clay mineralogy of all these sites might be more helpful in explaining the release and fixation of K upon drying than the determination of clay content of soil per se.

Ratio of (Ca + Mg)/ K was significantly correlated with KDry/KMoist with a correlation coefficient $r = 0.64$ ($p < 0.10$). A relationship between (Ca + Mg)/ K and KDry/KMoist was also reported by Barbagelata and Mal-

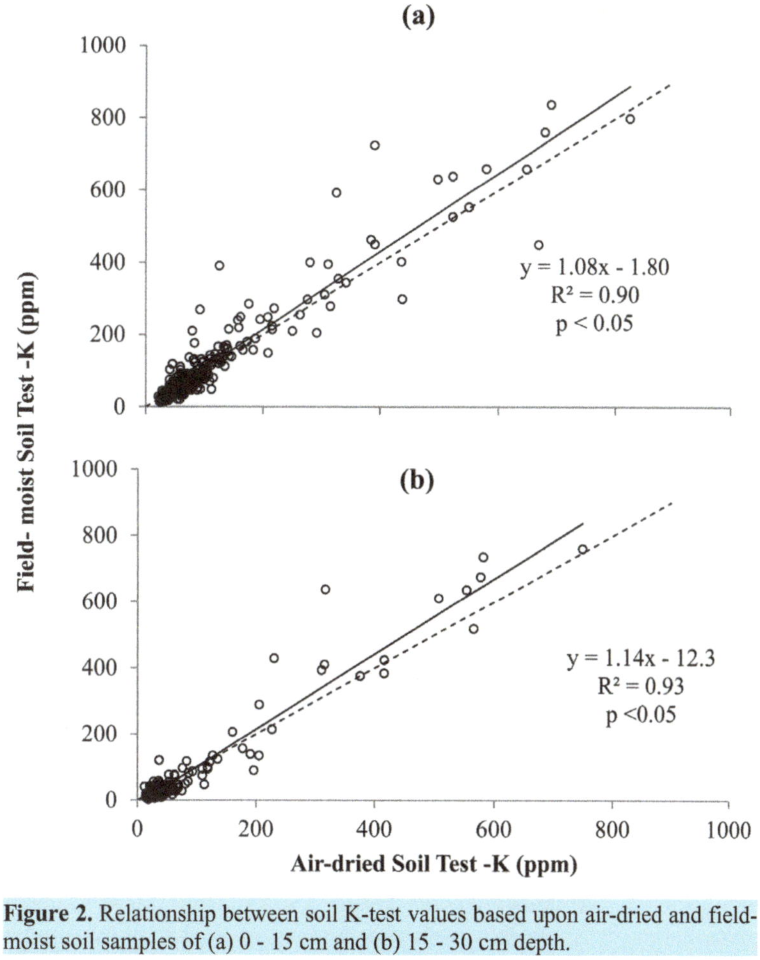

Figure 2. Relationship between soil K-test values based upon air-dried and field-moist soil samples of (a) 0 - 15 cm and (b) 15 - 30 cm depth.

Table 4. Soil test results of initial soil samples collected from 0 - 15 cm depth.

Soil properties	Number of observations (n)	Pearson correlation coefficient (r)
Initial soil samples		
pH	10	0.29
Organic matter (%)	10	0.61*
Cation exchange capacity ($cmol \cdot kg^{-1}$)	10	0.63*
Electrical conductivity ($dS \cdot m^{-1}$)	10	0.29
Clay (%)	10	0.45
Others		
(Ca + Mg)/K ratio†	40	0.64*
Soil moisture (%)‡	366	−0.02

*Significant at 90% confidence level, †Correlation of (Ca+Mg)/K ratio with KDry/KMoist ratio of soil samples collected in first fortnight of September, ‡Correlation of soil moisture (%) with KDry/KMoist ratio of all soil samples collected at fortnightly interval during the corn growing season.

larino [12]. It signifies that the concentration of cations present in soil solution can affect the release and fixation of K upon drying. It occurs because cations such as calcium which show high affinity for negative charged clays can compete with potassium ions for K fixation inducing wedge zones within clay interlayers which results in a release of K ions into the soil solution [27].

KDry and KMoist were significantly related for both depths (0 - 15 cm and 15 - 30 cm). Potassium levels of sub-soil samples were always lower in K compared to surface soil samples. Overall, sub-surface soils showed an appreciable increase in K levels in KDry compared to KMoist tests of surface soil samples. Since the sub-surface soils are less prone to weathering compared to surface soils, thereby, they show a high potential of release of K upon drying [10].

KDry compared to KMoist were significantly related in very low, low and very high category K soils (**Figure 3**). When the KDry content was below 120 ppm, K was released upon drying. Dry K analysis gave lower K values when the soils had >120 ppm initial K. Barbagelata and Mallarino results agree with these data where an exponential decrease of KDry/KMoist ratios was observed as soil K levels were increased [12].

Cation exchange capacity was correlated ($r = 0.63$, $p < 0.10$) with the KDry/KMoist ratio. The CEC of a soil partially depends upon the amount and type of clay minerals. CEC was observed to be positively related to the change of K levels in the soil samples when exposed to drying [28].

KDry/KMoist ratio was significantly related to organic matter content with a correlation coefficient of $r = 0.61$ ($p < 0.10$). The relationship of organic matter (non-volatile organic compounds) to the release of K from soils upon drying is also noted by Welch and Flannery [29] where organic compounds were found to retard the process of diffusion of K from interlayer of clay minerals.

As the season progressed, the difference between KDry and KMoist also changed (**Figures 4-6**). During April, with the exceptions of the Milnor and Arthur sites, KMoist levels were greater than KDry. By late September, this trend was reversed; KDry levels were greater K as compared to KMoist.

3.3. Effect of Time of Sampling on Soil K Test Results

Soil KDry levels of all sites decreased as the growing season progressed (**Figures 4-6**). This change was greater in Very high- K soils as compared to low K soils. There was a decrease of 265 ppm of K content at Valley City (Very high K-site) at the end of September as compared to those collected the previous April. In comparison, the decrease in K between April and September was only 25 ppm at Walcott West (Low K site). Greater variation of K levels in high K soils was also reported previously [30]. Temporal change of soil K level was significantly correlated with soil moisture content at three sites (Buffalo, Walcott East and Wyndmere) while temporal changes of K at other sites were poorly correlated with soil moisture content. An increase in non-exchangeable K was also observed by September in all sites except at Valley City. The temporal variation of soil K can at least be partially attributed to changing soil moisture and a reversion of exchangeable K to non-exchangeable forms. In addition, plant uptake during the growing season and leaching of K after physiological maturity until har-

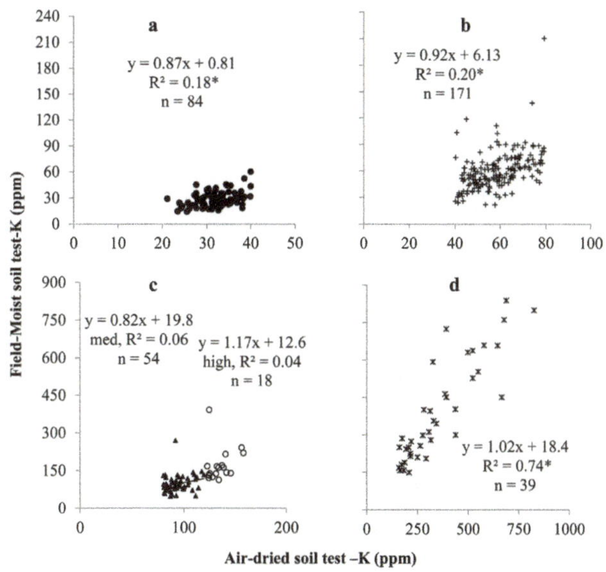

*refers to significant relation between soil test K results based upon air-dried and field-moist soils at 95% confidence level.

Figure 3. Relation of soil test K results based upon air-dried and field-moist soil samples of (a) Very low (0 - 40 ppm) soil K samples (b) Low (41 - 80 ppm) soil K sample (c) Medium (81 - 120 ppm) and high (121 - 160 ppm) soil K samples (d) Very high (>161 ppm) soil K samples.

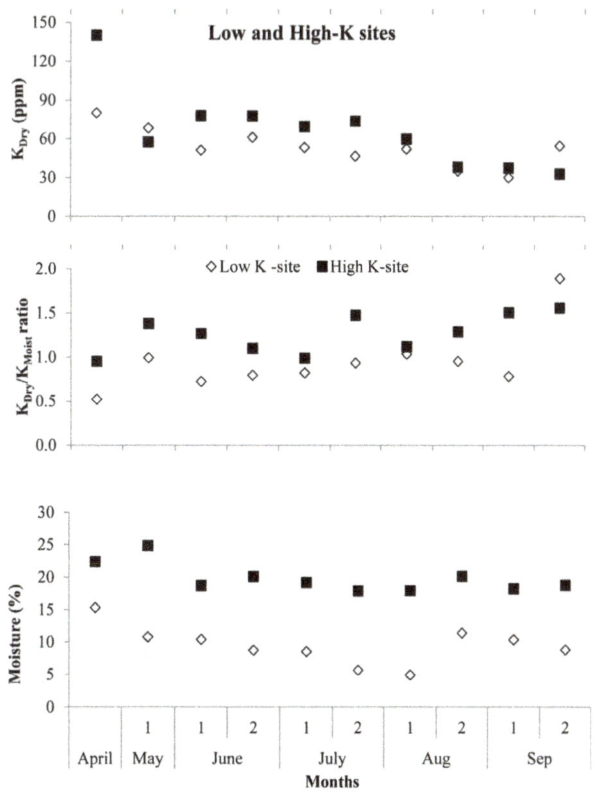

Figure 4. Effect of time of sampling on soil test-K (ppm), KDry/KMoist ratio and soil moisture (%) at Walcott W (low K site) and Fairmount (High K site).

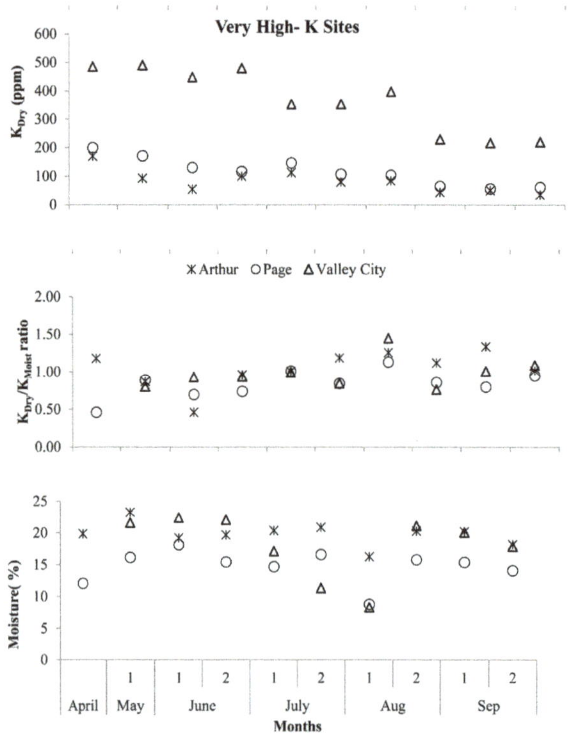

Figure 5. Effect of time of sampling on soil test-K (ppm), KDry/KMoist ratio and soil moisture (%) at Very high K testing sites (Arthur, Page and Valley City).

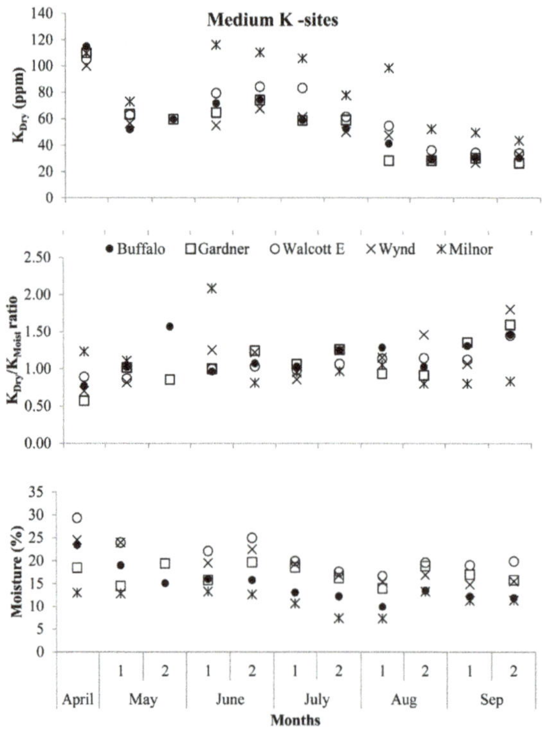

Figure 6. Effect of time of sampling on soil test-K (ppm), KDry/KMoist ratio and soil moisture (%) at medium K testing sites (Buffalo, Gardner, Walcott E, Wyndmere and Milnor).

vesting have been reported as the other possible factors responsible for temporal K variations [14].

Except at the Valley City site, soil K level of all sites dropped to Very low and Low categories with time (**Table 5**). Lower K levels during the fall may mislead farmers in applying fertilizer K rates for next year's crop. However, soil K levels usually recover during the winter season due to freezing and thawing effect and leaching of K from the crop residues, and comparatively higher exchangeable K is observed in April and May [31] [32]. It may be necessary to construct critical levels for early fall and June soil sampling, where the soil K levels are more stable over a practical length of time.

Among the KDry and KMoist soil test results, moist K soil levels were observed to be more variable within a corn growing season. Except for Arthur site, the coefficient of variation was greater for KMoist soil results compared with KDry for all other sites (**Table 6**). Some possible reasons for higher variation in KMoist results could be the manual error during molarity adjustments of extracting solution and while mixing of the moist samples to get a representative sample. This indicates that the current methodology used in determining soil K involving air-drying as a pre-treatment, have more potential in providing precise estimates of K levels over a growing season.

3.4. Corn Response to Applied K Fertilizer Rates

Experimental locations were quite variable in K-status, varying from 80 ppm to 485 ppm of plant available KDry levels. According to North Dakota's published K fertility categories [33], five of the sites had medium soil K level, three had soil K levels in the very high category while low and high categories were represented by one site each. Potassium in the profile was stratified; surface samples (0 - 15 cm) had higher K levels than the sub- surface layer (15 - 30 cm).

Corn grain yield was increased at four sites at the 10% probability level compared to plots receiving no K application. Maximum yield was obtained at 101 kg·ha^{-1} fertilizer rate at 5 sites and at 67 kg·ha^{-1} K rate over 4 out of 10 sites. None of the sites gave highest yield at maximum K fertilizer rate of 168 kg/ha of K. Only one site achieved maximum response at 134 kg/ha of K rate (**Table 7**). The present K category recommendations based on KDry predicted crop response at only 3 of 10 locations. The KMoist did not improve crop response prediction. In addition, the non-exchangeable K levels were not helpful in predicting crop response.

North Dakota experienced frequent rain in the spring and summer of 2014 (NDAWN, http://ndawn.ndsu.nodak.edu/) and good soil moisture conditions were maintained until August. Favorable soil moisture conditions promotes diffusion of K^+ ions [34]-[36] and may have resulted in comparable yields of control plots as that of plots receiving K-fertilizer.

Table 5. Changes in soil test K level between spring and fall soil sampling of control plots and its impact on soil test category.

Location	Change in soil K level†	Soil test category*	
	ppm	Spring	Fall
Buffalo	84.5 ± 5.35‡	Medium	Very low
Gardner	83.5 ± 4.44	Medium	Very low
Walcott E	71.1 ± 2.87	Medium	Very low
Wyndmere	67.2 ± 3.85	Medium	Very low
Fairmount	107 ± 3.85	High	Very low
Milnor	66.3 ± 2.53	Medium	Low
Walcott W	25.8 ± 10.6	Low	Low
Arthur	134 ± 9.66	Very high	very low
Page	139 ± 7.79	Very high	Very low
Valley City	265 ± 67.0	Very high	Very high

†Change in soil test K level calculated as spring minus fall sampling soil K test results. ‡Standard deviation of soil K change between four replications of a control plot (n = 4). *Soil test categories are given for corn in Franzen (2010) Extension Bulletin which include five categories as very low (0 - 40 ppm), low (41 - 80 ppm), medium (81 - 120 ppm), high (121 - 160 ppm) and very high (>161 ppm).

Table 6. Summary of soil K tests based on air-dried and field-moist soil samples during the growing season.

Location	Dry soil K test (ppm)			Moist soil K test (ppm)		
	Average	Std. Dev.	CV†	Average	Std. Dev.	CV
Buffalo	55.95	25.18	0.45	54.15	36.31	0.67
Gardner	54.78	25.39	0.46	59.93	47.59	0.79
Walcott E	63.50	24.39	0.38	63.59	29.73	0.47
Wyndmere	52.58	21.71	0.41	53.15	36.79	0.69
Fairmount	66.32	31.00	0.47	56.63	35.62	0.63
Milnor	83.70	28.03	0.33	81.45	30.56	0.38
Walcott W	53.06	14.68	0.28	64.04	35.24	0.55
Arthur	82.29	40.16	0.49	83.84	38.38	0.46
Page	115.9	47.96	0.41	156.9	109.2	0.70
Valley City	360.9	108.3	0.30	391.3	136.5	0.35

†CV refers to coefficient of variation.

Table 7. Summary of soil K tests based on air-dried and field-moist soil samples during the growing season.

Location	Treatment (kg K_2O ha^{-1})						
	0	33	67	101	134	168	LSD
				------(Grain yield, Mg·ha^{-1})------			
Buffalo	8.69 c†	9.64 a	8.88 bc	9.86 a	9.38 ab	9.31 ab	0.57*
Gardner	8.94 b	10.91 a	9.75 ab	11.0 a	10.1 ab	9.84 ab	1.43*
Walcott E	7.07	7.80	7.26	7.93	7.05	7.32	ns‡
Wyndmere	8.44	9.42	10.56	8.42	9.31	8.18	ns
Fairmount	10.1 b	10.9 ab	11.3 ab	11.0 ab	11.4 a	10.8 ab	1.27*
Walcott W	8.18	8.40	8.85	8.54	8.78	8.70	ns
Arthur	9.77 c	10.2 bc	11.3 ab	11.7 a	11.1 ab	10.6 abc	1.26*
Page	9.34 a	8.85 ab	9.74 a	9.49 a	9.03 ab	8.32 b	1.01*
Valley City	9.74	9.62	9.64	11.1	10.0	10.7	ns
	Treatment (kg K_2O ha^{-1})						
	0	67	134	202	269	336	
Milnor	12.3 ab	13.0 a	11.9 b	11.8 b	12.0 ab	12.8 ab	1.06*

*Significant at 90 % confidence level, †Different letters indicate significant differences at 0.10 significance level. ‡refers to non-significant corn yield response to applied K-treatments.

Based upon the observations of corn response to applied fertilizers, it can be concluded that a refined strategy is required to better predict corn yield response, or a different soil testing method is required for prediction improvement.

4. Summary and Conclusions

Air-drying of soil samples prior to soil analysis of plant-available K significantly affected soil K test results. Change of soil K test levels due to air-drying was not consistently increased or decreased, and was found to be significantly related to cation exchange capacity, organic matter and (Ca + Mg)/K ratio of the soil samples. Soil moisture content, clay content, pH and EC showed minimal influence over KDry/KMoist ratios. Time of soil sampling had considerable effects on soil K levels as well as KDry/KMoist ratios. Temporal K-variations of soil samples collected in fall and spring were large enough to change the soil test interpretation category of a site for making fertilizer recommendations, unless soil test interpretations were constructed for different sampling times. Corn response to applied K fertilizer was site specific and only related to initial soil K levels at three of ten sites.

Based upon these results, it can be concluded that air-drying of soil sample prior to soil K analysis alters the actual plant available-K levels, but KMoist is not a better predictor of corn yield response compared with KDry. The extent of K variation is dependent upon various factors and is likely to change over the time. Corn K response curves needs recalibration in North Dakota. Moreover, soil K levels along with time of sampling, soil moisture dynamics and plant's nutrient utilization potential should be taken into consideration when making K-fertilizer recommendations.

References

[1] National Agricultural Statistical Service (2011) Corn for Grain Yield Per Acre-North Dakota. U.S. Department of Agriculture, Washington DC. http://quickstats.nass.usda.gov/results/90C69DEC-38D6-31B4-9953-4C6EB5E82D79?pivot=short_desc

[2] Ransom, J., Franzen, D., Glogoza, P., Hellevang, K., Hofman, V., McMullen, M. and Zollinger, R. (2004) Basics of Corn Production in North Dakota. North Dakota Extension Service. A-834. http://www.ag.ndsu.edu/pubs/plantsci/rowcrops/a834.pdf

[3] Bender, R.R., Haegele, J.W. Ruffo, M.L. and Below, F.E. (2013) Modern Corn Hybrids' Nutrient Uptake Patterns. *Better Crops*, **97**, 7-10.

[4] Hanway, J.J. (2007) Iowa State University Research. *Integrated Crop*, **498**, 251-253. http://www.ipm.iastate.edu/ipm/icm/2007/8-6/nutrients.html

[5] Havlin. J.L., Beaton, J.D., Tisdale, S.M. and Nelson, W.L. (2005) Potassium. In: Havlin, J.L., et al., Eds., *Soil Fertility and Fertilizers*, Pearson Prentice Hall, New Jersey, 199-218.

[6] Hu, Y. and Schmidhater, U. (2005) Drought and Salinity: A Comparison of Their Effects on Mineral Nutrition of Plants. *Journal of Plant Nutrition and Soil Science*, **168**, 541-549. http://dx.doi.org/10.1002/jpln.200420516

[7] Cakmak, I. (2005) The Role of Potassium in Alleviating Detrimental Effects of Abiotic Stresses in Plants. *Journal of Plant Nutrition and Soil Science*, **168**, 521-530. http://dx.doi.org./10.1002/jpln.200420485

[8] Zorb, C., Senbayram, M. and Peiter, E. (2013) Potassium in Agriculture—Status and Perspectives. *Journal of Plant Physiology*, **171**, 656-669. http://dx.doi.org/10.1016/j.jplph.2013.08.008

[9] Haby, V.A., Russelle, M.P. and Skogley, E.O. (1990) Testing Soils for Potassium, Calcium and Magnesium. In: Westerman, R.L., Ed., *Soil Testing and Plant Analysis*, 3rd Edition, SSSA, Madison, 181-221.

[10] McLean, E.O. and Watson, M.E. (1985) Soil Measurements of Plant Available Potassium. In: Munson, R.D., Ed., *Potassium in Agriculture*, ASA, CSSA, and SSSA, Madison, 277-308.

[11] Wells, K.L. and Dollarhide, J.E. (2000) The Effect of Drying Soil Samples on Soil Test Potassium Values. *Agronomy Notes*, **32**. http://www.uky.edu/Ag/Tobacco/agr_notes/agvl32_5.pdf

[12] Barbagelata, P.A. and Mallarino, A.P. (2012) Field Correlation of Potassium Soil Test Methods Based on Dried and Field-Moist Soil Samples for Corn and Soybean. *Soil Science Society of America Journal*, **77**, 630-642.

[13] Franzen, D.W. (2011) Variability of Soil Test Potassium in Space and Time. *Proceedings of the North Central Extension-Industry Soil Fertility Conference*, Des Moines, 16-17 November 2011, 74-82.

[14] Murrell, T.S. (2011) Why Are Soil Test Potassium Levels So Variable over Time in the Corn Belt? http://fyi.uwex.edu/discoveryfarms/files/2011/06/murrell-k-variability.pdf

[15] Vitko, L.F., Laboski, C.A.M. and Andraski, T.W. (2010) Effect of Sample Time on Soil Test Potassium Levels. *Proceedings of Wisconsin Crop Management Conference*, **49**, 32-38.

[16] Elliott, E.T., Heil, J.W., Kelly, E.F. and Monger, H.C. (1999) Soil Structural and Other Physical Properties. In: Robertson, G.P., *et al.*, Eds., *Standard Soil Methods for Long Term Ecological Research*, Oxford University Press, Inc., New York, 74-88.

[17] Chapman, H.D. (1965) Cation-Exchange Capacity. In: Black, C.A., Ed., *Method of Soil Analysis*, *Part* 2: *Chemical and Microbiological Properties*, ASA, Madison, 891-900.

[18] Black, C.A. (1965) Methods of Soil Analysis: Part 1, Physical and Mineralogical Properties. American Society of Agronomy, Madison.

[19] SAS Institute (2013) SAS Version 9.3. SAS Inst., Cary.

[20] SAS Institute (2013) SAS Enterprise Guide 4.3. SAS Inst., Cary.

[21] Soil Survey Division Staff (1993) Soil Survey Manual. Soil Conservation Service. US Department of Agriculture

Handbook 18.

[22] Whitney, D.A. (1998) Soil Salinity. In: Brown, J.R., Ed., *Recommended Chemical Soil Test Procedures for the North Central Region*, NCR Publication No. 221, Missouri Agricultural Experiment Station, Columbia, 59-60.

[23] Combs, S.M. and Nathan, M.V. (1998) Soil Organic Matter. In: Brown, J.R., Ed., *Recommended Chemical Soil Test Procedures for the North Central Region*, NCR Publication No. 221, Missouri Agricultural Experiment Station SB 1001, University of Missouri-Columbia, Columbia, 53-58.

[24] Luebs, R.E., Stanfoed, G. and Scott, A.D. (1956) Relation of Available Potassium to Soil Moisture. *Soil Science Society of America Journal*, **20**, 45-50. http://dx.doi.org/10.2136/sssaj1956.03615995002000010011x

[25] Burns, A.F. and Barber, S.A. (1961) The Effect of Temperature and Moisture on Exchangeable Potassium. *Soil Science Society of America Proceedings*, **23**, 349-352. http://dx.doi.org/10.2136/sssaj1961.03615995002500050013x

[26] Dowdy, R.H. and Hutcheson, T.B. (1963) Effect of Exchangeable Potassium Level and Drying on Release and Fixation of Potassium by Soils as Related to Clay Mineralogy. *Soil Science Society of America Journal*, **27**, 31-34. http://dx.doi.org/10.2136/sssaj1963.03615995002700010014x

[27] Sparks, D.L. and Huang, P.M. (1985) Physical Chemistry of Soil Potassium. In: Munson, R.D., Ed., *Potassium in Agriculture*, ASA, CSSA and SSSA, Madison, 201-265.

[28] Barbagelata, P.A. (2006) Evaluation of Potassium Soil Tests and Methods for Mapping Soil Fertility Properties in Iowa Corn and Soybean Fields. Retrospective Thesis and Dissertations, Paper 1797, Ph.D. Dissertation, Iowa State University, Ames.

[29] Welch, L.F. and Flannery, R.L. (1985) Potassium Nutrition of Corn. In: Munson, D.R., Ed., *Potassium in Agriculture*, ASA, CSSA and SSSA, Madison, 647-664.

[30] Lockman, R.B. and Malloya, M.G. (1984) Seasonal Variation in Soil Test Results. *Communications in Soil Science and Plant Analysis*, **15**, 741-757. http://dx.doi.org/10.1080/00103628409367514

[31] Fine, L.O., Bailey, T.A. and Truog, E. (1940) Availability of Fixed Potassium as Influenced by Freezing and Thawing. *Soil Science Society of America Proceedings*, **5**, 183-186. http://dx.doi.org/10.2136/sssaj1941.036159950005000C0031x

[32] Mallarino, A.P., Barbagelata, P.A., Clover, M.W., Villavicencio, C.X. and Thompson, L.B. (2011) Factors Determining High Temporal Soil-Test Potassium Variation and Soil Sampling and Testing Alternatives. *Proceedings of the North Central Region Soil and Plant Analyst Workshop*, Bettendorf, IA, 23-24 February 2011, 3-14.

[33] Franzen, D.W. (2010) North Dakota Fertilizer Recommendation Tables and Equations. NDSU Extension Service, SF-882 (Revised), 7.

[34] Mackay, A.D. and Barber, S.A. (1985) Soil Moisture Effect on Potassium Uptake by Corn. *Agronomy Journal*, **77**, 524-527. http://dx.doi.org/10.2134/agronj1985.00021962007700040005x

[35] Schaff, B.E. and Skogley, E.O. (1982) Diffusion of Potassium, Calcium and Magnesium in Bozeman Silt Loam as Influenced by Temperature and Moisture. *Soil Science Society of America Journal*, **46**, 521-524.

[36] Zeng, Q. and Brown, P.H. (2000) Soil Potassium Mobility and Uptake by Corn under Differential Soil Moisture Regimes. *Plant and Soil*, **221**, 121-134.

9

Factors Affecting Tree Husbandry and Woodlots Establishment in Kilimanjaro Region, Tanzania

Revocatus Petro[1*], Francis Laswai[2], Mohammed Mijai[1], Geofrey Nyaradani[1], Chelestino Balama[2]

[1]Tanzania Forestry Research Institute, Moshi, Tanzania
[2]Tanzania Forestry Research Institute, Morogoro, Tanzania
Email: [*]mushumbuz2002@yahoo.co.uk

Abstract

The study on assessment of factors affecting tree husbandry and woodlots establishment was carried out between September and November 2014 in all seven districts of Kilimanjaro region, namely Hai, Siha Rombo, Mwanga, Same, Moshi Rural and Moshi Municipal Council (MMC). A purposive sampling design was employed whereby two wards per district were selected for the study. Household questionnaire survey was used in data collection. Results show that, factors affecting tree husbandry and woodlots establishment in Kilimanjaro region are gender, livelihood activities, access and ownership/land tenure, cultural factors, legal issues and by-laws insisting people to plant trees. About 32.4%, 32.4% and 30% of all respondents own land with size of 1.0 acre and below, 1.1 - 2.0 and 2.1 - 5.0 acres respectively. Respondents from Hai (55.3%), Same (26.3%), Rombo (11%) and MMC (8%) agreed that cultural issues affect tree husbandry. About 1%, 23%, 30% and 40.8% of respondents in Siha, Same, Hai and Rombo districts respectively agreed that some of by-laws insist people to plant trees on their environments. None of the respondents from any district reported labour availability to be one of the factors affecting tree husbandry. Other factors reported to affect tree husbandry in Kilimanjaro region were capacity building in tree planting and tending, insect pests, diseases and climate change. Formulation of by-laws which insist on tree planting and limit land degradation is recommended at village level. Every piece of land should be planned and used sustainably including area for tree planting. Custom and norms which inhibit women to plant trees should be prohibited through capacity building.

Keywords

Socio-Cultural Factors, Tree Husbandry, Woodlots, Kilimanjaro, Tanzania

[*]Corresponding author.

1. Introduction

Tanzania has a total area covering about 945,000 km^2 of which 33.5 million ha are forests and woodlands [1]. About 13 million ha have been gazetted as forest reserves. The main forest types are the extensive miombo woodlands in lowland areas across the central and southern parts of the country, the *Acacia* woodlands in the northern regions, the coastal forest/woodland mosaic in the east, mangrove forests along the Indian Ocean shoreline, and closed canopy forests on the ancient mountains of the Eastern Arc in the east [2]. Of these various forest types, 14.3 million ha are found within gazetted Forest Reserves, 2.5 million ha are proposed as Forest Reserves, and about 2 million ha are in Game Reserves or National Parks [2] [3]. The total area of forest plantations is estimated to be between 190,000 and 250,000 ha, made up of 85,000 ha of state managed industrial plantations, about 40,000 ha of private companies industrial plantations, and 80,000 to 140,000 ha of village and farm plantations [4].

Kilimanjaro region covers an area of 13,209 km^2 or 1.4 percent of the area of the entire Tanzania Mainland out of which 2181.38 km^2 are forest reserves, located mainly on high altitudes as water catchments forests [5]. Almost the entire region's population of 1.64 million by 2012 census [6] relies upon biomass fuels of firewood, charcoal, agricultural residues and biogas for domestic heating and cooking. However, wood fuel (firewood and charcoal) accounts for about 95% of the total energy consumed in the region [7]. Wood fuel is mainly used for cooking, water heating, house heating and lighting. Households are the most important category in wood energy consumption while the second highest consumer of wood fuel are the cottage industries which include brick making, tobacco curing, fish smoking, and bakeries [8]. Supply of the fuel wood in Kilimanjaro region is mainly from the thickets and bushes estimated to cover about 1000 km^2 of the land. The major tree species found in these forests falls into three generic groups, which are Acacias, Albizias and Combretums [9]. Trees like *Melicia excelsa* (mvule) and *Dalbergia melanoxylon* (mpingo) are also found in lower part of the region. Many important commercial trees in Kilimanjaro have been harvested with no programs for replanting which has led to widespread of deforestation which in turn causing erosion of the soil, then loss of soil fertility, loss of wildlife habitat and loss of biodiversity. Kilimanjaro communities are suffering from the results of these misfortunes and unsustainable land management. Agricultural activities all over the region have been affected due to shortage of rainfall and high temperature prevailing for quite long causing seasonal rivers and streams to dry out completely. Women, who traditionally collect firewood, spend longer hours to secure the same amounts of wood and many people are now using their precious hard-earned cash to purchase wood. The population increase in the region from 652,772 in 1967 to 1,640,087 in 2012 [5] [6] has caused an increase in competition of land use form for production of agricultural crops, livestock husbandry and tree husbandry.

Several institutions and individuals have been trying to reforestate the deforestated and degraded areas in order to rescue the current situation. Fast growing fuel wood, fruits, medicine and lumber producing tree species have been introduced to help families from spending time and money on these domestic products. Despite of all efforts made to rehabilitate the lost trees/forests in the region, it is not known as to why majority of the people in the region do not have tree woodlots. Therefore, this study aimed at identifying factors affecting tree husbandry and woodlot establishment in Kilimanjaro region. The results from this study will be used in making recommendations which will lead to overcome factors affecting tree planting in Kilimanjaro Region.

2. Methods

2.1. Description of the Study Areas

2.1.1. Location

The study was conducted in Kilimanjaro region, which is divided into seven districts, namely Hai, Siha, Rombo, Mwanga, Same, Moshi Rural and Moshi Municipality. Kilimanjaro region is located in the north eastern part of Tanzania Mainland. It lies between latitudes 2°25' and 4°15' south of the Equator and between longitude 36°25'30" and 38°10'45" east of Greenwich. The region has a common border with Kenya in the north, Tanga region in the southeast and Manyara and Arusha regions to the south and west (**Figure 1**).

2.1.2. Climate and Soils

In Kilimanjaro region, there are two rainy seasons and two dry seasons. Rainfall starts from September to November as short rainy season and March to May as long rainy season. The dry season starts from December to January and then from July to August every year. The mean annual rainfall varies from 500 mm in the lowlands

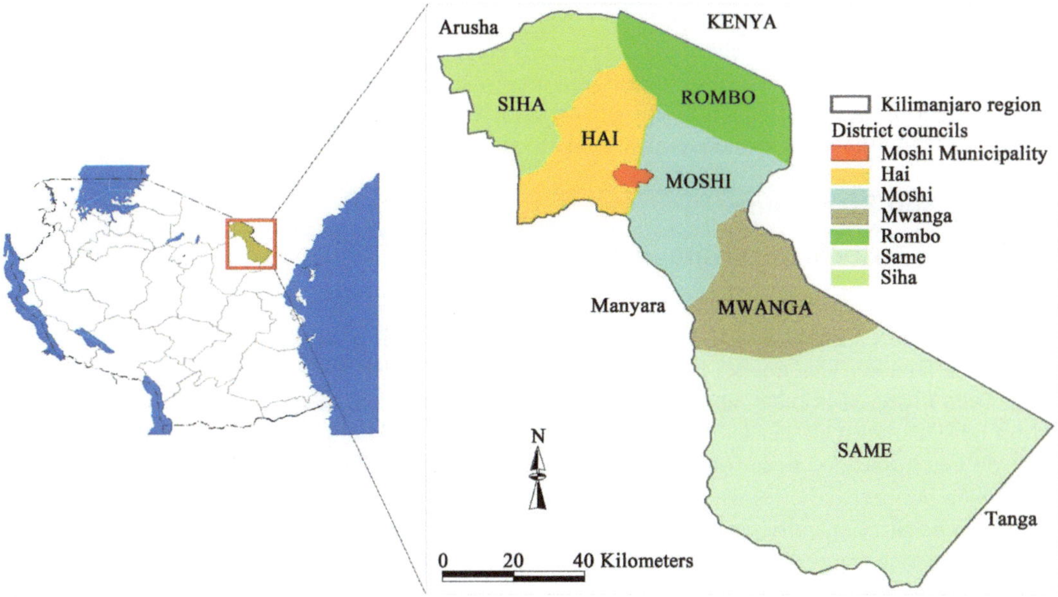

Figure 1. Location of Kilimanjaro region, Tanzania [10].

to over 2000 mm in highland areas located over 1600 meters above sea level. The average minimum and maximum temperatures of the region are 15˚C and 30˚C respectively in the mountainous areas. However, temperature goes up as far as 40˚C in the lowlands during hot period. Soils of the region vary; there are alluvial soils in lower Moshi and areas near Pangani and Mkomazi rivers. Areas in the foot of mountain ranges of Pare comprise colluvial soils and in the highland areas of Mount Kilimanjaro the soils are andosols, rich in nutrients, and are generally deep to very deep loam soil reddish in colour. Forest soils are loam with full of litter and the pH is 4 - 4.6 and less in this area [11] [12].

2.2. Data Collection

2.2.1. Sampling Design

A purposive sampling design was employed whereby two wards per district were selected. This method is a non probability sampling technique where by an experienced individual selects the sample based on his/her experience to judge the appropriate characteristics required for the sample. The criteria used included 1) presence of the sustainable land management (SLM) project, 2) topography of the area (upper, medium, and lower slopes so as to capture different weather conditions) and 3) household involvement in tree planting

2.2.2. Socio-Economic Survey

Socio-economic data were collected through questionnaire survey that intended to get information at household level independently. The questionnaire was structured and open-ended (**Appendix 1**). The structured questions were employed on issues where a specific range of known responses was expected. Such questions allow easier interpretation and analysis than open-ended questions. On the other hand, the open-ended questions were included so as to allow interviewees to construct their own version of experiences because the explanatory power of structured questions is limited [13].

2.2.3. Data Analysis

Quantitative data from questionnaire surveys were coded to facilitate their entry into the computer. They were analysed using the Statistical Package for Social Sciences (SPSS) Computer Programme Version 16 software and results were presented. Data on gender was aggregated as males and females for each district. Proportion (percentage) of each district was determined in relation to the total for the region. Education was grouped into primary, secondary, college and informal levels. Livelihood activities were grouped into crop production, livestock, crop and livestock production, business and employee & others. Numbers of trees owned by respondents were grouped in to five categories mainly; 1 - 50, 51 - 100, 101 - 200, 201 - 500 and >500 trees. Size of the land owned by respondents were classified in to four classes mainly; ≤1.0, 1.1 - 2.0, 2.1 - 5.0 and ≥5.1 acres. Percentages of each education level, livelihood activity, trees category, land classes were determined in relation to the total of all groups in the region. Proportions of cultural issues, legal issues, insects and diseases data per district were determined in relation to the total for the region. However, the number of respondents differed from one section to another as not all respondents answered all questions in the questionnaire.

3. Results and Discussion

3.1. Characteristics of the Respondents

3.1.1. Gender

The results in **Table 1** show that, the majority of the interviewed respondents (56.6%) were males and the rest (43.4%) were females. This implies that most of the households interviewed were male headed. The same situation has been observed in other studies conducted in other areas, e.g. in studies conducted in Morogoro Rural and Mufindi Districts male headed households were 73% and 62.5% respectively [14] [15]. Gender has an influence on tree husbandry since some of the activities are performed by men only while others by women and when women perform activities which are associated with men, then they are doing them on behalf of men not on their own behalf. Generally, majority of the activities in tree growing are for male dominated except weeding and watering which are performed by women in most of the Districts in Kilimanjaro region. Therefore, gender affects tree husbandry in the region.

3.1.2. Education Level of the Respondents

Majority of the respondents (72.3%) have attained primary level education (**Table 2**). However, few individuals

Table 1. Number of respondents and percentage distribution by gender, Kilimanjaro region, Tanzania.

Districts	Gender		Total Respondents	% of Total Sample
	Male	Female		
Hai	36	18	54	15.1
Siha	31	28	59	16.5
Same	27	24	51	14.3
Mwanga	32	28	60	16.8
Rombo	36	24	60	16.8
Moshi Rural	27	21	48	13.4
Moshi Municipal	13	12	25	7.1
Total	202	155	357	100
%	56.6	43.4	100	

Table 2. Number of respondents and percentage distribution by education level, Kilimanjaro region, Tanzania.

District	Education Level				Total
	Primary	Secondary	College	Informal	
Hai	34	10	5	2	51
Siha	37	12	1	8	58
Same	44	5	0	1	50
Mwanga	46	11	0	2	59
Rombo	46	9	2	1	58
Moshi Rural	27	19	0	0	46
Moshi Municipal	17	6	2	0	25
Total	251	72	10	14	347
%	72.3	20.8	2.9	4.0	100

(20.8%) have attained secondary education and minority 2.9% and 4.0% have attained college education and informal education respectively. Despite the low level of education observed in this study, majority of the respondents know how to plant and manage trees. This is probably due to the fact that in Tanzania, tree planting and management are taught even in primary schools. This is in contrary to the report [16] that people with high level of education stand a good chance of conserving natural resources.

3.1.3. Livelihood Activities of the Respondents

A combination of crop and livestock production was the most important (51%) economic activity followed by crop production alone (39%) (**Table 3**). The main cultivated crops were banana and coffee which are normally intercropped with trees. Other crops were maize, potatoes, vegetables, cowpeas, beans and sunflower. Coffee which was the main cash crop has been clear felled due to downfall of world prices. The removal of coffee tree affects much of tree husbandry in Kilimanjaro since shade trees were also removed to allow light intensity to other agricultural crops. Livestock keeping was the least important livelihood activity because of shortage of grazing land in highland areas of Kilimanjaro. Animals kept include cattle, goats, pork and poultry and were confined in a zero grazing system. The presence of livestock in a zero grazing system forced people to plant tree which are used as fodder. Livestock keeping in highland areas contribute positively in tree planting. Therefore livelihood activities affects tree husbandry in Kilimanjaro region.

3.2. Factors Affecting Tree Husbandry and Woodlot Establishment

3.2.1. Access and Ownership/Land Tenure

About 32.4%, 32.4% and 30% of all respondents owned land with size of 1.0 acre and below, 1.1 - 2.0 and 2.1 - 5.0 acres respectively (**Table 4**). The number of respondents who owned land in Moshi Municipal was very few compared to other districts because most of the respondents are living in rented houses hence they have no land. Siha and Same were the leading districts by having large number of respondents who reported to own land with

Table 3. Number of respondents and percentage distribution by livelihood activities, Kilimanjaro region.

District	Livelihood Activities					Total
	Crop production	Livestock	Crop & Livestock Production	Business	Employee & Others	
Hai	24	5	19	4	0	52
Siha	36	0	20	2	0	58
Same	17	0	32	1	1	51
Mwanga	13	2	41	1	1	58
Rombo	17	1	36	3	0	57
Moshi Rural	18	1	22	1	4	46
Moshi MC	11	1	6	6	0	24
Total	136	10	176	18	6	346
%	39.3	2.9	50.9	5.2	1.7	100

Table 4. Number of respondents in each land-size category, Kilimanjaro region, Tanzania.

District	Number of Respondents by Land Size (acre) Category				Total
	≤1.0	1.1 - 2.0	2.1 - 5.0	≥5.0	
Hai	29	14	9	0	52
Siha	16	19	21	3	59
Same	4	7	28	11	50
Mwanga	23	20	15	1	59
Rombo	20	22	16	0	58
Moshi Rural	13	19	12	3	47
Moshi MC	5	9	0	0	14
Total	110	110	101	18	339
%	32.4	32.4	30	5.2	

size of 2.1 - 5.0 acres (**Table 4**). Correspondingly, Same had the highest number of respondents who own land with size of above 5.0 acre. In the highland areas of Kilimanjaro, a family owns a land of an average of 0.5 of a hectare (1.25 acres) while in lowland area a family owns 1.5 hectares (3.75 acres) [5] and this land especially in the coffee belt or the Chagga home gardens is regulated by customary law, where transfer or change of ownership is ruled by inheritance patterns (ownership passing to sons). Environmental degradation in Kilimanjaro region seems to be increasing due to poor land husbandry practices such as lack of use of soil-erosion control methods [5]. In this regard, the region needs to intensify land management practices in order to improve land productivity per unit area since it affects tree husbandry.

The size of the land has a total reflection to the number of trees planted. More than 45 respondents in each district reported to have their own trees except Moshi Muncipality wherein only 17 respondents reported having own trees (**Table 5**). As stated earlier, most of respondents in Moshi Municipality are living in rented houses and tenants frequently change houses, hence commitment by this group to plant trees on the premises they live is limited. About 61% of all respondents owned 50 trees and below. Same, Siha and Rombo districts had relatively higher number of respondents who own 201 trees and more compared to other districts (**Table 5**). This is an indication that scarcity of land in these three districts is relatively less compared to other districts. Same and Rombo districts had large number of respondents (50% and 38% respectively) who had land specifically set aside for tree planting while Hai (11%) and Siha (1%) had lowest. Respondents from Mwanga, Moshi Rural and Moshi Municipal reported to have no specific area for tree planting because of insufficient land. Instead, they were mixing trees with other agricultural crops and other trees are planted in borders. Therefore access and ownership/land tenure were considered as factors affecting tree husbandry in the region.

3.2.2. Cultural Factors

Every society has its own socio-cultural factors which members of the society conforms to. These factors are important because a member of the society needs to know them in order to participate in various activities [17]. These socio-cultural factors can influence both men and women in a particular society. Though some of the factors cannot be explained scientifically, still the members of the society honour them. Results from this study revealed that, cultural factors, to some extent, affect tree husbandry in some districts of the region. 55%, 26%, 10.5% and 8% of all respondents from Hai, Same, Rombo and Moshi Municipal respectively agreed that cultural issues could be considered as one of factors affecting tree husbandry in Kilimanjaro region. However, all respondents in Siha, Mwanga and Moshi Rural totally disagreed that cultural issues influences tree planting and management. **Table 6** is an activity profile showing how culturally the society has allocated various activities to both women and men in tree growing. The profile shows that there are activities which are performed by men and others by women. However, when women perform activities which are associated with men, they are doing them on behalf of men but not on their own. Additionally, it was reported that some tree species like *Euphorbia triculli* and *Albizia coriaria* are not allowed to be grown by women. It is alleged that if a woman plants such tree species, she will become barren and her husband will die [18]. Generally, all activities in tree growing are done by men except weeding and watering which are done by women. The allocation of duties is discriminating against women. Therefore cultural aspects limit participation of women in afforestation and agro-forestry programs. Results from this study agree with the findings of Chavangi [18], that certain tree species may have culturally defined gender specific and ownership restrictions.

3.2.3. Legal Issues

About 23.4%, 30% and 40.8% of respondents in Same, Hai and Rombo districts respectively agreed that there are some by-laws which compel people to plant trees on the areas surrounding their environments. Very few percentages of respondents in Siha and Moshi Municipal agreed on the same (**Table 7**). The by-laws reported to compel people to plant trees including; planting of trees surrounding their areas, restrictions on cutting trees unnecessarily, planting trees in catchments and lowlands, restrictions on cutting tree without a permit, prohibiting unnecessary setting of fire and grazing animals in some areas. Some posters were observed in some areas of Hai and Same prohibiting cutting of trees unnecessarily, unnecessary setting of fire and prohibition on grazing of animals. Most of the by-laws are enforced by village, ward, and division officials. There are prescribed punishments for those who fail to abide by-laws and regulations. The kind of punishments reported includes verbal warning and fines. The findings further pointed out a number of complaints from various individuals regarding violation of existing by-laws by some of the officials in wards and villages including few dishonest forest offi-

Table 5. Number of respondents within each tree category, Kilimanjaro region, Tanzania.

District	Number of respondents by tree category					Total
	1 - 50	51 - 100	101 - 200	201 - 500	≥501	
Hai	33	12	2	2	2	51
Siha	28	19	2	9	1	59
Same	29	5	7	3	6	50
Mwanga	41	8	4	2	2	57
Rombo	26	8	13	7	4	58
Moshi Rural	33	7	2	3	1	46
Moshi Municipal	17	0	0	0	0	17
Total	207	59	30	26	16	338
%	61.2	17.5	8.9	7.7	4.7	

Table 6. Activity profile for women and men in tree growing, Kilimanjaro region, Tanzania.

Activities	Who does it?		
	Men	Females	Both
Preparation for the holes for transplanting tree seedlings	X		
Site selection where tree seedlings are planted	X		
Types of tree seedlings to be planted	X		
Actual planting of trees	X		
Weeding of tree seedlings		X	
Watering of tree seedlings		X	
Protecting the tree seedlings from destruction by putting protective devices	X		
Pruning of trees	X		

Table 7. Number of respondents who agreed that legal issues insist people to plant trees in their areas in Kilimanjaro region.

Districts	Number of Respondents	Percentage of Respondent
Hai	36	30.0
Siha	1	0.8
Same	28	23.4
Rombo	49	40.8
Moshi Municipal	6	5.0
Total	120	100

cials. Similar weaknesses have been observed in other areas in Africa [19]. Respondents from Mwanga and Moshi Rural districts, however, reported that they are unaware of the by-laws. Probably this unawareness explains why most of the respondents from these two districts owned few number of tress i.e. below 50 trees (**Table 5**).

3.2.4. Capacity Building

A high number of respondents in Hai (22%) and Rombo (23%) districts agreed that they have ever been trained on tree planting and tending. However, respondents though few in number; 7%, 8%, 12.5%, 13% and 15% in Moshi Municipal, Mwanga, Same, Siha and Moshi Rural districts respectively agreed to have been trained on tree planting and tending. More training is needed especially in Moshi Municipal and Mwanga districts. Some respondents from Rombo, Siha and Same districts reported to have never been trained by any institution but they know how to plant and tend trees from their parents. Some institutions reported to train people on tree planting and tending and issues concerning environmental conservation includes Hai Green Initiative (HGI), Kilimanjaro

National Park Authority (KINAPA), TENGERU, Sustainable Land Management Project (SLM), Foresters and environmentalists from Districts, Community Development Officers from Districts, Selian-Arusha, Tanzania Association of Foresters (TAF), Tanzania Forestry Research Institute (TAFORI), GTZ, PADEP and TASAF.

3.2.5. Labour Availability

None of the respondent from any district reported to employ any casual labour to assist in tree farm establishment and management. All the respondents reported to attend tree planting activities such as tree farm establishment, tree planting and tending on their own. This is because most of respondents do not have land specifically set aside for tree planting. Instead, trees are mixed on banana and coffee farms and on farm boundaries. Therefore, labour availability is not one of the factors affecting tree husbandry in Kilimanjaro region.

3.2.6. Other Factors

1) Insects and Diseases

Insect pests and diseases were among other factors affecting tree husbandry in Kilimanjaro region. Other pests reported to affect trees were livestock and human beings that destroy young trees especially near the roads. Rombo was the leading district in insect pests and diseases infestation while Moshi Rural was the least (**Table 8**). Insect pests and diseases infest the stressed trees which are weak. Some parts of Rombo, Same and Mwanga districts get insufficient rain which cannot sustain the growth of trees. The mean annual rainfall of Rombo and Mwanga districts are 500 and 400 mm (in lowland) respectively [22] [23]; In this case most of trees in these areas are stressed, hence prone to insects and diseases attack. Siha and Moshi Rural districts had shown low infestation of insect pests and diseases compared to other districts probably because of sufficient rain. The mean annual rainfall of Siha and Moshi Rural are about 900 and 800 mm [20] [21].

2) Climate

Large number of respondents from Siha (19.3%), Rombo (17.2%), Moshi Rural (16.8%), Mwanga (14%) and Hai (13.7%) reported that the existing climate allows good growth of most planted trees species. 10.5% and 8.8%, relatively few number of respondents in same and Moshi Municipal respectively, agreed that the existing climate does allow good growth of most planted trees species. Some respondents from Same and Mwanga reported a long period of drought and strong dry wind in lowland limit tree growth while respondents from Rombo reported stunted growth during strong cold season as the climatic factors hindering tree husbandry. Same and Mwanga districts are located in the semi-arid areas in Kilimanjaro region and they experiences 400 - 600 mm of rainfall per annum in low lands [22]. Temperatures range between an average of 14˚C during June-July and 32˚C usually during the month of January therefore tree growth and other agricultural crops are affected much by changing climate in these districts.

4. Conclusions and Recommendations

Factors affecting tree husbandry in Kilimanjaro region in Tanzania include access and ownership/land tenure, cultural issues, by-laws, capacity building, age, gender and education. Other factors are insect pests, diseases

Table 8. Insects and diseases reported to affect tree growth in Kilimanjaro region.

Districts	Number of Respondents	Percentage of Respondent	Insects	Disease
Hai	14	11.7	Termites, ants, white flies, catarpillars	Rotting of roots of Grevillea, frost, dropping of leaves
Siha	7	5.8	Termites	Die back, heart rot
Same	20	16.7	Termites	Die back, leaf spot
Mwanga	20	16.7	White flies, stem boares, termites, fruit boers	Damping off, die back, heart rot
Rombo	50	41.7	Termites, ants, white flies, catarpillars,	Frost
Moshi Rural	1	0.8	Termites	Leaf spots
Moshi Municipal	8	6.7	Spider, white flies, termites	
Total	120	100.0		

and climate change. It is therefore recommended that; strict implementation of by-laws in tree planting in the districts that are lagging behind, so that people can plant trees and reduce land degradation; insisting in the use of agro-forestry practices whereby agricultural crops are mixed with trees because of insufficient of land in the region; educating the community members to change their attitudes towards cultural factors which promote tree growing and conduct a study on the best way of controlling pests and diseases affecting tree husbandry in Kilimanjaro region.

Acknowledgements

Authors gratefully acknowledge the Government of Tanzania, Global Environment Facility (GEF) and UNDP through Sustainable Land Management (SLM) Project in Kilimanjaro for financing this study. Authors are also indebted to all Institutions and individuals who facilitated the study. Finally, the authors are thankful to the Project Technical Team from Project Coordination Unit (PCU) at the RAS-Kilimanjaro Office for their useful comments and suggestions.

References

[1] United Republic of Tanzania (URT) (1998) National Forest Policy. Government Printer, Dar-es-Salaam, 59 p.

[2] Akida, A. and Blomley, T. (2006) Trends in Forest Ownership, Forest Resources Tenure and Institutional Arrangements: Are They Contributing to Better Forest Management and Poverty Reduction? Case Study from Tanzania. Unpublished Report, FAO, Rome.

[3] United Republic of Tanzania (URT) (2001) National Forestry and Beekeeping Programme, 2001-2010. Forestry and Beekeeping Division, Ministry of Natural Resources and Tourism, 132 p.

[4] Ngaga, Y.M. (2011) Forest Plantations and Woodlots in Tanzania. African Forest Forum. Nairobi, 76 p.

[5] United Republic of Tanzania (URT) (1998) Kilimanjaro Region Social-Economic Profile. The Planning Commission, Dar es Salaam and Regional Commissioner's Office, Kilimanjaro, 230 p.

[6] United Republic of Tanzania (URT) (2013) 2012 Population and Housing Census. National Bureau of Statistics, Ministry of Finance, Dar es Salaam and Office of Chief Government Statistician President's Office, Finance, Economy and Development Planning, Zanzibar, 244 p.

[7] TAFORI (2010) Production and Consumption of Fuel Wood and Tree Species Preferences in Kilimanjaro Region. TAFORI Project Report, Moshi Timber Research Centre, Moshi, 21 p.

[8] Mugo, F.W. (2001) The Role of Woodfuel Conservation in Sustainable Supply of the Resource: Paper Presented at Charcoal Stakeholders. ICRAF Working Paper No. 20, Nairobi.

[9] Kimaryo, B.T. and Ngereza, K.I. (1993) Development and Promotion of Improved Traditional Earth Kilns for Charcoal Production in Kilimanjaro Region of Tanzania. A Consultancy Report Submitted to REDPU/MWEM, Dar es Salaam. A World Bank/IDA Supported Project, TAFORI, Morogoro.

[10] Sangeda, A.Z., Kahimba, F.C., Kashaga, R.A.L., Semu, E., Mahonge, C.P. and Mkanda, F.X. (2014) Testing of Decision Making Tools for Village Land Use Planning and Natural Resources Management in Kilimanjaro Region. *Open Journal of Soil Science*, **4**, 446-458.

[11] Lovett, J.C. and Pócs, T. (1993) Assessment of the Condition of the Catchment Forest Reserves, a Botanical Appraisal. Report Prepared for the Catchment Forestry Project.

[12] Lovett, J.C. (1993) Eastern Arc Moist Forest Flora. In: Lovett, J.C. and Wasser, S.K., Eds., *Biogeography and Ecology of the Rain Forest of Eastern Africa*, Cambridge University Press, Cambridge, 35-56. http://dx.doi.org/10.1017/cbo9780511895692.004

[13] Valentine, G. (1997) Tell Me about Using Interviews as a Research Methodology. In: Flowerdew, E. and Martin, D., Eds., *Methods in Human Geography: A Guide for Students Doing a Research Project*, Longman, London, 110-253.

[14] Lema, T.A. (2003) The Role of Non-Timber Forest Products in Household Food Security and Women Income in Morogoro Rural District, Tanzania. Master's Dissertation, Sokoine University of Agriculture, Morogoro.

[15] Raphael, T. and Swai, G. (2009) The Impact of Participatory Forest Management and Local People's Perceptions on Its Implementation at the Village Level in Mufindi Districts, Southern Tanzania Highlands. *Proceedings of the First Participatory Forest Management (PFM) Research Workshop: Participatory Forest Management for Improved Forest Quality, Livelihood and Governance*, Morogoro, 15-16 December 2008, 133-146.

[16] Lebora, H. (2006) Contribution of Participatory Forest Management to Sustainable Forest Management and Poverty Reduction in Lindi Rural District, Tanzania. Master's Dissertation, Sokoine University of Agriculture, Morogoro, 36-50.

[17] Oloo, J.O., Makenzi, P.M., Mwangi, J.G. and Abdulrazack, A. (2013) Influence of Traditions/Customs and Beliefs/ Norms on Women in Tree Growing in Siaya County, Kenya. *Global Journal of Environmental Science and Technology*, **1**, 1-6.

[18] Chavangi, N. (1984) Cultural Aspects of Fuel Wood Procurement in Kakamega District, Kenya. Working Paper No. 4, Kenya Fuel Wood Development Program/Beijer Institute, Nairobi.

[19] Anderson, A.B., May, P.H. and Balick, M.J. (1991) The Subsidy from Nature: Palm Forests, Peasantry, and Development in Amazon Countries. Columbia University Press, New York.

[20] Moshi District Council (2011) Moshi District Socio-Economic Profile. MoDC, Moshi, 22 p.

[21] Siha District Council (2013) Siha District Socio-Economic Profile. SDC, Siha, 93 p.

[22] Mwanga District Council (2011) Mwanga District Socio-Economic Profile. MDC, Mwanga, 19 p.

[23] Rombo District Council (2013) Rombo District Socio-Economic Profile. RDC, Rombo, 54 p.

Appendix I. Factors Affecting Tree Husbandry and Woodlots Establishment in Kilimanjaro Region, Tanzania

1. RESPONDENT'S GENERAL INFORMATION

District name__________ The interview date__________

Ward name__________ Village/Suburb/Street name__________

1.1. Name_______

1.2. Sex Male ☐ Female ☐ Age____ Years____

1.3. Education attained____

1.4. Marital status: Married ☐ Unmarried ☐ Divorce ☐ Widow ☐
Separated ☐ Widower ☐

1.5. Occupation: (1) Peasant (2) Livestock keeper (3) Peasant & Livestock keeper (4) Business
(5) Others state__________

2. ACCESS AND OWNERSHIP/LAND TENURE

2.1. Do you use firewood? Yes ☐ No ☐

2.2. Do you use charcoal for energy? Yes ☐ No ☐

2.3. If yes please state the source of your firewood/charcoal________________________

2.4. Do you own trees? Yes ☐ No ☐

2.5. If yes how many trees do you have?________________________________

2.6. Why don't you have more trees than what you have mentioned?

____

2.7. What are the tree species do you have and what are their uses?

____

2.8. What are the uses of the tree species that you have?

2.9. What is the growth status of your trees? Good ☐ Average ☐ Worse ☐

2.10. If not good what are the main problems? ________________________

2.11. Do you own land? Yes ☐ No ☐

2.12. If yes state how many hectares/acres________________________________

2.13. Have you set aside land for tree planting? Yes ☐ No ☐

2.14. If yes state how many hectares/acres________________________________

2.15. If no state why____

2.16. Where do you get seeds for tree planting exercise ________________________

2.17. Where do you get seedlings for tree planting exercise?____________________

2.18. What are your tree species preferences?________________________________

2.19. Please give reasons for their preference________________________________

3. CULTURAL ISSUES

3.1. Are there any cultural issues affecting land ownership? Yes ☐ No ☐

3.2. If yes state____

3.3. How do they affect land ownership? Yes ☐ No ☐

3.4. Are there any cultural issues affecting tree farm establishment? Yes ☐ No ☐

3.5. If yes state____

3.6. How do they affect tree farm establishment?________________________

4. LEGAL ISSUES

4.1. Are there any laws which insist you to plant trees on area around you? Yes ☐ No ☐

4.2. If yes state____

4.3. If yes, have you followed the law?________________________________

4.4. If no, can you give reasons for not following the law?____________________

5. CAPACITY BUILDING

5.1. Have you ever been trained on tree planting? Yes ☐ No ☐

5.2. If yes mention the organization involved in providing that training.

____

5.3. Have you ever been trained on tending of trees? Yes ☐ No ☐

5.4. If yes mention the organization involved in providing that training______________________

6. LABOUR AVAILABILITY

6.1. Have you employed any casual labourer to assist you in tree farm establishment? Yes ☐ No ☐

6.2. If yes how many casual labourers do have?______________________

6.3. What are the main duties/activities does the casual labourers assist you?______________________

6.4. Are casual labourers readily available? Yes ☐ No ☐

6.5. If yes, where/how do you get them?______________________

6.6. Do you pay them? Yes ☐ No ☐

6.7. If yes, how much do you pay them per day/week/month?______________________

6.8. Can scarcity of casual labourers affect establishment and tending of trees in your farm? Yes ☐ No ☐

7. OTHERS

7.1. Is there any disease that affects growth of your trees? Yes ☐ No ☐

7.2. If yes state type of disease______________________

7.3. Is there any insect that affects growth of your trees? Yes ☐ No ☐

7.4. If yes state type of insects______________________

7.5. Does the climate allow good growth of your trees? Yes ☐ No ☐

7.6. If no state why______________________

7.7. What should be done in order to speed up tree planting?______________________

10

Switchgrass Management Practice Effects on Near-Surface Soil Properties in West-Central Arkansas

Alayna Jacobs[1], Kristofor R. Brye[1], Randy King[2], Joel Douglas[3], Lisa S. Wood[1], Larry C. Purcell[1], Michael Looper[4]

[1]Department of Crop, Soil, and Environmental Sciences, University of Arkansas, Fayetteville, USA
[2]USDA-NRCS Plant Materials Center, Booneville, USA
[3]USDA-NRCS Central National Technology Support Center, Fort Worth, USA
[4]Department of Animal Science, University of Arkansas, Fayetteville, USA
Email: kbrye@uark.edu

Abstract

Agronomic management practices that maximize monoculture switchgrass (*Panicum virgatum* L.) yield are generally well understood; however, little is known about corresponding effects of differing switchgrass management practices on near-surface soil properties and processes. The objective of the study was to evaluate the effects of cultivar ("Alamo" and "Cave-in-Rock"), harvest frequency (1- and 2-cuts per year), fertilizer source (poultry litter and commercial fertilizer), and irrigation management (irrigated and non-irrigated) on near-surface soil properties and surface infiltration in a Leadvale silt loam (fine-silty, siliceous, semiactive, thermic, Typic Fragiudult) after four years (2008 through 2011) of consistent management in west-central Arkansas. Irrigating switchgrass increased ($P < 0.01$) soil bulk density in treatment combinations where poultry litter was applied (1.40 $g \cdot cm^{-3}$) compared to non-irrigated treatment combinations (1.33 $g \cdot cm^{-3}$). Root density was greater ($P = 0.031$) in irrigated (2.62 $kg \cdot cm^{-3}$) than in non-irrigated (1.65 $kg \cdot cm^{-3}$) treatments when averaged over all other treatment factors. The total infiltration rate under unsaturated conditions was greater ($P = 0.01$) in the 1-cut (33 $mm \cdot min^{-1}$) than 2-cut (23 $mm \cdot min^{-1}$) harvest treatment combinations when averaged over all other treatment factors, while the total infiltration rate under saturated conditions did not differ among treatment combinations ($P > 0.05$) and averaged 0.79 $mm \cdot min^{-1}$. Results from this study indicate that management decisions to maximize switchgrass biomass production affect soil properties over relatively short periods of time, and further research is needed to develop local best management practices to maximize yield while maintaining or improving soil quality.

Keywords

Soil Bulk Density, Infiltration, Extractable Soil Nutrients

1. Introduction

The United States (US) is the largest consumer of petroleum in the world at an estimated 18.6 million barrels per day [1]. Even though the US is also the third largest producer of petroleum, domestic production cannot keep pace with demand, as 40% of petroleum products consumed in the US are net imports from other countries [1]. Burning fossil fuels also releases carbon dioxide (CO_2). Promoting the generation and development of domestic fuel supplies, while reducing CO_2 emissions from fossil fuel combustion, has become a central issue in global policy.

Concerns related to the effects of global warming and possible shortages of finite fossil fuel sources have led to government-mandated regulations encouraging the development of alternative fuels. According to the Energy Policy Act of 2005 and the Renewable Fuel Standard Program of 2007, 36 billion gallons of renewable fuel must be blended into the nation's transportation fuel supply by 2022 [2]. However, only 12% of electricity generated in the US is currently generated from renewable sources, such as wind, hydroelectric, and biomass [3].

One potential solution is the conversion of plant biomass to fuel. Using plant biomass as a source of cellulosic biofuel as an alternative to fossil fuel is attractive because plants use sunlight for energy and capture existing CO_2 from the atmosphere. Though bioenergy derived from lignocellulosic sources may not provide a complete alternative to the massive energy needs of the US, lignocellulosic bioenergy sources may alleviate a portion of the negative impacts of burning fossil fuels and contribute to an extended supply of fossil fuels for the future [4].

Plant biomass can be utilized as a fuel in two major ways: by burning or mixing biomass with coal (*i.e.*, co-firing) to generate electricity [5] or conversion of biomass to ethanol. Co-firing is desirable because it reduces the amount of coal burned while maintaining a high level of energy output. Mixing renewable fuels with a non-renewable fuel like coal may also extend the supply of finite resources further into the future. One advantage of using plant biomass rather than corn (*Zea mays* L.) grain to produce ethanol is the avoidance of sacrificing food for fuel production. Corn is an important food crop for human and livestock consumption. Corn makes up over 95% of the grain rations for cattle, and the US is the world's largest producer of beef [6]. Use of a non-food cellulosic material for conversion to ethanol offers an attractive solution to the problem of competing with the livestock feed industry.

Corn is also an annual crop that must be replanted every year and requires large amounts of fertilizer. The extensive preparation required for conventional corn production is regarded as causing more total soil erosion than any other crop grown in the US [7]. Dedicated energy crops, or crops grown exclusively for conversion to renewable fuel sources, have emerged as potential sources of cellulose. These crops include short-rotation woody crops, such as willow (*Salix* spp.) and cottonwood (*Populus* spp.), annual crop residue from small-grain crops and biomass sorghum (*Sorghum bicolor* [L.] Moench), and perennial grass crops. Ideal energy crops from woody and herbaceous sources have low fertilizer requirements, are relatively cheap and easy to establish, have a perennial growth habit, are easily integrated with existing US farming practices, and are adaptable to sites unsuitable for food or cash crop production.

Switchgrass (*Panicum virgatum* L.) was recognized by the US Department of Energy in the 1990s as part of the Bioenergy Feedstock Development Program for its ability to produce large quantities of biomass on relatively poor sites. Annual yields of switchgrass in the US average 11.2 $Mg \cdot ha^{-1}$, ranging from 4.5 $Mg \cdot ha^{-1}$ in the northern plains to 23.0 $Mg \cdot ha^{-1}$ in Alabama [8]. Switchgrass is native to all of the US, except in California and the Pacific Northwest, and is adapted to a wide range of soil textures and soil drainage classes [9]. Another desirable aspect of switchgrass production is that the agricultural community is already familiar with perennial grass production, harvesting strategies, and harvesting equipment [8]. In order for producers to commit to growing dedicated energy crops for renewable fuel markets, detailed information about storage, handling, pricing, timing, and transport must be available [10]. Since most of that information does not exist for the majority of US producers, having the option of growing switchgrass as both a source of livestock forage or renewable fuel feedstock may provide added stability for long-term value to producers.

The western portion of Arkansas is largely unsuitable for traditional row-crop production due to steep topography and thin soil compared to the delta region of the Lower Mississippi River basin of eastern Arkansas. However, the average biomass yield for switchgrass grown in Booneville, Arkansas under a single annual harvest regime was 13.4 $Mg \cdot ha^{-1}$ over a four-year period, which was above the national average yield of 11.2 $Mg \cdot ha^{-1}$ estimated by McLaughlin and Kszos [8]. When using the ALMANAC yield prediction computer model, McLaughlin *et al.* [11] recognized the southeast and south-central US regions as being more suitable than the upper mid-western region for providing dependable future feedstock supplies of switchgrass biomass.

In addition to the projected increase in yield, western Arkansas is widely known for its forage and hay production and for its role in the beef cattle industry. Agricultural operators in western Arkansas are familiar with grass harvest, distribution, and storage and typically own conventional farm equipment that can be used to harvest switchgrass. The cost of producing one dry ton of switchgrass in Arkansas was estimated to be $26.73 in the third year of production, with an expected useful life of 12 years before having to replant [10]. Furthermore, Arkansas is the only US state to be ranked in the top 10 of broiler chicken (*Gallus gallusdomesticus*), turkey (*Meleagris gallopavo*), and egg production [12]. As a by-product of poultry production, poultry litter has been widely considered the most valuable manure produced by livestock due to its plant nutrient and organic matter content [13]. Consequently, many producers use dry, surface-applied poultry litter to fertilize hay and forage production pastures.

Past research has identified agronomic procedures for maximizing switchgrass biomass production using varying systems, including cultivar choice, fertilizer application, irrigation, harvest frequency, and row spacing [14]-[17]. Thorough evaluations of the implications associated with switchgrass production systems are needed to seriously determine if switchgrass provides a truly renewable and clean source of domestic energy. Therefore, the objective of this study was to evaluate the effects of switchgrass cultivar (*i.e.*, "Alamo" and "Cave-in-Rock"), water management (*i.e.*, irrigation and no irrigation), fertilizer source (*i.e.*, commercial or poultry litter fertilization), and harvest regime (*i.e.*, once- or twice-harvested per year) after four years of consistent management on near-surface soil properties (*i.e.*, bulk density, soil aggregation, soil pH and electrical conductivity, extractable soil nutrients, and root density) and surface infiltration in west-central Arkansas.

Since "Alamo" previously produced greater biomass yields in all treatment combinations [15], it was hypothesized that "Alamo" treatment combinations would have lower soil bulk density, greater root density, and greater surface infiltration rate than "Cave-in-Rock" combinations. Treatment combinations fertilized with poultry litter rather than commercial fertilizer were expected to have lower soil bulk density and greater surface infiltration rates as the addition of the organic material associated with the poultry litter would increase porosity. Irrigation was expected to decrease near-surface, mobile soil nutrient levels due to increased water available for leaching. Harvesting twice per year was expected to result in a lower root density, greater bulk density, and lower surface infiltration rates than harvesting once per year. Lastly, it was expected that the greatest-yielding treatment combinations would contain the lowest soil nutrient contents since a greater proportion of plant nutrients would have been removed from the soil in aboveground biomass production.

2. Materials and Methods

2.1. Site Description

The US Department of Agriculture's Natural Resource Conservation Service's Plant Materials Center (PMC) was established in Booneville, AR in 1987 and serves the plant material needs of the Southern Ozarks, Arkansas River Valley, Boston, and Ouachita Mountains (Major Land Resource Areas [MLRAs] 116A, 118A, 117, and 119, respectively). The PMC is located along the Petit Jean River in Logan County, AR at an elevation of 146 m [18] and lies along the north edge of the Ouachita National Forest in the eastern portion of the Arkansas Valley and Ridges (MLRA 118A). The 30-yr mean annual precipitation throughout the region is 127 cm and precipitation is delivered more or less evenly throughout the winter (23%), summer (21%), spring (29%), and autumn (27%) seasons [19]. The 30-yr mean annual air temperature for the region is 15.6˚C, with a winter minimum of 10.5˚C and a summer maximum of 32.4˚C [19].

The study site at the PMC (**Figure 1**) was located on a Leadvale silt loam (fine-silty, siliceous, semiactive, thermic Typic Fragiudult) [20]. A fragipan is located at a depth of between 30 and 60 cm from the surface such that the Leadvale soil is poorly drained in late winter and early spring due to the presence of a perched water table. The fragipan also causes plants to experience drought stress during the summer by limiting plant rooting

Figure 1. The field study was conducted at the United States Department of Agriculture's Natural Resource Conservation Service's Plant Materials Center (PMC) near Booneville, AR (*).

depth. The study site was a pasture dominated by tall fescue (*Schedonorus arundinaceus* [Schreb] Dumort., nom. Cons.) prior to being prepared for a yield study in 2006 [15]. No fertilizer was applied and the vegetation was kept mowed to a height of approximately 15 cm year-round. Based on soil sampling, soil nutrient levels were brought up to uniform, medium production values prior to planting according to recommended values for phosphorus (P) and potassium (K) for native warm-season grass establishment.

2.2. Initial Switchgrass Field and Treatment Establishment

"Alamo" and "Cave-in-Rock" switchgrass cultivars were planted in 12.2 by 12.2 m plots on 5 March 2007. Each plot contained one switchgrass cultivar and fertilizer source. Plots were seeded at a rate of 4.4 kg·ha^{-1} pure live seed (457 seed·m^{-2}) and planted with a no-tillage native grass drill (Sukup 2050 series, Jonesboro, AR). After drilling, the seedbed was rolled with a water-filled roller to establish good seed-to-soil contact. Temporary sprinkler irrigation was applied to all plots for initial seed emergence and establishment. A permanent sprinkler irrigation system was installed in replicated irrigation treatments in summer 2007. Each 12.2 by 12.2 m plot was divided in half, with one subplot harvested once per year and the other subplot harvested twice per year. There were a total of 48 plots encompassing three replications of each cultivar-irrigation-fertilizer source-harvest frequency treatment combination.

Rain gauges were placed in the irrigated plots to calibrate the irrigation delivery system. Irrigated plots received 2.54 cm of irrigation water from a nearby surface pond per week during June through August from 2008 to 2011. The average annual precipitation at the study site was 131 cm [18], which was slightly greater than the 30-yr average of 121 cm for the region [19]. Annual rainfall represented 127% in 2008, 124% in 2009, 70% in 2010, and 110% in 2011 of the 30-yr average rainfall [18] [19]. Average annual air temperature represented 90% in 2008, 89% in 2009, 93% in 2010, and 99% in 2011 of the 30-yr average [18] [19].

The study site was burned each year in early March to remove residue stubble, stimulate switchgrass seed production for wildlife, remove surface residue for native pollinator nesting habitat, and to create corridors for other ground-nesting wildlife species [21]. Fertilizer treatments were applied when green foliage appeared in early April each year (*i.e.*, 2008 to 2011). Poultry litter (PL) was applied at a rate of 4.5 Mg·ha^{-1}, while commercial fertilizer (CF) was applied to match nitrogen (N), P, and K levels in the applied PL. Litter batches were analyzed annually for nutrient concentrations prior to application.

Two harvest frequencies were imposed to test their effects on annual aboveground biomass production. A single harvest was made in November after the first killing frost for the 1-cut system. In the 2-cut system, harvests occurred twice per year. The first harvest occurred in June just prior to the boot stage when the seed head emerges and the second harvest occurred after a killing frost in November. Additional switchgrass harvesting details and results of the initial yield study were reported in Jacobs and King [15].

2.3. Soil, Plant, and Infiltration Measurements

In July 2012, soil samples were collected in all plots from the 0- to 10- and 10- to 20-cm depth intervals for bulk density, extractable soil nutrients, and soil particle-size analyses. In the top 10 cm, bulk density samples were collected manually with a 5-cm outside diameter, stainless steel core chamber and a slide hammer. For the 10- to 20-cm depth, bulk density samples were collected with a 5-cm diameter, stainless steel mechanical hydraulic probe. Samples from both depths were dried in a forced-air dryer at 70˚C for 48 hours, and then weighed for bulk density determinations. Soil from the bulk density samples was sieved through a 2-mm mesh screen and used to measure particle-size distribution using a modified 12-hour hydrometer method [22]. Mehlich-3 extractable soil nutrients (*i.e.*, P, K, Ca, Mg, S, Na, Fe, Mn, Zn, and Cu) were determined in a 1:2 soil-water extract [23] [24]. Soil pH and electrical conductivity (EC) potentiometrically in a 1:2 soil-water paste.

In August 2013, soil samples were collected from the 0- to 5- and 5- to 10-cm depths using a 4.8-cm diameter core chamber and slide hammer to measure aggregate stability (AS) (*i.e.*, water-stable aggregates > 0.25-mm diameter) using a wet-sieving procedure [25]. Soil core samples were collected from areas between switchgrass crowns in the switchgrass rows. Two replications were collected from each plot, mixed together for one sample per depth interval, and allowed to air dry for 15 days. Approximately 400 g of air-dried soil were used for the wet-sieving procedure, where soil was plunged in tap water at 30 cycles per minute for 5 minutes. Soil retained on the mesh openings of each of five sieve sizes (*i.e.*, 4, 2, 1, 0.5, and 0.25 mm) was rinsed into an aluminum container with tap water. Water was decanted from the container, and the remaining soil was weighed after being dried at 70˚C for 24 hours. The soil mass from each container was summed to calculate the total mass of water-stable aggregates from each plot.

One soil sample for root density was collected per plot in September 2013 from the 0- to 15-cm depth interval using a 7.3-cm diameter core chamber and slide hammer, and prepared according to the procedures followed by Brye and Riley [26]. Root samples were washed on a 2-mm mesh screen to collect the root material, dried at 55˚C for 24 hours, and weighed. Root density samples were collected from areas between switchgrass crowns in the switchgrass rows.

In November 2013, double-ring infiltrometers were used to measure surface infiltration rates two days after a soaking rainfall. Mature switchgrass was trimmed using hedge trimmers and residue carefully removed prior to placement of infiltrometers. One double-ring infiltrometer measurement was conducted in each cultivar-harvest frequency-fertilizer source treatment combination. In order to maximize potential infiltration differences among treatment combinations, infiltration measurements were only conducted in the non-irrigated treatment. Double-ring infiltrometers were placed between switchgrass crowns in the switchgrass rows after switchgrass was mowed to a height of 6 cm. The outer-ring diameter was 30 cm, the inner-ring diameter was 6 cm, and rings were 10 cm in height. The infiltrometer was inserted approximately 2 cm deep into the soil and the outer ring was filled with tap water to act as a buffer between dry soil outside the outer ring and saturated soil inside the inner ring. The inner ring was filled with tap water and the distance from the top of the soil to the water surface in the inner ring was measured at 0, 1, 2, 3, 4, 5, 8, 10, 12, 15, 18, and 20 minutes.

Double-ring infiltration measurements were paired with mini-disk tension infiltrometer (Decagon, Pullman, Washington) measurements, where infiltrometer tension was set at −2 cm for each measurement. Two mini-disk infiltrometer measurements were conducted in each plot, with one measurement collected from the center-ring area immediately following double-ring infiltrometer measurements and one measurement collected in a nearby location. The infiltrative surface of the mini-disk infiltrometer was 4.5 cm. The two mini-disk measurements were performed between switchgrass crowns in the switchgrass rows in each plot for a total of 48 measurements. The height of the water inside the 2.1-cm diameter mini-disk infiltrometer water reservoir was recorded at 0, 1, 2, 3, 4, 5, 8, 10, 12, 15, 18, and 20 minutes. The total surface infiltration rate was calculated for both sets of infiltration measurements by dividing the total amount of infiltrated water by 20 minutes.

2.4. Statistical Analyses

The effects of cultivar, harvest frequency, irrigation, fertilizer source, soil depth, and their interactions on soil bulk density, soil chemical properties, particle-size distributions, and AS were evaluated by analysis of variance (ANOVA) using PROC GLM in SAS (version 9.3, SAS Institute, Inc., Cary, NC). A separate ANOVA was conducted to evaluate the effects of cultivar, harvest frequency, irrigation, fertilizer source, and their interactions on root density using SAS. A separate ANOVA was also conducted to evaluate the effects of cultivar, harvest frequency, fertilizer source, and their interactions on total surface infiltration rate for both sets of infiltration data. All treatments were treated as fixed effects for all ANOVAs conducted. An analysis of covariance (ANCOVA) was conducted to evaluate the effects of cultivar, harvest frequency, and fertilizer source on the relationship between the natural logarithm of the infiltration rate and the natural logarithm of the mid-point of the measurement time interval. When appropriate, means were separated by least significant difference (LSD) at $\alpha = 0.05$.

3. Results

3.1. Soil Properties

3.1.1. Particle-Size Distribution

Throughout the study site, sand, silt, and clay in the top 20 cm varied somewhat. Though sand, silt, and clay distributions remained within the range of a silt-loam texture in all treatment combinations, sand, silt, and clay contents in the top 20 cm differed slightly among various treatment combinations ($P < 0.05$), with the largest differences occurring between soil depths, which was expected. In both depth intervals, sand content was greatest in the irrigated-"Alamo" treatment combinations (**Table 1**). Sand content was lowest in the top 10 cm in the non-irrigated-"Cave-in-Rock" (26.1%) and in the 10 - 20 cm depth interval in the non-irrigated-"Alamo" (27.5%) treatment combinations, which did not differ (**Table 1**). Averaged over all other treatment factors, silt content was greater ($P < 0.01$) in the top 10 cm than in the 10- to 20-cm depth interval (50.4 and 48.2%, respectively). Silt content was greater ($P = 0.04$) in the PL (50%) than in the CF (48.6%) treatment combinations. Clay content was greater ($P = 0.03$) in the CF (21.7%) than in the PL treatment combinations (20.6%). Silt content was greater ($P < 0.01$) in "Cave-in-Rock" (50.2%) than in "Alamo" (48.4%) treatments. Despite the significant fertilizer source effect on silt and clay and significant cultivar effect on silt content, all silt and clay differences between treatments were less than 2%. Small differences of this magnitude are not likely to affect agronomic management or productivity.

Averaged over harvest frequency, fertilizer source, and irrigation, clay content also differed between cultivars within soil depths ($P < 0.01$). As expected, clay content increased with increasing depth from the top 10 cm (19.5%) to the 10- to 20-cm depth interval (22.9%) for both cultivars (**Figure 2**). Clay content was greatest and least for "Alamo" treatment combinations in the 10- to 20-cm (22%) and in the 0- to 10-cm (17%) depth intervals, respectively (**Figure 2**).

The observed differences in sand, silt, and clay contents between soil depths were expected based on the reported textural classes of the top two horizons of the Leadvale silt-loam official series description [20]. Clay

Table 1. Summary of the effects of soil depth, cultivar, and irrigation on sand and extractable soil manganese (Mn) contents and electrical conductivity (EC).

Soil property	Soil depth (cm)	Alamo		Cave-in-Rock	
		Irrigated	Non-irrigated	Irrigated	Non-irrigated
Sand (%)	0 - 10	32.6 a*	31.4 ab	30.3 abc	26.1 d
	10 - 20	30.3 ab	27.5 cd	29.0 bc	28.9 bc
Mn ($kg·ha^{-1}$)	0 - 10	183.2 b*	170.9 b	179.3 b	169.4 b
	10 - 20	230.7 a	168.9 b	178.7 b	185.5 b
EC ($dS·m^{-1}$)	0 - 10	0.070 a*	0.072 a	0.071 a	0.070 a
	10 - 20	0.060 bc	0.053 c	0.053 c	0.067 ab

*For each soil property, means followed by different letters differ significantly at $P < 0.05$.

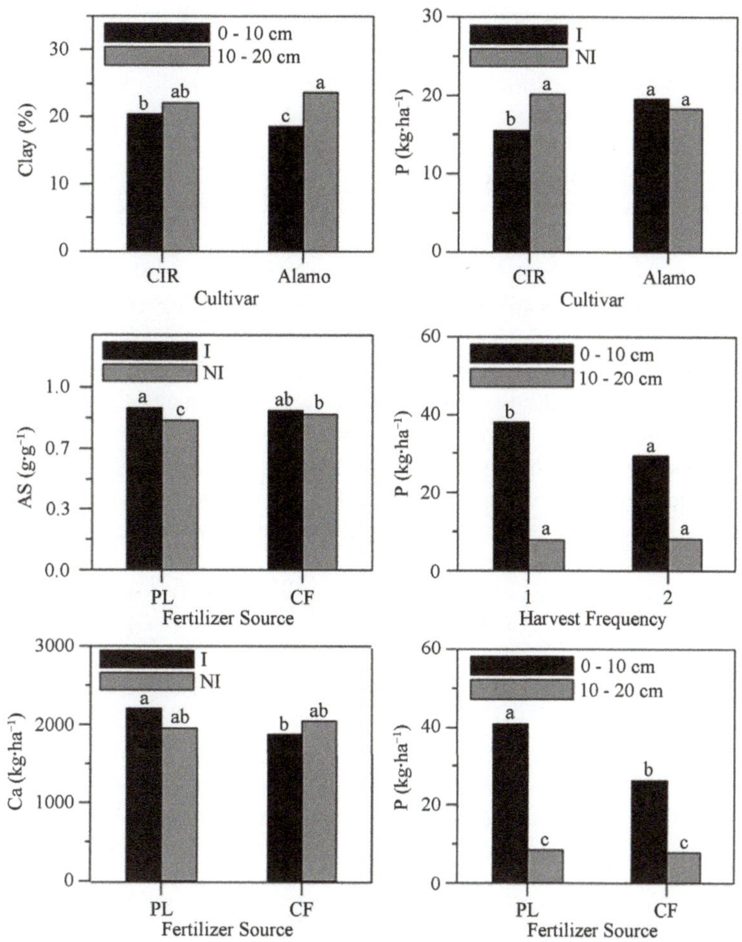

Figure 2. Soil clay content, aggregate stability (AS), and extractable soil phosphorus (P) and calcium (Ca) as affected by soil depth, irrigation [irrigated (I) and non-irrigated (NI)], cultivar ["Cave-in-Rock" (CIR) and "Alamo"], harvest frequency [one (1) harvest per year and two (2) harvests per year], and/or fertilizer source [poultry litter (PL) and commercial fertilizer (CF)]. Different letters atop bars within a soil property indicate a significant difference at $P < 0.05$.

accumulation with increasing depth was expected, as the surface Ap horizon grades to an argillic Bt horizon at approximately 20 cm [20]. Despite some relatively minor differences in particle-size distribution among various treatment combinations, differences were not large enough to change the soil taxonomic classification anywhere throughout the study site and would likely not influence practical agronomic management decisions related to switchgrass production.

3.1.2. Bulk Density

Bulk density was affected by several experimental treatment factors. As expected, averaged over all other treatment factors, soil bulk density was greater ($P < 0.001$) in the 10- to 20-cm depth (1.44 g·cm^{-3}) than in the top 10 cm (1.30 g·cm^{-3}). Averaged over soil depths and cultivars, soil bulk density was also greater ($P = 0.004$) in the irrigated-PL treatment combination with either harvest frequency (1.40 g·cm^{-3}) than the non-irrigated-1-cut treatment combination with either fertilizer source (1.33 g·cm^{-3}). Soil bulk density was unaffected ($P > 0.05$) by switchgrass cultivar.

3.1.3. Aggregate Stability

Similar to bulk density, aggregate stability was affected by several experimental treatment factors. Averaged over soil depth, cultivar, and harvest frequency, AS was greater ($P = 0.03$) under irrigated (0.93 g·g^{-1}) than non-irrigated (0.86 $g \cdot g^{-1}$) treatments with PL, but did not differ between irrigation treatments with CF (0.86 $g \cdot g^{-1}$; **Figure 2**). Averaged over irrigation, the greatest (P = 0.02) AS was measured in the "Alamo"-CF-2-cut treatment combination in the 10- to 20-cm depth (0.93 $g \cdot g^{-1}$), while the lowest AS was measured in the "Cave-in-Rock"-PL-1-cut treatment combination in the top 10 cm depth (0.87 $g \cdot g^{-1}$; **Table 2**).

3.1.4. Soil pH and Electrical Conductivity

Soil pH was unaffected by switchgrass cultivar, irrigation, harvest frequency, or soil depth (P > 0.05), but differed between fertilizer sources (P = 0.001). As expected, averaged over all other treatment factors, soil pH was greater in the PL (pH = 6.1) than in the CF (pH = 5.9) treatment. Though soil pH differed between fertilizer sources by only 0.02 units, nutrient availability and buffering capacity were likely different as well between fertilizer sources.

Similar to bulk density and aggregate stability, soil EC was affected by several experimental treatment factors. Averaged over fertilizer source and soil depth, soil EC was greatest (P = 0.016) in "Cave-in-Rock" (0.070 $dS \cdot m^{-1}$) and least in "Alamo" (0.056 $dS \cdot m^{-1}$) treatments without irrigation in the 2-cut harvest frequency. Averaged over harvest frequency and fertilizer source, soil EC was generally greater (P = 0.006) in the top 10 cm than in the 10- to 20-cm depth and was the lowest in the non-irrigated-"Alamo" and irrigated-"Cave-in-Rock" treatment combinations in the 10- to 20-cm depth interval, which did not differ (**Table 1**). Averaged over cultivar and irrigation, soil EC was also greatest (P = 0.020) in the 1-cut-PL treatment combination in the top 10 cm (0.081 $dS \cdot m^{-1}$) compared to all other treatment combinations (**Table 3**). Soil EC did not differ among harvest-frequency-fertilizer-source treatment combinations in the 10- to 20-cm depth and averaged 0.058 $dS \cdot m^{-1}$ (**Table 3**).

3.1.5. Extractable Soil Nutrients

Similar to bulk density, aggregate stability, and soil EC, numerous extractable soil nutrients were affected by multiple experimental treatment factors. Averaged over harvest frequency and fertilizer source, extractable soil Mn was greater (P = 0.026) in the irrigated-"Alamo" treatment combination in the 10- to 20-cm depth interval (230.2 $kg \cdot ha^{-1}$) than in all other treatment combinations, which did not differ and averaged 176.6 $kg \cdot ha^{-1}$

Table 2. Summary of the effects of soil depth, harvest frequency, cultivar, and fertilizer source on aggregate stability (AS).

Soil property	Soil depth (cm)	Harvest frequency	Alamo		Cave-in-Rock	
			PL†	CF†	PL	CF
AS ($g \cdot g^{-1}$)	0 - 10	1	0.90 bcde*	0.91 abc	0.87 f	0.89 cdef
		2	0.89 cdef	0.92 ab	0.91 abc	0.88 def
	10 - 20	1	0.90 bcde	0.92 ab	0.88 ef	0.89 cdef
		2	0.92 ab	0.93 a	0.90 bcde	0.90 bcd

†PL, poultry litter; CF, commercial fertilizer. *Means followed by different letters differ significantly at $P < 0.05$.

Table 3. Summary of the effects of harvest frequency, soil depth, and fertilizer source on extractable soil copper (Cu) content and electrical conductivity (EC).

Soil property	Harvest frequency	0 - 10 cm depth		10 - 20 cm depth	
		PL†	CF†	PL	CF
Cu ($kg \cdot ha^{-1}$)	1	2.9 a*	2.0 c	1.0 d	0.9 de
	2	2.5 b	2.01 c	1.0 de	0.8 e
EC ($dS \cdot m^{-1}$)	1	0.081 a*	0.067 b	0.059 c	0.059 c
	2	0.068 b	0.067 b	0.060 bc	0.055 c

†PL, poultry litter; CF, commercial fertilizer. *For each soil property, means followed by different letters differ significantly at $P < 0.05$.

(**Table 1**). Unexpectedly, averaged over all other factors, "Cave-in-Rock" treatments had greater ($P = 0.033$) extractable soil K (95 kg K ha^{-1}) and lower ($P = 0.019$) extractable soil Na (43 kg Na ha^{-1}) compared to "Alamo" treatments (79 kg K and 64 kg Na ha^{-1}, respectively). Averaged over all other treatment factors, extractable soil Ca was greater ($P = 0.031$) in the 10- to 20-cm depth (2079 kg·ha^{-1}) than in the top 10 cm (1965 kg Ca ha^{-1}). In contrast, extractable soil Fe was greater ($P < 0.001$) in the top 10 cm (216 kg Fe ha^{-1}) than in the 10- to 20-cm depth (164 kg Fe ha^{-1}).

Averaged over irrigation, cultivar, and harvest frequency, extractable soil P, K, Mg, S, Na, and Zn contents followed similar trends related to treatment combinations and were generally greater ($P < 0.04$) in the top 10 cm compared to the 10- to 20-cm depth. Extractable soil P (**Figure 2**), S, Zn (**Figure 3**), K, and Mg (**Figure 4**) contents were greatest in the PL treatment combinations in the top 10 cm and the lowest in the 10- to 20-cm depth regardless of fertilizer source, which did not differ. In contrast, extractable soil Na was generally greater in the 10- to 20-cm depth compared to the top 10 cm and was greatest in the 10- to 20-cm-depth-PL and lowest in the 0- to 10-cm CF treatment combination (**Figure 3**).

Averaged over irrigation, cultivar, and fertilizer source, extractable soil P, K, Mg, and Zn contents also differed between soil depths within harvest frequency. Extractable soil P (**Figure 2**), Zn (**Figure 3**), K, and Mg (**Figure 4**) contents followed similar trends related to treatment combinations and were generally greater ($P < 0.02$) in the top 10 cm compared to the 10- to 20-cm depth. Extractable soil P (**Figure 2**), Zn (**Figure 3**), K, and Mg (**Figure 4**) contents were greatest in the 1-cut treatment combination in the top 10 cm, which supported the

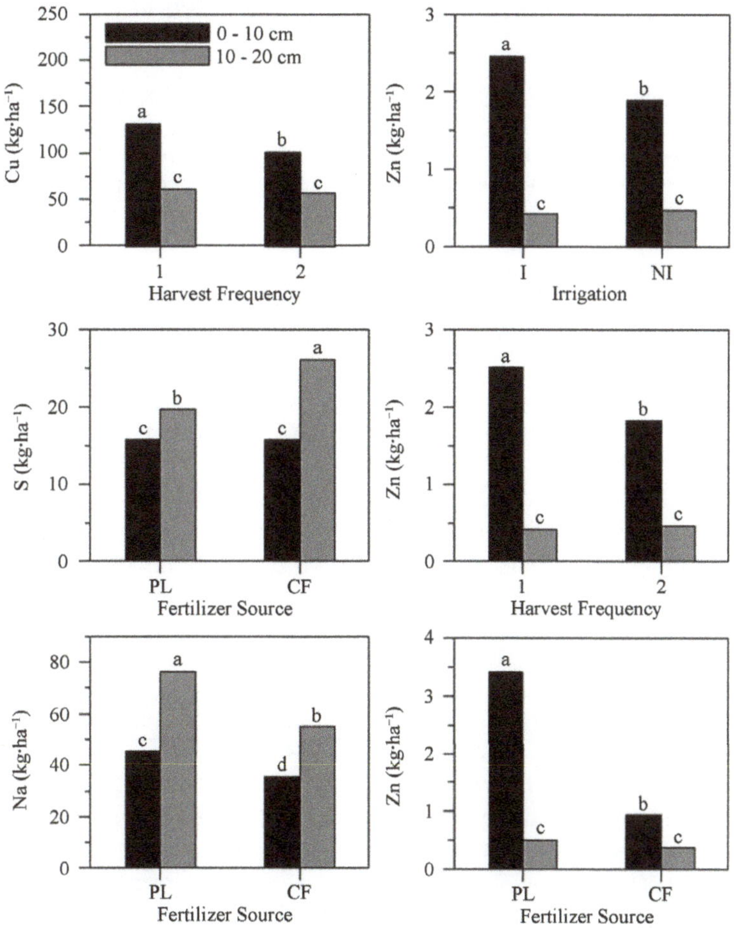

Figure 3. Extractable soil copper (Cu), zinc (Zn), sulfur (S), and sodium (Na) contents as affected by soil depth, irrigation [irrigated (I) and non-irrigated (NI)], harvest frequency [one (1) harvest per year and two (2) harvests per year], and/ or fertilizer source [poultry litter (PL) and commercial fertilizer (CF)]. Different letters atop bars within a soil property indicate a significant difference at $P < 0.05$.

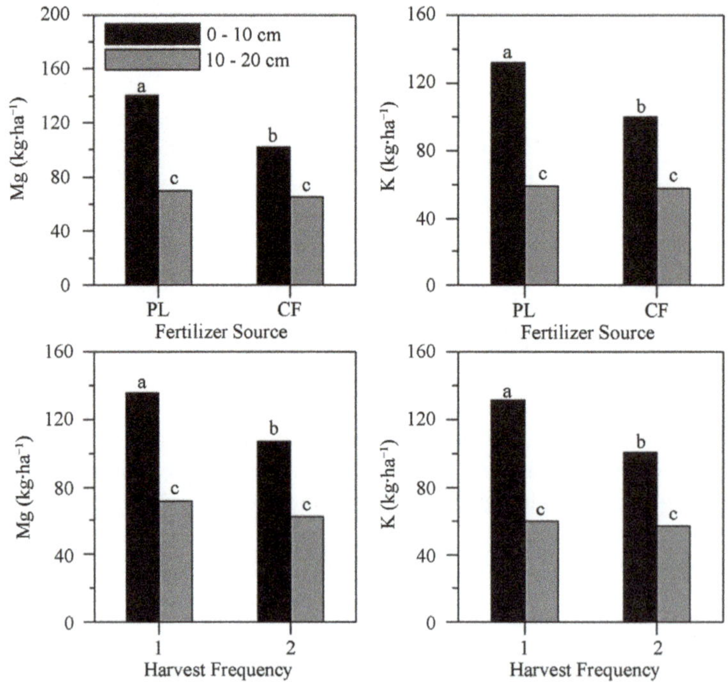

Figure 4. Extractable soil potassium (K) and magnesium (Mg) as affected by soil depth and fertilizer source [poultry litter (PL) and commercial fertilizer (CF)] or harvest frequency [one (1) harvest per year and two (2) harvests per year]. Different letters atop bars within a soil property indicate a significant difference at $P < 0.05$.

hypothesis that harvesting once after switchgrass senescence rather than twice per year allows greater retention of extractable soil nutrients. Extractable soil P (**Figure 2**), Zn (**Figure 3**), K, and Mg (**Figure 4**) contents were lower in the 10- to 20-cm, which did not differ between harvest frequencies, than in the 0- to 10-cm depth.

In addition, averaged over cultivar, harvest frequency, and soil depth, extractable soil Ca (**Figure 2**) and Cu (**Figure 3**) contents followed similar trends, as both were the lowest ($P < 0.03$) in the irrigated-CF and the greatest in the irrigated-PL treatment combination. Averaged over harvest frequency, fertilizer source, and soil depth, extractable soil P content was lower ($P = 0.016$) in the irrigated-"Cave-in-Rock" than in the "Alamo" of either irrigation treatment and in the non-irrigated-"Cave-in-Rock" treatment combination, which did not differ (**Figure 2**).

Averaged over cultivar, harvest frequency, and fertilizer source, extractable soil Zn content differed between irrigation treatments within soil depths. While the greatest ($P = 0.025$) extractable Zn content was in the irrigated-0- to 10-cm depth compared to all other treatment combinations, extractable Zn content was generally greater in the top 10 cm than in the 10- to 20-cm depth (**Figure 3**). Extractable soil Zn was lowest in the 10- to 20-cm depth of either irrigation treatment compared to all other treatment combinations (**Figure 3**). Averaged over soil depth, the greatest ($P = 0.043$) extractable Zn content was in the irrigated-"Alamo"-PL (2.3 $kg{\cdot}ha^{-1}$) treatment, but did not differ from the "Cave-in-Rock"-PL treatment of either irrigation combination. The lowest extractable Zn content was in the irrigated-"Alamo"-CF (0.6 $kg{\cdot}ha^{-1}$) treatment, but did not differ from all other CF treatments in combination with either irrigation or cultivar.

Averaged over irrigation and soil depth, extractable soil Cu ($P = 0.009$) was greatest in the "Alamo"-PL-1-cut and in the "Cave-in-Rock"-PL treatment combinations under both harvest frequencies (2.0 kg Cu ha^{-1}) compared to all other treatment combinations. Averaged over cultivar and irrigation, extractable soil Cu content was greater ($P = 0.038$) in the 1-cut-PL treatment combination in the top 10 cm (2.9 kg Cu ha^{-1}) than in all other treatment combinations, which were <2.0 kg Cu ha^{-1} (**Table 4**). Extractable soil Cu content was generally greater in the top 10 cm and in treatments that included PL. Averaged over harvest frequency and irrigation, extractable soil Cu content was greatest ($P = 0.032$) in the "Cave-in-Rock"-PL treatment combination in the top 10 cm (2.9 kg Cu ha^{-1}) than all treatment combinations in the 10- to 20-cm depth, which were all <1.1 kg Cu ha^{-1} and did not differ.

Table 4. Summary of the effects of harvest frequency, cultivar, and fertilizer source on extractable soil copper (Cu) and iron (Fe) contents.

Soil property	Harvest frequency	Alamo		Cave-in-Rock	
		PL[†]	CF[†]	PL	CF
Cu ($kg \cdot ha^{-1}$)	1	2.0 a[*]	1.5 b	1.9 a	1.5 b
	2	1.6 b	1.4 b	2.0 a	1.4 b
Fe ($kg \cdot ha^{-1}$)	1	212.5 a[*]	184.2 b	193.3 ab	189.5 b
	2	195.5 ab	188.8 b	192.6 ab	160.9 c

[†]PL, poultry litter; CF, commercial fertilizer. [*]For each soil property, means followed by different letters differ significantly at $P < 0.05$.

3.2. Root Density

In contrast to the initial hypothesis that irrigation would have minimal effect on belowground switchgrass growth in the climatic region of west-central Arkansas, averaged over all other treatment factors, switchgrass root density was greater ($P = 0.031$) in the irrigated (2.62 $kg \cdot m^{-3}$) than in the non-irrigated (1.65 $kg \cdot m^{-3}$) treatment.

It was also hypothesized that increasing the harvest frequency from 1 to 2 cuts per year would significantly decrease switchgrass root density. However, results did not support this initial hypothesis as root density was also unaffected ($P > 0.05$) by switchgrass cultivar and fertilizer source.

3.3. Infiltration Characteristics

3.3.1. Total Infiltration

Total water infiltration rates measured with the double-ring infiltrometer did not differ ($P > 0.05$) among any treatment combination. Total infiltration rates ranged from a low of 0.2 $mm \cdot min^{-1}$ in the non-irrigated-"Alamo" treatment combination with either fertilizer source to a high of 2.4 $mm \cdot min^{-1}$ in the non-irrigated-"Cave-in-Rock"-PL treatment combination and averaged 0.79 $mm \cdot min^{-1}$ across all treatment combinations.

Unlike total infiltration rates measured with the double-ring infiltrometer, total infiltration rates measured with the mini-disk infiltrometer at a tension of −2 cm were greater ($P = 0.034$) in the 1-cut (33 $mm \cdot min^{-1}$) than in the 2-cut (24 $mm \cdot min^{-1}$) treatment. The total tension infiltration rate was unaffected ($P > 0.05$) by cultivar, irrigation, or fertilizer source.

3.3.2. Relationship between Infiltration Rate and Time

1) Double-ring infiltrometer

Though total infiltration measured with the double-ring infiltrometer did not differ among experimental treatments, averaged over fertilizer sources, the slope of the linear regression characterizing the natural logarithm of the infiltration rate against the natural logarithm of the mid-point time was greatest ($P = 0.004$) for the "Cave-in-Rock"-2-cut and "Alamo"-1-cut treatment combinations, which did not differ (**Figure 5**). The slope was smallest for the "Cave-in-Rock"-1-cut and "Alamo"-2-cut treatment combinations, which did not differ (**Figure 5**). In addition, averaged over cultivar, the slope of the linear regression characterizing the natural logarithm of the infiltration rate against the natural logarithm of the mid-point time was greater ($P = 0.048$) for the CF-1-cut than the PL-1-cut treatment combinations (**Figure 5**).

2) Mini-disk infiltrometer

Unlike the double-ring infiltrometer, the slope of the linear regression equations characterizing the natural logarithm of the infiltration rate against the natural logarithm of the mid-point time was positive and numerically greatest ($P = 0.026$) for the "Alamo"-CF-2-cut and numerically smallest and negative for the "Alamo"-PL-2-cut and "Cave-in-Rock"-CF-2-cut treatment combinations, which did not differ (**Table 5**). The y-intercept of the linear relationship was positive and numerically greatest ($P = 0.049$) in the "Cave-in-Rock"-PL-1-cut and positive and numerically smallest in the "Alamo"-CF-2-cut treatment combination (**Table 5**).

Table 5. Summary of parameters that characterize the relationship between the natural logarithm of the mini-disk (tension) infiltration rate and time.

Cultivar	Fertilizer source	Harvest frequency	Slope (m)	Y-intercept
Alamo	Commercial fertilizer	1	−0.088 ab*	1.109 b
		2	0.047 a	0.253 c
	Poultry litter	1	−0.126 abcd	1.270 ab
		2	−0.212 cd	1.266 ab
Cave-in-Rock	Commercial fertilizer	1	0.131 abcd	1.190 ab
		2	−0.196 bcd	0.873 bc
	Poultry litter	1	−0.126 abcd	1.852 a
		2	−0.106 abc	0.909 bc

*Means followed by different letters differ significantly at $P < 0.05$.

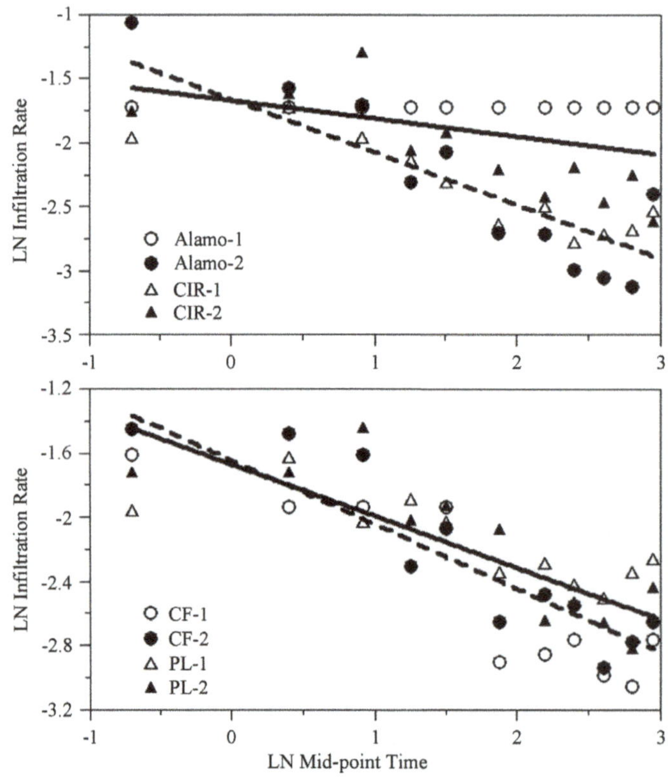

Figure 5. Natural logarithm (LN) of the double-ring infiltration rate per LN mid-point time as affected by cultivar ["Cave-in-Rock" (CIR) and "Alamo"] and harvest frequency [one (1) harvest per year and two (2) harvests per year] (top panel), or fertilizer source [Poultry Litter (PL) and Commercial Fertilizer (CF)] and harvest frequency (bottom panel). Solid regression line (top) includes CIR-2 and Alamo-1, for which slopes did not differ. Dashed regression line (top) includes CIR-1 and Alamo-2, for which slopes did not differ. Solid regression line (bottom) includes PL-1, CF-2, and PL-2, for which slopes did not differ. Dashed regression line (bottom) includes CF-1, PL-2, and CF-2, for which slopes did not differ.

4. Discussion

Overall, results of this study indicated treatments imposed on the switchgrass production system affected near-surface properties in the top 20 cm and surficial processes. However, treatment combinations that produced significant differences in the amount of aboveground biomass production noted previously by Jacobs and King [15] did not necessarily produce significant differences in the measured soil properties in this study.

4.1. Cultivar Effects

Switchgrass cultivar was not expected to affect most soil properties; however, AS and silt concentration, and Na, Fe, P, and K contents differed between cultivars when averaged over all other treatment factors. In addition, clay concentration differed by depth within cultivars when averaged over irrigation, harvest frequency, and fertilizer source. While the silt and clay differences were likely residual inherent differences from before the study was initiated, and assuming similar nutrient concentrations between cultivars, greater soil extractable P and K contents in "Cave-in-Rock" treatments may have resulted from lower nutrient removal rates according to lower biomass production compared to "Alamo" treatments [15]. Average biomass yields at this site over a four-year period (*i.e.*, 2008 through 2011) were significantly greater from "Alamo" than from "Cave-in-Rock" treatments [15]; thus, greater plant uptake and removal of macronutrients resulting in lower extractable soil P and K contents would be expected.

While extractable soil nutrient differences were relatively straightforward, AS depends on many factors related to soil, water, and plant relationships. Previous research in Arkansas has shown that aggregate stability in the top 10 cm is greater in undisturbed native prairie or prairie restorations than in cultivated, row-crop agroecosystems. Aggregate stability in row-crop production varied from 0.05 $g \cdot g^{-1}$ in a non-irrigated, wheat (*Triticum aestivum*)-soybean (*Glycine max*), double-cropped system managed consistently for 10 years [27] to 0.11 $g \cdot g^{-1}$ in irrigated and non-irrigated rice (*Oryza sativa*)-wheat double-cropped systems managed consistently for 6 years [28]. In contrast, Brye and Riley [26] reported total water-stable aggregate concentrations of 0.44 $g \cdot g^{-1}$ in an undisturbed native prairie in northwest Arkansas. Interestingly, AS measured in this study, which averaged 0.89 $g \cdot g^{-1}$ in the top 10 cm (**Table 2**), were two times greater than those measured in a native prairie ecosystem (0.44 $g \cdot g^{-1}$) [26]. Differences may be attributed to the intensive nature of treatment combinations imposed in this study to maximize aboveground biomass production, such as irrigation and fertilizer applications. Greater AS observed in this study may also be attributed to complex interactions involving plant roots and soil aggregation and show the similarity of perennial, monoculture switchgrass production to native grasslands.

Differences in root density could influence AS, but results from this study showed that root density did not differ between cultivars. Soil organic matter (SOM) may also influence AS, but SOM was not measured in this study. The 0- to 15-cm depth interval where root density was measured in this study may not have captured enough of each cultivar's unique rooting characteristics, which may explain why AS and not root density differed between cultivars when averaged over all other factors. Frank *et al.* [29] reported that 50% of switchgrass root mass occurred in the top 30 cm of soil. In addition, Jung *et al.* [30] reported that switchgrass root partitioning between fine and coarse roots is dynamic and may be altered with differing levels of soil fertility. Due to differing extractable soil nutrient contents between cultivars, the partitioning between fine and coarse roots may not have been accurately captured for all treatments by the 2-mm sieve size used for the root washing procedure.

4.2. Irrigation Effects

Similar to cultivar effects, it was hypothesized that irrigation treatments would have little influence on soil properties. However, irrigation treatments produced significant differences in root density. Though application of irrigation to the switchgrass crop did not significantly increase biomass yields [15], irrigation increased root density compared to the non-irrigated treatment when averaged over all other treatment factors. It was clear that irrigation directly stimulated root growth and, as a result, indirectly altered additional soil properties measured in this study, such as AS.

Averaged over all other treatment factors, AS did not differ between irrigation treatments. However, increased AS been linked to greater root density, as roots encourage the formation and stabilization of soil aggregates [31]. The lack of a significant irrigation effect on AS in this study was in contrast to Singer *et al.* [32], where the rewetting of dry soil during irrigation cycles decreased the concentration of soil macroaggregates compared to a non-irrigated site. The destruction of soil aggregates due to irrigation was attributed to an increase in air pressure inside dry soil macroaggregates during the rewetting process (*i.e.*, slaking) [32]. Soil aggregates may not have experienced sufficient wetting and drying for slaking to occur in west-central Arkansas, given irrigation did not significantly increase switchgrass biomass yield [15].

Whalen and Chung [33] compared soil aggregate stability with differing application rates of cattle manure on cropland in Canada. While Whalen and Chung [33] hypothesized that soil macroaggregate (>12.1 mm) stability would increase with increased manure application rate, results showed that the proportion of soil macroaggregates actually decreased with manure application compared to no manure application. The dispersive effects of monovalent cations and foreign soil present in cattle manure were cited as the explanation for their results [33]. Edaño [34] also determined that the application of manure fertilizer (*i.e.*, cattle manure or PL) compared to CF did not encourage macroaggregate formation in switchgrass plots harvested in both 1- and 2-cut systems.

The dispersive effects of monovalent cations typically present in PL, such as Na, may have affected soil properties evaluated in this study. For example, irrigation of plots fertilized with PL may have caused greater dispersion by further mobilizing monovalent cations. In this study, extractable soil Na was greater in the 10- to 20-cm depth than in the top 10 cm. Results from this study also further support the conclusions of Whalen and Chung [33] and Edaño [34] as soil bulk density was greater in the irrigated-PL treatment combination with either harvest frequency (1.41 $g \cdot cm^{-3}$) than in the non-irrigated-1-cut treatment combination with either fertilizer source (1.33 $g \cdot cm^{-3}$).

4.3. Fertilizer Source Effects

The application of two different fertilizer sources over a four-year period [15] produced significant differences in extractable soil nutrient levels. Even though significant differences between fertilizer sources were present among extractable soil nutrient levels after the four-year study period, biomass yield was unaffected by fertilizer source [15]. In contrast, Edaño [34] reported that switchgrass biomass yield was greater with PL compared to CF application, switchgrass seeded with legumes, and an unfertilized control.

Poultry litter was applied at a rate of 4.5 $Mg \cdot ha^{-1}$ and corresponding amounts of N, P, and K were applied as CF. Poultry litter contains more than just N, P, and K, and significant differences between the levels of extractable soil nutrients were apparent in this study. Only one soil nutrient tested (*i.e.*, Mn) did not differ between fertilizer sources when averaged across all other treatment factors. Soil collected from treatment combinations fertilized with PL contained significantly greater contents of extractable soil P, K, Ca, Mg, S, Na, Fe, Zn, and Cu than CF treatment combinations. Edaño [34] observed similar increases in extractable soil P when switchgrass was fertilized with cattle manure and PL instead of CF.

In addition, both soil pH and EC were significantly greater in PL than in CF treatment combinations, which were expected. Results from this study regarding soil pH changes in "Alamo" switchgrass from fertilizer sources were similar to those reported by Edaño [34] in Oklahoma. Soil pH slightly increased when fertilized with cattle manure or PL and slightly decreased when fertilized with CF compared to an unfertilized control [34]. The differences in pH would not likely affect agronomic management decisions, especially given that soil pH values of 5.5 to 6.6 are considered optimum for native, warm-season grass establishment in Arkansas [21].

4.4. Harvest Frequency Effects

The previous yield study by Jacobs and King [15] largely focused on the effects of harvest frequency on biomass yield due to the dual-purpose nature of switchgrass production as a potential livestock forage and as a potential renewable energy feedstock. Harvesting once per year in the fall after a frost (*i.e.*, the 1-cut system) produces biomass low in livestock forage quality, but desirable quality for conversion to renewable fuels. Conversely, harvesting twice per year (*i.e.*, the 2-cut system) increases switchgrass crop flexibility for landowners who may harvest the first cutting of switchgrass for hay or grazing when livestock forage quality is greatest [15]. Jacobs and King [15] noted that, averaged across all other treatment combinations, the 2-cut system produced significantly more biomass than the 1-cut system.

Despite the increased yield for switchgrass harvested in the 2-cut system, removing greater amounts of biomass also decreased near-surface extractable soil K and Mg levels. In addition, AS and mini-disk total infiltration rates were significantly lower in the 2-cut than in the 1-cut system. In a similar study, Edaño [34] noted switchgrass harvested in a 2-cut system had significantly lower extractable soil nitrate concentrations, total microbial biomass, and total mycorrhizal biomass in the top 10 cm than those in a 1-cut harvest system.

The lack of significant effects due to increased harvest frequency on root density are in contrast to harvest frequency effects on aboveground biomass yield. Jacobs and King [15] reported that mean biomass yield over the four-year study was significantly greater for the 2-cut (17.9 Mg·ha^{-1}) compared to the 1-cut (12.1 Mg·ha^{-1}) system when averaged over all other treatment factors. Data from above- and belowground plant production further demonstrate that increasing switchgrass yield by harvesting twice per year does not significantly decrease root density in the soil and climatic conditions of west-central Arkansas.

4.5. Infiltration Characteristics

In general, results indicated total infiltration rates were greater under tension when measured with the mini-disk infiltrometer compared to the double-ring infiltrometer because each method measured different mechanisms of infiltration. Infiltration was only measured in the inner ring for the double-ring infiltrometer while adjacent soil was nearly saturated throughout the 20-minute measurement period. Infiltration measured with the double-ring infiltrometer measured downward or vertical water infiltration (*i.e.*, one-dimensional water infiltration). In contrast, soil around the mini-disk infiltrometer was unsaturated throughout the 20-minute measurement period; thus infiltration was likely both vertically downward and lateral (*i.e.*, more three-dimensional water infiltration).

Due to the differences associated with measuring infiltration in unsaturated and nearly saturated conditions, only mini-disk infiltration rates were directly affected by harvest frequency. The mini-disk infiltration rate was greater in the 1-cut than in the 2-cut treatment, which supported the hypothesis that more intensive harvest frequencies decrease water infiltration through soil micropores. Increases in wheel traffic from tractors and other harvesting equipment in 2-cut systems may not consistently increase bulk density, as previous data showed, but may reduce soil micropore abundance and volume and consequently hydraulic conductivity. Removal of the switchgrass canopy cover during the growing season (*i.e.*, at the June harvest in the 2-cut system) likely exposed more soil than in the 1-cut treatment and may have allowed soil micropores to become sealed or clogged by temporary crusting from precipitation or irrigation.

The slope of the linear regression relating the natural logarithm of the infiltration rate and the natural logarithm of time provided additional information about what could happen to surface water (*i.e.*, precipitation or irrigation) in each treatment combination. Large (*i.e.*, steep) slopes represented greater surface infiltration per unit time, while smaller (*i.e.*, flatter) slopes represented slower surface infiltration per unit time. This relationship was readily apparent when comparing the slopes averaged over fertilizer source for double-ring infiltration, which differed by cultivar within harvest frequency (**Figure 5**). The treatment combinations with modest infiltration rates and a smaller slope characterizing the relationship between the natural logarithm of the infiltration rate and the natural logarithm of time included the "Cave-in-Rock"-2-cut and "Alamo"-1-cut combinations, which did not differ (**Figure 5**). Conversely, the slopes characterizing the relationship between the natural logarithm of the infiltration rate and the natural logarithm of time associated with the "Cave-in-Rock"-1-cut and "Alamo"-2-cut treatment combinations were larger and represented greater surface infiltration per unit time (**Figure 5**). A similar relationship was observed when the treatments were averaged over cultivar and differed by fertilizer source within harvest frequency, as the slope of the PL-1-cut treatment combination was smaller compared to the slope of the CF-1-cut treatment combination (**Figure 5**). One reason the slope was smaller for the PL treatment combination could be related to the presence of monovalent cations in the PL, particularly Na, which could have caused more surface soil dispersion than CF. The ecological implications for greater infiltration and storage of surface water in the soil profile over shorter periods are vast and may affect the amount of plant available water for crop growth, the amount of off-site nutrient loss, and the potential for erosion of topsoil.

The mean double-ring infiltration rate into a silt-loam soil, which was measured in this switchgrass production study after four years of continuous and consistent management, was less than that reported by Bonin et al. [35] for switchgrass harvested in a 1-cut system in Ohio (2.8 mm·min^{-1}). Decreased infiltration rates recorded in this study may be due to a variety of site differences among the two studies, including different switchgrass plant population densities resulting from differing initial seeding rates (*i.e.*, 11.2 and 4.4 kg pure live seed·ha^{-1} in Ohio and for this study, respectively), differences in soil surface texture (*i.e.*, silty clay loam in Ohio and silt loam in this study), and/or differences in switchgrass stand age (*i.e.*, seven years in Ohio and four years in this study) [35]. Another factor that may have contributed to decreased infiltration rates at this site compared to the Ohio site [35] could have been the hydrophobic properties of ash from burning as part of annual stand management activities. Switchgrass stubble was burned each March throughout the four-year treatment application period in this study, while the Ohio site was not burned at all during their study period [35]. Consequently, reducing the frequency of prescribed burns may increase infiltration rates.

Studies that evaluate soil aggregate stability may give ancillary information toward interpreting water infiltration rates because soil aggregate size influences soil pore space. In this study, though AS did not systematically differ between harvest frequencies, the total tension infiltration rate was greater in the 1-cut than in the 2-cut system, which supports the supposition of Edaño [34] that harvesting switchgrass in a 1-cut system has the potential to increase water infiltration rates. Results from this study indicate water infiltration rates do not consistently improve (*i.e.*, increase) with greater AS. In addition, results show that treatment effects may be overlooked if only saturated water infiltration is evaluated.

4.6. Practical Applications

Producers and other private landowners engaged in feedstock production must make informed management decisions to maximize production while protecting soil and water resources. A first step in providing information to producers and landowners lies in pairing aboveground management strategies to maximize biomass yield, such as cultivar choice, irrigation management, harvest frequency, and choice of fertilizer, with belowground consequences.

Similar to the results of this study, Jacobs and King [15] also noted that each switchgrass cultivar responded differently to the 2-cut harvest frequency, even though mean annual production over the four-year period did not differ significantly between cultivars within harvest frequency averaged over fertilizer source. The "Cave-in-Rock"-2-cut treatment combination produced an average of 62% more biomass compared to the 1-cut system, while the "Alamo"-2-cut treatment combination only produced an average of 35% more biomass compared to the 1-cut system. The results regarding cultivar response to the 2-cut harvest frequency were similar to those reported by Fike *et al.* [36], where maximum switchgrass biomass yield was achieved by harvesting lowland cultivars (e.g., "Alamo") in a 1-cut harvest frequency and upland cultivars (e.g., "Cave-in-Rock") in a 2-cut harvest frequency [36]. This study also supported the conclusion of Edaño [34] that lowland cultivars are better suited to dedicated energy crop production, while upland cultivars are better suited as dual-use livestock forage and energy crop production.

Biomass yield information coupled with soil hydraulic properties provide valuable information to landowners for making management decisions. For example, landowners may choose to harvest twice per year only if "Cave-in-Rock" is used, knowing that increased biomass yield could be balanced by the trade-off of decreased water infiltration rates. Similarly, landowners may only harvest "Alamo" once per year, as harvesting twice per year may only increase yields by 35%, while also decreasing infiltration rates. Landowners who are interested in converting highly erodible cropland to switchgrass production may choose to plant "Alamo" instead of "Cave-in-Rock" to decrease the potential for soil erosion due to more water-stable soil aggregation. Subtleties associated with these management strategies may become important if switchgrass is grown on the vast areas needed to support industrial facilities for renewable fuel production.

5. Conclusions

Results showed that four years of consistent agronomic management with various strategies to maximize switchgrass production in west-central Arkansas produced significant soil property differences, some of which were small and of limited practical use, while other differences were larger and had practical management implications. Results from this study also illuminated new opportunities for producers and landowners to customize switchgrass management systems to address specific natural resource goals. In general, "Alamo" produced greater biomass yields and greater AS, which resulted in greater depletion of extractable soil P and K contents over time compared with "Cave-in-Rock". Irrigating switchgrass, though not currently cost-effective or recommended for marginal sites, significantly increased AS and switchgrass root densities when averaged over all other treatment factors. Harvesting switchgrass in a 2-cut system significantly decreased AS, extractable soil K and Mg contents, and total tension infiltration rates. Fertilizing switchgrass with PL rather than CF generally increased soil pH, EC, and extractable soil P, K, Ca and Mg contents, though both PL and CF appear to be viable options for landowners interested in switchgrass production.

Further investigations are needed to evaluate the effect of management strategies when no fertilizer is applied to assess soil property change if or when the bulk of fertilizers are diverted for growing food crops to feed a growing world population. More information is also needed about the fate of surface water and infiltration characteristics associated with switchgrass production strategies, as the importance of surface water and groundwater recharge only increase with time. Finally, this study demonstrated the need to enlarge the scope of discussion when assessing the feasibility and natural resource consequences of producing energy crops in the US.

Acknowledgements

The authors are grateful for the field and laboratory assistance provided by Bryan Jacobs, Eddie Pratt, Debbie Orick, Dale Goff, Michele Helton, Julie Osborne, and Taylor Adams.

References

[1] Energy Information Administration (EIA) (2013) How Dependent Are We on Foreign Oil? http://www.eia.gov/energy_in_brief/article/foreign_oil_dependence.cfm

[2] United States Environmental Protection Agency (USEPA) (2013) Renewable Fuel Standard. http://www.epa.gov/otaq/fuels/renewablefuels/

[3] Energy Information Administration (EIA) (2013) Monthly Energy Review. http://www.eia.gov/totalenergy/data/monthly/pdf/mer.pdf

[4] Parrish, D.J. and Fike, J.H. (2005) The Biology and Agronomy of Switchgrass for Biofuels. *Critical Reviews in Plant Science*, **24**, 423-459. http://dx.doi.org/10.1080/07352680500316433

[5] Tillman, D.A. (2000) Biomass Cofiring: The Technology, the Experience, the Combustion Consequences. *Biomass & Bioenergy*, **19**, 365-384. http://dx.doi.org/10.1016/S0961-9534(00)00049-0

[6] United States Department of Agriculture, Economic Research Service (USDA-ERS) (2012) Cattle and Beef Production Overview. http://www.ers.usda.gov/topics/animal-products/cattle-beef.aspx#.U-5by1bxVcw

[7] Pimentel, D., Harvey, C., Resosidarmo, P., Sinclair, K., Kurz, D., McNair, M., Crist, S., Sphritz, L., Fitton, L., Saffouri, R. and Blair, R. (1995) Environmental and Economic Costs of Soil Erosion and Conservation Benefits. *Science*, **276**, 1117-1123. http://dx.doi.org/10.1126/science.267.5201.1117

[8] McLaughlin, S.B. and Kszos, L.A. (2005) Development of Switchgrass (*Panicum virgatum*) as a Bioenergy Feedstock in the United States. *Biomass & Bioenergy*, **28**, 515-535. http://dx.doi.org/10.1016/j.biombioe.2004.05.006

[9] Stubbendick, J., Hatch, S.L. and Butterfield, C.H. (1998) North American Range Plants. 4th Edition, University of Nebraska Press, Lincoln.

[10] Popp, M.P. (2007) Assessment of Alternative Fuel Production from Switchgrass: An Example from Arkansas. *American Journal of Agricultural Economics*, **39**, 373-380.

[11] McLaughlin, S.B., Kiniry, J.R., Taliaferro, C.M. and De La Torre Ugarte, D. (2006) Projecting Yield and Utilization Potential of Switchgrass as an Energy Crop. *Advances in Agronomy*, **90**, 267-297. http://dx.doi.org/10.1016/S0065-2113(06)90007-8

[12] United States Department of Agriculture, National Agricultural Statistics Service (USDA-NASS) (2014) Rank of States: Production of Livestock, Arkansas, 2013. http://www.nass.usda.gov/Statistics_by_State/Arkansas/Publications/Statistical_Bulletin/Livestock_Data/arliverank.pdf

[13] Mitchell, C.C. and Donald, J.O. (1999) The Value and Use of Poultry Manures as Fertilizer. Alabama Cooperative Extension System, Circular ANR-244, Alabama A&M University, Auburn University, Auburn.

[14] Adler, P.R., Sanderson, M.A., Boateng, A.A., Weimer, P.J. and Jung, H.J.G. (2006) Biomass Yield and Biofuel Quality of Switchgrass Harvested in Fall and Spring. *Agronomy Journal*, **98**, 1518-1525. http://dx.doi.org/10.2134/agronj2005.0351

[15] Jacobs, A.A. and King, J.R. (2012) Using Fertilization, Irrigation, and Harvest Strategies to Maximize "Alamo" and "Cave-in-Rock" Switchgrass Biomass Yield in the Southern Ozarks. http://www.plant-materials.nrcs.usda.gov/pubs/arpmctr11197.pdf

[16] Kering, M.K., Butler, T.J., Biermacher, J.T., Mosali, J. and Guretzky, J.A. (2012) Effect of Potassium and Nitrogen Fertilizer on Switchgrass Productivity and Nutrient Removal Rates under Two Harvest Systems on a Low Potassium Soil. *BioEnergy Research*, **6**, 329-335. http://dx.doi.org/10.1007/s12155-012-9261-8

[17] Bransby, D. and Huang, P. (2014) Twenty-Year Biomass Yields of Eight Switchgrass Cultivars in Alabama. *BioEnergy Research*, **7**, 1186-1190. http://dx.doi.org/10.1007/s12155-014-9448-2

[18] Burner, D.M. (2012) Weather Summary for the Dale Bumpers Small Farms Research Center 2005-2012. Unpublished Data, Booneville.

[19] National Oceanic and Atmospheric Administration (NOAA) (2010) Climatography of the United States, Annual and Seasonal Normals of Temperature and Precipitation 1981-2010: Booneville, Arkansas. United States Department of Commerce, National Climatic Data Center, Asheville.

[20] United States Department of Agriculture, Natural Resource Conservation Service (USDA-NRCS) (2003) Leadvale Soil Series Official Description. https://soilseries.sc.egov.usda.gov/OSD_Docs/L/LEADVALE.html

[21] United States Department of Agriculture, Natural Resource Conservation Service (USDA-NRCS) (2009) Establishing Native Warm Season Grass Mixtures Fact Sheet. http://efotg.sc.egov.usda.gov/references/public/AR/native_warm_season_grass.pdf

[22] Bouyoucos, G.J. (1927) The Hydrometer as a New Method for the Mechanical Analysis of Soils. *Soil Science*, **23**, 343-353. http://dx.doi.org/10.1097/00010694-192705000-00002

[23] Mehlich, A. (1984) Mehlich 3 Soil Test Extractant: A Modification of Mehlich 2 Extractant. *Communications in Soil Science & Plant Analysis*, **15**, 1409-1419. http://dx.doi.org/10.1080/00103628409367568

[24] Tucker, M.R. (1992) Determination of Phosphorous by Mehlich 3 Extraction. In: Donohue, S.J., Ed., *Reference Soil and Media Diagnostic Procedure for the Southern Region of the United States*, Southern Cooperative Series Bulletin 374, Virginia Agricultural Experiment Station, Blacksburg, 9-12.

[25] Yoder, R.E. (1936) A Direct Method of Aggregate Analysis of Soils and a Study of the Physical Nature of Erosion Losses. *Agronomy Journal*, **28**, 337-351. http://dx.doi.org/10.2134/agronj1936.00021962002800050001x

[26] Brye, K.R. and Riley, T.L. (2009) Soil and Plant Property Differences across a Chronosequence of Humid-Temperate Tallgrass Prairie Restorations. *Soil Science*, **174**, 346-357. http://dx.doi.org/10.1097/SS.0b013e3181a93daa

[27] Smith, S.F., Brye, K.R., Chen, P., Gbur, E. and Korth, K. (2014) Residue and Water Management Effects on Aggregate Stability and Aggregate-Associated Carbon and Nitrogen in a Wheat-Soybean, Double-Crop System. *Soil Science Society of America Journal*, **78**, 1378-1391. http://dx.doi.org/10.2136/sssaj2013.12.0534

[28] Anders, M.M., Brye, K.R., Olk, D.C. and Schmid, B.T. (2010) Rice Rotation and Tillage Effects on Soil Aggregation and Aggregate Carbon and Nitrogen Dynamics. *Soil Science Society of America Journal*, **76**, 994-1004. http://dx.doi.org/10.2136/sssaj2010.0436

[29] Frank, A.B., Berdahl, D.J., Hanson, J.D., Liebig, M.A. and Johnson, H.A. (2004) Biomass and Carbon Partitioning in Switchgrass. *Crop Science*, **44**, 1391-1396. http://dx.doi.org/10.2135/cropsci2004.1391

[30] Jung, J.Y., Lal, R., Jastrow, J.D. and Tyler, D.D. (2011) Nitrogenous Fertilizer Effects on Soil Structural Properties under Switchgrass. *Agriculture, Ecosystems & Environment*, **141**, 215-220. http://dx.doi.org/10.1016/j.agee.2011.01.016

[31] Six, J., Bossuyt, H., Degryze, D. and Denef, K. (2004) A History of Research on the Link between (Micro)aggregates, Soil Biota, and Soil Organic Matter Dynamics. *Soil & Tillage Research*, **79**, 7-31. http://dx.doi.org/10.1016/j.still.2004.03.008

[32] Singer, M.J., Southard, R.J., Warrington, D.N. and Janitzky, P. (1992) Stability of Synthetic Sand-Clay Aggregates after Wetting and Drying Cycles. *Soil Science Society of America Journal*, **56**, 1843-1848. http://dx.doi.org/10.2136/sssaj1992.03615995005600060032x

[33] Whalen, J.K. and Chang, C. (2002) Macroaggregate Characteristics in Cultivated Soils after 25 Annual Manure Applications. *Soil Science Society of America Journal*, **66**, 1637-1647. http://dx.doi.org/10.2136/sssaj2002.1637

[34] Edaño, M.L.S. (2009) Effects of Nutrient Sources, Harvest Frequency and Environmental Conditions in Switchgrass Production and Soil Properties. Ph.D. Dissertation, Oklahoma State University, Stillwater.

[35] Bonin, C., Lai, R., Schmitz, M. and Wullschleger, S. (2012) Soil Physical and Hydrological Properties under Three Biofuel Crops in Ohio. *Acta Agriculturae Scandinavica*, **62**, 595-603.

[36] Fike, J.H., Parrish, D.J., Wolf, D.D., Balasko, J.A., Green, J.T., Rasnake, M. and Reynolds, J.H. (2006) Switchgrass Production for the Upper Southeastern USA: Influence of Cultivar and Cutting Frequency on Biomass Yields. *Biomass & Bioenergy*, **30**, 207-213. http://dx.doi.org/10.1016/j.biombioe.2005.10.008

11

Impacts of Forest Harvest on Active Carbon and Microbial Properties of a Volcanic Ash Cap Soil in Northern Idaho

Deborah S. Page-Dumroese[1*], Matt D. Busse[2], Steven T. Overby[3], Brian D. Gardner[4], Joanne M. Tirocke[1]

[1]USDA Forest Service, Rocky Mountain Research Station, Moscow, USA
[2]USDA Forest Service, Pacific Southwest Research Station, Davis, USA
[3]USDA Forest Service, Rocky Mountain Research Station, Flagstaff, USA
[4]USDA Natural Resource Conservation Service, Moscow, USA
Email: [*]ddumroese@fs.fed.us

Abstract

Soil quality assessments are essential for determining impacts on belowground microbial community structure and function. We evaluated the suitability of active carbon (C), a rapid field test, as an indicator of soil biological quality in five paired forest stands (clear cut harvested 40 years prior and unharvested) growing on volcanic ash-cap soils in northern Idaho. Active C was compared with several traditional measures of soil microbial properties (microbial biomass, respiration, fungal hyphal biomass, bacterial number and biomass and PLFA community structure). Despite the significant differences in forest vegetation between paired stands, no differences in active C and only a few significant differences in microbial properties were detected. Total bacteria (microscope counts) and PLFA signatures (gram positive bacteria, gram negative bacteria, actinomycetes) were significantly higher in the managed stands. Our results indicate that either mineral soil biological properties in managed stands were relatively unaffected at the time of harvest or some biological recovery occurred 40 years later. Additionally, volcanic ash-cap soils in moist ecosystems could be highly resilient to the impacts of harvest operations and therefore few significant biological changes could be detected.

Keywords

Soil Quality, Forest Soil, Soil Monitoring, Soil Organic Matter

[*]Corresponding author.

1. Introduction

Changes in the labile fraction of the soil organic matter pool can provide an indication of both short- and long-term changes in soil quality that are a direct response to land management practices. Although there are a myriad of soil physical, chemical or biological properties that can be tested for soil quality assessments [1] to maintain forest sustainability, there are few consistently used measures that can be easily applied [2]. In addition to the limited accessibility and applicability of many field and laboratory tests, application of many testing procedures to forests and forest soils can also be a challenge [3]. Of particular concern is an appropriate surrogate test for biological activity after a given management activity. Indicators include potentially mineralized nutrients, decomposition, microbial biomass, total or active carbon, or organic matter content.

Assessments of soil quality often include a measure of soil organic matter, because it is critical for maintaining site productivity. In forest soils, organic matter content has been linked to soil water availability, nutrient supply, soil aggregation, and disease incidence or prevention [4] and contributes to numerous soil functions [3] [5]. In addition, forest management practices that impact organic matter accumulation and decomposition could affect future tree growth and site productivity [6].

Forest harvest or site preparation activities that remove the surface organic horizons may cause changes in soil organic matter that affect long-term productivity. For instance, loss of soil organic matter may alter soil productivity by leading to changes in the hydrologic function, erosion rates, soil water retention, or microbial activity [3] [7]. Total soil organic matter, however, is not a ready indicator of how management might affect the more labile fraction of soil organic carbon and it is therefore important to use methods that expresses a change in organic matter and other biological properties [8] and that forest soil scientists or other land managers can use to quickly assess potential impacts.

Tephra-derived forest soils have unique physical and chemical characteristics that affect their responses to harvest operations. Volcanic materials have undeveloped crystalline structure and weather rapidly in humid environments, producing soils that are characterized by high water holding capacity, cation exchange capacity, and their ability to stabilize organic matter [9]. These soil types are widely distributed in northern Idaho and support highly productive forests [10].

Forest harvesting can result in changes in the active fractions of organic matter and result in alteration of microbial properties or biomass or changes in C and N content [1] [11]. The active C pool can be a sensitive indicator of changes in other biological properties such as microbial biomass C [12], particulate organic matter [13] and soil carbohydrates [14]; however, it has not been widely tested in forest soils. Given that a rapid assessment of soil microbial response to land management is needed and that a measure of biological activity should reflect organic matter quality, our objectives were to 1) test the rapid field measure of active C as an indicator of soil biological quality, and 2) determine the long-term impact of forest harvesting on soil microbial and chemical properties.

2. Methods and Materials

2.1. Site Description

Soils were collected from the forest floor and surface 0 - 5 cm of the mineral soil in 5 different stands each of a 40-year-old managed forest and an unharvested control in northern Idaho during a 3-week time period in June. The unharvested stands were a mixed conifer forest type with an overstory consisting primarily of grand fir (*Abies grandis* (Douglas ex D. Don) Lindl.), western redcedar (*Thuja plicata* Donn ex D. Don) and western hemlock (*Tsuga heterophylla* (Raf.) Sarg.) with a small amount of Douglas-fir (*Pseudotsuga menziesii* (Mirb.) Franco (**Table 1**). Canopy coverage was visually estimated, diameter was measured at 1.4 m from ground level, and height was measured from ground level to the top of the terminal leader.Managed stands were clearcut harvested, dozer piled, and the piles burned; a typical harvest system from 40 yrs ago. Species composition for the managed stands consisted of grand fir, Douglas-fir, lodgepole pine (*Pinus contorta* (Douglas ex Loudon)), grand fir, western white pine (*Pinus monticola* (Douglas ex D. Don)), and ponderosa pine (*Pinus ponderosa* Lawson & C. Lawson). Both the harvested and unharvested stands were selected because they occur on the Threebear soil series (Medial over loamy, amorphic over mixed, superactive, Frigid Alfic Udivitrand) [15]. The Threebear series is an extensive soil in northern Idaho, (USDA-NRCS Major Land Resource Area 43A) and occurs on structural benches, hills and mountain foots lopes formed in eolian and alluvial material with a thick (>30 cm) mantle

Table 1. Overstory species and mean canopy coverage, diameter, and height for managed and unharvested stands. NM indicates that this species was not measured for diameter or height.

Tree species	Managed stand			Unharvested stand		
	Canopy coverage (%)	Diameter (cm)	Height (m)	Canopy coverage (%)	Diameter (cm)	Height (m)
A. grandis	20	23.8	16.5	64	50.3	36.8
L. occidentalis	0	0	0	<1	NM	NM
P. contorta	28	25.1	12.4	0	0	0
P. monticola	7	23.5	15.6	0	0	0
P. ponderosa	5	32.5	15.1	0	0	0
P. menziesii	40	29.4	15.7	3	45.2	30.8
T. plicata	0	0	0	17	69.8	28.6
T. heterophylla	0	0	0	16	70.5	33.6

of volcanic ash. The soil is mapped at elevations ranging from 2800 to 3500 feet and is one of the most productive soils in the forested lands of northern Idaho, and as such is of great interest to land managers as a unit to manage and monitor for changes in soil quality.

2.2. Soil Sampling

In each forest stand (5 paired managed and unharvested stands), 3 randomly selected locations within each forest site (5 sites × 3 soil samples = 15 total samples) were selected and cores (2 cm dia.) were collected to a depth of 5 cm in the mineral soil. Soils were immediately placed in a cooler until the end of the day when they were cooled to 4˚C until processing. For each forest condition we collected 15 soil cores. Moist soil samples were subsampled for all analyses. Bulk density was calculated as dry mass per volume of the cylinder. No rocks were present in this soil; therefore total bulk density is presented. Soil pH was conducted on a 2:1 water:soil paste. Total C and N in the forest floor and mineral soil were analyzed on a LECO CN analyzer (LECO Corp, St. Joseph, MI). Cation exchange capacity (CEC) was determined with ammonium acetate extract and base saturation was calculated as the sum of the cations [16]. Active carbon was determined with a field kit using the modified permanganate oxidizable C method [17] [18] that is sensitive to land management changes. Sample and standard absorbance was read on a Hach (Hach Company, Boulder, CO) colorimeter at 550 nm. In this method, the bleaching of the purple permanganate color is proportional to the amount of oxidizable C in the soil (*i.e.*, the greater the color loss, the greater the amount of oxidizable C).

2.3. Microbial Biomass and Basal Respiration

Microbial biomasswas measured by the substrate-induced respiration method (SIR) [19] using 25 g soil and 5 $g \cdot kg^{-1}$ glucose. CO_2 production was measured within 3 h of glucose addition using a LI-6200 infra-red gas analyzer. Separate 25 g soil samples were wetted to approximately 60% of total porosity and incubated at 20˚C for 8 days before measuring basal respiration (24 h CO_2 production) with an infra-red gas analyzer.

2.4. Fungi and Bacteria (Microscope)

Fungal hyphal length and biomass were estimated by epifluorescent microscopy [20]. Briefly, hyphae were extracted from 10 g soil (dry wt. equivalent) by shaking in 95 ml dilute Ringers solution for 10 minutes. The samples were then diluted in 0.3 M phosphate buffer to a final dilution of 1 mg soil ml^{-1}. Diluted samples (1 ml) were stained with 0.5 ml fluorescent brightener 28 (Sigma-Aldrich, St. Louis, MO) for 30 min. at room temperature in the dark, filtered onto 0.2 μm blackened Millipore filters, and viewed at 400× magnification with a 01 filter set. Hyphal lengths and widths were measured on 20 fields per sample, and converted to biomass [21].

Bacteria were stained with SYBR Green I (Invitrogen Corp., Carlsbad, CA) to estimate total number and biomass. Soil extractions followed the same method as those for determining fungal biomass. Samples (1 ml) were filtered onto 0.02 μm what mananodisc filters using <80 mm Hg pressure. Each filter was then stained with 15 μl SYBR Green I (10^{-3} dilution in filter-sterilized H_2O), incubated in the dark for 15 min, and viewed using an epifluorescent microscope with a 09 filter set. Cell numbers and diameters were measured on 10 fields per sample, averaging 20 - 50 cells per field. Total biomass was calculated from the microscope measurements

2.5. Phospholipid Fatty Acid (PLFA)

Freeze-dried mineral soil subsamples were utilized to assess microbial community structure. Mass spectral analysis identified numerous compounds between C14 to C22 in C chain length, of which 16 were used as microbial biomarkers [21]-[23]. Total microbial biomass was calculated by summing these 16 biomarkers. Subsets of the biomarkers were used to estimate gram-negative bacteria (cy17:0, cy19:0, 16:1ω9, 16:1ω7, 18:1ω5c, and 18:1ω7) and gram-positive bacteria biomarkers (i15:0, a15:0, i16:0, i17:0, a17:0, and 10me16:0) discriminate between groups. Two isomers of C18:2n6 (*trans* and *cis*) were used to estimate the fungal group [22] and 10me 16:0 were used to represent the actinobacteria [23].

PLFA extractions were performed using 5 g of freeze-dried soil with a single-phase mixture of chloroform, methanol, and phosphate buffer [24] then fractionated into neutral, glyco-, and phospholipids [25]. The phospholipids are then esterfied and reconstituted in hexane prior to analysis [26]. A capillary column was used for compound separation. Identification and quantification of the each specific fatty acid methylester (FAME) was established using commercially available standards (Accustandard, Sigma-Aldrich, St. Louis, Missouri, USA) through electron ionization on a quadrapole mass selective detector (Agilent 6890N/5973N gas chromatograph/ mass spectrometer, Santa Clara, CA, USA). Quantification (μmol of PLFA/kg oven-dry material) of samples is based on calibration curves derived from individual FAME standards.

2.6. Statistical Analyses

Analysis of variance using the mixed model (SAS Inc., Cary, NC) was used to test for differences in soil properties among harvested and unharvested control stands. The least significant difference (LSD) test of significance was used to assess the differences of means among the clear cut harvested and uncut treatments at $P < 0.10$. Linear regression analysis was used to explore the relationships of active C and individual soil chemical or microbial properties (SAS Inc., Cary, NC).

3. Results

In general, soil properties still reflect the impact of land management 40 years after harvest and site preparation operations. The surface organic horizons had lower C and N contents, but only N was significantly lower in the managed stands as compared to the unharvested stands (**Table 2**). Total bulk density was slightly (8%), yet significantly higher in the managed stand (**Table 2**). In addition, the surface (0 - 5 cm) mineral soil C, N, CEC, and base saturation were lower in the managed stand than in the control, but only CEC values were significantly different among these two stand types.

The microbial measurements showed mixed responses to legacy stand management. Although not significant, active C was less in the managed stand; whereas SIR, respiration at day 8, bacteria and hyphal biomass were greater 40 years after management, but not significantly (**Table 3**). Total bacteria were significantly greater in the managed stand as compared to the unharvested stands.

Our objective for this study was to determine if active C could be used as a low-cost, time efficient field method for determining organic matter quality in lieu of using lab- and cost-intensive measurements. For each stand type (managed and unharvested) we used a simple linear regression to evaluate the relationships. Active C was positively correlated with several microbial measures (**Table 4**). Interestingly, only SIR, basal respiration, and the microscope fungi: bacteria ratios were significantly correlated with active C in the unharvested and managed stands whereas total bacteria, bacterial biomass, hyphal length, hyphal biomass, actinomycetes (PFLA), and fungi (PFLA) were only significantly correlated with active C in the managed stands. As expected, the best relationship with active C was with total C ($P \leq 0.0001$ for both the managed and unharvested stands; **Figures 1(a)-(b)**).

4. Discussion

Using the unharvested stand as a reference assumes that both stand types were identical when the managed

Table 2. Mean surface organic horizon and mineral soil (0 - 5 cm depth) characteristics in managed and unharvested stands in northern Idaho. Values in parentheses are the standard error of the mean.

Soil property	Stand type		
	Managed	Unharvested	*P*
Surface organic horizons			
Carbon (Mg/ha)	48.6 (4.9)	59.4 (3.2)	0.3855
Nitrogen (kg/ha)	1095 (130)	1597 (170)	0.0152
Mineral soil (0 - 5 *cm*)			
Carbon (Mg/ha)	18.0 (2.1)	21.4 (2.4)	0.4258
Nitrogen (kg/ha)	2838 (215)	2913 (232)	0.4153
pH	6.2 (0.1)	6.3 (0.1)	0.5253
Bulk density (Mg/m^3)	0.72 (0.08)	0.67 (0.05)	0.0152
Cation exchange capacity (cmols/kg)	24.4 (3.5)	28.5 (3.9)	0.0401
Base saturation (%)	66.1 (4.1)	71.4 (5.3)	0.2081

Table 3. Mean soil microbial measurements in the mineral soil (0 - 5 cm depth) in managed and unharvested stands in northern Idaho.Values in parentheses are the standard error of the mean.

Soil property	Stand type		
	Managed	Unharvested	*P*
Active C (mg/kg)	842.0 (251.5)	934.8 (221.7)	0.2930
SIR (mg/kg)	500.5 (235.1)	416.5 (125.2)	0.2858
Basal respiration (μg/h/g soil)	0.061 (0.030)	0.049 (0.011)	0.2400
Total bacteria ($\times 10^9$ cells/g soil)	4.23 (0.21)	2.69 (0.14)	0.0443
Bacterial biomass (μg/g soil)	458.4 (176.2)	374.8 (178.4)	0.2081
Hyphal biomass (μg/g soil)	2914.7 (974.3)	2657.6 (1321.2)	0.5159
Hyphal length (m/g soil)	925.4 (292.5)	950.5 (473.6)	0.8544
Fungi: bacteria (microscope)	9.5 (3.0)	7.9 (1.0)	0.2672
Gram positive (μmol of PFLA/kg)	50.89 (17.84)	34.57 (10.52)	0.0100
Gram negative (μmol of PFLA/kg)	33.17 (9.76)	25.56 (6.89)	0.0221
Actinomycetes (μmol of PFLA/kg)	1.62 (0.44)	1.35 (0.34)	0.0911
Fungi (μmol of PFLA/kg)	10.82 (3.48)	8.95 (2.98)	0.1315

stands were harvested and had not been selectively affected by biological factors since that time [27] [28]. In this area of northern Idaho, past timber sale records indicate that this assumption is likely correct (Idaho Panhandle National Forest Timber Stand Management Database). We also assumed that since this was a relatively narrow geographic area, that climate had not changed among our stand types. Clearcut harvesting alters hydrological processes [29] [30], surface organic matter, and carbon [7]. Using the unharvested stand as a baseline reference for changes in the managed stand, the impacts of clearcut harvesting 40 yrs ago on soil properties were still evident, though the effects were not consistent among properties nor were they particularly sizable when detected (**Table 2**). Forest floor N content showed the greatest difference due to harvesting (31% lower in managed compared to unharvested stands). We attribute this to the slow development of the Oe and Oa horizons during stand regeneration in this forest type or that the incremental tree growth in the managed stand exceeds that of the unharvested stand thereby moving more N into the aboveground portion of the stand [29]. Cation ex-

Table 4. Linear regression relationship of microbial measurements to active C in the mineral soil (0 - 5 cm depth) in managed and unharvested stands in northern Idaho. Values in bold indicate a significant relationship between the variable and active C at $P < 0.20$.

Soil property	Managed stand		Unharvested stand	
	F	*P*	F	*P*
SIR (mg/kg)	3.629316	0.067473	4.186894	0.063296
Basal respiration (μg/h/g soil)	2.508301	0.137263	4.329197	0.059552
Total bacteria ($\times 10^9$ cells/g soil)	2.896061	0.112574	0.787941	0.392168
Bacterial biomass (μg/g soil)	2.021855	0.178601	0.001147	0.973536
Hyphal biomass (μg/g soil)	0.00911	0.925416	0.037679	0.849395
Hyphal length (m/g soil)	3.02617	0.105536	0.084167	0.77668
Fungi: bacteria (microscope)	4.40091	0.056021	1.510192	0.05601
Gram positive (PFLA)	1.19006	0.295124	0.495088	0.494077
Gram negative (PFLA)	1.420241	0.254666	0.775937	0.394381
Actinomycetes (PFLA)	2.887063	0.113082	1.781503	0.204871
Fungi (PFLA)	2.276916	0.155228	0.887178	0.363418

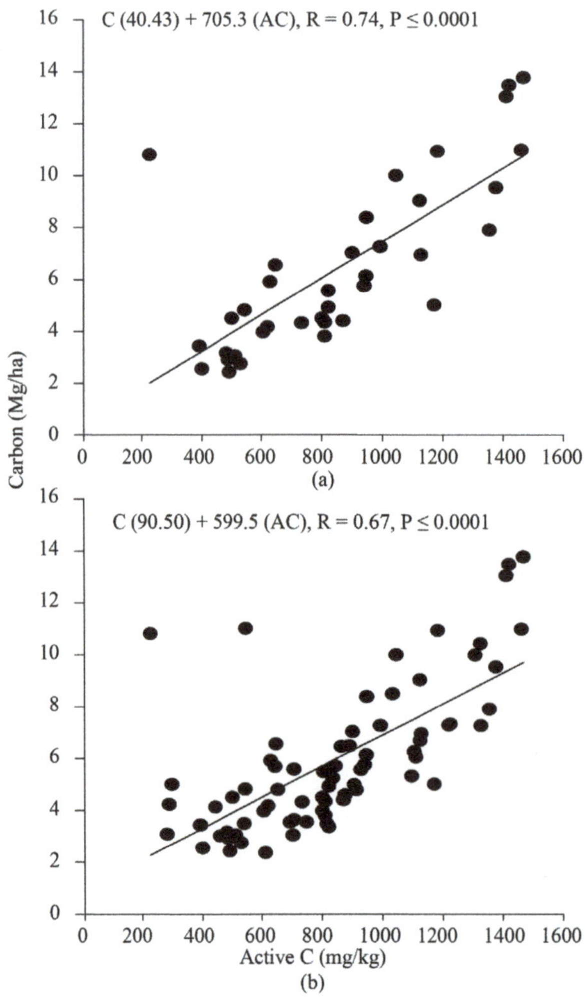

Figure 1. Correlation of C (Mg/ha) to active C (mg/kg) in (a) managed and (b) unharvested forest soil (0 - 5 cm) in northern Idaho.

change capacity was 14% lower in managed compared to unharvested stands yet was still well above the threshold level for ash-cap soil. In addition, bulk density remained significantly higher in the managed stand than unmanaged stand as a consequence of clearcut harvest operations, piling slash for burn piles, the susceptibility of volcanic ash-cap soil to compaction, and the lack of a freeze/thaw cycle in this area [31]. However, we anticipate the small increase in bulk density with harvesting (8% higher compared to unharvested stands) will have a nominal effect on soil and plant processes [32].

Microorganisms are an essential constituent of soil quality and productivity, particularly after harvest operations [4] [33] [34]. Microbes are responsible for mediating wood decomposition, nitrogen conversion, and mycorrhizal activity, thereby affecting plant growth [35] and they depend on organic matter for energy to facilitate the transformation of nutrients making them available for plant uptake [36]. Harvesting can alter microbial activity by changing the amount and type of organic matter, altering soil temperature and moisture, and increasing soil bulk density [37]. After 40 years, microbial measures on our harvested sites had recovered to, or exceeded, levels in the unharvested stand. Stand recovery after harvesting in the Oregon Cascades indicated that increasing time after harvesting resulted in soil properties (soil organic matter and C), but not litter depth or soil respiration, shifting towards unharvested conditions after 40 years [38]. We assume this observation reflects both the tolerance of microbial communities to infrequent disturbance [20] and the positive effect of the aggrading forest floor on soil temperature, moisture, nutrient and carbon inputs and, consequently, microbial community resilience. Interestingly, bacterial numbers and PLFA content (gram positive and gram negative bacterial signatures) were significantly higher in managed compared to unharvested stands. Reasons for this observation and whether ecological significance can be attached to them are unclear.

One of our objectives was to test the effectiveness of active C as a rapid method for determining management impacts on forest soils. Field testing of active C is a simple and inexpensive procedure and has been shown to be effective indicator of soil organic matter content and biological activity in other ecosystem types [12] [17]. Importantly, we tested active C across a wide range of soil C contents and microbial conditions. For example, there was a 7-fold range in total soil C among samples (2 - 14 Mg/ha) and microbial biomass ranged from approximately 50 to 1200 mg C/kg. Active C followed a similar trend as total C in both stand types (**Figures 1(a)-(b)**). Active C has also been shown to have greater sensitivity to land management or environmental gradients than particulate organic C, microbial biomass C, and soil organic C [39].

5. Conclusion

In our 40-year-old harvested stand, moderate changes in soil bulk density, cation exchange capacity and some PFLA measurements indicate that the soil environment is still exhibiting chemical, biological and physical changes due to harvesting. Some of the changes may be associated with the shift in overstory species, but likely the residual impacts of soil compaction and reduced total C are influential on the changes in microbial and active C response. We have shown that active C can be used to estimate the most readily oxidizable (active) forms of soil C and that this measure is related to other microbial measures such as SIR, basal respiration, PFLA fungi and the total C pool. This method determining the labile soil C pool provides a low-cost tool for forest land managers who are interested in determining changes in soil quality from harvest or site preparation practices. The active C field-going method offers a way to gather routine measurements. For example, in forest stands that will be harvested, measurements of active C may indicate the need to leave more residual slash or to limit burning operations. Alternatively, in stands that already have been harvested, active C may be a tool to determine site preparation or restoration needs.

Acknowledgements

The authors wish to thank Tom Rice (retired, Rocky Mountain Research Station), Dr. Mark Kimsey (University of Idaho Intermountain Forest Tree Nutrition Cooperative), Frank Gariglio, Scott Bare, Dave Hoover, Hal Swenson, Rod Kyar, Arlene Tugel, Francine Lauritie, Allyson Young and Charles Kome (USDA-NRCS) for their assistance with field data collection. We also thank Anita Falen (University of Idaho) for her assistance in the field and laboratory.

References

[1] Doran, J.W. and Parkin, T.B. (1994) Defining and Assessing Soil Quality. In: Doran, J.W., Coleman, D.C., Bezdicek,

D.F. and Stewart, B.A., Eds., *Defining Soil Quality for a Sustainable Environment*, Soil Science Society of America, Madison, 3-21.

[2] Page-Dumroese, D., Jurgensen, M., Elliot, W., Rice, T., Nesser, J., Collins, T. and Meurisse, R. (2000) Soil Quality Standards and Guidelines for Forest Sustainability in Northwestern North America. *Forest Ecology and Management*, **138**, 445-462. http://dx.doi.org/10.1016/S0378-1127(00)00430-8

[3] Burger, J.A. and Kelting, D.L. (1998) Soil Quality Monitoring for Assessing Sustainable Forest Management. In: Adams, M.B., Ramakrishna, K. and Davidson, E.A., Eds., *The Contributions of Soil Science to the Development of and Implementation of Criteria and Indicators of Sustainable Forest Management*, Soil Science Society of America, Madison.

[4] Harvey, A.E., Larsen, M.J. and Jurgensen, M.F. (1980) Biological Implication of Increasing Harvest Intensity on the Maintenance and Productivity of Forest Soils. In: *Environmental Consequences of Timber Harvesting in Rocky Mountain Coniferous Ecosystems*, Intermountain Research Station, Ogden, 180-210.

[5] Powers, R.F. (1997) Are We Maintaining the Productivity of Forest Lands? Establishing Guidelines through a Network of Long-Term Studies. In: Harvey, A.E. and Neuwenschwander, L., Eds., *Management and Productivity of Western-Montane Forest Soils*, Intermountain Research Station, Ogden, 70-81.

[6] Grigal, D.F. (2000) Effects of Extensive Forest Management on Soil Productivity. *Forest Ecology and Management*, **138**, 167-185.http://dx.doi.org/10.1016/S0378-1127(00)00395-9

[7] Powers, R.F., Scott, D.A., Sanchez, F.G., Voldseth, R.A., Page-Dumroese, D., Elioff, J.D. and Stone, D.M. (2005) The North American Long-Term Soil Productivity Experiment: Findings from the First Decade of Research. *Forest Ecology and Management*, **220**, 31-50. http://dx.doi.org/10.1016/j.foreco.2005.08.003

[8] Franzluebbers, A.J., Haney, R.L., Honeycutt, C.W., Schomberg, H.H. and Hons, F.M. (2000) Flush of Carbon Dioxide Following Rewetting of Dried Soil Relates to Active Organic Pools. *Soil Science Society of America Journal*, **64**, 613-623. http://dx.doi.org/10.2136/sssaj2000.642613x

[9] Moldrup, P., Yoshikawa, S., Olesen, T., Komatsu, T. and Rolston, D.E. (2003) Gas Diffusivity in Undisturbed Volcanic Ash Soils: Test of Soil-Water-Characteristic-Based Prediction Models. *Soil Science Society of America Journal*, **67**, 41-51.http://dx.doi.org/10.2136/sssaj2003.0041

[10] Gardner, B. (2007). Field Identification of Andic Soil Properties for Soils of North-Central Idaho. In: Page-Dumroese, D.S., Miller, R.E., Mital, J., McDaniel, P. and Miller, D., Eds., *Volcanic-Ash-Derived Forest Soils of the Inland Northwest: Properties and Implications for Management and Restoration*. Rocky Mountain Research Station, Ft. Collins, 23-30.

[11] Page-Dumroese, D.S., Jurgensen, M.F. and Terry, T.A. (2010) Maintaining Soil Productivity during Biomass Harvesting in Western Forests. *Western Journal Applied Forestry*, **25**, 5-10.

[12] Islam, K.R. and Weil, R.R. (2000) Soil Quality Indicator Properties in Mid-Atlantic Soils as Influenced by Conservation Management. *Journal of Soil and Water Conservation*, **55**, 69-78.

[13] Wander, M.M. and Bidart, M.G. (2000) Tillage Practice Influences on the Physical Protection, Bioavailability and Composition of Particulate Organic Matter. *Biology and Fertility of Soils*, **32**, 360-367. http://dx.doi.org/10.1007/s003740000260

[14] DeLuca, T.H. and Keeney, D.R. (1993) Soluble Organics and Extractable Nitrogen in Paired Prairie and Cultivated Soil of Central Iowa. *Soil Science*, **155**, 219-228. http://dx.doi.org/10.1097/00010694-199303000-00008

[15] Soil Survey Staff (2006) Soil Taxonomy.10th Edition, US Department of Agriculture, Natural Resources Conservation, Washington DC.

[16] Sumner, M.E. and Miller, W.P. (1996) Cation Exchange Capacity. In: Sparks, D.L., Ed., *Methods of Soil Analysis*, *Part* 3: *Chemical Methods*, Soil Science Society of America, Madison, 1201-1230.

[17] Weil, R.R., Kankikar, I.R., Stine, M.A., Gruver, J.B. and Samson-Liebig, S.E. (2003) Estimating Active Carbon for Soil Quality Assessment: A Simplified Method for Laboratory and Field Use. *American Journal of Alternative Agriculture*, **18**, 3-17. http://dx.doi.org/10.1079/AJAA2003003

[18] Blair, G.J., Lefroy, R.D.B. and Lise, L. (1995) Soil Carbon Fractions Based on Their Degree of Oxidation, and the Development of a Carbon Management Index for Agricultural Systems. *Australian Journal of Agricultural Research*, **46**, 1459-1466. http://dx.doi.org/10.1071/AR9951459

[19] Anderson, J.P.E. and Domsch, K.H. (1978) A Physiological Method for the Quantitative Measurement of Microbial Biomass in Soil. *Soil Biology and Biochemistry*, **11**, 215-221. http://dx.doi.org/10.1016/0038-0717(78)90099-8

[20] Busse, M.D., Sanchez, F.G., Ratcliff, A.W., Butnor, J.R., Carter, E.A. and Powers, R.F. (2009) Soil Carbon Sequestration and Changes in Fungal and Bacterial Biomass Following Incorporation of Forest Residues. *Soil Biology and Biochemistry*, **41**, 220-227. http://dx.doi.org/10.1016/j.soilbio.2008.10.012

[21] Bottomley, P.J. (1994) Light Microscopic Methods for Studying Soil Microorganisms. In: Weaver, R.W., *et al.*, Eds., *Methods of Soil Analysis*, *Part* 2: *Microbiological and Biochemical Properties*, Soil Science Society of America, Mad-

ison, 81-105.

[22] Frostegard, A. and Bååth, E. (1996) The Use of Phospholipid Fatty Acid Analysis to Estimate Bacterial and Fungal Biomass in Soil. *Biology and Fertility of Soils*, **22**, 59-65. http://dx.doi.org/10.1007/BF00384433

[23] Zelles, L., Bai, Q.Y., Beck, T. and Beese, F. (1992) Signature Fatty Acids in Phospholipid and Lipopolysaccharides as Indicators of Microbial Biomass and Community Structure in Agricultural Soils. *Soil Biology and Biochemistry*, **24**, 317-323. http://dx.doi.org/10.1016/0038-0717(92)90191-Y

[24] White, D.C., Meadows, P., Eglinton, G. and Coleman, M.L. (1993) *In Situ* Measurement of Microbial Biomass, Community Structure and Nutritional Status. *Philosophical Transactions: Physical Sciences and Engineering*, **344**, 59-67. http://dx.doi.org/10.1098/rsta.1993.0075

[25] Frostegard, A., Tunlid, A. and Bååth, E. (1991) Microbial Biomass Measured as Total Lipid Phosphate in Soils of Different Organic Content. *Journal of Microbiological Methods*, **14**, 151-163. http://dx.doi.org/10.1016/0167-7012(91)90018-L

[26] Frostegard, A., Bååth, E. and Tunlid, A. (1993) Shifts in the Structure of Soil Microbial Communities in Limed Forests as Revealed by Phospholipid Fatty Acid Analysis. *Soil Biology and Biochemistry*, **25**, 723-730. http://dx.doi.org/10.1016/0038-0717(93)90113-P

[27] Dyck, W.J. and Cole, D.W. (1994) Strategies for Determining Consequences of Harvesting and Associated Practices on Long-Term Productivity. In: Dyck, W.J., Cole, D.W. and Comerford, N.B., Eds., *Impacts of Forest Harvesting on Long-Term Site Producitivity*, Chapman and Hall, London, 13-40. http://dx.doi.org/10.1007/978-94-011-1270-3_2

[28] Pennock, D.J. and van Kessel, C. (1997) Clear-Cut Forest Harvest Impacts on Soil Quality Indicators in the Mixedwood Forest of Saskatchewan, Canada. *Geoderma*, **75**, 13-32. http://dx.doi.org/10.1016/S0016-7061(96)00075-4

[29] Seidel, K.W. (1980) Diameter and Height Growth of Suppressed Grand Fir Saplings after Overstory Removal. US Department of Agriculture, Forest Service Research Paper, PNW-RP-275.

[30] Neary, D.G. and Hornbeck, J.W. (1994) Impacts of Harvesting and Associated Practices on Off-Site Environmental Quality. In: Dyck, W.J., Cole, D.W and Comerford, N.B., Eds., *Impacts of Forest Harvesting on Long-Term Site Productivity*, Chapman and Hall, London, 81-118. http://dx.doi.org/10.1007/978-94-011-1270-3_4

[31] Page-Dumroese, D.S., Jurgensen, M.F., Tiarks, A.E., Ponder Jr., F., Sanchez, F.G., Fleming, R.L., *et al.* (2006) Soil Physical Property Changes at the North American Long-Term Soil Productivity Sites: 1 and 5 Years after Compaction. *Canadian Journal of Forest Research*, **36**, 551-564. http://dx.doi.org/10.1139/x05-273

[32] Ponder Jr., F., Fleming, R.L., Berch, S., Busse, M.D., Elioff, J.D., Hazlett, P.W., *et al.* (2012) Effects of Organic Matter Removal, Soil Compaction and Vegetation Control on 10th Year Biomass and Foliar Nutrition: LTSP Continent-Wide Comparisons. *Forest Ecology and Management*, **278**, 35-54. http://dx.doi.org/10.1016/j.foreco.2012.04.014

[33] Larsen, M.J., Harvey, A.E. and Jurgensen, M.F. (1980) Residue Decay Processes and Associated Environmental Functions in Northern Rocky Mountain Forests. In: *Environmental Consequences of Timber Harvesting in Rocky Mountain Coniferous Ecosystems*, Intermountain Research Station, Ogden, 157- 174.

[34] Mroz, G.D., Jurgensen, M.F., Harvey, A.E. and Larsen, M.J. (1980) Effects of Fire on Nitrogen and Forest Floor Horizons. *Soil Science Society of America Journal*, **44**, 395-400. http://dx.doi.org/10.2136/sssaj1980.03615995004400020038x

[35] Jurgensen, M.F., Harvey, A.E., Graham, R.T., Page-Dumroese, D.S., Tonn, J.R., Larsen, M.J. and Jain, T.B. (1997) Impacts of Timber Harvesting on Soil Organic Matter, Nitrogen, Productivity, and Health of Inland Northwest Forest. *Forest Science*, **43**, 234-251.

[36] Entry, J.A., Stark, N.M. and Loewenstein, H. (1986) Effect of Timber Harvesting on Microbial Biomass Fluxes in a Northern Rocky Mountain Forest Soil. *Canadian Journal of Forest Research*, **16**, 1076-1081. http://dx.doi.org/10.1139/x86-186

[37] Bailey, V.L., Smith, J.L. and Bolton, H. (2002) Fungal-to-Bacterial Rations in Soils Investigated for Enhanced C Sequestration. *Soil Biology and Biochemistry*, **34**, 393-395. http://dx.doi.org/10.1016/S0038-0717(02)00033-0

[38] Griffiths, R.P. and Swanson, A.K. (2001) Forest Soil Characteristics in a Chronosequence of Harvested Douglas-Fir forests. *Canadian Journal of Forest Research*, **31**, 1871-1879. http://dx.doi.org/10.1139/cjfr-31-11-1871

[39] Culman, S.W., Snapp, S.S., Freeman, M.A., Schipanski, M.E., Beniston, J., Lal, R., Drinkwater, L.E., Franzluebbers, A.J., Glover, J.D., Grandy, A.S., Lee, J., Six, J., Maul, J.E., Mirksy, S.B., Spargo, J.T. and Wander, M.M. (2012) Permanganate Oxidizable Carbon Reflects a Processed Soil Fraction That Is Sensitive to Management. *Soil Science Society of America Journal*, **76**, 494-504. http://dx.doi.org/sssaj2011.0286

12

Soil Respiration in Drying of an Organic Soil

Zofia Fischer[1], Pavel Blažka[2*]

[1]Department of Biotechnology and Environment, John Paul II, Catholic University of Lublin, Lublin, Poland
[2]Department of Science, University of South Bohemia, České Budějovice, Czech
Email: *pavelblazka@seznam.cz, zofiaf@plusnet.pl

Abstract

Drying of soil was linearly related to time, soil volume decreased and ratio of air within the soils increased. Respiration was related with decreasing humidity, storage of CO_2 in soil water results in RQ < 0.5 in the larger soil items at least for a while. Rate of drying decreased in the second part of the process. RQ increased as the CO_2 stored was aerated when its solvent-water evaporated and access of air into the soil increased; eventually RQ = 1.0 in the last days of the experiment. Respiration of the experimental soil stopped when GWC reached 0.15. ΣRQ for the whole process is about 0.7, a bit higher in experiments with less soil suggesting less anoxia.

Keywords

RQ, Anoxia, Storage of CO_2, Changes of Soil Volume

1. Introduction

In the present contribution our original idea was to study process of drying more in detail. Soils in natural settings are changing; this may be approached not only in field work, but also in lab experiments, where the conditions may be set more accurately and less rely on unpredictable weather etc. and show how they are reflected in the process, particularly in soil respiration. During the work we realized however, that some concepts are generally not quite clear or understood and we propose some clarification, both by pointing to literature not frequently referred in soil science and/or through experimental design.

Respiration was/is considered as a general measure of soil activity (somewhat mystical statement, preferably of soil organisms), recirculation of nutrients from plant residues, while drying of soil is associated with wet and dry periods, seasons, hence more recently with global changes.

Soil respiration was studied many times; the literature on it is quite extensive [1]-[3]. It is based both on field measurements and on laboratory experimentation. The latter is mostly associated with sieving to get more re-

*Corresponding author.

producible results, and also with mixing and thus aeration of the soil material. We tried to overcome the sieving effect by slightly compressing the sieved soil by hand [4]. We have seen no other reference on the effect of compression in literature. We want to however first make some introductory comments to the subject respiration. The word respiration is historically (since 18th century) associated primarily with oxygen consumption by men and animals, more recently in biochemistry with oxidation of protons to water in the oxidative phosphorylation pathway as part of the general scheme of energy metabolism in most organisms, prokaryotes and eukaryotes; plants, animals and fungi—except only obligatory anaerobes [5]-[7]. Respiration is a heterotrophic process and also root respiration is not autotrophic e.g. [3], [8] though it is run by organisms, which largely live autotrophically; all plants are however heterotrophs at night. Soil respiration, meaning respiration of the whole soil community, was introduced later (first half of 20th century—Suchtelen [9] was probably among the first). Respiration is anyway a biological process; all biological principles are to be included, not discarded, but it is also a part of soil biology, a composite science as soil physics, biochemistry, geology etc.; all parts have to be observed with the same scrutiny, accuracy and respect to achieve real progress in soil science as a whole.

The sequence/wording "anaerobic respiration" (meaning reduction of e.g. nitrate, nitrite or sulfate replacing dioxygen as proton acceptor) is used quite frequently, though it is not particularly accurate, nitrate and sulfate reduction or proton acceptance would be more to the point, but may be perhaps tolerated.

Respiratory production of carbon dioxide occurs largely in the tricarboxylic acids cycle, is associated through oxidative phosphorylation with oxygen consumption, their ratio in terms of absolute values of moles or volumes is the respiratory quotient (RQ = [production CO_2]/[consumption O_2]).

Introduction of reliable IR non-dispersive spectrophotometers (In fact already 1850 paper based on CO_2) enabling measurements of small concentrations and changes of carbon dioxide concentrations to soil science provoked some shifts in terminology; soil biological activity—measured as production of carbon dioxide—is now quite frequently referred as soil respiration e.g. [2] [8] [10], most clearly formulated by Gaumont-Guay *et al.* [11] ("The efflux of carbon dioxide (CO_2) from the soil, referred to hereafter as soil respiration…"). There are however quite a few sources of carbon dioxide in soils, besides processes involving primarily inorganic substrates, e.g. acidification of carbonates by acids [12], there is also a number of fermentation types occurring when oxygen availability is limited. Both respiration and fermentation are the only processes providing energy to organisms (as macroergic P-bonds [~P], largely ATP). Fermentation in soils contemporary with respiration is apparently quite frequent, particularly when soils become wet and both involve also consumption of some substrate(s); it is therefore surprising, if there is no information on SOM given, particularly in dry environment where very low SOM may be expected e.g. [13]. Carbon dioxide produced in fermentation is associated with much smaller gain in useful energy (about 3 to 4-times) compared to respiration. Respiration supplemented by fermentation is likely the most frequent situation in soils with inadequate oxygen availability; though even low partial tensions of O_2 are mostly adequate for microorganisms. This results in RQ $\geq$ 1 and decreasing production of ~P. On the other hand, many products of fermentations may be readily available substrates in respiration, if dioxygen becomes available again. This is also reflected in the interpretation of the RQ. Very often limits for RQ are given e.g. within 0.7 till 1.3. In fact in anaerobes (obligatory or facultative, microbial or eukaryotic) RQ may be much higher (e.g. >10), in strict anoxia it cannot be defined at all, as dividing by 0 (O_2 consumption) is a prohibited math operation. Two most important types of fermentation for soil may be lactic and ethanolic (alcoholic). While respiration of lactate and many other products of fermentation (including e.g. acetate) results in RQ = 1, like most carbohydrates; for catabolism of ethanol RQ = 0.57; in reduction of methane RQ = 0.50. Efflux of CO_2 is in fact a mix of respiration, fermentation and abiotic processes leading to absorption or release of CO_2. Field measurements of CO_2 release to the atmosphere are useful particularly for understanding of the role of soils in the global carbon cycle. Experimental conditions (e.g., participation of autotrophs, soil humidity, differences of temperature between the "dome" and the neighboring free area…), are however to be thoroughly specified and the CO_2 efflux is not considered as soil respiration.

In soil literature retention of CO_2 in soil water due its high solubility (distribution coefficient air/water = approx 1 compared to dioxygen 0.03) is mentioned only exceptionally. In soils with pH > 4 dissolved bicarbonate becomes the dominating form of the carbonate system [14]. What we measure by any instrument is the apparent production of CO_2 and RQ modified by diffusion and solubility interference. RQ is a ratio, relationship between the differences of the two components and their ratio are thus non-linear, rather exponential. Jenkins and Adams [15] are thus not correct in saying "RQ is free of empirical assumptions…", Araujo *et al.* [16] in accepting without comment measured RQ in the range RQ = 0.2 till 1.5 depending on the soil degradation level. On the low side remediation of hydrocarbon degradation of soils results in low RQ [17], again below the theoretical limit 0.65 for hexadecane; methane is the most reduced hydrocarbon, its oxidation (assuming only one process) results in RQ = 0.5; this should be considered as the low limit of acceptable RQ data. On the other side of possible range synthesis of lipids from carbohydrates are sometimes considered with RQ ≈ 1.3 as possible maximum. It is probably correct for respiration; we must however expect hypoxic or anaerobic conditions particularly in wet soils. Soil organisms are forced to get additional energy to what they get from decreased or even stopped respiration through some fermentation; with prevailing fermentation RQ may be well over 10. In deep anoxia RQ cannot be calculated at all, as CO_2 is produced, but there is no oxygen consumption, division by 0 is a prohibited math operation. Let us just remind summed equation of ethanolic fermentation:

$$C_6H_{12}O_6 = 2\ CH_3 \cdot CH_2OH + 2\ CO_2$$

Being aware of interferences in measurements is particularly important in studies of processes, like various kinds of acclimation to changing temperature and humidity or in the budget of a process.

We want to analyze particular parts of the drying process, not only desiccation, but also changes of soil volume, access of air/oxygen to soil aggregates, storage capacity for CO_2 within the soil etc. and their impact on measured respiration data. This applies both to laboratory and field setups and involves size of soil items in both. Our soil is to some extent exceptional in high SOM content (around 50% of dry weight), many relationships studied in less organic soils need not to be applicable, also some of our conclusions are necessarily not of general use.

Drying of soils is a natural part of regular seasonal events, in observations and/or experiments many papers tend to limit themselves to just momentarily flashes or steady states; we tried however to estimate the budget of the drying process to go through different situations, states of the soil, get rid of many obstacles in getting real metabolic data. Incidentally, we came across low rates, their statistical treatment and time scales.

2. Material and Methods

2.1. The Site and Soil Sampling

The location is on the northern outskirts of the city of Warsaw (Poland). Geographic coordinates are: 52°20'212 N, 020°51'260 E, elevation 81 m. Refer to Blazka and Fischer [18]. The aim of the study is understands of mechanisms of drying through laboratory experimentation and not description of a particular locality. We used the soil for experiments as a laboratory model. The soil is highly organic; nearly 50% LOI, most of the mineral fraction is sand. It comes from a wet site in a dense forest, from the very active A1 zone, depth about 10 cm below surface, it was 5 mm sieved. We got our samples respecting suggestions by Carter and Gregorich [19].

2.2. Laboratory Analyses

2.2.1. Humidity

(GWC—gravimetric water content, WHC—water holding capacity, dm—dry mass).

Our procedure was based essentially on Grace *et al.*, Priha and Smolander [20] [21] with minor modifications [18]. GWC and other contents are given as ratios, but not %. Dry mass was determined by drying at 110°C to constant weight.

2.2.2. Soil Organic Matter (SOM and Density)

SOM was estimated as loss of weight in ignition to constant weight (LOI) at 500°C on selected samples. Mean densities of SOM (ρ = 1.3) and of the mineral fraction of dry weight (ρ = 2.65), are from Ilsted *et al.* [4] and many others.

Organic matter = LOI was between 0.45 and 0.50.

ρ(soil) = (LOI * 1.30 + (1 − LOI) * 2.65 = 1.96.

2.2.3. Respiration

(OCD—oxygen consumption, PCO_2—production of CO_2, RQ—respiratory quotient).

We used the volumetric method with the Scholander type respirometer to get both the oxygen consumption and carbon dioxide production. The capacity of its cylindrical flasks is 50 ml (inside radius r = 2 cm, V = 4 cm^2) and we refer data as $\mu l \cdot g^{-1}$(dry mass)$\cdot h^{-1}$. Each experimental series consisted of two parts. In the first, without hydroxide in the flasks, changes in the volumes were recorded mostly for 5 hours, this gave the sum of CO_2 changes and O_2 consumption, in the second part after placing filter paper soaked with NaOH solution into the respirometric chamber, we recorded the oxygen consumption (OCD) for another 5 hours interval, readings each 30 minutes. This involves assumption of slow rates of metabolic changes. The bath temperature was 17.5°C. Sensitivity of the volumetric respirometer is effectively not much worse (<0.1 μl gas·g^{-1}·$hour^{-1}$) than of the IR meters, though rates of providing data are not comparable.

While data of the second part were the OCD, PCO_2 was calculated as first reading (without hydroxide) + second reading (with hydroxide), respecting the signs in readings and calculations. We then plotted the results against time of readings, the slope coefficient of the regression (β_1) was the rate (O_2 consumption-OCD or PCO_2); the respective regression coefficients were all $r > 0.95$, they were later corrected for pressure and temperature, multiplied by the calibration factor and related per gram dry soil and hour. Weights of respiration vials with soil were taken regularly; tare was subtracted to get rates of drying.

Later in Results and discussion we will consider to what extent the data measured might be biased by diffusion or other transport problems within the soil items, the measured data may then be classified as apparent data (AOCD, $APCO_2$ and ARQ).

2.2.4. Soil (Vs), Air (Va), Pore Volume (Vpo)

Wet soil was gently pressed before start, there were no fissures between soil and wall of respirometric flask, they appeared however later during drying. We measured them by a series of blades (used in fine mechanics to measure narrow crevices, range 0.05 till 2 mm) on four sides of the vial, their mean was subtracted from r = 2 cm to get the radius of the base. The height (he) of the cylinders was estimated using a plate put on the top of the open vial and fixed to center by 4 screws projecting down on the outside of the flask, there were 5 evenly distributed holes on the plate, through which the distance from the top of the plate to soil top was measured by a rod, read by caliper and subtracted from the distance bottom of flask-top of plate. Mean of the 5 readings was height of the soil (he). Final calculation as cylinder volume:

$$\text{Soil volume } (\text{Vs}): \text{Vs} = \pi * r^2 * \text{he} \quad (1)$$

$$\text{Pore volume}(\text{porosity})\,\text{Vpo} = \text{Vs} - ((\text{Vs} - \text{Ws})/\rho) = \text{Vs} - ((\text{Vs} - H_2O)/\rho) \quad (2)$$

Ws is mass of wet soil.

$$\text{Volume of air Va} = \text{Vpo} - H_2O \quad (3)$$

2.3. Experimental

Changes of GWC in experiment are included as background for other parameters of the soil studied. We refer four principal experiments, which were preceded by several preliminary checks or pilot experiments, all done on one and the same kind of soil collected from one spot. All were aimed towards analysis of changes in soil respiration (OCD, PCO_2 and RQ) in drying of the soil. The main difference between the experiments was in the amount of soil in flasks and therefore were coded as LS (less soil) and MS (more soil) respectively. Outline of the experiments is in **Table 1**. Drying proceeded in MS1 and MS2 in two rates, up to about half of the experiments the rate (a) was somewhat faster, then later (rate b). Drying was terminated when oxygen uptake and CO_2 production dropped below the sensitivity of the instrument, except LS1 terminated earlier. There were 12 (LS1, MS1) or later 6 (LS2, MS2) parallel respirometers in each series. The basic data differ by an order of magnitude and change during the experiment; therefore we compared the coefficients of variance-cv's. It turned out that the cv's of the means based on six replicates are smaller than those of 12 replicates (**Table 2**). As standard deviations (second roots of variance) are involved in calculations of the t-test, we consider the data as comparable (**Table 2**). Weights of vials for GWC estimates were taken not necessarily on the day of respiration measurement. Therefore GWC data for comparison with respiration were interpolated from the respective regression of GWC to time (all $r > 0.99$).

After each 10 days WHC was estimated in 10 of the vials in the first series (MS1, LS1). In experiment MS2 and LS2 we estimated WHC at the beginning and at termination of the experiment.

Additionally we made a 14-days drying experiment (MP—mud pies) starting at GWC = 1.0 and finished at

Table 1. Outline of the experiments.

	LS1	LS2	MS1	MS2
dry mass [g] mean; se	3.8; 0.0294	3.26; 0.025	15.22; 0.0751	14.61; 0.147
days of experiment	24	13	52	42
starts of experiment	Aug 25, 14	Feb 23, 15	Nov 11, 14	Feb 23, 15
1st linear section [d]	24	13	32	29
°C air	17.2 - 19.0	18.8 - 19.7	17.0 - 18.8	18.4 - 19.7
β_1 GWC rate a [m]	−1.845	−4.43	−1.632	−1.257
β_1 GWC rate b [m]	x	x	−0.721	−0.412
mud-pies formation [d]	8	6	16	20
humidity air %	50 - 65	40	45 - 62	40 - 45
GWC start	2.4	1.54	2.34	1.58
GWC end	0.644	0.12	0.15	0.16
GWC estimates N	7	5	15	15
GWC replications	12	6	12	6

GWC rates are β_1-coefficients in linear regressions per month, a stands for the first, b for the second linear section. In the LS was one section. LS1 was finished prematurely.

Table 2. Cv's of means of cv's in MS1 and MS2 experiments for OCD, PCO_2 and GWC to compare variation in data with 12 or 6 replicates.

	OCD	PCO_2	GWC	replicates	determinations
MS1	0.2919	0.3581	0.1275	12	14
MS2	0.1373	0.1401	0.0511	6	13

GWC = 0.13 at the respiration stop. There were two modes of soil drying, each in six respirometers: a) soil (350 g dry weight) dried freely distributed on a plate, aliquots were taken for each respirometric measurements (per 6 * 5 g dry weight), returned to the plate and mixed; b) In the second mode wet soil was pressed by hand into the respirometric vials (per 5 g to each vial), it was left to dry freely between measurements each second day.

We also checked respiration in strict anoxia (SA) comparing respiration of a fresh collected soil with respiration after flushing the respirometers by pure dinitrogen (Air products.com, purity 5.2).

2.4. Statistics

Was rather basic, it involved averages = means (m), standard deviations (s), coefficients of variation (cv) and standard errors of the mean (se) calculated for all series of data obtained, referred here or not. Two sided t-test was applied for checks of differences of starting and closing data of series and for checks of differences in neighboring data within series. In final evaluation principal (**Table 2**) rates were related to month: (30days per month).

This came out from the discrepancy in evaluation of rates in changes in GWC, Vs, Vpo related to day which was our working time unit. We were unable to reject the 0-hypothesis (slope-rate of change of GWC is not different from zero—(from horizontal plane line), whereas paired T-test showed significant differences not only between start and end of experiment ($p < 0.001$), but also between most of neighboring data in the series (**Table 3**). In general the choice of reference units data is frequently intuitional, comparison of two or more statistical treatments might be a more general tool.

3. Results and Discussion

Results are largely based on the four drying experiments specified in Methods, Experimental, **Table 1** and in **Figure 1**. Additionally we performed two checks to get respiration data of the hard mud pies and the small

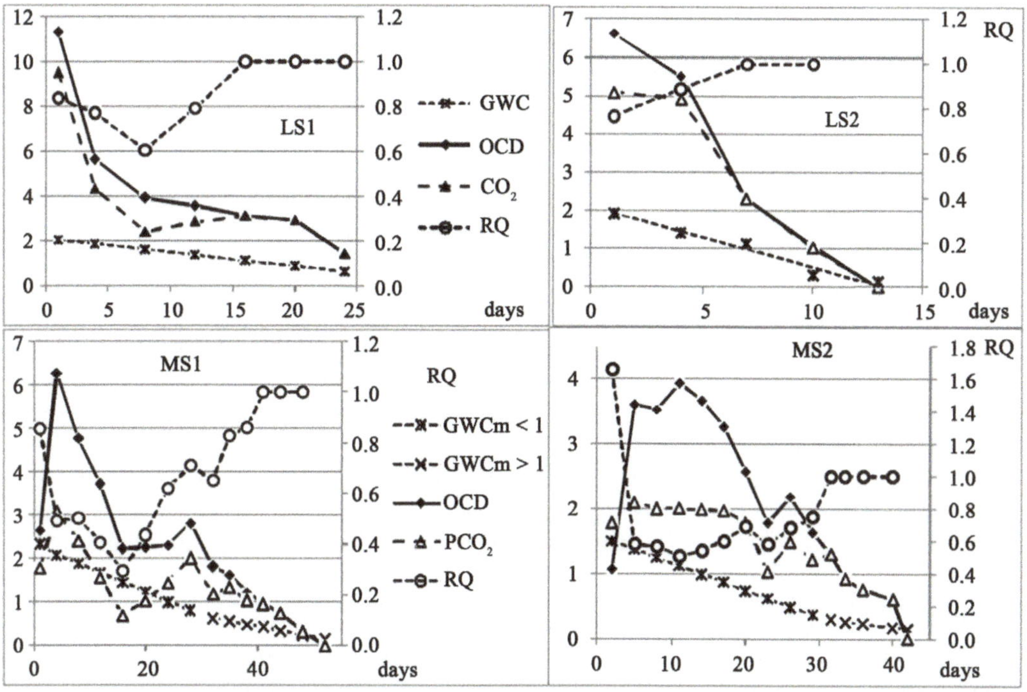

Figure 1. Changes of the gravimetric water content-GWC (as ratios), OCD (oxygen consumption), PCO_2 (production of carbon dioxide)—both as $\mu l \cdot g\ [dm]^{-1} \cdot h^{-1}$, and RQ (alternative scales) in the four experiments. Rates of changes in GWC are in **Table 1** and **Table 3**, rates of OCD and PCO_2 changes as related to GWC are in **Table 5**, overall budget of the experiments in **Table 6**. The LS1 experiment was finished before respiration dropped to zero. All standard errors are within the point icons. Specification of symbols for LS2 and MS2 experiments are in parts for LS1 and MS1 respectively.

Table 3. Rates (slopes) GWC, Vs, OCD per week and month for MS2 and LS2 compared to 2-sided t-test for start-end and neighboring days.

MS2	β_1 (slope)/day	N = 14 all p	slopes/month	p	T-test of pairs start-end	T-test of neighboring days			
						$p > 0.05$	$p < 0.05$	$p < 0.01$	$p < 0.001$
GWC	−0.0361	>0.05	−1.083	<0.005	$p < 0.001$	0		0	14
Vs	−0.4255	>0.05	−12.76	<0.001	$p < 0.001$	6	5	1	2
OCD	−0.0733	>0.05	−2.319	<0.001	$p < 0.001$	3		1	10
LS2		N = 5 for GWC and Vs; N = 4 for OCD							
GWC	−0.1138	>0.05	−3.41	<0.005	$p < 0.001$	0	0	0	5
Vs	−0.3963	>0.05	−11.89	<0.001	$p < 0.001$	0	0	2	3
OCD	−0.5904	>0.05	−17.71	<0.001	$p < 0.001$	0	1	0	3

Significance of slopes related to day and month, compared to 2-sided T-test for data pairs start-end of experiment and neighboring pairs during the experiment.

sandy like aggregates and demonstrate CO_2 production in strict anoxia. Also we assured us that the hard dry mud pies are not just gypsum or carbonate.

3.1. Humidity

Relationships of humidity (GWC) are therefore the first to be examined more in detail. Its changes related to time in terms of days[d] proved to be controversial; this resulted in use of month [m] as time reference unit in some selected relationships (**Table 3**).

GWC decreased during the experiments linearly with time for approximately 2/3 of the course, towards the end in the MS experiments, there was a significant drop in rate (**Table 1**). The course was a single linear function with respect to time in the shorter LS experiments. A slowing of the function is indicated in LS2, there are, however, too few data to get a significant picture (**Figure 1**).

3.2. Soil Volume

So far cracks in dry soils were often referred from the field in the literature, but there was/is much less interest in experiments on changes of soil volume in lab. There are more recent works on the impact of repeated drying and rewetting cycles on characteristic of soils resulting so far in non-complete understanding of these processes. We registered shrinking of soil during LS1 and MS1 experiments and estimated it with assumptions of linearity; nevertheless these estimates for MS1 and LS1 are in good agreement with the data from the second series of experiments. All four experiments have shown significant shrinking and changes of pores (porosity). The LS2 experiment is less successful due to high rate of drying and GWC decrease, therefore it was short and only few data were obtained (**Table 2** and **Table 3**).

Soil humidity (GWC) appears to be the governing parameter in shrinking of soil and respiration of soil biota. Shrinking [soil volume Vs, cm^3] and volume of calculated pores, both Vpo and Va, are linearly related with GWC, only Va in LS2 is not significant (**Table 4**). Regressions of Vs to time [d] were significant.

3.3. Changes of Soil Consistency

Formation of "mud-pies", hard stone—like concretions of dried soil, occurred in all four experiments. In MS1 and MS2 it begun approximately in 1/4 of the drying time, resulted from shrinking of soil, fissures appeared between soil and the walls of respirometric flasks, thus more of the soil surface was opened to the gas phase in

Table 4. Soil volume (Vs), porosity (Vpo) and air filled porosity (Va), their % share in Vs, rates of change (β_1, r), probability of no change (p).

	LS1	LS2	MS1	MS2
Vs; start	9.79 (d8)	8.042	37.7 (d16)	32.735
N; se	9;	9; 0.33	13;	24; 0.509
Vs end	8.17 (d20)	3.035	25.4	15.987
se	x	0.222	x	0.245
Vs-GWC; β_1	1.57	3.74	7.45	11.73
−r	0.944	0.977	0.941	0.990
$p<$	0.025	0.001	0.001	0.001
%Vpo in Vs start	84.2	83.2	87.5	86.38
%Vpo in Vs end	62.3	56.5	50.1	58.32
Vpo-GWC; β_1	2.1775	3.15	10.696	13.252
−r	0.955	0.9947	0.995	0.8767
$p<$	0.001	0.001	0.001	0.001
%Va in Vs start	15.18	16.13	10.3	15.13
%Va in Vs end	36.20	41.79	45.01	68.75
Va-GWC; β_1	−1.13	−0.139	−3.572	−1.468
−r	0.983	0.347	0.964	0.877
p	<0.001	>0.05	<0.001	<0.001

Va did not change significantly in drying, ratio to Vs changed just due to decrease of Vs.

the flasks. The rate of drying was also related to the amount of soil in the flasks. In the MS1 and MS2 series there is a higher soil layer, longer diffusion path and supposed humidity zonation. In the LS series drying and course of the experiments were faster (**Table 1**); the "mud pies" eventually formed just a narrow plate of dried soil.

All these processes result in increasing access of air into the soil preparation, in a much better ventilation of the dried soils than is in the wet soils and contributed to formation of the "mud pies" during the drying experiments.

It's a common knowledge, that wet soils are soft, pastry; easily adopt form of container under only very slight pressure. When dried they become hard, stone-like, if slightly pressed together before drying; they resist pressures approximately like wall nuts, larger "mud pies" are more resistant than small ones. Sieved soil loosely distributed on a plate acquires in drying a sandy character; its total surface is much greater than that of the "mud pie".

To check the character of the hardiness of the mud pies we chopped off 5.0 g material from the top of the 25 g "mud pie" and took an aliquot of the same weight from the sandy like dried soil. Both were suspended in 50 ml of pure water and extracted for 2 hours and 30 minutes. Both soils looked hydrophobic at the beginning, after approx. 45 minutes they turned more polar. Some air bubbles were attached to dark, presumably light organic soil particles. The extracts were separated from the rests of soils by filtration through Macherey-Nagel filter paper MN 640 m. Both filtrates were yellowish brown, the "mud pie" extract was distinctly darker, but not brown, both very slightly colloidal. Most of the residues remained in the Erlenmeyer's with the heavy sand fraction, much less on the filters. We measured pH, conductivity and redox potential in both extracts: pH = 4.48 and 4.53, conductivity = 54 and 116 $\mu S \cdot cm^{-1}$ in "sand" and "mud pie" respectively, redox was E = 320 mV in both. Addition of 1 ml 0.2 molar solution of $BaCl_2$ to each of the extracts did not result in formation of any precipitate. We conclude, that the top of the "mud pie" probably contained more organic matter than "the sandy" fraction (more intense humic coloration of the extract), there was certainly no carbonate nor gypsum formed in the drying soils, could not been responsible for formation of the "mud pies". The two major components of our soil are the postglacial sand, which however remained unchanged in all the procedures as a separate and non-reactive fraction. The second major component is SOM.

We want suggest replacement of H-bonds (water bridges) in wet soil by some other weak interactions (non-covalent bonds) during drying, possibly by hydrophobic weak bonds. A slight pressure and greater mass of wet soil brings soil aggregates closer together and enables formation of other particle interactions when O-(water-) bridges are abolished in drying.

For considerations of bonding within soil aggregates a change in scale of the aggregates might be useful. Remind that there are 6E23 single molecules per each mole (Avogadro number), thus in 1 μg of SOM 6E14 molecules of organic compounds of mean molecular mass 1000 may be accommodated. Soil science does not seem to be generally ready for thinking in such scales (unfortunate Table 2.1 in Voroney [22], containing organic molecules [10^{-9} m] larger than viruses, $\Psi > -1500$ kPa not only in Chowdhury [13] etc.).

3.4. Respiration

3.4.1. Relationships to GWC

Relationships of gas exchange (OCD, PCO_2) with time and GWC are more complex, the respective regression coefficients are lower compared to regressions of GWC and time; the overall tendency of respiration (OCD and PCO_2) is a decrease with progressing drying as water is a primary essential-even substrate-of life (**Table 5**, **Figure 1**). Data on respiration, both OCD and PCO_2, follow common trends: high values immediately at the begin-

Table 5. Regression coefficients (β_1, r) for respiration to GWC.

r/GWC	LS1	LS2	MS1 all	MS1 after d2	MS2 all	MS2 after d3
OCD β_1	−5.104	−3.612	−1.970	−2.560	−2.000	−2.950
r	0.833	0.956	0.837	0.936	0.712	0.949
PCO_2 β_1	−3.704	−0.877	−0.816	−0.978	−1.149	−1.351
r	0.735	0.953	0.732	0.765	0.842	0.881

For MS experiments coefficients are for all data and also for dates after d2 and d3 due to the delayed Birch effect.

ning of experiments in both LS, or after three or five days in the MS series. There is a more or less stable respiration in the central part of the course in MS1 and MS2 (**Table 2** and **Table 4**). Basic statistics of respiration data show small variance, se's in all presented and vast majority of data not shown are within the point icons. The data in figures are of similar magnitude with those published by selected authors for forest soils considering problems in recalculation of area related data to mass related. Butler *et al.* [8] consider "rainfall as main driver of soil CO_2 efflux and its components".

3.4.2. CO_2 Storage

Distribution coefficient between gas and water for CO_2 is very close to one at the temperature of the experiments, storage capacity for CO_2 in soil water is therefore equal to GWC * dry mass of soil. It is the volume of water in respirometer (microcosm), about 20 ml in the first days of the MS experiments or 5 ml in the both LS; apparent production in two or three days.

Another estimate of CO_2 stored may result from the difference between AOCD and $APCO_2$ during the flush of respiration when soil is still wet, in MS1 days 3 to12. Assuming real RQ = 1.0 it is 2.81 ml CO_2 stored in 28.8 ml H_2O, the storage capacity is thus used in less than 10%.

Similar data for the MS2 experiment is for days 5 to 12 and show 2.79 ml CO_2 stored in 12.7 ml water, storage capacity used in approximately 22%. Assuming real RQ = 0.7 result in 1.97 ml CO_2 stored in MS1 and 1.95 ml CO_2 in MS2 respectively. Details of calculations are not shown.

The storage capacity for CO_2 was falling in experiments with progressing drying. Angert *et al.* [23] compared measured data of RQ (ARQ) with soil air analysis showing relatively low pCO_2. Their assumption on equilibrium between headspace (soil air) and wet acidic soil might not be fulfilled; there might be physical separation between the sections of the soil gas and the watered section, where CO_2 is primarily produced. Also a careful acidification and mixing of a tightly closed dilute solution of bicarbonate without gas phase need not result in production of gas bubbles. This may apply to a wet soil, where CO_2 is produced by microorganisms. ARQ data of Angert *et al.* [23] agree with our RQ in the respiration budget, there is also agreement on denial of carbonate chemistry as cause in the background of RQ < 1, we propose oxidation of methane and particularly ethanol produced in fermentation as hypothetic substrates in respiration with RQ < 1. We also seem to agree in accent on processes in studies. Sierra and Renault [14] conclude, that production is negligible compared to dissolved CO_2, this does not apply to acidic soils (pH < 4), where bicarbonate is practically missing. Therefore we did not consider bicarbonate storage, as estimates in our soil were always pH < 4.0. In acid soils however even a small volume stored may result in serious shifts in RQ.

3.4.3. Anoxia in Soil

To show gas exchange in strictly anoxic soil we placed 21 g fresh collected sieved wet soil (7.3 g dm) into each of four respirometers. Two of the respirometers were flushed with pure dinitrogen 50 $ml \cdot min^{-1}$, for 10 minutes; we registered O_2 consumption and CO_2 production [with and without NaOH trap inside]. The other pair of respirometers was a control to see regular soil respiration. There were 3.83 and 3.89 $\mu l \cdot g^{-1}(dm) \cdot h^{-1}$ respectively gas produced in the N_2 flushed respirometers without the hydroxide trap. There was however no gas production, nor consumption when hydroxide trap was placed in both respirometers; this also shows that the gas produced without the trap is absorbed by hydroxide, it is most probably CO_2 and not CH_4. RQ cannot be determined as shown in Introduction. The regular respiratory control in the second pair of respirometers provided 8.92 and 8.56 $\mu l\ O_2 \cdot g^{-1}(dm) \cdot h^{-1}$ respectively and 4.51 $\mu l\ CO_2 \cdot g^{-1}(dm) \cdot h^{-1}$ in both respirometers, this is consistent with data in **Figure 1**.

3.4.4. Respiration and Size of "Mud Pies"

Another pilot experiment aimed to differences of respiration in highly dried soils as "mud pies" or small sandy like aggregates. In the "mud pies" both OCD and PCO_2 are some 20% lower than in the sandy sieved small aggregates. Fast drops of simultaneous decrease of OCD and PCO_2 by approximately 0.6 - 0.8 $\mu l \cdot g^{-1} \cdot h^{-1}$ occurred at 0.63 GWC in "mud pies", at 0.43 in "sand". Drying in "mud pies" and associated decrease of respiration is slower in "mud pies". There is no contradiction to **Figure 1**.

3.4.5. "Birch Effect"

High respiration data occurring after changed conditions (sieving, watering…) are referred as the Birch effect

[24]. In our experiments it occurred in LS1 on d1, in MS1 on d 5 - 14 and also at lower GWC = 1.41 starting value in LS2 and MS2 thus at a somewhat lower moisture throughout the drying experiment (**Figure 1**). The delayed respiration flush in both MS experiments resulted in distorsions of the relationships of respiration to GWC, this was overcome by skipping the d 1 or d 2 data from the relationships in MS2 and MS1 (**Table 6**). The delay of flush in both MS experiments was probably caused by pressing the greater amount of soil at start resulting in formation of solid "mud pies", while the small amounts of soil in both LS resulted in a thin layer of soil with minor limitation in gas traffic in the first days. The slightly smaller humidity in MS2 compared to MS1 was of minor importance in this range. We however feel that the impression of extraordinarily of high starting RQ's in MS2 is partly misleading, it comes rather from general characteristics of ratios; OCD and PCO_2 data are of similar magnitude (1.78 and 1.07 respectively), not strikingly different indeed.

High flushes of respiration following change of conditions are debated for last 50 years [24]. So far they were attributed to increased availability of respiration substrates from organisms which died in preceding dries, even if the drop in water content was small [25] or to repeated drying and rewetting (e.g. [26]) or simply to watering of soil (e.g. [13]). Even the detailed study by Unger [27] which includes interaction of soil and the vegetation using highly advanced methodology does not seem to be the last word in the debate. Flush took place in all four experiments in fresh and sieved samples, but not watered, of course sampling was at the very beginning (**Figure 1**, **Table 5**, first section of the course). Respiration decreased dramatically in the next readings. Therefore we want to propose a hypothesis not relying primarily on the experimental conditions, but rather on the organisms themselves: Before the collection of soil organisms were adapted to low respiration rates in their natural settings, supplemented by fermentation. Sieving (or in other papers published-water added) allows more air to enter the soil; previously accumulated CO_2 is released to air or absorbed in water. The overall tendency during the experiments is decrease of AOCD and $APCO_2$ with decreasing GWC, the data of the first part of experiments are however influenced by CO_2 produced and stored in soil water and Va available, at this stage it was not included in the changes of gas contents measured. Some loss of humidity (GWC) in sieving is highly probable. Soil biota may require time to adapt to available dioxygen, presumably by activation/reconstruction of respiratory mechanisms (e.g. enzyme complexes of oxidative phosphorylation pathway) which appear to be of small use in more enzyme copies in wet soils in natural settings with limited oxygen availability.

3.4.6. Summing up OCD and PCO_2 in the Process

Summed OCD and PCO_2 for the whole drying experiment showed in all four experiments RQ around 0.7 (**Table 6**). ΣRQ is somewhat higher in the faster drying experiments (LS), suggesting less hypoxia. This may support the hypothesis presented in introduction on a mixed status of organisms in wet soils, respiration contemporarily with fermentation and use of the fermentation products (e.g. ethanol) as substrates in respiration when O_2 becomes available.

Studies of some process which take into account all data, characterizing the variable soil condition during a process may provide more information, than short measurements or flashes of soil metabolism in ± constant conditions. Type of instrumentation may be of secondary importance only–modern methods of O_2 estimation are used only exceptionally [27].

Nevertheless if there is O_2 consumption, anoxia is not in the total mass of soil, possibly in a limited area in center of the soil bloc.

3.4.7. Respiration Ends at Limiting GWC

There is a small, but significant peak of respiration in at the beginning of the decreasing phase at d 26 in MS1; in MS2 at d 26. A similar peak was registered by Thomson *et al.* [10] on the 5-th day of their 7-days experiment,

Table 6. Sums of OCD, COP, and ΣRQ in four experiments [ml].

	LS1	LS2	MS1	MS2
ΣOCD	7.67	15.14	42.53	33.13
ΣCOP	6.31	13.14	23.97	22.41
ΣRQ	0.82	0.87	0.56	0.68

ΣRQ is the ratio of the two sums. Duration of experiments in **Table 3**.

which did not continue to a respiration stop. We propose to interpret as increased respiration due to increasing access of air to the drying soil, in our experiments shortly overrun but proceeding desiccation of the soil and confirmed by fast drops afterwards.

All four series (including the prematurely finished LS1) end up with RQ = 1 when GWC is approximately = 1 in LS experiments, but GWC ≈ 0.6 in MS (**Figure 1**).

CO_2 retained in the first section of experiments in water is completely aerated from the soil towards the end when soil is very dry; when respiration seizes at GWC ca 0.15, there is no measurable PCO_2, as is no OCD.

All drying experiments were terminated when GWC dropped to 0.15 and respiration stopped. The experiment LS1 was in fact finished prematurely, nevertheless its final RQ = 1. Also in a preliminary experiment not reported here (summer 2014) respiration (both OCD and PCO_2) seized at GWC = 0.15 and RQ = 1. The respiration limit at GWC = 0.15 need not be a universal value for many soils, Chowdhury *et al.* [13] reported respiration in Australian soils at recalculated GWC = 0.025.

4. Conclusions

There is a distinct peak of respiration rate at the beginning of the experiment (Birch effect); it occurred with some delay in the larger soil blocks. The effect was not associated with increased humidity, rather with increased availability of oxygen within the soil; therefore we suggest interpreting as reaction of the enzymatic systems of the soil organisms. The respiration rate decreased linearly with decreasing humidity of the soil; RQ dropped below 0.5 in both larger soil blocks for some time; reason is evidently in storage of CO_2 produced in soil water. Rate of respiration registered is also related to consistency of soil; hard "mud pies" respire some 20% less than the small aggregates of sandy character.

Respiration ceased when GWC dropped down to 0.15. RQ was increasing in the second part of the experiments; a few days before the end it reached 1.00 and remained so till the last moment. We think it comes from CO_2 stored in the earlier parts of the drying process which is aerated when amount of water within the soil decreased and air in soil got higher. ΣRQ for the whole budget of the experiments is about 0.7; it is somewhat higher for the smaller soil items (LS) suggesting less hypoxia and may be consistent with partial use of fermentation products like ethanol (RQ = 0.57).

References

[1] Davidson, E.A., Savage, K., Verchot, L.V. and Navarro, R. (2002) Minimizing Artifacts and Biases in Chamber-Based Measurements of Soil Respiration. *Agricultural and Forest Meteorology*, **113**, 21-37. http://dx.doi.org/10.1016/S0168-1923(02)00100-4

[2] Luo, Y. and Zhou, X. (2010) Soil Respiration and the Environment. Academic Press, Burlington.

[3] Luan, J., Liu, S., Zhu, X., Wang, J. and Liu, K. (2012). Roles of Biotic and Abiotic Variables in Determining Spatial Variation of Soil Respiration in Secondary Oak and Planted Pine Forests. *Soil Biology and Biochemistry*, **44**, 143-150. http://dx.doi.org/10.1016/j.soilbio.2011.08.012

[4] Ilstedt, U., Nordgren, A. and Malmer, A. (2000) Optimum Soil Water for Soil Respiration before and after Amendment with Glucose in Humid Tropical Acrisols and a Boreal Mor Layer. *Soil Biology & Biochemistry*, **32**, 1591-1599.

[5] Campbell, N.A. (1993) Biology. 3rd Edition, Benjamin/Cummings, Francis Group, Redwood City.

[6] Alberts, B. and Johnson, A. (2002) The Molecular Biology of the Cell. Garland, Taylor and Francis Group, New York.

[7] McGill, W.B. (2007) The Physiology and Biochemistry of Soil Organisms. In: Paul, E.A., Ed., *Soil Microbiology Ecology, and Biochemistry*, 3rd Edition, Academic Press, Burlington. http://dx.doi.org/10.1016/b978-0-08-047514-1.50013-5

[8] Butler, A., Meir, P., Saiz, G., Maracahipes, L., Marimon, B.S. and Grace, J. (2012) Annual Variation in Soil Respiration and Its Component Parts in Two Structurally Contrasting Woody Savannas in Central Brazil. *Plant Soil*, **352**, 129-142. http://dx.doi.org/10.1016/b978-0-08-047514-1.50013-5

[9] van Suchtelen, H. (1910) Uber die Messung der Lebenstätigkeit der aerobiotischen Bakterienim Boden durch die Kohlensaureproduktion. *Centrablattfür Bakteriologie*, **28**, 46-89. Cited from: Lundegardh, H. (1926) Carbon Dioxide Evolution of Soil and Crop Growth. *Soil Science*, **23**, 417-453.

[10] Thomson, B.C., Ostle, N.J., McNamara, N.P., Whiteley, A.S. and Griffith, R.I. (2010) Effects of Sieving, Drying and Rewetting upon Soil Bacterial Community Structure and Respiration Rates. *Journal of Microbiological Methods*, **83**, 69-73.

[11] Gaumont-Guay, D.T., Black, T.A., Tim, J., Griffis, T.J., Alan, G., Barr, A.G., Jassal, R.S. and Nesic, Z. (2006) Interpreting the Dependence of Soil Respiration on Soil Temperature and Water Content in a Boreal Aspen Stand. *Agricultural and Forest Meteorology*, **140**, 220-235. http://dx.doi.org/10.1016/j.agrformet.2006.08.003

[12] Walker, T.S., Harsh, P.B., Grotewold, E. and Vivanco, J.M. (2003) Update on Root Exudation and Rhizosphere Biology. *Plant Physiology*, **132**, 44-51. http://dx.doi.org/10.1104/pp.102.019661

[13] Chowdhury, N., Burns, R.G. and Marschner, P. (2011) Recovery of Soil Respiration after Drying. *Plant Soil*, **348**, 269-279. http://dx.doi.org/10.1007/s11104-011-0871-2

[14] Sierra, J. and Renault, P. (1995) Oxygen Consumption by Soil Microorganisms as Affected by Oxygen and Carbon Dioxide Levels. *Applied Soil Ecology*, **2**, 175-184. http://dx.doi.org/10.1016/0929-1393(95)00051-L

[15] Jenkins, M.E. and Adams, M.A. (2011) Respiratory Quotients and Q10 of Soil Respiration in Sub-Alpine Australia Reflect Influences of Vegetation Types. *Soil Biology & Biochemistry*, **43**, 1266-1274. http://dx.doi.org/10.1016/j.soilbio.2011.02.017

[16] Araujo, A.S.F., Cesarz, S., Leite, L.F.C., Borges, C.D., Tsai, S.M. and Eisenhauer, N. (2013) Soil Microbial Properties and Temporal Stability in Degraded and Restored Lands of Northeast Brazil. *Soil Biology & Biochemistry*, **55**, 175-181. http://dx.doi.org/10.1016/j.soilbio.2013.07.013

[17] Lamy, E., Tran, T.C., Mottelet, S., Pauss, A. and Schoefs, O. (2013) Relationships of Respiratory Quotient to Microbial Biomass and Hydrocarbon Contaminant Degradation during Soil Bioremediation. *International Biodeterioration & Biodegradation*, **83**, 85-91. http://dx.doi.org/10.1016/j.ibiod.2013.04.015

[18] Blazka, P. and Fischer, Z. (2014) Moisture, Water Holding, Drying and Wetting in Forest Soils. *Open Journal of Soil Science*, **4**, 174-184. http://dx.doi.org/10.4236/ojss.2014.45021

[19] Carter, M.R. and Gregorich, E.G., Eds. (2008) Soil Sampling and Methods of Analysis. CRC Press, Taylor & Francis Group, Boca Raton.

[20] Grace, C., Hart, M. and Brookes P.C. (2006) Laboratory Manual of the Soil Microbial Biomass Group. Rothamsted Research.

[21] Priha, O. and Smolander, A. (2003) Short-Term Uptake of $^{15}NH_4$ into Soil Microbes and Seedlings of Pine, Spruce and Birch in Potted Soils. *Biology and Fertility of Soils*, **37**, 324-327.

[22] Voroney, R.P. (2007) The Soil Habitat. In: Paul, E.A., Ed., *Soil Microbiology Ecology and Biochemistry*, 3rd Edition, Elsevier, Academic Press, London, 25-49. http://dx.doi.org/10.1016/b978-0-08-047514-1.50006-8

[23] Angert, A., Yakir, D., Rodeghiero, M., Preisler, Y., Davidson, E.A. and Weiner, T. (2014) Using O_2 to Study the Relationships between Soil CO_2 Efflux and Soil Respiration. *Biogeosciences Discussions*, **11**, 12039-12068. http://dx.doi.org/10.5194/bgd-11-12039-2014

[24] Birch, H.F. (1964) Mineralisation of Plant Nitrogen Following Alternate Wet and Dry Conditions. *Plant Soil*, **20**, 43-49. http://dx.doi.org/10.1007/BF01378096

[25] Fischer, T. (2009) Substantial Rewetting Phenomena on Soil Respiration Can Be Observed at Low Water Availability. *Soil Biology & Biochemistry*, **41**, 1577-1579. http://dx.doi.org/10.1016/j.soilbio.2009.04.009

[26] Shi, A. and Marschner, P. (2014) Drying and Rewetting Frequency Influences Cumulative Respiration and Its Distribution over Time in Two Soils with Contrasting Management. *Soil Biology & Biochemistry*, **72**, 172-179. http://dx.doi.org/10.1016/j.soilbio.2014.02.001

[27] Unger, S., Maguas, C., Pereira, J.S., David, T.S. and Werner, C. (2012) Interpreting Post-Drought Rewetting Effects on Soil and Ecosystem Carbon Dynamics in a Mediterranean Oak Savannah. *Agricultural and Forest Meteorology*, **154-155**, 9-18. http://dx.doi.org/10.1016/j.agrformet.2011.10.007

13

Microbe—Chloroacetanilide Herbicide Interaction across Soil Type

Sutapa Paul[1], Arunabha Chakravarty[2], Prasanta Kumar Patra[1*], Niladri Paul[1], Premasis Sukul[2], Debatosh Mukherjee[1]

[1]Department of Agricultural Chemistry and Soil Science, Bidhan Chandra Krishi Viswavidyalaya, Mohanpur, India

[2]Department of Agricultural Bio-Chemistry, Bidhan Chandra Krishi Viswavidyalaya, Mohanpur, India

Email: *drpatrapk@yahoo.co.in

Abstract

An investigation was carried out under laboratory conditions to study the persistence of butachlor applied at recommended dose (2 kg ai/ha) along with its impact on microbial activity as well as growth of colonial bacteria and fungi in alluvial (*Typic Haplaquent*), lateritic (*Typic Haplustalf*) and coastal (*Typic Haplaquept*) soils. Butachlor caused a significant increment in microbial activity following an initial diminution in between 10 to 22 days of incubation depending on the type of soil. The herbicide resulted in a significant shrinkage in bacterial community during later stages of incubation in lateritic and coastal soils in spite of a significant swelling on the 15th day in lateritic and alluvial soils. Fungal community significantly expanded at the initial stage in lateritic soil and during later stages in alluvial soil by the application of butachlor but shriveled during later stages in the lateritic soil, intermediate stage in coastal soil and initial stages in alluvial soil. Alluvial soil reared the highest population of colonial bacteria and exhibited highest microbial activity while coastal soil significantly pressed them down to the lowest. However, lateritic soil was the best niche for fungal community. Co-metabolism was the main phenomenon in butachlor metabolism particularly in coastal soil, though zymogenous microbes including bacteria and fungi also participated in both lateritic and alluvial soil at certain stages. The persistence of butachlor was the lowest in alluvial soil followed by lateritic and coastal soil, respectively. Among the soil types application of butachlor is safe in alluvial soil.

Keywords

Alluvial Soil, Lateritic Soil, Coastal Soil, Bacteria, Fungi, Butachlor, Persistence

*Corresponding author.

1. Introduction

Since man first began to cultivate crops, undesirable plants, called weeds, have been a problem causing reduction in agricultural production in several ways [1]. In India, as a whole, weeds cause an estimated 37% reduction of agricultural produce annually [2] and on global basis, a 10% reduction of crop yield [3]. Herbicides are used to kill or stunt weed infestation allowing crop plants to grow and gain a competitive advantage [4] and offer one of the most effective means to reduce labor costs on weed control. Herbicide usage in rice now accounts for about 30% of the total consumption of pesticides [5], and is expected to increase dramatically in India in future [6]. Butachlor (**Table 1**), available in granular as well as emulsifiable concentrate formulations, is a systemic pre-emergent herbicide controlling most of the annual grasses [7] [8]. Soil, the ultimate recipient of herbicides, comprises biotic and abiotic components. The abiotic components determine growth and activities of the biotic

Table 1. General characteristics of butachlor and test soils.

A. Properties of butachlor
Chemical structure

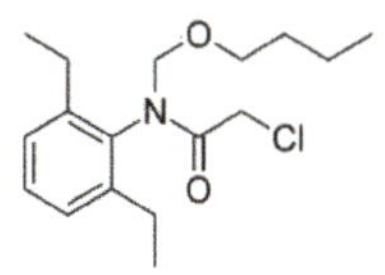

Chemical class	Chloroacetanilide
Chemical name	[N-(butoxymethyl)-2-chloro-N-(2,6-diethylphenyl)acetamide]
Molecular formula	$C_{17}H_{26}ClNO_2$
Molar mass	311.85 g·mol^{-1}
Solubility in water	20 mg l^{-1} (20˚C)
Vapour pressure	1.8 × 10^{-6} mm Hg (25˚C)
LD_{50}	1740 mg·kg^{-1} (oral, rat)

B. Soil properties	Laterite	Soil type coastal	Alluvial
Water holding capacity (%)	30.0	54.0	49.0
pH (1:2.5)	5.5	6.7	7.7
Electrical conductivity (dsm^{-1})	0.06	1.55	0.48
Cation exchange capacity [c mol (p$^+$) Kg^{-1}]	9.1	4.7	16.9
Total nitrogen (%)	0.03	0.11	0.07
Available nitrogen (%)	0.0015	0.031	0.0196
Available Phosphorus (ppm Olsen P)	4.3	60.0	17.1
Organic carbon (%)	0.49	0.64	0.58
C:N ratio	17.5:1	5.7:1	8.5:1
Mechanical analysis			
Sand (%)	72.2	17.6	22.2
Silt (%)	8.4	35.8	35.4
Clay (%)	19.4	46.6	42.4
Ca [c mol (p$^+$) kg^{-1}]	2.5	8.2	10.4
Mg [c mol (p$^+$) kg^{-1}]	0.6	2.3	1.0
Origin	Regional Research Farm, Jhargram, West Medinipur	No. 1 Farm, ICAR, Canning, 24 Pgs (S)	Univ. Research Farm, Mohanpur, Nadia
Soil taxonomy	*Typic Haplustalf*	*Typic Haplaquept*	*Typic Haplaquent*

components, which, in turn, regulates the physical, chemical and biological properties of soil. The living organisms exhibit different types of inter-relationships leading to establishment of a biological equilibrium in soil. Among different internal and external factors affecting the equilibrium, soil pH, and level of organic matter and electrical conductivity of soil have drawn maximum attention [9]. Generally neutral to slightly alkaline soil reaction support higher growth of bacteria [10], while acidic soils harbor more fungal propagules [11]. Level of organic matter in soil also determines the growth as well as activities of microorganisms [12]. Soil salinity on the contrary exerts diminutive impact on the preponderance of microbes in soil [12]. Soil micro flora exists normally in a state of starvation due to low energy substrate supply [13] which can be changed temporarily into their active growth and metabolism on availability of energy and nutrients from applied herbicides [14]. As a consequence, the herbicides are degraded in soils by biotic agencies. Though herbicides effectively control weeds, they also cause qualitative and quantitative alterations in the soil microbial populations and their enzyme activities [15]-[17], resulting in decline in crop productivity [18]. So herbicides are aptly considered as "two-edged sword" [19] and viewed as a serious threat to global environment [20]. They kill species of bacteria, fungi and protozoa combating pathogenic microorganisms and thus upset the balance between pathogens and beneficial organisms [21]. Biochemical transformation is the principal avenue of pesticide metabolism in addition to physical and chemical transformations [22] wherein micro-organisms are mainly involved [23] [24]. Among them, bacteria and fungi exhibit maximum capability of degrading pesticides [25]. Some of them utilize herbicide as energy and nutrient sources [26] while others degrade pesticide without exploiting them as energy and nutrient sources [22] by a process called co-metabolism [27]. Reports on negative effect of butachlor on the growth and activities of microbes are also available [28]. However, Information regarding fate and behavior of butachlor under varying soil conditions affecting growth and activities of micro-organism is scanty. An attempt was thus made to study the persistence of butachlor at the recommended dose in alluvial, lateritic and coastal soils along with its impact on microbial activity as well as growth of colonial bacteria and fungi.

2. Materials and Methods

2.1. Chemicals

Butachlor of analytical grade (97.7%) was obtained from Sigma-Aldrich (Germany). All solvents and chemicals were of A. R. grade.

2.2. Soil Collection and Analysis

The investigation was based on three simultaneous experiments in the laboratory of Department of Agricultural Chemistry and Soil Science, Bidhan Chandra Krishi Viswavidyalaya, Mohanpur, Nadia, West Bengal, India and the experiments were conducted with three soils from three different geographical origins possessing a wide variation in physicochemical properties (**Table 1**). Surface soil samples (0 - 10 cm), collected from the monoculture (*kharif* rice) cultivated fields of all three different locations, were kept in a green house at 25˚C ± 5˚C at 60% - 80% of maximum water holding capacity. Shortly before the study, soils were air-dried, ground and passed through a 80 mesh sieve, thoroughly homogenized and analyzed for pH (1:2.5 soil-water ratio) [29]; organic carbon by 1 N $K_2Cr_2O_7$ solution method [30]; total nitrogen by modified Kjeldahl method [29] and available P_2O_5 by extraction with alkaline $NaHCO_3$ (pH 8.5) method [31]. Mechanical analysis for determining the texture of the soil was done by Hydrometer method [32] [33]. Carbon-di-oxide evolved per 100 gm soil following the method of Pramer and Schimidt [34] by absorbing the evolved carbon-di-oxide in NaOH solution and back titrating with hydrochloric acid using Barium chloride and phenolphthalein as indicator.

2.3. Experiments

2.3.1. Experiment 1

A set comprising three earthen pots were filled with 2 kg for each of the three soil types left as such to be considered as control. Another set of similar pots were thoroughly blended with butachlor at the rate of 2 kg a.i./ha-separately for each soil type in a tumbling mixture (30 ± 2 rpm) for 1 h. The moisture content of the soils was maintained at 50% of the water holding capacity and checked gravimetrically every week throughout the experimental period. The pots were covered with black polyethylene sheet to avoid photo degradation of the herbicide and to minimize moisture loss. The pots were then arranged in a Completely Randomized Design with three replications of each of the two treatments viz. control and treated, and were incubated at 30˚C ± 1˚C for 60 days.

2.3.2. Experiment 2

In order to study the influence of butachlor on microbial activity across soil type another experiment was conducted in one litter conical flasks containing 100 g soil with the same treatments (with or without butachlor at 2.0 kg a.i./ha), replications (3), soil types (alluvial, lateritic and coastal) and environmental conditions (moisture 50% of water holding capacity and incubation temperature 30˚C ± 1˚C). Microbial activity was determined at periodic intervals in terms of mg of carbon-di-oxide evolved per 100 gm soil following the method of Pramer and Schimidt [36] by absorbing the evolved carbon-di-oxide in NaOH solution and back titrating with hydrochloric acid using Barium chloride and phenolopthalein as indicator.

2.3.3. Experiment 3

Third experiment was carried out in 50 ml conical flask containing 10 gm of soil with the same treatments (with or without butachlor at 2.0 kg a.i./ha), replications (3), soil types (alluvial, lateritic and coastal) and environmental conditions (moisture: 50% of water holding capacity, incubation temperature 30˚C ± 1˚C and incubation period 60 d) to determine the dissipation pattern of butachlor in each soil type in absence of microorganisms. For this one batch of soil sample was sterilized by autoclaving at 15 psi and 121˚C [35]. The experimental set-up for sterilized soils was performed aseptically.

2.4. Sample Collection and Analysis

Soil sample was drawn from the respective pots periodically (5th, 10th, 15th, 30th, 45th and 60th days after incubation) from the replicated pots of each treatment. The moist soil samples were analyzed to count the colony forming units (CFU) of total bacteria and fungi following serial dilution and pour plate technique [36] in asparagine-mannitol agar [37] and dextrose-rose bengal agar [38], respectively. Soils (sterile and non-sterile) were also analyzed for the presence of butachlor by extracting the soils following the procedure described by Debnath *et al.*, [39]. For residue analysis, soil samples (sterile and non-sterile) were also collected on "0" day (1 h) in addition to the periods stated above. Herbicide residues were analyzed by GLC [Hewlett Packard (USA), model 5890A] coupled with a Ni^{63}—electron capture detector, a glass column ($6^{/} \times 2$ mm) packed with 3% OV-101 on 80 - 100 mesh chromosorb-W (HP), and 3329A integrator. During analysis following gas chromatographic parameters were maintained: injection temperature, 250˚C; oven temperature, 200˚C; detector temperature, 250˚C; carrier gas, nitrogen having 65 mL/min flow rate. The residue data were processed to calculate the half-life ($T_{1/2}$) following the method of Hoskins [40].

2.5. Statistical Analysis

The results were evaluated by analysis of variance (ANOVA), and the statistical significance ($P = 0.05$) of difference between means within factors (herbicide, soil type and incubation time) was computed using Completely Randomized Design following the method of Gomez and Gomez [41].

3. Results and Discussion

3.1. Butachlor Dissipation

Sorption and degradation are two fundamental processes which govern predicting the environmental fate and behavior of organic contaminants in soil, which are affected by many factors like interaction with microorganisms, chemical and soil constituents [42]. A study on butachlor dissipation was carried out in soil in presence and absence of biological activity under laboratory condition. The dissipation and thus persistence of butachlor varied across soil types and prevailing conditions (sterilized and non-sterilized). After 60 days of incubation, 45.9% - 49.5% and 83.9% - 90.5% attenuation of butachlor residues were noticed in three soils under sterilized and non-sterilized condition, respectively (**Table 2**). When log residues in all cases were plotted against time, a linear relationship (r^2, 0.85 - 0.98) was observed suggesting a first order reaction in the dissipation behavior of the herbicide (**Figure 1**), which conforms earlier findings [43] [44]. The half-life ($T_{1/2}$) values were calculated by means of log 2/K, where K represents to the slope of the straight line regression. Under non-sterilized condition

Table 2. Dissipation pattern of butachlor in sterilized and non-sterilized soil.

Days after application	Sterilized		Non-sterilized		% loss due to biotic forces
	Residues* (ppm)	% loss due to abiotic forces	Residues* (ppm)	% total loss	
			Laterite soil		
0	0.961 ± 0.009	-	0.965 ± 0.015	-	0
5	0.839 ± 0.006	12.64	0.719 ± 0.019	25.49	12.85
10	0.748 ± 0.008	22.16	0.576 ± 0.016	40.31	18.15
15	0.639 ± 0.009	33.51	0.411 ± 0.011	57.41	23.9
30	0.621 ± 0.004	35.38	0.292 ± 0.01	69.74	34.36
45	0.547 ± 0.007	43.08	0.208 ± 0.008	78.44	35.36
60	0.509 ± 0.009	47.03	0.118 ± 0.018	87.77	40.74
Regression equation	Y = 1.91 − 0.004X		Y = 1.89 − 0.014X		
r^2	0.867		0.975		
RL_{50} (days)	75.25		21.50		
			Coastal soil		
0	0.953 ± 0.018	-	0.956 ± 0.021	-	0
5	0.84 ± 0.100	11.86	0.735 ± 0.100	23.12	11.26
10	0.746 ± 0.020	21.72	0.586 ± 0.010	38.7	16.98
15	0.648 ± 0.020	32	0.442 ± 0.020	53.76	21.76
30	0.584 ± 0.040	38.71	0.316 ± 0.016	66.94	28.23
45	0.563 ± 0.020	40.92	0.23 ± 0.030	75.94	35.02
60	0.516 ± 0.016	45.85	0.154 ± 0.020	83.89	38.04
Regression equation	Y = 1.89 − 0.003X		Y = 1.89 − 0.012X		
r^2	0.850		0.971		
RL_{50} (days)	100.33		25.08		
			Alluvial soil		
0	0.98 ± 0.080	-	0.982 ± 0.080	-	0
5	0.844 ± 0.020	13.88	0.698 ± 0.012	28.92	15.04
10	0.736 ± 0.020	24.9	0.546 ± 0.020	44.39	19.49
15	0.626 ± 0.020	31.12	0.354 ± 0.020	63.95	32.83
30	0.591 ± 0.009	39.69	0.252 ± 0.022	74.33	34.64
45	0.535 ± 0.015	45.41	0.172 ± 0.012	82.48	37.07
60	0.493 ± 0.007	49.49	0.093 ± 0.007	90.52	41.03
Regression equation	Y = 1.91 − 0.004X		Y = 1.87 − 0.015X		
r^2	0.851		0.967		
RL_{50} (days)	75.25		20.06		

*Mean value of three replications.

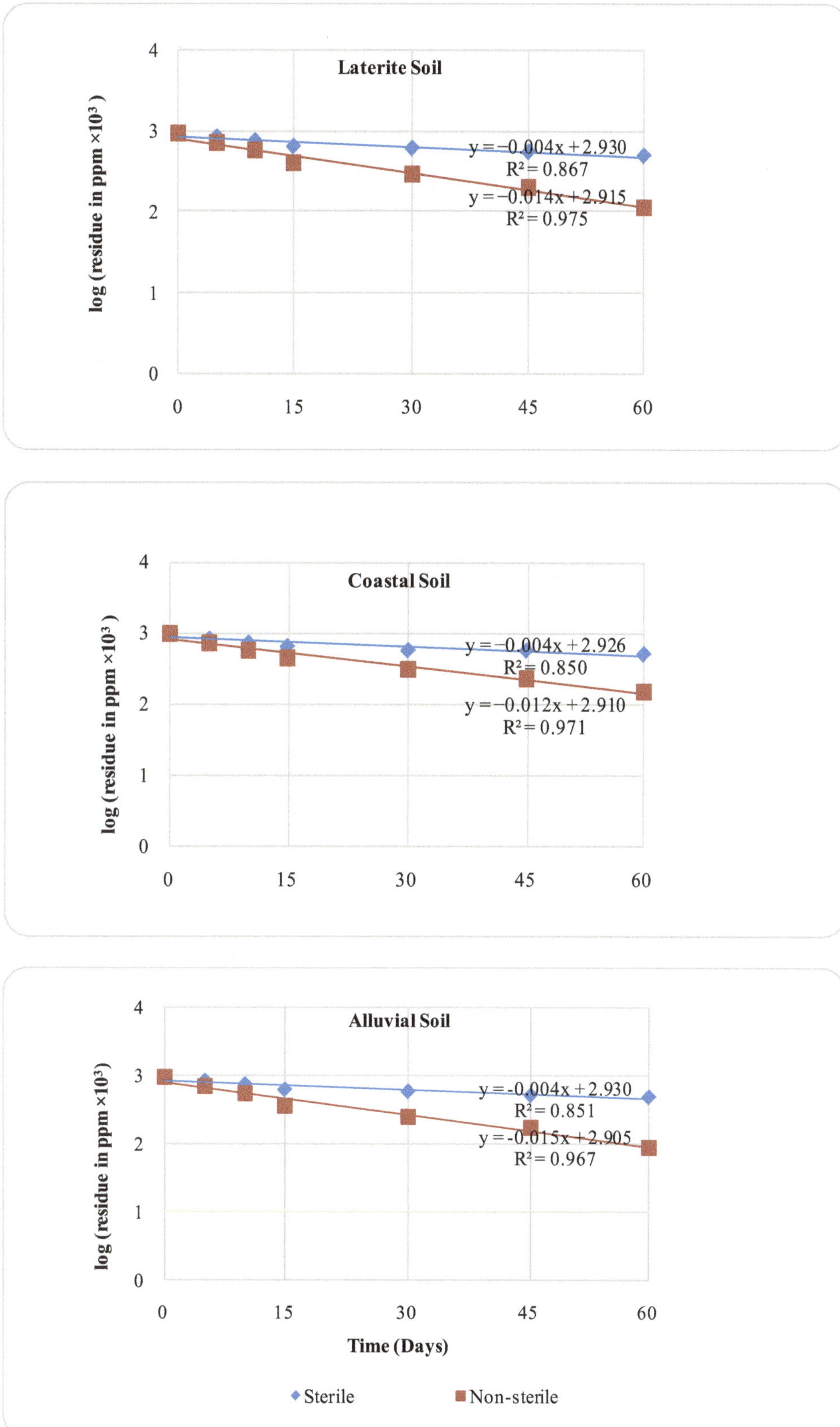

Figure 1. Linear plot for first order reaction of butachlor dissipation in sterile and non-sterile laterite, coastal and alluvial soils.

rate of attenuation was highest in alluvial soil ($T_{1/2}$, 20.1 d) followed by lateritic soil ($T_{1/2}$, 21.5 d) and coastal ($T_{1/2}$, 25.1 d) soils. However, under sterilized soil the trend was as follows: alluvial soil ($T_{1/2}$, 75.3 d) = lateritic soil ($T_{1/2}$, 75.3 d) < coastal soil ($T_{1/2}$, 100.3 d). The half-life found in the present investigation under non-sterilized condition is quite comparable with the half-life values found earlier ($T_{1/2}$, 16 - 19 d, [45]) under field study, using application rate similar to our present study (2 kg ai/ha). However, the observed minor difference might be due to the differences in experimental conditions in two studies. Opposite to our experiment they conducted the experiment in rice grown soil under field condition with and without organic manure treatment, which attributed to difference in physiological metabolic activities in soil. The biochemical activity is also encouraged by root exudates [46]. However, quality and quantity of root exudates depend on plant types and microorganisms also have a choice on root exudates utilization [47]. In a separate experiment half-life value of butachlor in the rhizosphere soil containing three different plants ranged from 5 - 10 d [28], which also reflects the same explanation. Faster degradation of butachlor in alluvial than that of coastal soil, observed in the present study, was also previously reported [48]. The low persistence of the chemical in alluvial soil was earlier demonstrated due to the most congenial environmental factors particularly pH [44], favoring microbial degradation process through enhanced activities of micro-organisms [49]. On the other hand, it was highly persistent in coastal soil due to higher salt content [50] adversely affecting growth and activities of micro-organisms as evidenced from the least microbial activity across soil type. Commendable dissipation was observed in sterile soil. This indicates the involvement of abiotic forces (sorption and chemical factors) other than photo degradation as the soils were kept under dark during the entire experimental period. In non-sterilized soil herbicide dissipation was higher than that in sterilized soil, confirming the possible microbial role in addition to abiotic factors. It is also interesting to note that irrespective of soil types and days of incubation there was a general trend of more dissipation of the chemical due to abiotic forces (49.5% loss after 60 d) than biotic forces (41.0% loss after 60 d) (**Table 2**). Sorption plays a vital and influencing role on organic contaminants fate and behavior in soil because the process exerts tremendous impact over the bioavailability of the chemical for degradation [51]. Adsorption of butachlor in soil is well established phenomenon and well documented [52]-[54].

3.2. Effect of Butachlor on Microbial Activity across Soil Types

Substantiating the results of Min *et al.*, [15] and Barman *et al.*, [55] butachlor effectuated significant detrimental influence on carbon di-oxide production, the virtue of microbial activity, from the beginning up to 10, 13 and 22 d of incubation in coastal, lateritic and alluvial soil, respectively (**Table 3**). The initial detrimental influence reflected the lag phase or adaptive phase of microbial community for synthesizing required enzymes [56] to degrade the herbicide with the formation of new strains through mutation [57] and then multiply their population to the required level not only for the detoxication [26] but also the utilization of either herbicide [58], the dead susceptible biotic agents [59] or both as energy and nutrient sources. Consequently, supporting the reports of Min *et al.*, ([15] there was a significant boost in microbial activity from 13th, 16th and 28th day of incubation in coastal, lateritic and alluvial soil, respectively, persisting thereafter throughout. Among the three soil types, significant higher amount of carbon di-oxide was released from alluvial soil followed by lateritic and coastal soil, respectively. The lowest microbial activity in coastal soil was due to the hypertonic soil environment [60] while

Table 3. Effect of butachlor on the evolution of CO_2 (mg 100 g^{-1} dry soil) across soil type (values are mean of 5 replications) Con = Control, Tre = Butachlor treated at 2 kg a.i. per ha.

Soil type		1	2	3	4	5	6	7	10	13	16	19	22	25	28	35	42	49	56	63
Laterite	Con	8.06	16.02	23.70	36.13	47.53	58.54	69.32	80.91	93.58	106.25	119.81	133.90	148.36	163.19	178.86	195.49	213.45	232.68	254.13
	Tre	5.56	12.82	20.15	32.25	44.10	55.53	66.59	79.37	92.75	106.67	121.10	136.10	151.40	166.85	182.95	200.45	219.35	240.35	264.85
Coastal	Con	9.10	17.15	25.47	34.13	42.11	49.34	55.35	61.64	69.07	77.27	86.37	96.48	107.40	108.63	124.25	141.35	160.18	182.57	207.80
	Tre	6.06	13.22	21.42	31.08	40.04	47.82	53.92	61.27	69.40	78.40	88.90	100.15	112.03	124.78	142.35	163.00	185.05	211.51	241.26
Alluvial	Con	17.00	33.13	48.41	61.94	73.14	84.04	95.57	109.16	124.44	144.03	168.52	195.41	221.36	244.27	266.12	286.70	305.70	322.80	339.55
	Tre	13.06	25.64	36.88	47.17	57.12	69.02	81.34	96.37	105.15	137.69	165.32	195.17	221.24	244.76	267.88	290.21	309.31	327.26	344.09
CD (P = 0.05)		0.003	0.009	0.005	0.003	0.003	0.001	0.003	0.001	0.005	0.009	0.017	0.014	0.015	0.016	0.018	0.014	0.019	0.049	0.036

the highest microbial activity in alluvial soil [61] was due to the most favorable environmental condition particularly the pH [44]. Consequently during a period of 63 d, the cumulative release of carbon di-oxide from alluvial soil was 33.6% higher than that of lateritic soil which, in turn was 22.3% higher than that of coastal soil.

3.3. Effect of Butachlor on Colonial Microbes across Soil Types

Butachlor brought about differential influence on colonial micro-organisms at different stages of incubation which, in turn, varied in different soil types. The herbicide caused an initial non-significant influence on bacterial population in each soil type—10 d for lateritic as well as alluvial soils and 45 d for coastal soil (**Figure 2**). For lateritic and alluvial soil the non-significant influence reflected stationary phase. Whereas, significant stimulation of bacteria by application of butachlor on the 15th day of incubation in spite of significant detrimental influence on microbial activity during initial 13 and 22 d of incubation in lateritic and alluvial soil, respectively, ensured that the bacterial flora derived their energy and nutrients from herbicide for their growth and multiplication [62]. Consequently the population of bacteria significantly increased in lateritic and alluvial soil by 22.7 and 12.4%, respectively over their respective control on the 15th day. The stimulation was more pronounced in lateritic soil. That was due to the additive effect of energy and nutrient from the susceptible dead cells as evidenced from significant enhancement in microbial activity on the 16th day of incubation. Then there was another stationery phase persisting up to 30th day and 60th day for lateritic and alluvial soil, respectively. Butachlor at the later stages in lateritic soil, however, induced significant detrimental influence on bacterial population manifesting the activation process through which a putative herbicide was converted to toxic molecules [26]. As a result, there was a progressive activation of butachlor resulting in significant gradual reduction in colonial bacteria from 16.3% on the 45th day to 30.0% on the 60th day of incubation as compared to that of untreated control in

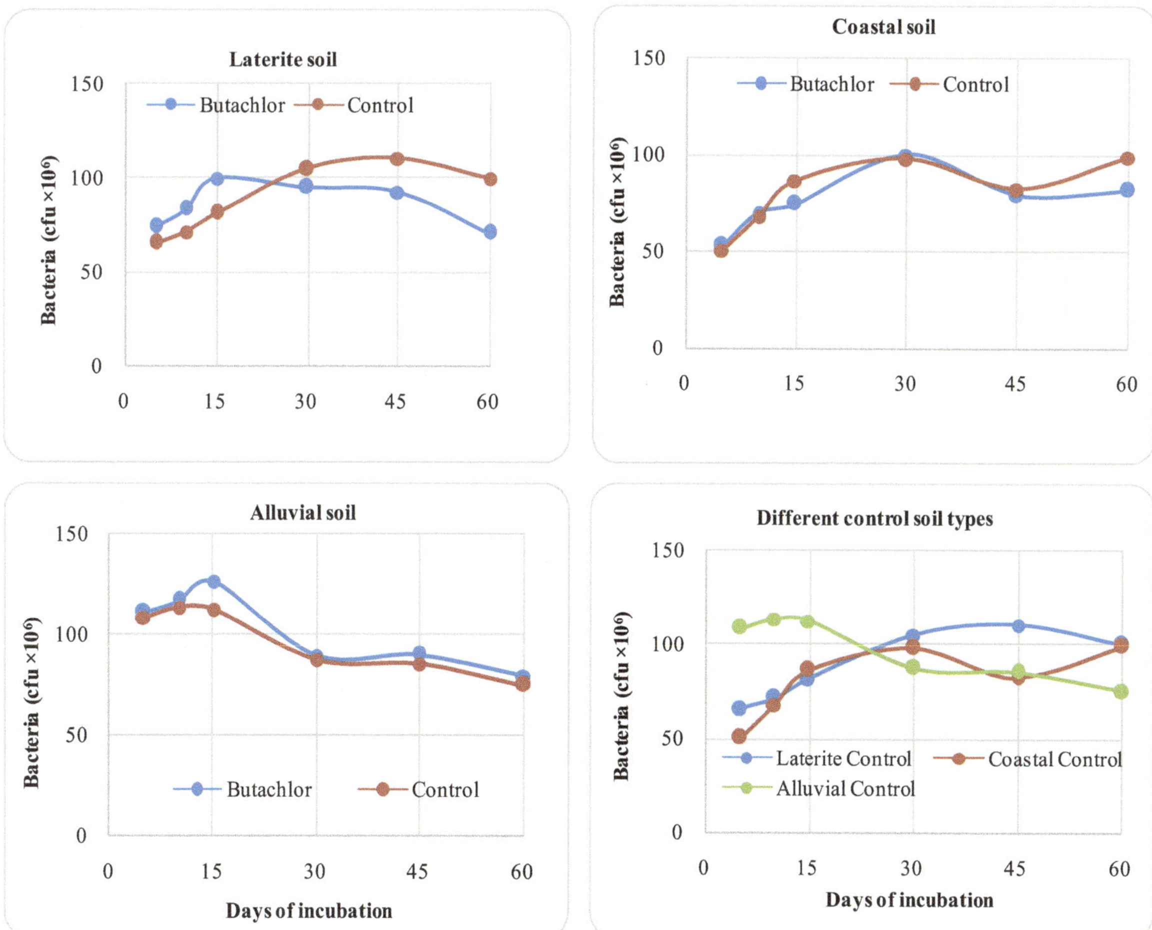

Figure 2. Effect of butachlor application on bacterial population in three soils at different periods of incubation.

lateritic soil. Butachlor also brought about significant reduction in colonial bacteria by 17.2% as compared to the untreated control in the coastal soil on the 60th day of incubation. Sustaining the reports of Min *et al.*, [63] fungal propagules were significantly increased on the 5th day of incubation by the application of butachlor over that of untreated control in lateritic soil illustrating the utilization of the chemical as nutrient and energy sources by the chemo heterotrophs (**Figure 3**). Then there was manifestation of the stationery phase on the 10th day before the

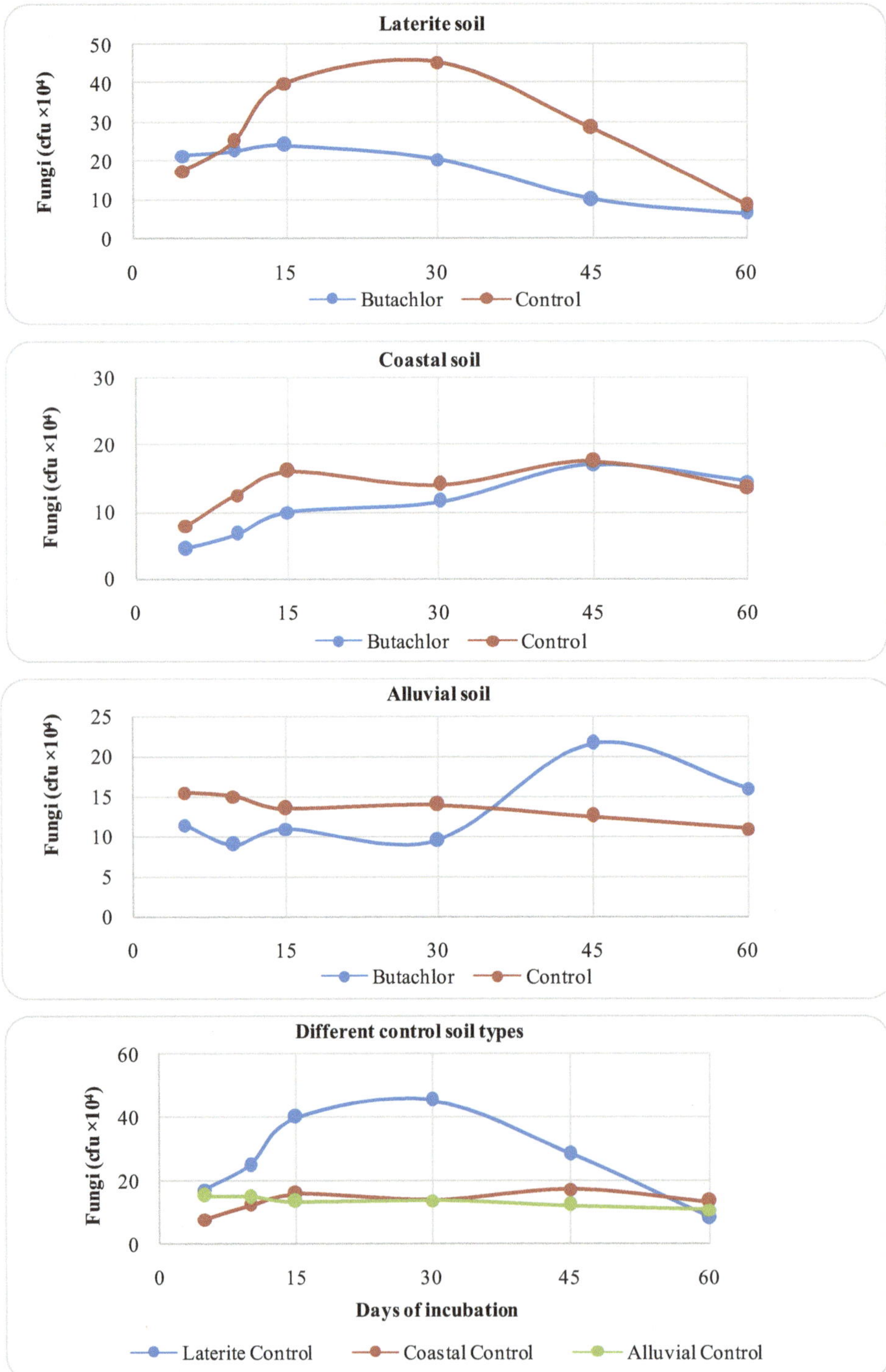

Anova (P value at 0.05): Treatment (Tr) 1.25, Days (D) 1.20, Interaction (Tr x D) 3.81

Figure 3. Effect of butachlor application on fungal population in three soils at different periods of incubation.

influence of activated compounds. Thereafter, there was a significant progressive devastating impact of activated herbicide molecules resulting in sharp decline in fungal propagules from 40% on the 15th day to 64% on the 45th day of incubation. But on the 60th day there was again detoxification process as evidenced from the non-significant influence in the lateritic soil. In contrast, butachlor resulted in the stationary phase of fungal propagules on the 5th day in coastal soil before induction of activated compounds. Then activated compounds of butachlor resulted in significant reduction in fungal propagules from 46.4% on the 10th day to 37.5% on the 15th day revealing a significant gradual declining impact of activated compounds of butachlor and/or butachlor as such, which were detoxified thereafter from 30th to 60th day of incubation as evidenced from their non-significant influence. This conformed earlier finding [62]. Whereas, butachlor resulted in significant reduction in fungal propagules in alluvial soil as compared to untreated control on the 5th day of incubation. The impact was more severe as evidenced from the 40% reduction in fungal propagules on the 10th day. Following a mysterious non-significant influence on the 15th day the activated butachlor molecules again resulted in a significant fall in fungal propagules by 32.1% as compared to untreated control on the 30th day. Thereafter, butachlor resulted in a significant enhancement in fungal propagules during later stages of incubation in alluvial soil, commemorating the findings of Yang *et al.*, [28]. The extent of increment, however, gradually decreased from 75% on the 45th day to 45.5% on the 60th day as compared to untreated control in alluvial soil. So far as soil types are concerned, alluvial soil, owing to most favorable environmental conditions particularly the pH, was the best niche for the maximum proliferation of colonial bacteria followed by lateritic and coastal soils, respectively. On the other hand, lateritic soil due to favorable acidic pH harbored the highest fungal propagules followed by alluvial and coastal soil, respectively. Coastal soil reared the least colonial bacteria and fungi due to higher osmotic soil environment. The results substantiate the findings of [14]. Butachlor was subjected to co-metabolism by commensal microbes during initial 10 days of incubation and 19.5% residues were dissipated as evidenced from non-significant influence of herbicide on colonial microbes in alluvial soil (**Figure 2** & **Figure 3**, **Table 2**). Then the significant increment in colonial bacteria on the 15th day pointed out that 13.3% butachlor residues were utilized by zymogenous [64] bacteria as energy and nutrient sources besides commensal microorganisms from the 10th to 15th day of incubation. Again there was 1.8% of residual loss in butachlor through co-metabolism from the 15th to 30th days. But during later stages, 6.4% of butachlor residues were utilized by colonial fungi in alluvial soil along with co-metabolic micro-organisms. On the other hand, co-metabolism was the only phenomenon prevailing in coastal soil and 38.0% of butachlor residues were co-metabolized in coastal soil by commensal microbes (**Figure 2** & **Figure 3**, **Table 2**). However, along with co-metabolic microbes, colonial fungi were also the responsible agents during initial 5 d in lateritic soil and 12.9% of butachlor residues were utilized by them as nutrients and energy sources as evidenced from significant increment in fungal propagules. Then there was co-metabolism of 5.3% of butachlor residues from 5 to 10 d of incubation. But again from the 10th to 15th days of incubation about 5.7% of butachlor residues were metabolized by zymogenous bacteria together with co-metabolic ones as evidenced from their significant stimulation. Thereafter, from 30th to 60th day about 10.4% of butachlor was co-metabolized in soil.

4. Conclusion

Albeit considering the importance and significance of abiotic forces in dissipating butachlor in soil, it is understood that highest microbial activity and maximum enlargement of microbial community resulted in the least persistence of butachlor in alluvial soil. On the other hand, the herbicide induced highest persistence in coastal soil due to inimical influence on the growth and the activity of micro-organisms. So among the three soil types alluvial soil furnishes its suitability for butachlor application.

References

[1] Klingman, G.C., Ashton, F.M. and Noordhoff, L.J. (1982) Weed Science: Principles and Practices. Wiley, New York.

[2] Yaduraju, N.T., Prasad Babu, M.B.B. and Chandla, P. (2006) Herbicide Use. In: Swaminathan, M.S., Chadha, K.L. (Eds). *Agriculture and Environment.* Malhotra Publishing House, New Delhi, 192-210.

[3] Froud-Williams, R.J. (2002) Weed Competition. In: Naylor, R.L. (Ed), *Weed Management Handbook*, Blackwell Science, 16-38. http://dx.doi.org/10.1002/9780470751039.ch2

[4] Monaco, T.J., Weller, S.C. and Ashton, F.M. (2002) Weed Science: Principles and Practices. 4th Edition, John Wiley & Sons, Inc., New York.

[5] Yadaraju, N.T. and Mishra, J.S. (2002) Herbicides-Boon or Bane? *Pestology*, **26**, 43-46.

[6] Bhan, V.M. and Mishra, J.S. (2001) Herbicides in Relation to Food Security and Environment in India. *Pestic Inform*, **12**, 28-33.

[7] Jena, P.K., Adhya, T.K. and Rao, V.R. (1987) Influence of Carbaryl on Nitrogenase Activity and Combination of Butachlor and Carbofuran on Nitrogen-Fixing Microorganisms in Paddy Soils. *Pesticide Science*, **19**, 179-184. http://dx.doi.org/10.1002/ps.2780190303

[8] Yu, Y.L., Chen, Y.X., Luo, Y.M., Pan, X.D., He, Y.F. and Wong, M.H. (2003) Rapid Degradation of Butachlor in Wheat Rhizosphere Soil. *Chemosphere*, **50**, 771-774. http://dx.doi.org/10.1016/S0045-6535(02)00218-7

[9] Lauber, C.L., Hamady, M., Knight, R. and Fierer, N. (2009) Pyrosequencing-Based Assessment of Soil pH as a Predictor of Soil Bacterial Community Composition at the Continental Scale. *Applied and Environmental Microbiology*, **75**, 5111-5120. http://dx.doi.org/10.1128/AEM.00335-09

[10] Rousk, J., Bååth, P.C., Lauber, C.L., Lozupone, C., Caporaso, J.G., Knight, R. and Fierer, N. (2010) Soil Bacterial and Fungal Communities across a pH Gradient in an Arable Soil. *ISME Journal*, 1-12.

[11] Rousk, J., Brookes, P.C. and Baath, E. (2009) Contrasting Soil pH Effects on Fungal and Bacterial Growth Suggests Functional Redundancy in Carbon Mineralization. *Applied and Environmental Microbiology*, **75**, 1589-1596. http://dx.doi.org/10.1128/AEM.02775-08

[12] Lauber, C.L., Strickland, M.S., Bradford, M.A. and Fierer, N. (2008) The Influence of Soil Properties on the Structure of Bacterial and Fungal Communities across Land-Use Types. *Soil Biology and Biochemistry*, **40**, 2407-2415. http://dx.doi.org/10.1016/j.soilbio.2008.05.021

[13] Grossbard, E. (1976) Effect on the Soil Microflora. In: Audus, L.J., Ed., *Herbicides Physiology*, *Biochemistry*, *Ecology*, Academic Press, London, 99-147.

[14] Mukherjee, D., Praharaj, A.K., Saha, N. and Chakravarty, A. (1999) Influence of Basalin on the Preponderance of Soil Microflora. *Journal of Interacademicia*, **3**, 172-177.

[15] Min, H., Ye, Y.F., Chen, Z.Y., Wu, W.X. and Du, Y.F. (2002) Effects of Butachlor on Microbial Enzyme Activities in Paddy Soil. *Journal of Environmental Sciences*, **14**, 413-417.

[16] Saeki, M. and Toyota, K. (2004) Effect of Bensulfuron-Methyl (a Sulfonyurea Herbicide) on the Soil Bacterial Community of a Paddy Soil Microcosm. *Biology and Fertility of Soils*, **40**, 110-118. http://dx.doi.org/10.1007/s00374-004-0747-1

[17] Mandal, B.B., Bandyopadhyaya, P. and Maity, S.K. (2007) Effect of Some Preemergence Herbicides on Soil Microflora in Direct Seeded Rice. *Indian Agriculturist*, **31**, 19-23.

[18] Reichardt, W., Dobermann, A. and George, T. (1998) Intensification of Rice Production Systems: Opportunities and Limits. In: Dowling, N.G., Greenfield, S.M. and Fisher, K.S., Eds., *Sustainability of Rice in the Global Food System*, Pacific Basin Study Centre and IRRI Publ., Davis, California, US; Manila, Philippines, 127-144.

[19] Kudsk, P. and Streibig, J.C. (2003) Herbicides—A Two-Edged Sword. *Weed Research*, **43**, 90-102. http://dx.doi.org/10.1046/j.1365-3180.2003.00328.x

[20] Olofsdotter, M., Watson, A. and Piggin, C. (1998) Weeds: A Looming Problem in Modern Rice Production. In: Dowling, N.G., Greenfield, S.M. and Fisher, K.S., Eds., *Sustainability of Rice in the Global Food System*, Pacific Basin Study Center & IRRI Publ., Davis, California, US; Manila, Philippines, 165-173.

[21] Kalia, A. and Gupta, R.P. (2004) Disruption of Food Web by Pesticides. *Indian Journal of Ecology*, **31**, 85-92.

[22] Van Ferd, L., Hoagland, R.E., Zablotowicz, R.M. and Hall, J.C. (2003) Pesticide Metabolism in Plants and Microorganisms. *Weed Science*, **51**, 472-495. http://dx.doi.org/10.1614/0043-1745(2003)051[0472:PMIPAM]2.0.CO;2

[23] Beestman, G.B. and Deming, J.M. (1974) Dissipation of Acetanilide Herbicides from Soils. *Agronomy Journal*, **66**, 308-311. http://dx.doi.org/10.2134/agronj1974.00021962006600020035x

[24] Chen, Y.L. and Chen, J.S. (1979) Degradation and Dissipation of Herbicide Butachlor in Paddy Fields. *Journal of Pesticide Science*, **4**, 431-438. http://dx.doi.org/10.1584/jpestics.4.431

[25] Abd-Airahman, S.H. and Salem-Benkhit, M.M. (2013) Microbial Biodegradation of Butachlor Pollution (Obsolete Pesticide Machete 60% EC). *African Journal of Microbiology Research*, **7**, 330-335.

[26] Alexander, M. (1977) Introduction to Soil Microbiology. 2nd Edition, John Wiley Eastern Limited, New Delhi, 467.

[27] Horvath, R. (1972) Microbial Co-Metabolism and the Degradation of Organic Compounds in Nature. *Bacteriology Reviews*, **36**, 146-155.

[28] Yang, C.M., Wang, M., Chen, H. and Li, J. (2011) Responses of Butachlor Degradation and Microbial Properties in a Riparian Soil to the Cultivation of Three Different Plants. *Journal of Environmental Sciences*, **23**, 1437-1444. http://dx.doi.org/10.1016/S1001-0742(10)60604-3

[29] Jackson, M.L. (1973) Soil Chemical Analysis. Prentice Hall of India Pvt. Ltd., New Delhi, 498.

[30] Walkley, A.J. and Black, I.A. (1934) Estimation of Soil Organic Carbon by the Chromic Acid Titration Method. *Soil Science*, **37**, 29-38.

[31] Olsen, S.R., Cole, C.V., Watanabe, F.S. and Dean, L.A. (1954) Estimation of Available Phosphorus in Soils by Extraction with Sodium Bicarbonate. United States Department of Agriculture Circular 19, Washington DC, 939.

[32] Bouyoucos, G.J. (1922) The Hydrometer as a New Method for the Mechanical Analysis of Soils. *Soil Science*, **23**, 343-354. http://dx.doi.org/10.1097/00010694-192705000-00002

[33] Bouyoucos, G.J. (1962) Hydrometer Method Improved for Making Particle-Size Analysis of Soils. *Agronomy Journal*, **54**, 464-465. http://dx.doi.org/10.2134/agronj1962.00021962005400050028x

[34] Pramer, D. and Schmidt, E.L. (1965) Experimental Soil Microbiology. Burgess Publishing Co., Minneapolis.

[35] Salle, A.J. (1973) Laboratory Manual of Fundamental Principles of Bacteriology. McGraw-Hill Book Company, New York.

[36] Thornton, H.G. (1922) On the Development of a Standardized Agar Media for Counting Soil Bacteria with a Special Regards to the Repression of Spreading Colonies. *Annals of Applied Biology*, **9**, 241-274. http://dx.doi.org/10.1111/j.1744-7348.1922.tb05958.x

[37] Martin, J.P. (1950) Use of Acid, Rose Bengal and Streptomycin in the Plate Method for Estimating Soil Fungi. *Soil Science*, **69**, 215-232. http://dx.doi.org/10.1097/00010694-195003000-00006

[38] Stroetmann, I., Kämpfer, P. and Dott, W. (1994) The Efficiency of Sterilized Methods for Different Soils. *Zentralblatt für Hygiene und Umweltmedizin*, **195**, 111-120.

[39] Debnath, A., Das, A.C. and Mukherjee, D. (2002) Persistence and Effect of Butachlor and Basalin on the Activities of Phosphate Solubilizing Microorganisms in Wet Land Rice Soil. *Bulletin of Environmental Contamination and Toxicology*, **68**, 766-770. http://dx.doi.org/10.1007/s001280319

[40] Hoskins, W.M. (1961) Mathematical Treatment of Loss of Pesticide Residues. *FAO Plant Protection Bulletin*, **9**, 163-168.

[41] Gomez, K.A. and Gomez, A.A. (1984) Statistical Procedures for Agricultural Research. John Wiley and Sons, New York.

[42] Maheswari, S.T. and Ramesh, A. (2011) Adsorption and Degradation of Anilophos in Different Soils and Its Environmental Impact in Soils. *International Journal of Environmental Research*, **6**, 451-456.

[43] Prakash, N.B. and Suseela, D.L. (2000) Persistance of Butachlor in Soils under Different Moisture Regimes. *Journal of the Indian Society of Soil Science*, **48**, 249-256.

[44] Prakash, N.B., Suseeladevi, L. and Siddaramappa, R. (2000) Effect of Organic Amendments on Persistence of Butachlor and Pendimethalin in Soil. *Karnataka Journal of Agricultural Sciences*, **13**, 575-580.

[45] Rao, P.C., Rama Laksmi, C.S., Madhavi, M., Swapna, G. and Sireesha, A. (2012) Butachlor Dissipation in Rice Grown Soil and Its Residues in Grain. *Indian Journal of Weed Science*, **44**, 84-87.

[46] Raaijmakers, J.M., Paulitz, C.T., Steinberg, C., Alabourette, C. and Moenne-Loccoz, Y. (2009) The Rhizosphere: A Playground and Battlefield for Soilborne Pathogens and Beneficial Microorganisms. *Plant and Soil*, **321**, 341-361. http://dx.doi.org/10.1007/s11104-008-9568-6

[47] Garbeva, P., van Elsan, J.D. and van Veen, J.A. (2008) Rhizosphere Microbial Community and Its Response to Plant Species and Soil History. *Plant and Soil*, **302**, 19-32. http://dx.doi.org/10.1007/s11104-007-9432-0

[48] Pal, R., Das, P., Chakrabarty, K., Chakravarty, A. and Chowdhury, A. (2006) Butachlor Degradation in Tropical Soils: Effect of Application Rate, Biotic-Abiotic Interactions and Soil Conditions. *Journal of Environmental Science and Health, Part B*, **41**, 1103-1113. http://dx.doi.org/10.1080/03601230600851141

[49] Chen, Y.L. (1980) Degradation of Butachlor in Paddy Fields. In: *Weeds and Weed Control in Asia*, FFTC Book Series No. 20, Food and Fertiliser Technology Centre, Taiwan, 121-142.

[50] Tripathi, S., Kumari, S., Chakraborty, A., Gupta, A., Chakrabarti, K. and Bandyapadhyay, B.K. (2006) Microbial Biomass and Its Activities in Salt-Affected Coastal Soils. *Biology and Fertility of Soils*, **42**, 273-277. http://dx.doi.org/10.1007/s00374-005-0037-6

[51] Regitano, J.B., Koskinen, W.C. and Sadowsky, M.J. (2006) Influence of Soil Aging on Sorption and Bioavailability of Simazine. *Journal of Agricultural and Food Chemistry*, **54**, 1373-1379. http://dx.doi.org/10.1021/jf052343s

[52] Yu, Y.L., Wu, X.M., Li, S.N., Fang, H., Zhan, H.Y. and Yu, J.Q. (2006) An Exploration of the Relationship between Adsorption and Bioavailability of Pesticides in Soil to Earthworm. *Environmental Pollution*, **141**, 428-433. http://dx.doi.org/10.1016/j.envpol.2005.08.058

[53] Liu, Z., Ding, N., Hayat, T., He, Y., Xu, J. and Wang, H. (2010) Butachlor Adsorption in Organically Rich Soil

Particles. *Soil Science Society of America Journal*, **74**, 2032-2038. http://dx.doi.org/10.2136/sssaj2010.0032

[54] Liu, Z., He, Y., Xu, J. and Zeng, F. (2013) How Do Amorphous Sesquioxides Affect and Contribute to Butachlor Retention in Soils? *Journal of Soils and Sediments*, **13**, 617-628. http://dx.doi.org/10.1007/s11368-012-0638-2

[55] Barman, K.K., Srivastava, E. and Varshney, J.G. (2009) Effect of Butachlor on Total Microbial Activities: Azotobacter and Phosphate Solubilizing Fungal Population. *Indian Journal of Weed Science*, **41**, 27-31.

[56] Elsas, J.D.V., Jansson, J.K. and Trevors, J.T. (2007) Modern Soil Microbiology. 2nd Edition, Taylor and Francis Group, London.

[57] Kalyanasundaram, D. and Kavitha, S. (2012) Effect of Butachlor on the Microbial Population of Direct Sown Rice. World Academy of Science, Engineering and Technology, Vol. 6, 853-855.

[58] Baruah, M. and Mishra, R.R. (1986) Effect of Herbicids Butachlor, 2,4-D and Oxyfluorfen on Enzyme Activities and CO_2 Evolution in Submerged Paddy Field Soil. *Plant and Soil*, **96**, 287-291. http://dx.doi.org/10.1007/BF02374772

[59] Agnihotri, V.P., Sinha, A.P. and Singh, K. (1981) Influence of Insecticides on Soil Microorganisms and Their Biochemical Activity. *Pesticides*, **15**, 16-24.

[60] Setia, R., Marschner, P., Baldock, J. and Chittleborough, D. (2010) Is CO_2 Evolution in Saline Soil Affected by an Osmotic Effect and Calcium Carbonate? *Biology and Fertility of Soils*, **46**, 781-792. http://dx.doi.org/10.1007/s00374-010-0479-3

[61] Praharaj, A.K., Saha, N., Chakravarty, A. and Mukherjee, D. (1997) Effect of Basalin on Microbiological Activities in Soil. *Journal of Interacademicia*, **1**, 43-47.

[62] Bera, S. and Ghosh, R.K. (2013) Soil Physico-Chemical Properties and Microflora as Influenced by Bispyribac Sodium 10% SC in Transplanted *Kharif* Rice. *Rice Science*, **20**, 298-302. http://dx.doi.org/10.1016/S1672-6308(13)60148-1

[63] Min, H., Ye, Y.F., Chen, Z.Y., Wu, W.X. and Du, Y.F. (2001) Effects of Butachlor on Microbial Populations and Enzyme Activities in Paddy Soil. *Journal of Environmental Science and Health, Part B*, **36**, 581-595. http://dx.doi.org/10.1081/PFC-100106187

[64] Winogradsky, S. (1924) Sur la microflora autochtone de la terre arable. *Comptes rendus hebdomadaires des seances de l'Academie des Sciences (Paris) D*, **178**, 1236-1239.

14

Physicochemical Property Testing of a Novel Maize Seed Coating Agent and Its Antibacterial Mechanism Research

Qingxiang Deng, Defang Zeng*

School of Resource and Environmental Engineering, Wuhan University of Technology, Wuhan, China
Email: *13971066854@163.com

Abstract

We used the no pollution natural polymer polysaccharide chitosan as the main raw material, which is supplemented by other additives, to prepare a high-efficiency environmental friendly maize seed coating agent. By detecting the seed quality, maize seedling indexes, which include root activity, chlorophyll content and malondialdehyde (MDA) content, and through the results we found that this seed coating agent could improve the photosynthetic capacity and enhance seedling resistance. Laboratory bacteriostasis test and study showed that the antibacterial rate of this seed coating agent has reached more than 88%. Meanwhile, we analyzed and studied the antimicrobial mechanism of the seed coating agent.

Keywords

Maize Seed Coating Agent, Antibacterial, Mechanism Analysis

1. Introduction

The history of seed coating agent and seed coating technology research, development, demonstration, popularization and application in China has lasted for over 20 years. The function of seed coating agent played in crop protection has been gradually recognized by people [1] [2]. Seed coating agent is of obvious benefits to control underground pest, kill seed and seedling diseases, promote seedling healthy growth, improve the crop quality and seed germination rate, reduce the use amount of seeds and increase the output and so on [3].

Maize is not only one of China's three major grain crops, but also an important feed and industrial raw materials [4]. The harm of maize leaf blight, sheath blight, head smut and other harmful organisms have great impact

*Corresponding author.

on maize seedling, yield and quality [5]. The seed coated by maize seed coating is an effective method to control the northern maize leaf blight, sheath blight and head smut [6] [7]. Seed coating agent can promote seed germination, prevent and control diseases and insect pests and supply essential nutrients to achieve the purpose of strong seedlings. It can also create a good micro ecological environment [8].

In view of the above problems, we designed a novel maize seed coating agent to replace the traditional maize seed coating agent. We tested the indexes of this seed coating agent to check if it has meet the relevant standards. Then we measured maize seedlings indexes after the maize seed had been coated with the novel maize seed coating agent and the traditional seed coating agent Shileshi respectively, which include root activity, chlorophyll content and MDA content to characterize the maize seedling indexes of life. We also conducted laboratory experiments to analysis the mechanism of the seed coating agent in promoting seedling growth of maize seedlings, enhancing physical indexes, preventing and controlling of maize leaf blight, sheath blight and maize head smut.

2. Methods and Materials

2.1. Experiment Instruments and Materials

2.1.1. Main Experimental Instruments

Electronic analytical balance (JA1003N, Tianjin Shunte Electronic Instrument Co., Ltd., China), electronic constant speed blender (GS-12B, Shanghai Lushuo Real Co., Ltd., China), constant temperature and humidity incubator (LHP-160L, Changzhou Runhua Electrical Appliances Co., Ltd., China), Petri dish (Φ90 mm, Shanghai Yuejin Medical Instrument Factory, China), ultrasonic cleaner (KQ-700DE, Shanghai Qiangqiang Instruments Equipment Co., Ltd., China), high temperature blast drying oven (DHG-9078A, Shanghai Shanzhi Instrument Equipment Co., Ltd., China), electrothermal pressure steam sterilizer (XFH-30CA, Dingxinyi Experimental Equipment Co., Ltd., China), microporous test sieve (0.04 mm, Endecotts, Shanghai Guowang Chemical Co., Ltd., China), vacuum microporous filter (TAPS, Beijing Jinghuikaiye Technology Co., Ltd., China), Uv-vis spectrophotometer (UV-3, Shanghai Meipuda Instrument Co., Ltd., China), digital rotational viscometer (NXS-11, Chengdu Instrument Factory, China), spectrophotometer (721S, Shanghai Edge Light Technology Co., Ltd., China), digital electro-thermostatic water bath (HH-4, Honghua Instrument Factory, China).

2.1.2. Main Experimental Materials

Chitosan (homemade), film-forming additives (homemade), antifreeze (homemade), microelement nutrient solution (homemade), violet pigments (Hebei Baiwei Biological Technology Co., Ltd., China), deionized water (homemade), maize seeds (Shandong Pingyuanshennong seed industry Co., Ltd., China), Shileshi (traditional maize seed coating agent, Swiss Syngenta Crop Protection Co., Ltd., China), maize northern leaf blight fungus, sheath blight fungus, head smut fungus (Hubei Academy of Agricultural Sciences, China).

2.2. Preparation of Environmental Friendly Maize Seed Coating Agent

The above ingredients first mixed with deionized water respectively to be aqueous solution of a certain concentration, then mixed them according to the designed sequence and proportion, finally, stirred completely. This mixed liquid is the product of environmental friendly maize seed coating agent, we marked it as 4#-1.

2.3. Physicochemical Properties Test of the Maize Seed Coating Agent

2.3.1. Physicochemical Properties Requirements of Maize Seed Coating Agent

According to China's Pesticide Industry Standards, the physicochemical properties and parameters requirements of seed coating agent are in **Table 1** [9].

2.3.2. Testing Methods for Physicohemical Properties of Seed Coating Agent

1) pH Value Test

Refer to pH Value Measurement Method of GB/T 1601-1993 Pesticide.

2) Floating Rate Test

Refer to Measurement Method of GB/T 17768-1999 Standards Specification for the Suspended Seed Coating Agents.

Table 1. The physicochemical property requirements of seed coating agent.

Indicators	Requirements
pH Value	3.8 - 7.2
Floating Rate/%	No less than 90%
Sieve Analysis/%	No less than 95%
Viscosity	200 - 800 mPa·s
Film-Forming Property	qualified
Coating Uniformity	No less than 90%
Coating Peeling Rate	No less than 8%
Storage Stability	qualified

3) Sieve Analysis Test

Refer to Measurement Method of GB/T 16150-1995 Pesticide Powder, Wettable Powder Fineness.

4) Viscosity Test

Refer to Measurement Method of NY621-2002 Agriculture Industry Standards of Duofuke Suspended Seed Coating Agent.

5) Film-Forming Property Test

Refer to Measurement Method of GB/T 17768-1999 Standards Specifications for Suspended Seed Coating Agents.

6) Coating Uniformity Test

Refer to Measurement Method of GB/T 17768-1999 Standards Specifications for Suspended Seed Coating Agents

7) Coating Peeling Rate Test

Refer to Measurement Method of GB/T 17768-1999 Standards Specifications for Suspended Seed Coating Agents.

8) Storage Stability Test

Refer to Measurement Method of GB/T 17768-1999 Standards Specifications for Suspended Seed Coating Agents.

2.4. Testing of Different Seed Coating Agents on Maize Seedling Index

2.4.1. Testing of Different Seed Coating Agents on Maize Seedling Root Activity

Plant roots are active absorbing and synthetic organ, the growth and activity level of root influence maize nutrition status and the yield level above ground directly, root activity is an amount that characterize whether plant root is healthy.

The root activity was determined by TTC (three phenyl tetrazolium chloride) method. The testing steps about TTC reduction amount were as follows: took sample of maize root tip for 0.5 g, put the root immersed in a mixed solution containing of 0.4% TTC and phosphate buffer, then kept it preserving in 37°C for three hours, after that, added 2 mL·mol/L sulfuric acid to terminate reaction. Made a blank contrast test at the same time, added sulfuric acid firstly, then added apica sample, other operations were as the same above. Then removed the root, carefully blotted moisture, put it together with 3 - 4 mL of ethyl acetate solution into a mortar, added a small amount of quartz sand for full grinding to extract the three phenyl methyl hydrazone. After filtering the extracting liquid, put it into a 10 mL volumetric flask, then washed with a small amount of ethyl acetate solution for 3 times, put it into test tube to the scale at last. By using spectrophotometer at a wavelength of 485 nm colorimetric to read the absorbance of OD with blank experiment as reference, through examine the standard curve, we can find the reduction amount of TTC [10]. Finally, the root activity can be calculated by formula.

The calculation formula is as follows:

$$\text{Root activity}\left(\text{mg}\cdot\text{g}\cdot\text{h}^{-1}\right)=\text{TTC reduction amount}\left(\text{mg}\right)/\left[\text{root weight}\left(\text{g}\right)\times\text{time}\left(\text{h}\right)\right] \quad (1)$$

2.4.2. Testing of Different Seed Coating Agents on Maize Seedling Chlorophyll Content

Chlorophyll is one of the most important pigments related with photosynthesis. It absorbs energy from the light, then the energy is used to convert carbon dioxide into carbohydrates. Photosynthesis is the synthesis process that chlorophyll absorbs light energy to synthesis organic compounds, meanwhile the light energy converts to chemical energy.

Chlorophyll content was determined by spectrophotometry. The testing steps were as follows: 1) The extraction of pigment. Weighed 2 g fresh maize leaves, put it into a mortar, added a mixture of acetone ethanol (1:1) 10 mL, then added a little calcium carbonate and quartz sand, grinded into homogenate. Then added 80% acetone 5 mL, put the homogenate into a centrifuge tube, used 80% acetone to wash mortar. After centrifugation, discarded the pellet, used 80% acetone to set the supernatant constant volume to 20 mL; 2) The determination of absorbance. Took the above pigment extract 1mL, added 4 mL 80% acetone to dilute the extract, then put them into the cuvette, set 80% acetone as control, determined the absorbance value at 663 nm and 645 nm. Finally, the root activity can be calculated by formula.

The calculation formula is as follows:

$$\text{Total chlorophyll content}\left(\text{mg}\cdot\text{g}^{-1}\text{FW}\right)=\left(20.2\times A645+8.02\times A663\right)V/1000\times W \quad (2)$$

where $A663$ and $A645$ are respectively the absorbance under the corresponding wave; V for extraction liquid volume; W for leaf weight.

2.4.3. Testing of Different Seed Coating Agents on Malondialdehyde (MDA) Content

MDA is one of the main products of the cell membrane lipid peroxidation, its accumulation is the performance of free radical toxicity [11], MDA quantity can not only indirectly reflect the plant tissue antioxidant capacity, but also represent the degree of membrane lipid peroxidation in a certain extent. Therefore, MDA content is a very important index in plant senescence physiology and physiological research.

The testing steps of MDA content were as follows: weighed 1 g leaves of maize seedlings, cut it into shear, then added 2 mL 10% three chloroacetic acid (TCA) and a small amount of quartz sand, grinded into homogenate, then added 8 mL TCA for further grinding. The homogenate used centrifuge to centrifugal 10 min in 4000 rpm, took the supernatant as the sample extraction. Drew 2 mL the centrifugal supernatant (contrast part put 2 mL of distilled water), added 2 mL 0.6% thiobarbituric acid (TBA) solution, after shaking it, put it in a boiling water bath 15 min for fully react. Then rapid cooled it after centrifugation 10 min, took out the supernatant to determine the absorbance in 532 nm, 600 nm and 450 nm wavelength, calculated MDA content.

The calculation formula is as follows:

$$\text{Malondialdehyde}\left(\text{MDA}\right)\ \text{content}=6.45\times\left(D532-D600\right)-0.56\times D450 \quad (3)$$

where $D532$, $D600$ and $D450$ are respectively the absorbance under the corresponding wave.

2.5. Bacteriostatic Experiments of Seed Coating Agent in Lab

Large plaque bacteria, sheath blight fungus and maize head smut fungus are several main harmful bacteria that restrict maize yield [12] [13]. Maize large leaf spot occurred mainly on the growth of maize seedling leaves. The pathogens harm seedling leaf, leaf sheath and outer bract. Leaves appear green gray spots on early onset, multi spots connected pieces later, then the leaves turn yellow and withered, it seriously affect the normal growth of maize seedlings. Sheath blight occurred mainly in the development of the whole process of maize growth, the pathogens harm seedlings leaf sheaths, bract and ear. Firstly, the leaf sheath get disease, then appear green spots, when it become serious, it can cause stalk rot, ear rot and lodging, which seriously affect maize yield. Head smut occurred mainly in panicle stage of maize growth. The pathogens harm tassel and ear, the tillering were clustered, plants were dwarf, shortened internodes, leaf color became dark green, when it get seriously, the whole plant won't have any harvest.

In this test, we put maize large plaque bacteria, sheath blight fungus and maize head smut fungus as the research objects, used the medium culture them to determine the maize seedling growth rate respectively, which can measure the bacteriostatic effect of the novel environmental friendly maize seed coating agent 4#-1 and the traditional one Shileshi. The maize seeds are coated with the self-made maize seed coating agent and Shileshi respectively according to 1:50 (w/w) proportion, then put them into PDA medium. The experiment set up a blank group as contrast, which marked as CK. After the Petri dish being treated with high temperature and high pressure, we put the culture medium into them, then we inoculate equal amount of maize large plaque bacteria, sheath blight fungus, maize head smut fungus into each Petri dishes. Finally, put the Petri dish into temperature and humidity incubator, whose temperature at (27 ± 1)˚C and humidity at 85%, to culture the seed. By using colony cross method to measure the diameter, we could calculate the antibaterial rate through the diameter [14], the calculation formula is as follows:

$$AR(\%) = \frac{A-B}{A} \times 100\% \quad (3)$$

where A and B are the colony diameter of the contrast group and the treatment group respectively.

3. Results

3.1. Physicochemical Property Testing Results of Maize Seed Coating Agent

The physicochemical property testing results of the novel maize seed coating agent were shown in **Table 2**. It showed that all the indexes of this environment friendly maize seed coating agent have reached the qualified level.

3.2. Effect of Different Seed Coating Agents on Several Major Maize Seedling Activity Indexes

3.2.1. Effect of Different Seed Coating Agents on Maize Seedling Root Activity

The root system of plant is not only the main absorptive organ of water and fertilizer nutrients, but also the important assimilate organ of substance transformation and synthesis, its growth condition is directly related to the plant life activities [15].

Effects of different seed coating agents on root activity were shown in **Figure 1**. The results showed that comparing with the blank group, the root activity of maize seedling was improved via the seed coated by different seed coating agents. The maize seedlings root activity of 4#-1 on the third day, the fifth day and the seventh day were 2.2, 2.3, 2.5, the Shileshi group were respectively 2, 2.1, 2.2, root activity of 4#-1 were relatively higher than the Shileshi group about 10%, 9.52%, 13.64%. The improvement of maize seedling root activity may be caused by this self-made polysaccharide polymer, it can change root cell membrane structure to reduce the permeability of cell membrane and enhance the cell membrane stability. It can also activate the enzyme activity that is related to absorpt mineral nutrition in cell membrane, which can improve the ability of absorbing mineral elements. Maize growth is built on the basis of various nutrients absorbed from the roots and this can improve its various life activities in return [10].

3.2.2. Effect of Different Seed Coating Agents on Maize Seedling Chlorophyll Content

Effects of different seed coating agents on chlorophyll content were shown in **Figure 2**. We can see from **Figure 2** that the total chlorophyll in maize seedlings leaves of 4#-1 on the fifth day, the seventh day were improved in different degrees compared with the Shileshi group. The maize seedlings chlorophyll content of 4#-1 on the fifth day and seventh day were 1.9, 2.3, while the Shileshi's were respectively 1.8 and 2, the chlorophyll content of 4#-1 were relatively higher than the Shileshi group about 5.56% and 15%. The increasing of chlorophyll content improved the light to capture pigment of chloroplast in unit leaf area and the center pigment quantity of the photosynthetic reaction. With the content of light capture pigment increasing, the leaves absorb more light, so the increase of chlorophyll content were in favor of maize leaf photochemical efficiency and photosynthesis, it promoted the accumulation of photosynthetic products and seedling growth in return [10].

3.2.3. Effect of Different Seed Coating Agents on Maize Seedling Malondialdehyde (MDA) Content

The effect of different seed coating agents results on maize seedling MDA content were shown in **Figure 3**. It can be seen from **Figure 3** that the 4#-1 coating agent can effectively reduce the MDA content in newborn root cells. The maize seedlings MDA content of 4#-1 on the third day, the fifth day and the seventh day were 1.9, 2.7, 3.2, the Shileshi's were respectively 2.2, 3.1, 3.9. The MDA content of 4#-1 compared with Shileshi's were relatively decreased about 13.64%, 12.90%, 17.95%. MDA is a typical index of cell membrane lipid peroxidation,

Table 2. Physicochemical property testing results of maize seed coating agent.

Indicators	Requirements
pH Value	6.0
Floating Rate/%	95
Sieve Analysis/%	99
Viscosity	380
Film-Forming Property	qualified
Coating Uniformity	95
Coating Peeling Rate	2.5
Storage Stability	qualified

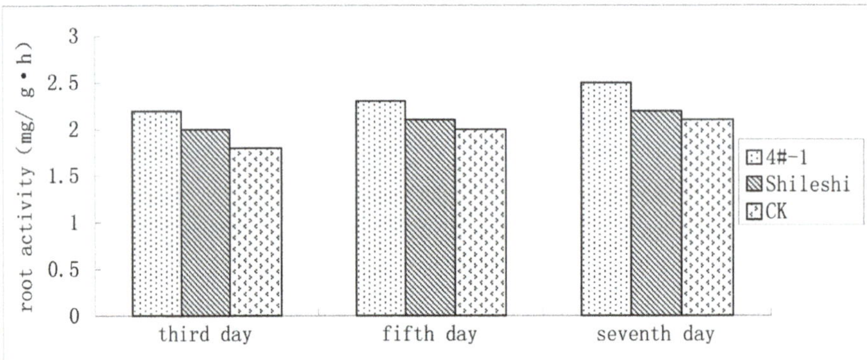

Figure 1. Effect of different seed coating agents on root activity of maize seedling.

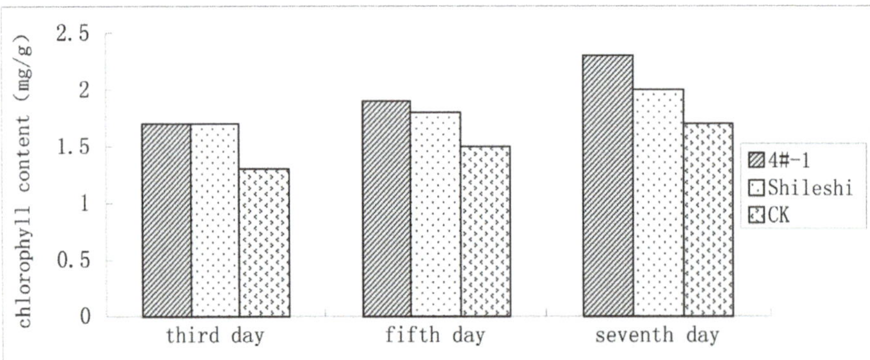

Figure 2. Effect of different seed coating agents on maize seedling chlorophyll content.

its content level is closely related to the stability of cell membrane structure, membrane material exchanges in and outside the cell membrane and the activity level of each enzyme on the cell membrane. The decrease of MDA content can reduce the lipid peroxidation products and protect biological membrane systems. It is extremely useful for alleviate some of the damage [10].

3.3. Bacteriostatic Experiment Results of Different Seed Coating Agents in Lab

The experiment results of 4#-1 and Shileshi seed coating agent on maize big spot pathogen, sheath blight fungus and head smut fungus were shown in **Table 3**. From the table, we knew that both 4#-1 and Shileshi had different degrees of improvement in inhibiting those funguses. The antibacterial effect of 4#-1 is higher than the Shileshi, its inhibition rate of maize big spot pathogen, sheath blight fungus and head smut fungus have reached 88.12%,

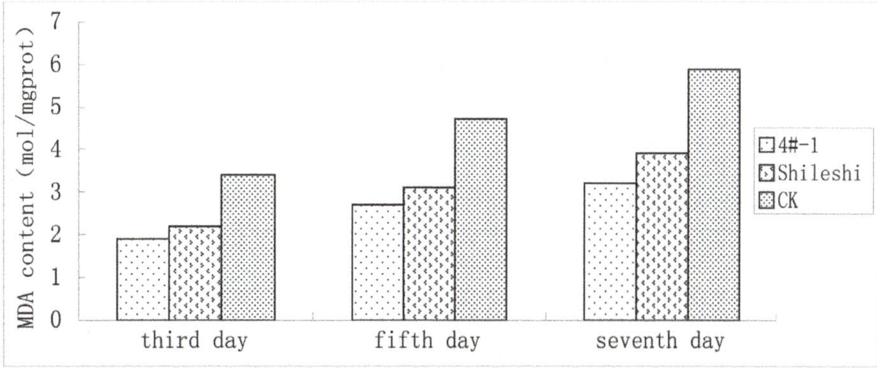

Figure 3. Effect of different seed coating agents on maize seedling MDA content.

Table 3. Bacteriostasis experiment results of different seed coating agents in lab.

Treatment Group	Large Plaque Bacteria		Sheath Blight Fungus		Head Smut Fungus	
	Colony Diameter/mm	Bacteriostatic Rate/%	Colony Diameter/mm	Bacteriostatic Rate/%	Colony Diameter/mm	Bacteriostatic Rate/%
4#-1	5.3	88.12	4.6	89.38	4.1	90.21
Shileshi	9.7	78.25	9.1	78.98	9.6	77.09
CK	44.6	-	43.3	-	41.9	-

89.38%, 90.21% respectively. It was visible that the self-made maize seed coating agent 4#-1 can effectively restrain harmful fungus in maize growth and reduce the incidence rate of seeds and plants.

3.4. Antibacterial Mechanism Analysis

When compared with traditional maize seed coating agent Shileshi, this novel environmental friendly maize seed coating agent 4#-1 had obvious advantage in improving antibacterial rate. This was mainly due to chitosan was added into this novel seed coating agent. By studying and summarizing, we got to know that there were two mechanism analysis of antimicrobial activity of chitosan.

First, the bacteriostasis effect was achieved by chitosan interfer with cell membrane. Chitosan form a very thick protective layer outside of seed cell when it was in high concentration, which hinder the cells to absorb nutrients [16], it can combine negative ions on the cell wall when in low concentration, such as fat polysaccharide and protein, which make cells in low stability and result in a leakage of intracellular nutrients [17]. In addition, the low degree of polymerization and low molecular chitosan are able to penetrate into the cell membrane, combine with cytoplasm, hinder the transcription of mRNA, which make the bacteriostatic function become true [18].

Second, the chitosan's destruction to the cell wall. The chitinase activity of pathogenic fungi will be stimulated by chitosan. When the chitosan concentration is very high, chitinase of fungus will be excessive expressed, which leads to chitin degradation of its own cell wall, thereby damages cell wall [19].

4. Conclusion

This study is built on a series of contrast experiments on the self-made environmental friendly maize seed coating agent 4#-1 and the traditional maize seed coating agent Shileshi. We tested the maize seedling root activity and some other parameters after the maize seeds were coated with different kinds of maize seed coating agent. The results showed that the self-made maize seed coating agent 4#-1 can effectively promote the maize root growth, increase maize seedlings chlorophyll content, reduce MDA content and inhibit a variety of maize bacteria. The environmental friendly maize seed coating agent can obviously increase maize ability to adapt outside environment and promote healthy growth of maize seedlings.

References

[1] Liu, Z.H., Xu, Y.W. and Shang, Y.J. (2010) Discuss on How to Improve the Quality of Pesticide Exposure. *Seed Science and Technology*, **8**, 9-11.

[2] Liu, P.F., Liu, X.L., Mu, K.G., Bai, J.J., Wu, H.X. and Wu, H.Y. (2000) Selection and Effects of Different Bentonite as Thickening Agent of Seed Coating Formulation. *Journal of Pesticide Science*, **2**, 62-67.

[3] Wu, X.H., Liu, X.L., Wang, H.M. and Gao, R.J. (2003) The Progress and Study of Seed Coating Agent in China. *Journal of Pesticides*, 5, 26-28.

[4] Li, L. (2008) Different Seed Coating Agent on the Influence of Corn Head Smut. *Seeds World*, **11**, 24.

[5] Wang, X.M., Dai, F.C. and Liao, Q. (2002) Manual of Corn Diseases and Insect Pests in the Field. China Agricultural Science and Technology Press, Beijing, 22-23, 95-95.

[6] Sha, H.L., Sha, H.Z., Fang, S.Q., Jin, Q.M. and Li, H. (2004) Control Effect of 6.9% Corn Seed Coating of Isofenphos-Methyl Triadimefon and Tebuconazole on Hypogeous Pests and Corn Smut. *Journal of Jilin Agricultural University*, **26**, 438-440, 444.

[7] He, X.Y., Kong, Q.Q. and Xi, X.M. (2008) Security Testing of 20% ke·fu·cuo Seed Coating Agent on Different Maize Seeds. *Inner Mongolia Agricultural Science and Technology*, **5**, 60-61.

[8] Lei, X.T., Pan, F.X. and Zhang, Y.M. (2007) Effects of Maibao Seed Clothing Agent on the Wheat Seed Vigor and Seedling Character. *Journal of Henan Institute of Science and Technology*, **35**, 8-10.

[9] TeKrony, D.M. (2006) Seeds: The Delivery System for Crop Science. *Crop Science*, **46**, 2263-2269. http://dx.doi.org/10.2135/cropsci2005.12.0445

[10] Zeng, D.F. and Tu, R.J. (2011) Preparation and Study of an Environmentally Friendly Seed-Coating Agent for Cucumber. *Frontiers of Agriculture in China*, **50**, 328-332. http://dx.doi.org/10.1007/s11703-011-1103-5

[11] Wang, J.S., Guo, C.R. and Zhang, Y. (1997) Agricultural Biological Chemical Technology. Science and Technology Press, Taiyuan.

[12] Lei, C.L. and Wang, M.Q. (2008) Central China Insect Research. China Agricultural Science and Technology Press, Beijing, 258-264.

[13] Wan, B.M. (2009) Summer Maize and Its Common Diseases Prevention and Treatment. *Journal of Seed Industry Tribune*, **8**, 34.

[14] Zhao, B. and He, S.J. (2002) Microbiology Experiment. Science Press.

[15] Benhamou, N. (1992) Ultrastructural and Cytochemical Aspects of Chitosan of *Fusarium oxysporum f. sp. radicis-lycopersici*, Agent of Tomato Crown and Root Rot. *Phytopathology*, **82**, 1185-1193.

[16] EI-Ghaouth, A., Arul, J., Asselin, A. and Benhamou, N. (1992) Antifungal Activity of Chitosan on Two Post-Harvest Pathogen of Strawberry Fruits. *Phytooath*, **82**, 398-402.

[17] Young, D., Köhle, H. and Kauss, H. (1982) Effect of Chitosan on Membrane Permeability of Suspension-Cultured *Glycine max* and *Phaseolus vulgaris* Cells. *Plant Physiology*, **70**, 1449-1454. http://dx.doi.org/10.1104/pp.70.5.1449

[18] Hadwinger, L.A., Kendra, D.F., Fristensky, B.W. and Wagoner, W. (1986) Chitin in Nature and Technology. Springer US, New York.

[19] Li, J.Y. and Li, H.Y. (2002) Bacteriostasis of Chitosan to Peach Brown Rot Fungus. *Journal of Electron Microscopy*, **21**, 38-140.

15

Biochemical Fractionation of Soil Organic Matter after Incorporation of Organic Residues

Serge-Étienne Parent, Léon E. Parent*

Department of Soils and Agrifood Engineering, Université Laval, Québec, Canada
Email: *leon-etienne.parent@fsaa.ulaval.ca

Abstract

Soil organic matter (SOM) is a key factor for building and maintaining soil quality. The SOM quality is commonly assessed using densitometric and sieving separation methods, but such methods do not inform on the biochemical composition of SOM. Our objective was to evaluate the van Soest extraction procedure for soluble (SOL), holocellulose (HOLO) and lignin/cutin (LIC) fractions of SOM after incorporating crop residues and animal wastes into a C-depleted loamy sand. Millet cuttings, oat straw, fresh cattle manure and cattle manure compost were dried, sieved to obtain 53 - 250 and 250 - 2000 μm size fractions and characterized biochemically using a modified NDF-ADF-ADL van Soest method. Soil was also sieved into 53 - 250 and 250 - 2000 μm fractions. On a dry mass basis, crop residues contained 60% - 70% holocellulose while animal wastes contained more than 40% ash. Each soil fraction was combined with three rates of the corresponding organic fraction (2, 4, and 6 $Mg \cdot ha^{-1}$ millet forage cuttings or oat straw and 5, 10, and 15 $Mg \cdot ha^{-1}$ of cattle manure or cattle manure compost). Changes in soil biochemical components were analyzed using the balance method of compositional data analysis. Amendment, application rate and size fraction influenced significantly ($p < 0.05$) the [SOL | HOLO] balance but did not significantly affect the [SOL,HOLO | LIC] balance. The [SOL | HOLO] increased linearly with addition rate of crop residues, and decreased linearly with addition rate of animal wastes. This approach of balancing biochemical SOM components is a promising method to monitor the changes in SOM quality after the incorporation of organic residues and to elaborate beneficial practices for managing crop residues and animal wastes in agro-ecosystems.

Keywords

Soil Organic Matter, Van Soest, Holocellulose, Lignin, Compositional Data Analysis

*Corresponding author.

1. Introduction

Soil organic carbon (SOC) is the primary soil quality indicator driving soil chemical, biological, and physical functions [1]-[3]. The SOC decomposition rate depends on physical, chemical and biochemical protection mechanisms [4]. The physical protection to SOC against decomposition is provided by soil micro-aggregates (<250 μm in size) that are compact assemblages of mineral and organic matter [5]. The SOC models generally consider two or three biochemical C pools [6] [7]. The most labile SOC pools are sensitive indicators of soil biological activity [8] and structural stability [9] [10] resulting from agro-ecosystem management [11]-[14].

Labile and recalcitrant soil organic matter (SOM) fractions can be separated by size [15]. The particulate organic matter (POM) fraction is 53 to 2000 μm in size and provides an estimate of labile C [16]. The POM often has a bulk density less than one [17] [18], hence facilitating densimetric separation, is high in C [19], and contains plant debris [20], hyphae, spores, seeds, faunal skeletons, microbial biomass and partially humified materials [21] that can be separated by sieving. The biochemical composition of POM requires analyzing POM using a biochemical fractionation procedure.

In general, organic matter added to soil is analyzed before incorporation. Its biochemical stability in soil is measured by incubation [7] [22]. The van Soest method [23] [24] which is a routine biochemical fractionation method of fiber analysis for neutral detergent fiber (NDF), acid detergent fiber (ADF) and acid detergent lignin (ADL) has been modified to analyze the biochemical composition of organic residues and composts before soil incorporation [22] [25] and can be expanded to analyze the biochemical composition of POM after incorporation of organic amendments into soil.

Our objective was to test the capacity of a modified van Soest method to characterize the biochemical composition of soil C after incorporation of organic amendments into soil with special attention to the 53 - 250 and 250 - 2000 μm POM fractions. In this experiment, we incorporate four organic sources at increasing rates into soil and determine the biochemical composition of mixtures.

2. Material and Methods

We sampled an Orthic Humo-ferric Podzol [26] sandy loam of the Ivry soil series collected on an experimental field at Saint-Ubalde de Portneuf, Quebec, Canada (46°45'23"N, 123 72°19'57"W), containing 1.6% SOM (550°C, 16 h) in the 0 - 17 cm layer; pH ($CaCl_2$ 0.01 M) was 5.2. In potato (*Solanum tuberosum* L.) crop systems, millet cuttings and oat straw, as well as cattle manure and manure compost, are often incorporated into soil. Samples of pearl millet (*Pennisetum glaucum*) forage cuttings, oat (*Avena sativa* L.) straw, cattle manure (feedlot) and cattle manure compost (field stored manure) were ground to ≤ 2 mm and sieved to obtain 53 - 250 and 250 - 2000 μm fractions of POM [27]. Millet forage has lowest LIC content due to mid-season sampling [28]. Cattle manure compost is presumably the most altered material. Soil particles were also separated into 53 - 250 and 250 - 2000 μm sieve-size fractions after shaking a mixture of soil and sodium hexametaphosphate (5 $g \cdot L^{-1}$) solution for 16 h at 360 strokes min^{-1}. Each size fraction of organic residues was mixed with the same soil sieve-size fraction to represent the amount of soil mineral and organic fractions that could be recovered on the same sieve. Based on a soil bulk density of 1.32 $g \cdot mL^{-1}$ and an incorporation depth of 17 cm, crop residues were added at rates of 2, 4, and 6 $Mg \cdot ha^{-1}$ (0.0267, 0.0534 and 0.8010 g of plant residue per 30 g of soil, respectively) and animal wastes were incorporated at rates of 5, 10, and 15 $Mg \cdot ha^{-1}$ (0.0677, 0.1344 and 0.2017 g per 30 g of soil), on a dry mass basis.

Fibers were extracted following the French standard XPU 44-162 of the van Soest procedure adapted to compost biochemical fractionation [22]. Briefly, 2.00 ± 0.05 g of the soil/organics mixtures were weighed in F57 filter bags and placed into an Ankom$^{200/220}$ fiber analyzer. At each extraction step, the ash content of residual matter was quantified. The proportion of soluble substances (SOL) was computed as %SOM - %NDF (neutral detergent fibers) and that of holocelullose as %NDF - %ADL; the residual lignin plus cutin fraction was reported as %ADL (acid-digestible lignin). The recalcitrant LIC fraction was computed as %SOM - %SOL - % HOLO. Because the LIC fraction is computed by difference, experimental errors must accumulate in the LIC fraction.

3. Compositional Data Analysis

Because biochemical compositions are constrained to 100% and one component is computed by difference, they convey important properties as follows: 1) the normality assumption—which scans the real space—is inappropriate because confidence intervals may reach beyond the compositional space, and 2) the covariance matrix is biased (if the proportion of a component increases, at least another must decrease) and spurious (the covariance depends on scale). As compositional data transformation method, the additive log dual ratio (alr) or log contrast between two proportions (c_1 and c_2) has been first proposed based on the logistic distribution as follows [29]:

$$alr = \ln\left(\frac{c_1}{c_2}\right) = \ln(c_1) - \ln(c_2) \quad (1)$$

A D-part compositional vector has D-1 degrees of freedom [30] and can be expanded to a multiplicative logistic distribution [29]. To avoid biases when analyzing the biochemical fractions of maize mulch residues, [31] used orthonormal balances [32]. The isometric log ratio (ilr) provides D-1 orthonormal balances that allow avoiding biases when analyzing compositional data [32]. The ilr transformation is based on a system of balances formalized as a sequential binary partition (SBP), *i.e.* a (*D*-1) ×*D* matrix, in which parts labeled "+1" (group numerator) are contrasted with parts labeled "−1" (group denominator) in each ordered row. A part labeled "0" is excluded from the balance. The composition is partitioned sequentially at every ordered row into two contrasts until the (+1) and (−1) non-overlapping subsets each contain a single part. The ilr is a ratio of the geometric means of the (+) and (–) groups normalized by a coefficient, computed as follows [32]:

$$ilr_k = \sqrt{\frac{r_k s_k}{r_k + s_k}} \ln\left(\frac{\left(c_{i_1} c_{i_2} \times \ldots \times c_{i_r}\right)^{\frac{1}{r_k}}}{\left(c_{j_1} c_{j_2} \times \ldots \times c_{j_s}\right)^{\frac{1}{s_k}}}\right) \quad (2)$$

where, in the kth row of the SBP, r_k is the number of proportions in the (+) group, s_k is the number of proportions in the minus (–) group, c_{i_n} is the nth component in the (+) group and c_{i_n} is the nth component in the (–) group. For example, the balance between the more labile fractions (SOL and HOLO) is computed as follows:

$$[SOL|HOLO] = \sqrt{\frac{1x1}{1+1}} \ln\left(\frac{HOLO}{SOL}\right) \quad (3)$$

The balance between the more labile (SOL and HOLO) and the more recalcitrant (LIC) fractions is computed as follows:

$$[SOL, HOLO|LIC] = \sqrt{\frac{2x1}{2+1}} \ln\left(\frac{LIC}{\sqrt{SOLxHOLO}}\right) \quad (4)$$

where SOL, HOLO, and LIC are proportions of each biochemical fractions of SOM expressed on a common basis (e.g. dry soil mass or organic mass). Note that balances are subcompositionally coherent, *i.e.* results are independent whether results are expressed relatively to the fresh, dry or organic mass. This is an important advantage of the balance over percentages that vary with scale (moist, wet or organic basis), because the analytical results can be interpreted coherently whatever the scale referential [33] [34]. Fractions on the left are at denominator and those on the right at numerator (Equation (1)) because in algebra, numbers are generally row-listed from the most negative to the most positive.

In this paper, we contrasted the labile SOM fractions (SOL, HOLO) with the more recalcitrant one (LIC), then SOL and HOLO fractions as follows: [HOLO,SOL | LIC] and [SOL | HOLO] (**Figure 1**).

4. Statistical Analysis

For each of the two soil fractions (53 - 250 and 250 - 2000 μm), there were three rates of organic residues of the corresponding sieve-size fraction. Each treatment was replicated nine times. The statistical setup was a completely random factorial design. The statistical analysis was performed using the [HOLO,SOL | LIC] and [SOL | HOLO] balances as response variables along a gradient of POM additions. Fixed descriptors were size fraction (53 - 250 and 250 - 2000 μm), source of organic residue (millet, oat manure, compost) and application rate (continuous variable). The model was computed in the R statistical environment [35] with the lme4 [36] and lmer Test [37] packages. Prior to analysis, outliers were discarded at the 0.05 level of significance using the voutlier package [38]. The compositions package [39] was used to transform percentages into ilrs, hence providing D-1 degrees of

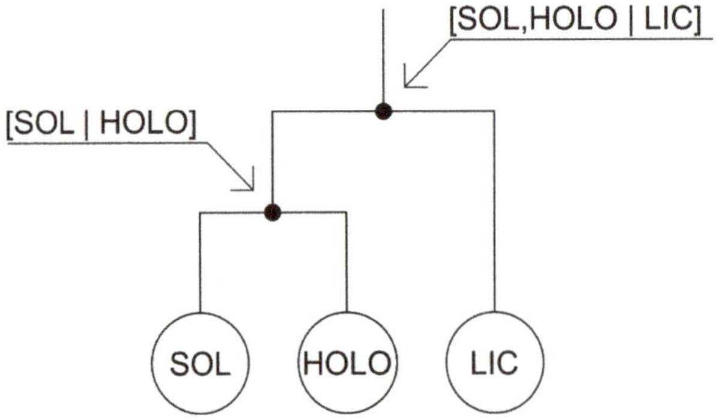

Figure 1. Balances between biochemical components.

freedom [40].

5. Results

5.1. Organic Residues

There was a large variation in biochemical fractions among organic residues and their size fractions expressed on either dry or organic mass basis while balances between biochemical components remained constant (**Table 1**). The correlation between biochemical fractions expressed on a soil or an organic mass varied from 0.47 (ns) for SOL to 0.75 ($p < 0.05$) for LIC and 0.936 ($p < 0.01$) for HOLO, indicating the large influence of ash content on the compositional vector. Balances thus avoid incoherence. Proportions computed on different bases that obviously return different results are referred to as "subcompositional incoherence" or scale dependency. However, the influence of the basis (dry soil or organic mass) is avoided by ratioing the proportions (e.g. [HOLO/SOM]/[SOL/SOM] = HOLO/SOL) as shown on the right-hand side of **Table 1**.

Balances between biochemical fractions expressed as isometric log ratios (Equations (2)-(4)) varied little in animal wastes, indicating that the manure compost was still immature despite partial composting in field piles. Indeed, the ilr transformation maps the dry mass composition from the three initial biochemical components (SOL, HOLO, LIC) in the compositional space to two orthonormal balances in the Euclidean space. Ash content was much lower in crop residues than animal wastes. Ash content was highest in the fine POM fraction of cattle manure, indicating mixture of manure with soil in the feedlot. Hence, the amounts of biochemical components added to soil were smaller in the fine than the coarse POM fractions. The advantage of orthonormal balances is shown here by the partitioning of organic and mineral components into linearly independent subsets.

The proportion of SOL was higher in the fine than the coarse POM fractions of crop residues, indicating that biochemical components were sorted out upon sieving. Compared to oat straw that was collected in the fall, the SOL fraction was higher and HOLO smaller in millet forage cuttings due to in-season sampling far from plant-maturity. Compared to their coarse POM fraction, the fine POM fraction of cattle manure and manure compost also appeared higher in SOL content.

5.2. Soil-Residue Mixtures

There were significant differences ($p \leq 0.05$) in biochemical composition of the mixtures between soil and organic amendments (**Table 2**). The SOL component varied with the size fraction. The HOLO fraction varied with the source of organic amendment, application rate as well as the size fraction and its interaction with the organic source. The LIC fraction varied with the source of organic amendment, the rate of application, the interaction between the source of amendment and the rate of application, and the interaction between the size fraction and the rate of application. However, as we recall from compositional data theory, the variance analyses computed across the inter-related SOL, HOLO and LIC proportions that have D-1 matrix rank are not statistically valid: the statistics should be interpreted from the balances between components, *i.e.* [SOL,HOLO | LIC] and [SOL| HOLO], at fulcrum (**Figure 1**), as isometric log ratios between two geometric means (Equation (1)).

The [SOL,HOLO | LIC] balance of size fractions did not differ significantly ($p < 0.05$) among treatments (**Table 2**). The [SOL,HOLO | LIC] balance thus appeared to be a poor indicator of SOM quality after incorporation

Table 1. Average biochemical compositions and compositional balances of four organic residues.

Organic residue	Sieve size	Ash	SOL	HOLO	LIC	Balances	
	µm					[SOL,HOLO | LIC]	[SOL | HOLO]
			% on dry matter basis				
Millet forage	2000 - 250	16.6	18.1	60.8	4.5	1.632	−0.858
	250 - 53	17.3	20.9	58.8	3.0	1.997	−0.732
Oatstraw	2000 - 250	12.5	3.4	71.2	12.9	0.151	−2.150
	250 - 53	12.7	8.9	69.8	8.5	0.877	−1.455
Cattlemanure	2000 - 250	52.4	14.3	18.2	15.1	0.052	−0.168
	250 - 53	87.6	6.4	3.3	2.7	0.436	0.461
Manure compost	2000 - 250	41.2	16.9	22.8	19.1	0.024	−0.213
	250 - 53	47.8	27.5	12.1	12.5	0.308	0.579
			% on organicmatter basis				
Millet forage	2000 - 250	-	21.7	72.9	5.4	1.632	−0.858
	250 - 53	-	25.2	71.1	3.7	1.997	−0.732
Oatstraw	2000 - 250	-	3.9	81.3	14.7	0.151	−2.150
	250 - 53	-	10.2	80.0	9.8	0.877	−1.455
Cattlemanure	2000 - 250	-	30.1	38.1	31.8	0.052	−0.168
	250 - 53	-	51.4	26.8	21.8	0.436	0.461
Manure compost	2000 - 250	-	28.7	38.8	32.4	0.024	−0.213
	250 - 53	-	52.7	23.3	24.0	0.308	0.579

Table 2. Effects of organic residues, rate of application and particle size on proportions of organic and mineral pools and on compositional balances between biochemical pools. SOL = soluble; HOLO = holocellulose; LIC = lignin + cutin; DMB = dry mass basis.

Source of variation	Proportion on DMB			Balance	
	SOL	HOLO	LIC	[SOL,HOLO | LIC]	[SOL | HOLO]
Organics	ns	***	*	ns	*
Rate	ns	***	**	ns	*
Size	**	*	ns	ns	***
Organics: Rate	ns	***	*	ns	***
Organics: Size	ns	ns	ns	ns	ns
Rate: Size	ns	ns	*	ns	ns
Organics: Rate: Size	ns	ns	ns	ns	ns

ns, * and **: non-significant and significant at the 0.10, 0.05 and 0.01 levels, respectively.

of organic residues into soil. On the other hand, the organic amendment, application rate, size fraction and the interaction between the organic amendment and application rate influenced the [SOL | HOLO] balance significantly ($p \leq 0.05$). The SOL weighted more in the fine than the larger fraction as shown by a more negative [SOL | HOLO] balance in the finer fraction.

The [SOL | HOLO] balance facilitated interpreting the results as shown by well-defined linear response trends upward for crop residues and downward for animal wastes (**Figure 2**). The [SOL | HOLO] balance decreased linearly with the addition rate of animal wastes due to a combination of higher SOL and lower HOLO contents in animal wastes (**Table 1**) resulting from feed digestion in the animal tract and partial decomposition in manure piles. Hence, compared to current laboratory methods and the ensuing statistical analysis distorted by raw compositions, a combination of the modified van Soest extraction method with tools of compositional data analysis proved to be useful in interpreting the changes in SOM quality after incorporating organic residues into soil.

6. Discussion

Soil organic carbon is protected through physical, chemical, and biochemical mechanisms [4]. This paper is a first attempt to provide a biochemical indicator of SOM quality measured as a log ratio between two operational pools using a van Soest procedure that has been modified to assess compost biological stability. Adding crop and animal residues to soil changed the operational biochemical pools of the new mixture. After incorporation of organic residues into SOM, the HOLO pool and the [SOL | HOLO] balance were found to be highly contrasting across size fractions but the direction of the change depended on the organic residue and the application rate. On the other hand, the functionality of biochemical pools must depend on the chemical make-up of their organic polymers and molecules and could be addressed in future research.

Organic residues and SOM are complex mixtures of organic molecules [41]. Organic C from fresh organic materials is labile and a small fraction accumulates as recalcitrant humus [42]. There are also products of intermediate decomposability or highly resistant to decomposition [43]. The SOL fraction of crop residues is a mixture of proteins, starch, lipids and other easily decomposable components while other fractions are specific to hemicelluloses, cellulose and lignin plus cutin, respectively [44]. The NDF may extract a large variety of soluble products [45] like phenolic compounds that color the extracting solution (yellow to brown color) and that may resist to decomposition. As a result, the SOL fraction could comprise both labile and recalcitrant C. The LIC fraction in the present study is an elusive component computed by difference that may contain plant lignin and cutin as well as recalcitrant humic substances and humin. Although humic substances are generally associated with the soil mineral fractions [8] [14] [46], NDF soluble humic substances from SOM may swamp out NDF soluble materials from organic residues. The HOLO fraction is the most specific one. The HOLO and the labile part of SOL could be considered as proxies for labile C pools. However, the van Soest method did not allow separating the labile from the recalcitrant C forms in the SOL fraction while a SOM quality index must take into account the compositional nature of biochemical fractions. Balances are computed as isometric log ratios to avoid redundancy (three components closed to 100% SOM are reduced to two linearly independent variables) and confusing scale dependency (the ratios eliminate the basis as soil dry mass or organic mass that return different proportions from the same analytical results). The [SOL | HOLO] balance computed using Equation 3 proved to be a biochemical POM index that can be detected by the modified van Soest method. Because a single balance proved to be significant, the distance between two compositions can be computed as a difference between two log ratios (Equations 1 or 3) rather than a Euclidean distance across two or more orthonormal balances.

Existing fractionation methods are divided into physical and chemical fractionation, which are based on different scopes regarding the mechanisms of SOC stabilization [47]. The POM is a SOM quality indicator sensi-

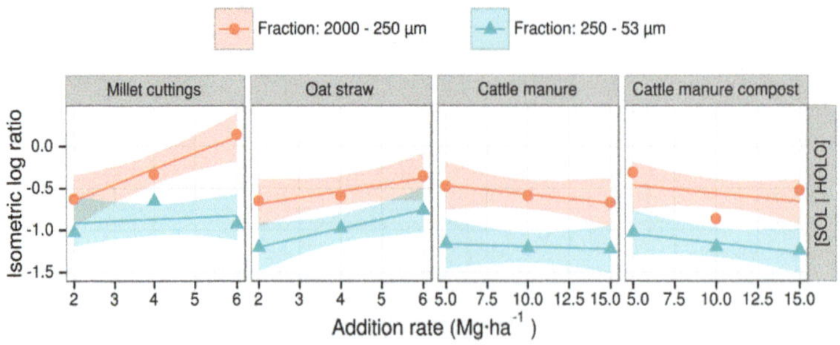

Figure 2. Change inthe [SOL | HOLO] balance with addition rates of organic residues in soil size fractions. Error ribbons reflect the precision of the isometric log ratio prediction at a 0.05 level.

tive to changes in C inputs [48]. In a mechanistic approach to soil C sequestration, [4] [49] suggested that SOC fractions have their own specific saturation capacity in the soil. In support of this theory, [50] found evidence of soil C saturation of chemically and biologically protected C pools far from their theoretical C saturation, while saturation of physically protected C occurred closer to their C saturation level. However, a mechanistic explanation of a saturation level is still required for the unprotected and the biochemically protected C pools [49]. Loose terms such as "recalcitrance', the vague definitions for fractions such as particulate organic matter (POM) or mineral-associated organic matter (MOM) and poorly defined fractionation procedures have confused the discussion on SOM [47]. Although this research did not address the functionality of the operationally derived [SOL | HOLO] balance, it showed that the van Soest method modified for compost analysis can detect the changes in SOM quality and thus has the potential to assist modeling changes in biochemical C pools in soil particle-size fractions as driven by crop residues and animal wastes management.

7. Conclusions

A sensitive biochemical SOM quality index supported by a laboratory method could contribute to elaborating theories on C sequestration and cycling in soils and beneficial practices for managing crop residues and animal wastes in agro-ecosystems. The van Soest method detected changes in SOM quality in the 53 - 250 and 250 - 2000 μm fractions after mixing a C-depleted acid sandy loam soil with 2 to 6 $Mg \cdot ha^{-1}$ of crop residues or 5 to 15 $Mg \cdot ha^{-1}$ of animal wastes. The [SOL | HOLO] balance computed as an isometric log ratio was found to respond linearly to additions of crop residues and animal wastes across size fractions.

The van Soest method could be tested on other soils using a larger spectrum of organic products and cultural practices because such method had the potential to monitor changes in SOM quality following residue management practices. The change in [SOL | HOLO] balance could be related to soil properties and functions to improve SOM management in agro-ecosystems.

Acknowledgements

This project is part of a more extensive project entitled "Implementing Means to Increase Potato Ecosystem Services" supported financially by the Natural Sciences and Engineering Council of Canada (CRDPJ 385199-09) and the following Canadian farm partners: Cultures Dolbec Inc., St-Ubalde, Québec; Groupe Gosselin FG Inc., Pont Rouge, Québec; Agriparmentier Inc. and Prochamps Inc., Notre-Dame-du-Bon-Conseil, Québec; Ferme Daniel Bolduc et Fils Inc., Péribonka, Québec.

References

[1] Doran, J.W. and Parkin, T.B. (1994) Defining and Assessing Soil Quality. In: Doran, J.W., Coleman, D.C., Bezdicek, D.F.and Stewart, B.A., Eds., *Defining Soil Quality for a Sustainable Environment*, Soil Science Society of America Journal, Madison, 3-21. http://dx.doi.org/10.2136/sssaspecpub35.c1

[2] Louwagie, G., Gay, S.H. and Burrell, A. (2009) Addressing Soil Degradation in EU Agriculture: Relevant Processes, Practices and Policies. Report on the Project Sustainable Agriculture and Soil Conservation (SoCo), European Communities. http://eusoils.jrc.ec.europa.eu/esdb_archive/eusoils_docs/other/EUR23767.pdf

[3] Weil, R.R. and Magdoff, F. (2004) Significance of Soil Organic Matter to Soil Quality and Health. In: Magdoff, F. and Weil, R.R., Eds., *Soil Organic Matter in Sustainable Agriculture*, CRC Press, Boca Raton, 1-43. http://dx.doi.org/10.1201/9780203496374.ch1

[4] Stewart, C.E., Paustian, K., Conant, R.T., Plante, A.F. and Six, J. (2007) Soil Carbon Saturation: Concept, Evidence and Evaluation. *Biogeochemistry*, **86**, 19-31. http://dx.doi.org/10.1007/s10533-007-9140-0

[5] Tisdall, J.M. and Oades, J.M. (1982) Organic Matter and Water-Stable Aggregates in Soils. *Journal of Soil Science*, **33**, 141-163. http://dx.doi.org/10.1111/j.1365-2389.1982.tb01755.x

[6] Andrén, O. and Kätterer, T. (1997) ICBM: The Introductory Carbon Balance Model for Exploration of Soil Carbon Balances. *Ecological Applications*, **7**, 1226-1236. http://dx.doi.org/10.1890/1051-0761(1997)007[1226:ITICBM]2.0.CO;2

[7] Thuriès, L., Pansu, M., Feller, C., Herrmann, P. and Rémy, J.C. (2001) Kinetics of Added Organic Matter Decomposition in a Mediterranean Sandy Soil. *Soil Biology and Biochemistry*, **33**, 997-1010. http://dx.doi.org/10.1016/S0038-0717(01)00003-7

[8] Gregorich, E.G. and Janzen, H.H. (1996) Storage of Soil Carbon in the Light Fraction and Macro-Organic Matter. In: Carter, M.R. and Stewart, B.A., Eds., *Structure and Soil Organic Matter Storage in Agricultural Soils*, CRC Press, Boca Raton, 167-190.

[9] Haynes, R.J. and Beare, M.H. (1996) Aggregation and Organic Matter Storage in Mesothermal, Humid Soils. In: Carter, M.R. and Stewart, B.A., Eds., *Structure and Soil Organic Matter Storage in Agricultural Soils*, CRC Press, Boca Raton, 213-262.

[10] Janzen, H.H., Campbell, C.A., Ellert, B.H. and Bremer, E. (1997) Soil Organic Matter Dynamics and Their Relationship to Soil Quality. In: Gregorich, E.G. and Carter, M.R., Eds., *Soil Quality for Crop Production and Ecosystem Health*, Elsevier Science, Amsterdam, 277-291. http://dx.doi.org/10.1016/s0166-2481(97)80039-6

[11] Cambardella, C.A. and Elliott, E.T. (1992) Particulate Soil Organic-Matter Changes across a Grassland Cultivation Sequence. *Soil Science Society of America Journal*, **56**, 777-783. http://dx.doi.org/10.2136/sssaj1992.03615995005600030017x

[12] Campbell, C.A., Biederbeck, V.O., Wen, G., Zentner, R.P., Schoenau, J. and Hahn, D. (1999) Seasonal Trends in Selected Soil Biochemical Attributes: Effects of Crop Rotation in the Semiarid Prairie. *Canadian Journal of Soil Science*, **79**, 73-84. http://dx.doi.org/10.4141/S98-008

[13] Graham, M.H., Haynes, R.J. and Meyer, J.H. (2002) Soil Organic Matter Content and Quality: Effects of Fertilizer Applications, Burning and Trash Retention on a Long-Term Sugarcane Experiment in South Africa. *Soil Biology and Biochemistry*, **34**, 93-102. http://dx.doi.org/10.1016/S0038-0717(01)00160-2

[14] Wander, M. and Nissen, T. (2004) Value of Soil Organic Carbon in Agricultural Lands. *Mitigation and Adaptation Strategies for Global Change*, **9**, 417-431. http://dx.doi.org/10.1023/B:MITI.0000038847.30124.77

[15] Balesdent, J. (1996) Un point sur l'évolution des réserves organiques des sols de France. *Étude et Gestion des Sols*, **3**, 245-260.

[16] Cambardella, C.A., Gajda, A.M., Doran, J.W., Wienhold, B.J. and Kettler, T.A. (2001) Estimation of Particulate and Total Organic Matter by Weight Loss-on-Ignition. In: Lal, R., Kimble, J.M., Follett, R.F. and Stewart, B.A., Eds., *Assessment Methods for Soil Carbon, Advances in Soil Science*, CRC Press, Boca Raton, 349-359.

[17] Guggenberger, G. and Zech, W. (1999) Soil Organic Matter Composition under Primary Forest, Pasture, and Secondary Forest Succession, Region Huetar Norte, Costa Rica. *Forest Ecology and Management*, **124**, 93-104. http://dx.doi.org/10.1016/S0378-1127(99)00055-9

[18] Six, J., Elliott, E.T., Paustian, K. and Doran, J.W. (1998) Aggregation and Soil Organic Matter Accumulation in Cultivated and Native Grassland Soils. *Soil Science Society of America Journal*, **62**, 1367-1377. http://dx.doi.org/10.2136/sssaj1998.03615995006200050032x

[19] Carter, M.R., Gregorich, E.G., Angers, D.A., Donald, R.G. and Bolinder, M.A. (1998) Organic C and N Storage, and Organic C Fractions, in Adjacent Cultivated and Forested Soils of Eastern Canada. *Soil and Tillage Research*, **47**, 253-261. http://dx.doi.org/10.1016/S0167-1987(98)00114-7

[20] Skjemstad, J.O., Lefeuvre, R.P. and Prebble, R.E. (1990) Turnover of Soil Organic Matter under Pasture as Determined by ^{13}C Natural Abundance. *Australian Journal of Soil Research*, **28**, 267-276. http://dx.doi.org/10.1071/SR9900267

[21] Baldock, J.A., Oades, J.M., Waters, A.G., Peng, X., Vassallo, A.M. and Wilson, M.A. (1992) Aspects of the Chemical Structure of Soil Organic Materials as Revealed by Solid-State Carbon-13 NMR Spectroscopy. *Biogeochemistry*, **16**, 1-42. http://dx.doi.org/10.1007/BF02402261

[22] Lashermes, G., Nicolardot, B., Parnaudeau, V., Thuriès, L., Chaussod, R., Guillotin, M.L., Linères, M., Mary, B., Metzger, L., Morvan, T., Tricaud, A., Villette, C. and Houot, S. (2009) Indicator of Potential Residual Carbon in Soils after Exogenous Organic Matter Application. *European Journal of Soil Science*, **60**, 297-310. http://dx.doi.org/10.1111/j.1365-2389.2008.01110.x

[23] Van Soest, P.J. and Wine, R.H. (1967) Use of Detergents in the Analysis of Fibrous Feeds. IV. Determination of Plant Cell-Wall Constituents. *Journal of the Association of Official Analytical Chemists*, **50**, 50-55.

[24] Van Soest, P.J. and Wine, R.H. (1968) Determination of Lignin and Cellulose in Acid-Detergent Fiber with Permanganate. *Journal of the Association of Official Analytical Chemists*, **51**, 780-785.

[25] Gabrielle, B., Da-Silveira, J., Houot, S. and Francou, C. (2004) Simulating Urban Waste Compost Effects on Carbon and Nitrogen Dynamics Using a Biochemical Index. *Journal of Environmental Quality*, **33**, 2333-2342. http://dx.doi.org/10.2134/jeq2004.2333

[26] Soil Classification Working Group (1998) The Canadian System of Soil Classification. 3rd Edition, Agriculture and Agri-Food Canada Publications, Ottawa, 1646.

[27] Gregorich, E.G. and Beare, M.H. (2007) Physically Uncomplexed Organic Matter. In: Carter, M.R. and Gregorich, E.G., Eds., *Soil Sampling and Methods of Analysis*, 2nd Edition, CRC Press, Boca Raton, 1262.

[28] Morales, J.U., Alatorre, A.H., Escalante, A.A., Lopez, S.B., Vazquez, H.G. and Gomez, M.O.D. (2011) Nutritional Characteristics of Silage and Hay of Pearl Millet at Four Phonological Stages. *Journal of Animal and Veterinary Advances*, **10**, 1378-1382. http://dx.doi.org/10.3923/javaa.2011.1378.1382

[29] Aitchison, J. (1986) The Statistical Analysis of Compositional Data. Chapman and Hall, London. http://dx.doi.org/10.1007/978-94-009-4109-0

[30] Aitchison, J. and Greenacre, M. (2002) Biplots of Compositional Data. *Journal of the Royal Statistical Society: Series C (Applied Statistics)*, **51**, 375-392. http://dx.doi.org/10.1111/1467-9876.00275

[31] Aslam, S., Garnier, P., Rumpel, C., Parent, S.-É. and Benoit, P. (2013) Adsorption and Desorption Behavior of Selected Pesticides as Influenced by Decomposition of Maize Mulch. *Chemosphere*, **91**, 1447-1455. http://dx.doi.org/10.1016/j.chemosphere.2012.12.005

[32] Egozcue, J.J. and Pawlowsky-Glahn, V. (2005) Groups of Parts and Their Balances in Compositional Data Analysis. *Mathematical Geology*, **37**, 795-828. http://dx.doi.org/10.1007/s11004-005-7381-9

[33] Abdi, D., Cade-Menun, B.J., Ziadi, N. and Parent, L.-É. (2015) Compositional Statistical Analysis of Soil 31P-NMR Forms. *Geoderma*, in Press.

[34] Parent, S.-É., Parent, L.E., Rozane, D.-E., Hernandes, A. and Natale, W. (2012) Nutrient Balance as Paradigm of Soil and Plant Chemometrics. In: Issaka, R.N., Ed., *Soil Fertility*, InTech, Rijeka, Croatia, 83-114.

[35] R Core Team (2015) R: A Language and Environment for Statistical Computing.

[36] Bates, D., Maechler, M., Bolker, B. and Walker, S. (2014). Lme4: Linear Mixed-Effects Models Using Eigen and S4. *Journal of Statistical Software*, E-Print.

[37] Kuznetsova, A., Brockhoff, P.B. and Bojesen Christensen, R.H. (2015) LmerTest: Tests in Linear Mixed Effects Models.

[38] Filzmoser, P. and Gschwandtner, M. (2015) Mvoutlier: Multivariate Outlier Detection Based on Robust Methods.

[39] Van den Boogaart, K.G., Tolosana-Delgado, R. and Bren, M. (2014) Compositions: Compositional Data Analysis.

[40] Greenacre, M. (2011) Measuring Subcompositional Incoherence. *Mathematical Geosciences*, **43**, 681-693. http://dx.doi.org/10.1007/s11004-011-9338-5

[41] Paul, E.A. and Clark, F.E. (1996) Soil Microbiology and Biochemistry. 2nd Edition, Academic Press, London.

[42] Robert, M. (2001) Soil Carbon Sequestration for Improved Land Management. Rome. ftp://ftp.fao.org/agl/agll/docs/wsrr96e.pdf

[43] Haynes, R.J. (2005) Labile Organic Matter Fractions as Central Components of the Quality of Agricultural Soils: An Overview. *Advances in Agronomy*, **85**, 221-268. http://dx.doi.org/10.1016/s0065-2113(04)85005-3

[44] Francou, C. (2003) Stabilisation de la matière organqiue au cours du compostage de déchets urbains: Influence de la nature des déchets et du procédé de compostage—Recherche d'indicateurs pertinents. Ec. Dr. ABIES. Institut National Agronomique Paris-Grignon. https://tel.archives-ouvertes.fr/file/index/docid/47348/filename/tel-00007519.pdf

[45] Robin, D. (1997) Intérêt de la caractérisation biochimique pour l'évaluation de la proportion de matière organique stable après décomposition dans le sol et la classification des produits organominéraux. *Agronomie*, **17**, 157-171. http://dx.doi.org/10.1051/agro:19970303

[46] Hayes, M.H.B. and Clapp, C.E. (2001) Humic Substances: Considerations of Compositions, Aspects of Structure, and Environmental Influences. *Soil Science*, **166**, 723-737. http://dx.doi.org/10.1097/00010694-200111000-00002

[47] Poeplau, C., Don, A., Dondini, M., Leifeld, J., Nemo, R., Schumacher, J., Senafati, N. and Wiesmeier, M. (2013) Reproductibility of a Soil Organic Carbon Fractionation Method to Derive RothC Carbon Pools. *European Journal of Soil Science*, **64**, 735-746. http://dx.doi.org/10.1111/ejss.12088

[48] Carter, M.R. (2002) Soil Quality for Sustainable Land Management: Organic Matter and Aggregation Interactions That Maintain Soil Functions. *Agronomy Journal*, **94**, 38-47. http://dx.doi.org/10.2134/agronj2002.0038

[49] Six, J., Conant, R.T., Paul, E.A. and Paustian, K. (2002) Stabilization Mechanisms of Soil Organic Matter: Implications for C-Saturation of Soils. *Plant and Soil*, **241**, 155-176. http://dx.doi.org/10.1023/A:1016125726789

[50] Stewart, C.E., Plante, A.F., Paustian, K., Conant, R.T. and Six, J. (2008) Soil Carbon Saturation: Linking Concept and Measurable Carbon Pools. *Soil Science Society of America Journal*, **72**, 379-392. http://dx.doi.org/10.2136/sssaj2007.0104

16

Comparison of the Effect of Vermicompost and Inorganic Fertilizers on Vegetative Growth and Fruit Production of Tomato (*Solanum lycopersicum* L.)

Md. Abul Kashem*, **Ashoka Sarker, Imam Hossain, Md. Shoffikul Islam**

Department of Soil Science, University of Chittagong, Chittagong, Bangladesh
Email: *kashem00@yahoo.com

Abstract

This study attempted to compare the effect of cow manure vermicompost and inorganic fertilizers on the vegetative growth and fruits of tomato plant (*Solanum lycopersicum* L.). An air dried sandy loam soil was mixed with five rates of vermicompost equivalent to 0 (control), 5, 10, 15 and 20 t ha^{-1} and three rates of NPK fertilizer equivalent to 50% (N-P-K = 69-16-35 kg ha^{-1}), 100% (N-P-K = 137-32-70 kg ha^{-1}) and 200% (N-P-K = 274-64-140 kg ha^{-1}). The treatments were replicated three times. The data revealed that shoot length, number of leaves, dry matter weight of shoots and roots, fruit number and fruit weight were influenced significantly ($P < 0.05$) by the application of vermicompost and NPK fertilizer in the growth media. The highest dose of vermicompost of 20 t ha^{-1} increased dry weight of shoot of 52 folds and root of 115 folds, number of fruit(s)/plant of 6 folds and mean fruit weight of 29 folds while the highest rate of NPK fertilizer of 200% increased dry weight of shoot of 35 folds and root of 80 folds, number of fruit(s)/plant of 4 folds and mean fruit weight of 18 folds over the control treatment. The growth performance of tomato was better in the vermicompost amended soil pots than the plants grown in the inorganic fertilizer amended soil pots. This study suggested that the vermicompost served as a potential source of nutrients for plant growth.

Keywords

Vermicompost, NPK Fertilizer, Tomato, Growth, Yield

*Corresponding author.

1. Introduction

Application of chemical fertilizers in the crop field contributes greatly to the deterioration of the environment, loss of soil fertility, less agricultural productivity and soil degradation [1]. The price of chemical fertilizers is very high and sometimes it is not available in the market as well for which the farmers sometimes fail to apply these fertilizers to the crop field in time. Compare to inorganic fertilizer, organic manure is readily available to the farmers and the price is also low [2]. In addition to that, application of manure may reduce the environmental pollution that is caused while manufacturing the chemical fertilizers in the industry [3].

It is a well known fact that organic fertilizers supply nutrients to the plants in adequate amount for optimum growth of plant and may increase the uptake of nutrients, assimilation capacity and the hormonal activity [4] [5]. Cow manure vermicompost could be used as an excellent soil amendment for main fields and nursery beds and has been reported to be useful in raising nursery species plants [6]. Vermicomposting means the use of earthworms for composting organic residues. Earthworms can eat as much as their own body weight per day. It is estimated that 1000 tons of moist organic manure can be converted into 300 tons of compost by earthworms [7]. Vermicompost could release nutrients slowly and steadily into the system and enables the plants to absorb these nutrients over time [8]. Edwards [9] reported that vermicompost could promote early and vigorous growth of seedlings. Vermicompost has been found to effectively enhance the root formation, elongation of stem and production of biomass, vegetables, ornamental plants, etc. [5] [10]-[14]. More available plant nutrients and microbial metabolites may be released into the growth media because earthworms may stimulate microbial activities and metabolism and also influence microbial populations [15]. Edwards and Burrows [16] reported that animal waste vermicomposts usually contained more nutrient elements than chemical plant growth media and composts released nutrients gradually in available forms that could be readily taken up by the plants.

The objective of the present study was to compare the response of different doses of cow manure vermicompost and inorganic fertilizer on the growth and fruits of tomato plant.

2. Materials and Methods

2.1. Plant Growth Experiment

A greenhouse pot experiment was carried out in the Department of Soil Science, University of Chittagong, Bangladesh, using a sandy loam surface soil (0 - 15 cm). Soil sample was air dried and passed through a 4-mm sieve for using the soil in the pots for plant growth experiment. A sub sample was air dried and passed through a 2-mm sieve and stored for laboratory analysis, Soil pH of 5.07 was measured at 1:2.5 soil to water ratio, organic carbon of 0.93% was measured by Walkley and Black [17] and CEC of 4.01 cmol kg^{-1} was measured with 1 M NH_4OAc extraction [18]. The percentages of sand (73%), silt (13%) and clay (14%) were measured by hydrometer method [19]. The vermicompost used in this research was resulted from processing on the cow manure with the help of Eisenia fetida worms. The pH of vermicompost was of 8.23. Five rates of vermicompost equivalent to 0 (control), 5, 10, 15 and 20 t ha^{-1} and three rates of NPK fertilizers equivalent to 50% (N-P-K = 69-16-35 kg ha^{-1}), 100% (N-P-K = 137-32-70 kg ha^{-1}) and 200% (N-P-K = 274-64-140 kg ha^{-1}) were applied separately in each pot containing four (4) kg soil. The pots were arranged in a completely randomized design (CRD) with three replications.

Twenty one (21) days old and uniform seedlings of tomato (*Solanumly copersicum* L.) having 2-3 leaves were transplanted to each pot. Water was applied to each pot regularly up to the field capacity. The height (cm) and number of leaves of the plants were recorded at 15 days intervals after transplanting (15, 30, 45, 60 and 75 DAT). The number of fruit(s) were counted and weighed after harvesting. The plants were harvested at 75 DAT and separated into shoots and roots. The roots were cleaned carefully to remove adhering soil particles. The shoots and roots were air dried for several days and oven dried at 65˚C for 72 hours and dry weights were recorded.

2.2. Statistical Analysis

Microsoft Excel and Minitab program [20] were used for analysis of variance (ANOVA).

3. Results

Application of different rates of vermicompost (5, 10, 15 and 20 t ha^{-1}) and NPK fertilizer (50%, 100% and 200%) in tomato plant significantly ($P < 0.05$) affected the growth and development of plants, number of fruits and mean fresh weight of fruit and yield during the study period of 75 days.

3.1 Effect of Vermicompost and NPK Fertilizer on Growth Parameters of Tomato

3.1.1. Shoot Length

From 15 to 75 DAT, the plant height ranged from 11.0 to 32.0 cm in 5 t ha^{-1} and 13.33 to 52.67 cm in 20 t ha^{-1} of vermicompost. Whereas, at the same DAT, the plant height ranged from 12.0 to 24.33 cm in 50% and 14.67 to 39.67 cm in 200% of NPK fertilizer and from 10.33 to 16.33 cm in control treatment (**Figure 1(a)**). Significant increase were observed among different treatments for shoot length (F = 70.41, $P < 0.001$) during different growth intervals. At harvest, application of highest rate of vermicompost (20 t ha^{-1}) and NPK fertilizer (200%) showed an increase of 36.34 cm and 23.34 cm of shoot length respectively, as compared to control. Vermicompost dose of 20 t ha^{-1} resulted in maximum plant height of 52.67 cm.

3.1.2. Number of Leaves

The number of leaves from 15 to 75 DAT ranged from 21 to 44 in 5 t ha^{-1} and 20 to 104 in 20 t ha^{-1} of vermicompost. Whereas, at the same DAT, the number of leaves ranged from 17 to 51 in 50% and 26 to 74 in 200% of NPK fertilizer and from 14 to 20 in control treatment (**Figure 1(b)**). The effect of application of vermicompost and NPK fertilizer was significant (F = 60.99, $P < 0.001$) on the number of leaves among different treatments. At harvest, application of highest rate of vermicompost (20 t·ha^{-1}) and NPK fertilizer (200%) showed an increase of 84 (4 folds) and 54 (3 folds) of leaves per plant respectively, as compared to control.

3.2. Effect of Different Doses of Vermicompost and NPK Fertilizer on Yield Parameters of Tomato

3.2.1. Dry Shoot Weight

Dry shoot weight ranged from 1.40 g in 5 t ha^{-1} to 4.25 g in 20 t ha^{-1} of vermicompost and 0.99 g in 50% to 2.90 g in 200% of NPK fertilizer treated pots respectively, and in control it was 0.08 g (**Figure 2(a)**). There is a significant increase in dry shoot weight (F = 135, $P < 0.001$) among the treatments and the highest rate of vermicompost (20 t ha^{-1}) and NPK fertilizer (200%) resulted an increase in dry shoot weight in respective treatments by 4.17 g (52 folds) and 2.82 g (35 folds) as compared to control. However, vermicompost application at the rate of 15 t ha^{-1} showed better results (3.61 g) than the highest rate (200%) of NPK fertilizer. The results of 5 and 10 t ha^{-1} of vermicompost were comparable with the results of 50% and 100% of NPK fertilizer treatments, respectively (**Figure 2(a)**).

3.2.2. Dry Root Weight

Dry root weight ranged from 0.91 g in 5 t ha^{-1} to 3.44 g in 20 t ha^{-1} of vermicompost and 0.76 g in 50% to 2.43 g in 200% of NPK fertilizer treated pots respectively, and in control it was 0.03 g (**Figure 2(a)**). There is a significant increase in dry root weight (F = 582, $P = 0.000$) among the treatments and the highest rate of vermicompost (20 t ha^{-1}) and NPK fertilizer (200%) resulted an increase in dry root weight in respective treatments by 3.41 g (115 folds) and 2.40 g (80 folds) as compared to control. However, vermicompost application at the rate of 15 t ha^{-1} showed better results (2.85 g) than that of the highest rate of NPK fertilizer (200%). The results obtained from the rate of 5 and 10 t ha^{-1} of vermicompost showed no significant difference ($P < 0.05$) with that of the results obtained from the rates of 50% and 100% of NPK fertilizer, respectively (**Figure 2(b)**).

3.3. Effect of Different Doses of Vermicompost and NPK Fertilizer on Quantity of Fruit of Tomato

3.3.1. Number of Fruit(s)/Plant

The numbers of fruit(s)/plant were ranged from 2 in 5 t ha^{-1} to 7 in t ha^{-1} of vermicompost and 1 in 50% to 4 in 200% of NPK fertilizer treated pots respectively, and in control it was 1 g (**Figure 3(a)**). Rates of vermicompost and NPK application has resulted in significant increase (F = 20, $P < 0.001$) in number of fruit(s)/plant and the increase in the highest rate of vermicompost (20 t ha^{-1}) and NPK fertilizer (200%) treatments was of 6 (6 folds) and 4 (4 folds) fruit(s)/plant as compared to control.

3.3.2. Mean Fresh Fruit Weight

Mean fresh fruit weight ranged from 7.9 g in 5 t ha^{-1} to 40 g in 20 t ha^{-1} of vermicompost and 6.1 g in 50% to 25 g

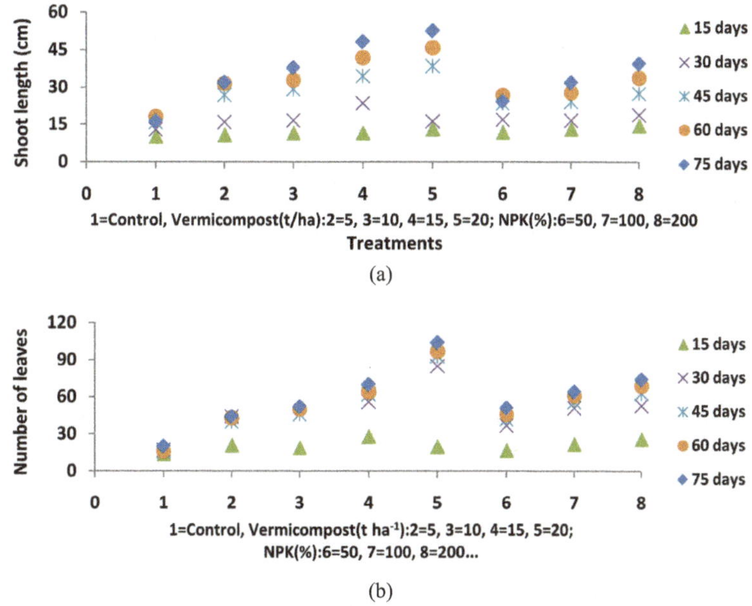

Figure 1. Application of vermicompost and inorganic fertilizer on growth parameters of tomato (*Solanum lycopersicum* L.) (mean ± standard error).

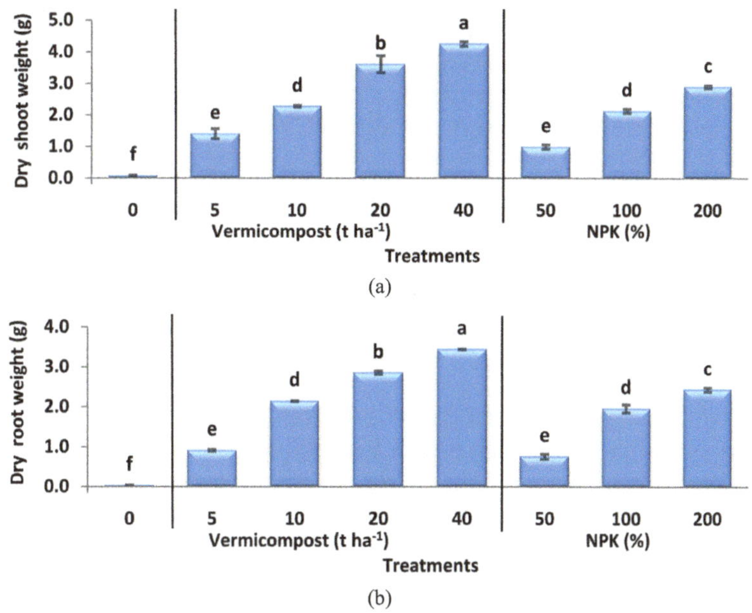

Figure 2. Application of vermicompost and inorganic fertilizer on dry shoot and root weight of tomato (*Solanum lycopersicum* L.) (mean ± standard error).

in 200% of NPK fertilizer treated pots respectively, and in control it was 1.34 g (**Figure 3(b)**). There is a significant increase in mean fresh fruit weight ($F = 74$, $P < 0.001$) among the treatments and the highest rate of vermicompost (20 t ha^{-1}) and NPK fertilizer (200%) resulted an increase in mean fresh fruit weight in respective treatments by 39 g (29 folds) and 24 g (18 folds) as compared to control treatment.

4. Discussion

The results of this experiment showed that the treatments of vermicompost and NPK fertilizer significantly increased the production of tomato in pots.

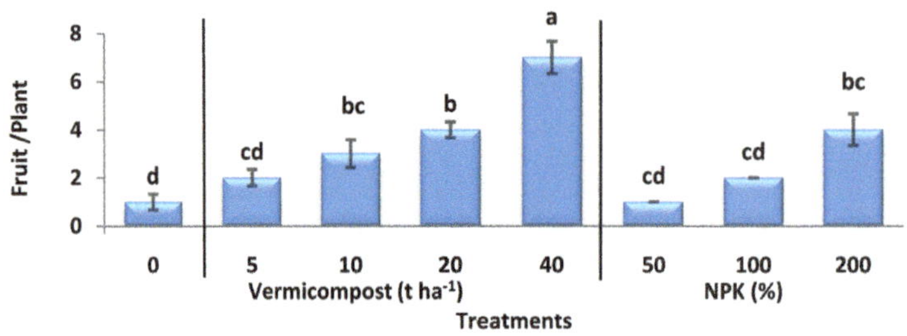

(a)

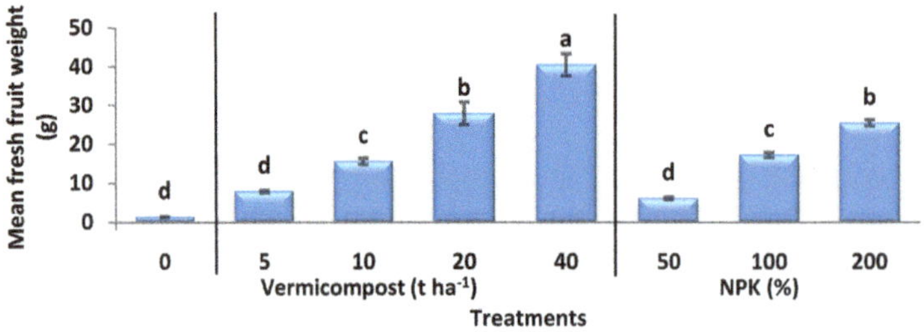

(b)

Figure 3. Application of vermicompost and inorganic fertilizer on number of fruit/plant and mean fresh fruit weight of tomato (*Solanum lycopersicum* L.) (mean ± standard error).

Plant growth parameters such as shoot length, number of leaves, shoot and root dry weights, number of fruits and fresh fruit weight were better in vermicompost treated pots than the NPK fertilizer treated pots. Among the vermicompost rates, the 20 t ha^{-1} showed the highest response in tomato plants. The differential response of plants to different doses of vermicompost may be due to release of variable amount of available nutrients and growth-promoting substances [4] [5]. Our results corroborate with the findings of other investigators [4] [21] [22]. They [4] [21] [22] reported significant increase in growth parameters of plants after application of vermicompost in the growth media. Scientists of different agricultural fields agreed that plant growth and development are strictly dependent on biological fertility factors. Earthworms may stimulate microbial activities and metabolism and also influence microbial populations. As a consequence, more available nutrients and microbial metabolites are released into the soil [15].

Moreover, in terms of plant growth and soil health, vermicompost plays an important role in improving soil texture, aeration, soil compaction and thus enhances more water and nutrients uptake by plants from their surrounding areas of root zone. There is much evidence that the activity of earthworms accelerates organic matter mineralization, decomposition of polysaccharides, increase the humus in the soil, and oppositely reduce the availability of toxic heavy elements to plants [23]. Macronutrients play important role in crop yield based on their role in activation of enzymes for chlorophyll synthesis, growth, fruit ripening and maintenance of the plant's enzyme system [24]. Therefore, the use of vermicompost constitutes an important alternative source of fertilizer that has environmental benefits, productivity and crop quality as well as compared to inorganic fertilizer.

5. Conclusion

From our results, it can be concluded that cow manure vermicompost application to soil increased growth parameters and fruit numbers of tomato plants (*Solanum lycopersicum* L.) over the inorganic fertilizer (NPK fertilizer) and hence could be useful for different crops production instead of inorganic fertilizers. The use of cow manure vermicompost could be a better option and farmers need to be educated about the importance of vermicomposting. More field trials with different crops need to be conducted in the future.

Acknowledgements

The authors wish to express their warm gratitude to the Research and Planning Division of Chittagong University, Bangladesh, for providing fund through the University Grants Commission of Bangladesh (UGC) to conduct this research.

References

[1] Inbar, Y., Hadar, Y. and Chen, Y. (1993) Recycling of Cattle Manure: The Composting Process and Characterization of Maturity. *Journal of Environmental Quality*, **22**, 857-863. http://dx.doi.org/10.2134/jeq1993.00472425002200040032x

[2] Alam, M.N., Jahan, M.S., Ali, M.K., Islam, M.S. and Khandaker, S.M.A.T. (2007) Effect of Vermicompost and NPKS Fertilizers on Growth, Yield and Yield Components of Red Amaranth. *Australian Journal of Basic and Applied Sciences*, **1**, 706-716.

[3] Rathier, T.M. and Frink. C.R. (1989) Nitrate in Runoff Water from Container-Grown Juniper and Alberta Spruce under Different Irrigation and N Fertilization Regimes. *Journal of Environmental Horticulture*, **7**, 32-35.

[4] Tomati, U., Galli, E., Grappelli, A. and Di Lena, G. (1990) Effect of Earthworm Casts on Protein Synthesis in Radish (*Raphanus sativum*) and Lettuce (*Lactuca sativa*) Seedlings. *Biology and Fertility of Soils*, **9**, 288-289. http://dx.doi.org/10.1007/BF00634102

[5] Grappelli, A., Tomati, U. and Galli, E. (1985) Earthworm Casting in Plant Propagation. *Horticultural Science*, **20**, 874- 876.

[6] Chanda, G.K., Bhunia, G. and Chakraborty, S.K. (2011) The Effect of Vermicompost and Other Fertilizers on Cultivation of Tomato Plants. *Journal of Horticulture and Forestry*, **3**, 42-45.

[7] Sajnanath, K. and Sushama, P.K. (2004) Recycling of Bio-Wastes through Vermicomposting. *Agrobios News Letter*, **3**, 33-34.

[8] Sharma, A.K. (2003) Biofertilizers for Sustainable Agriculture. Updesh Purohit for Agrobios, Jodhpur, 41-46.

[9] Edwards, C.A. and Burrows, I. (1988) The Potential of Earthworm Composts as Plant Growth Media. In: Edwards, C.A. and Neuhauser, E.F., Eds., *Earthworms in Environmental and Waste Management*, SPB Academic Publishing, The Hague, 211-220.

[10] Kale, R.D. and Bano, K. (1986) Field Trials with Vermicompost (Vee Comp. E.83 UAS) an Organic Fertilizer. In: Dash, M.C., Senapati, B.K. and Mishra, P.C., Eds., *Procedings of the National Seminar on Organic Waste Utilization, Vermicompost, Part B: Verms and Vermicomposting*, Five Star Printing Press, Burla, 151-157.

[11] Kale, R.D., Bano, K., Sreenivasa, M.N. and Bagyaraj, D.J. (1987) Influence of Worm Cast (Vee Comp. E.83 UAS) on the Growth and Mycorrhizal Colonization of Two Ornamental Plants. *South Indian Horticulture*, **35**, 433-437.

[12] Kale, R.D. (1998) Earthworm: Cinderella of Organic Farming. Prism Books, Bangalore.

[13] Bano, K., Kale, R.D. and Satyavathi, G.P. (1993) Vermicompost as Fertilizer for Ornamental Plants. In: Rajagopal, D., Kale, R.D. and Bano, K., Eds., *Proceedings of 4th National Symposium on Soil Pleurotus ostreatus against Infectious Bursal Biology and Ecology*, ISSBE, UAS, Bangalore, 165-168.

[14] Atiyeh, R.M., Subler, S., Edwards, C.A. and Metzger, J. (1999) Growth of Tomato Plants in Horticulture Potting Media Amended with Vermicompost. *Pedobiologia*, **43**, 724-728.

[15] Tomati, U., Grapppelli, A. and Galli, E. (1988) The Hormone-Like Effect of Earthworm Casts on Plant Growth. *Biology and Fertility of Soils*, **5**, 288-294. http://dx.doi.org/10.1007/BF00262133

[16] Edwards, C.A. (1988) Use of Earthworms in Breadown and Management of Organic Wastes. In: Edwards, C.A., Ed., *Earthworm Ecology*, CRC Press LLC, Boca Raton, 327-354.

[17] Walkley, A. and Black, I.A. (1934) An Examination of the Degtjareff Method for Determining Soil Organic Matter and a Proposed Modification of the Chromic Acid Titration Method. *Soil Science*, **37**, 29-38. http://dx.doi.org/10.1097/00010694-193401000-00003

[18] Soil Survey Laboratory Staff (1992) Soil Survey Laboratory Methods Manual. Soil Survey Laboratory Investigations Report No. 42, USDA-SCS, Washington DC.

[19] Bouyoucos, G.J. (1962) Hydrometer Method Improved for Making Particle Size Analysis of Soils. *Agronomy Journal*, **54**, 464-465. http://dx.doi.org/10.2134/agronj1962.00021962005400050028x

[20] Minitab, Inc. (1996) Minitab User Guide Release 11. Minitab, State College.

[21] Joshi, R. and Vig, A.P. (2010) Effect of Vermicompost on Growth, Yield and Quality of Tomato (*Solanum lycopersicum* L.). *African Journal of Basic Applied Sciences*, **2**, 117-123.

[22] Bachman, G.R. and Metzger, J.D. (2008) Growth of Bedding Plants in Commercial Potting Substrate Amended with Vermicompost. *Bioresource Technology*, 99, 3155-3161.

[23] Dominguez, J., Edwards, C.A. and Subler, S. (1997) A Comparison of Vermicomposting and Composting. *Biocycle*, **38**, 57-59.

[24] Grusak, M.A. and DellaPenna, D. (1999) Improving the Nutrient Composition of Plants to Enhance Human Nutrition and Health. *Annual Review of Plant Physiology and Plant Molecular Biology*, **50**, 133-161. http://dx.doi.org/10.1146/annurev.arplant.50.1.133

17

Comparative Assessment of Cow Manure Vermicompost and NPK Fertilizers and on the Growth and Production of Zinnia (*Zinnia elegans*) Flower

Shahin Sultana[1], Md. Abul Kashem[1,2*], Abul Kalam Mohammad Moniruzzaman Mollah[1]

[1]Asian University for Women, Chittagong, Bangladesh
[2]Department of Soil Science, Chittagong University, Chittagong, Bangladesh
Email: *abul.kashem@auw.edu.bd

Abstract

A pot experiment was conducted to investigate the comparative effect of cow manure vermicompost and NPK fertilizers on the growth and flower production of Zinnia (*Zinnia elegans*). An air dried sandy loam soil was mixed with four rates of NPK fertilizer equivalent to 0 (control), 0.5 dose (N-P-K = 69-16-35 kg·ha^{-1}), 1 dose (recommended) (N-P-K = 137-32-70 kg·ha^{-1}), 2 doses (N-P-K = 274-64-140 kg·ha^{-1}) and 4 doses (N-P-K = 548-128-280 kg·ha^{-1}) and three rates of vermicompost equivalent to 5%, 10%, 20% by oven dry weight. Two plants were grown in each pot. After blooming of flowers, the plants were harvested at 50 days of growth and leaves and stems were separated. The growth parameters (shoot height, root length, leaf number, total number of flower, flower diameter, fresh and dry weight of flower) of Zinnia plant increased by the application of vermicompost and the effect of NPK fertilizer was not found effective. The growth performance was similar between 10% and 20% and hence 10% vermicompost should be taken into consideration when ornamental plants are grown in potting media.

Keywords

Organic Fertilizer, Soil, Rates of Fertilizer, Flower

1. Introduction

Vermiculture is basically the science of breeding and raising earthworms. It defines the thrilling potential for

*Corresponding author.

waste reduction, fertilizer production, as well as an assortment of possible uses for the future [1]. Vermicompost does not have any adverse effect on soil, plant and environment. It improves soil aeration and structure thereby reducing soil compaction. It improves water retention capacity of soil because of its high organic matter content. It also promotes better root growth and nutrient absorption and improves nutrient status of soil, both macro-nutrients and micro-nutrients [2].

Generally, farmers use chemical fertilizers to improve soil fertility and hence increase the yield of their crops. However, the use of chemical fertilizers causes a great impact on the soil quality and the surrounding environment. Vermicomposting is a non-thermophilic and simple biotechnological process of composting, in which certain species of earthworms and microorganisms are used for biological degradation of organic waste [3] [4]. Vermicompost is aerobically decomposed products of organic wastes such as cattle dung and animal droppings, farm and forest wastes, vegetative waste, municipal solid wastes (MSW). However, for this study, cow manure was used as source of vermicompost because it is available in rural areas. Vermicomposting is a low-cost technology, environmentally-friendly process used to treat degradable organic waste. It is a plant nutrient-rich, microbiologically-active organic amendment that results from the interactions between earthworms and microorganisms during the decomposition process of organic waste [5]. Vermicomposting generally converts organic matter to a more uniform size, which gives the final substrate a characteristic earthy appearance, whereas the material resulting from composting usually has a more heterogeneous appearance [6] [7].

This organic fertilizer is therefore increasingly considered in agriculture and horticulture as a promising alternative to chemical fertilizers and/or peat in greenhouse potting media. However, the effects of vermicompost on the growth of horticultural plants are not yet fully understood.

Therefore, the objective of this study was to assess the effect of cow manure vermicompost and NPK Fertilizer on the growth and production of Zinnia *(Zinnia elegans)* flower.

2. Materials and Methods

A pot experiment was carried out in the roof top of a building of Asian University for women in Chittagong, Bangladesh. A sandy loam surface soil (0 - 15 cm) was used. Soil sample was air dried and passed through a 4-mm sieve for using the soil in the pots for plant growth experiment. A sub sample was air dried and passed through a 2-mm sieve and stored for laboratory analysis, Soil pH of 5.1 was measured at 1:2.5 soil to water ratio, soil organic carbon of 0.93% was measured by Walkley and Black [8] and CEC of 4.01 cmol·kg^{-1} was measured with 1 M NH_4OAc extraction [9]. The percentages of sand (73%), silt (13%) and clay (14%) were measured by hydrometer method [10]. The vermicompost used in this research was resulted from processing on the cow manure with the help of Eisenia fetida worms. The pH of vermicompost was of 7.17 and organic carbon was of 17.8% measured by ignition method. Other chemical properties (N, K, Ca, Mg, Fe, Mn, Zn and Cu) in the vermicompost were determined by Atomic Absorption Spectrophotometry (AAS) and P by Flame Photometry after digestion with H_2O_2-H_2SO_4 acids mixture [11] (**Table 1**). Four rates of NPK fertilizers equivalent to 0 (control),

Table 1. Nutrient status of vermicompost used in the experiment.

Property	Vermicompost
pH	7.17
C (%)	17.8
N (%)	1.78
P (%)	0.77
K (%)	0.94
Ca (%)	1.5
Mg (%)	0.53
Fe (%)	0.44
Mn (mg·kg^{-1})	180
Zn (mg·kg^{-1})	260
Cu (mg·kg^{-1})	354

0.5 dose (N-P-K = 69-16-35 kg·ha^{-1}), 1 dose (recommended) (N-P-K = 137-32-70 kg·ha^{-1}), 2 doses (N-P-K = 274-64-140 kg·ha^{-1}) and 4 doses (N-P-K = 548-128-280 kg·ha^{-1}) and three rates of vermicompost equivalent to 5%, 10%, and 20% by oven dry weight were applied separately in each pot containing five (5) kg soil. The 1 dose (N-P-K = 137-32-70 kg·ha^{-1}) fertilizer rate is consider as normal dose for most crops grown in this soil [12]. The pots were arranged in a completely randomized design (CRD) with three replications. The PK fertilizers were applied in the form of tripple super phosphate (TSP) and murate of potash (MP) during soil preparation prior to transplanting and N was applied as urea after 15 days of transplanting the Zinnia *(Zinnia elegans)* plant.

Two seedlings Zinnia of 20 days old were grown in each pot and water was maintained at field capacity. The plants were harvested at 50 days of growth after transplanting and flowers were collected during this growth period. At harvest, growth parameters (shoot height, root length, total leaf number, leaf length, total number of flower, flower diameter, fresh and dry weight of flowers) were recorded. To get dry weight of flowers, the fresh flowers were kept in oven at 60˚C for 72 hours.

Statistical Analysis

Microsoft Excel and Minitab program [13] were used for analysis of variance (ANOVA).

3. Results

Variance Analysis of data showed that the impact of growth media treated with vermicompost and NPK fertilizers on shoot height, root length, total leaf number, leaf length, total number of flower, flower diameter, fresh and dry weights of flowers were highly significant at 0.1 percent level ($p < 0.001$) (**Table 2**).

3.1. Shoot Height

At 50 days of growth after transplanting, maximum height (41 cm) of the shoot was related to 20% vermicompost and minimum height (15 cm) of the shoot was obtained in the control. The heights of shoots in the various rates of NPK fertilizers varied from 15 cm to 20 cm and in the vermicompost treated pots varied from 29 cm to 41 cm. The impact of NPK fertilizers on the shoot heights of Zinnia plant were found similar with that of control. Shoot heights of plant increased with the rates of vermicompost application but the values obtained in the 10% and 20% was not significantly different (**Table 3**).

3.2. Root Length

There was a strong positive correlation between the shoot and root heights of Zinnia plant regardless of treatments. The rates of NPK fertilizers and vermicompost on root height showed similar response as did for shoot heights. The heights of roots were between 5 cm and 7 cm in the NPK fertilizers treated pots and between 9 cm and 11 cm in the vermicompost treated pots (**Table 3**).

Table 2. Analysis of variance from ANOVA in various parameters.

Parameters	F
Shoot height	29.99*
Roots length	57.62*
Leaves number	60.09*
Leaf length	45.15*
Total number of flowers	101.23*
Flowers diameter	37.12*
Fresh weight of flowers	56.16*
Dry weight of flowers	35.41*

*indicates $p < 0.001$ level.

Table 3. Effect of vermicompost and inorganic fertilizers on the growth parameters of Zinnia.

Treatment	Shoot height (cm)	Root length (cm)	Total leaf number/plant	Leaf length (cm)	Total number of flowers/plant	Flowers diameter (cm)	Fresh weight (g) of flowers/plant	Dry weight (g) of flowers/plant
0 (Control)	14.9 c	5.1 c	8 d	3.3 c	1.5 d	1.62 d	0.09 d	0.03 d
0.5 dose NPK	15.5 c	5.4 c	9 d	3.9 c	2.5 c	1.83 d	1.08 c	0.08 c
1 dose NPK	18.4 c	6.2 c	9 d	4.4 bc	2.5 c	1.96 cd	1.35 c	0.07 c
2 doses NPK	19.7 c	7.0 c	10 d	5.1 bc	2.5 c	2.31 c	1.38 c	0.08 c
4 doses NPK	20.5 c	7.0 c	14 c	5.6 b	3.0 c	2.83 c	1.43 c	0.13 b
5% VC	28.7 b	9.2 b	21 b	7.8 a	4.0 b	4.43 a	1.97 b	0.25 a
10% VC	39.8 a	11.1 a	29 a	8.5 a	6.0 a	5.13 a	2.46 a	0.30 a
20%VC	41.3 a	11.2 a	32 a	9.1 a	7.0 a	5.61 a	2.65 a	0.27 a

Means followed by the same letter(s) in a column is not significantly different at $p < 0.01$ level. VC denotes vermicompost.

3.3. Number of Leaf

Total number of leaf per plant varied from 8 in the control to 32 in the 20% in the vermicompost treated pot. The influence of NPK fertilizers and vermicompost on number of leaf production was found similar as was observed for shoot height and root length. Vermicompost produced four and three folds higher leaf number compare to control and NPK fertilizers, respectively (**Table 3**).

3.4. Leaf Length

The length of leaf increased slightly from 3.3 to 5.6 cm with the rates of NPK fertilizer and from 7.8 cm to 9.1 cm with vermicompost applications in the growth media. The results showed that the rate effects within NPK fertilizers and vermicompost treatments were not found significant; however, the best performance was recorded in the plants grown with 20% vermicompost (**Table 3**).

3.5. Total Number of Flower

The data of **Table 3** showed that total number of flowers per plant was of 1.5 in the control, 2.5 to 3.0 with NPK fertilizers and 4 to 7 in the vermicompost treated pots. The rate effect of NPK fertilizers was not found significant in the production of flowers but it increased with rates of vermicompost application significantly from 5% to 10%. The rate effect was not significant between 10% and 20% vermicompost applied in the growth media (**Table 3**). We also observed that zinnia plants flowered earlier when grown in vermicomposts amended soil pots compared to those grown in NPK fertilizer treated media.

3.6. Flowers Diameter

The average diameter flower was the lowest of 1.6 cm in the control and the height of 5.33 cm in the plants grown in 20% vermicompost application pot. Flowers diameter increased with the rates of NPK fertilizers gradually from 1.8 to 2.3 cm and from 4.3 to 5.3 cm with the rates of vermicompost application in the growth media (**Figure 1**).

3.7. Fresh and Dry Weight of Flower

Flowers diameter reflected the fresh and dry weights of flowers irrespective of amendments. These three (Diameter, fresh and dry weights flowers) parameters of flowers were strongly and positively correlated with each other. Like other growth parameters of Zinnia plant, the lowest fresh (0.09 g) and dry weights (0.03 g) of flowers obtained in the control pot and highest fresh (2.65 g) and dry weights (0.27 g) was in the 20% vermicompost treated pots (**Table 3**, **Figure 1**).

Figure 1. The effect of NPK fertilizers and vermicompost on Zinnia flower.

4. Discussion

The response of Zinnia plant growth and flower production to vermicompost showed much better results compare to NPK fertilizers. The NPK fertilizers were not effective to enhance flower production of Zinnia plants and their rate effects were not also noticeable (**Table 3**). Similar our results, Edwards and Burrows [14] found that some ornamental plants such as *Eleagnus pungens*, *Cotoneaster conspicua*, *Pyracantha*, *Viburnum bodnantense*, *Chaemaecyparis lawsonia*, *Cupressocyparis leylandii and Juniperus communis* grew better in vermicomposts amended soil than in a commercial plant growth medium. They also reported that chrysanthemums, salvias and petunias plants flowered earlier in vermicomposts compared to those grown in a commercial planting media. In our experiment, Zinnia grew better at 10% vrmicompost amended soil and 20% vermicompost did not show any significant result. Wilson and Carlile [15] were reported to be best at substitution into soils at rates of 8% - 10% vermicompost for the growth of tomatoes, lettuces, and peppers. Zinnia plants produced big size of flowers in the vermicompost treated pots which is an important indicator of landscape. Our results corroborate with the findings of Shandanpour *et al.* [16], they studied the effect of cow manure vermicompost as the planting medium on the growth marigold.

Vermicompost can improve physical, chemical and biological and processes of soil which have their bearings on plant's growth. Goutam *et al.* [17] found that only organic fertilizer treated tomato plants showed more branching than inorganic fertilizer treated plants. It is assured that other factors, such as the presence of beneficial microorganisms or biologically active plant growth influencing substances such as phytohormone are released by beneficial microorganisms present in the vermicompost amended soil [18] [19]. Increased root biomass, enhanced plant growth and development of plant morphology are among the most frequently claimed effects of vermicompost treatment [20]. Researchers of various disciplines agree that plant growth and development are strictly dependent on biological fertility factors. Earthworms stimulate microbial activities and metabolism and also influence microbial populations. As a consequence more available nutrients and microbial metabolites are released into the soil and hence promote plant growth [20].

5. Conclusion

To avoid the negative impacts on soil, plant growth and surrounding environment, the uses of vermicompost needs to be increased as they are cheap, production sources are available, organic, reduce the pool of waste from our environment and keep our environment clean and healthy. Vermicompost plays an important role for improving soil physical and chemical properties and hence promote plant growth. For growing ornamental plants, the use of vermicompost should be given priority instead of inorganic fertilizer and the rate of vermicompost 10% should be taken into consideration.

References

[1] Entre Pinoys (2012) Vermiculture, the Management of Worms. http://www.mixph.com/2006/12/vermiculture-themanagement-of worms.html

[2] Lazcano, C. and Dominguez, J. (2010) Effects of Vermicompost as a Potting Amendment of Two Commercially-Grown Ornamental Plant Species. *Spanish Journal of Agricultural Research*, **8**, 1260-1270. http://dx.doi.org/10.5424/sjar/2010084-1412

[3] Khan, A. and Fouzia, I. (2011) Chemical Nutrient Analysis of Different Composts (Vermicompost and Pitcompost) and Their Effect on the Growth of a Vegetative Crop *Pisum sativum*. *Asian Journal of Plant Science and Research*, **1**, 116-130.

[4] Arancon, N.Q. and Edwards, C.A. (2005) Effects of Vermicomposts on Plant Growth. Soil Ecology Laboratory, The Ohio State University, Columbus.

[5] Dominguez, J. (2004) State-of-the-Art and New Perspectives on Vermicomposting Research. In: Edwards. C.A., Ed., *Earthworm Ecology*, CRC Press LLC, Boca Raton, 401-424. http://dx.doi.org/10.1201/9781420039719.ch20

[6] Ndegwa, P.M. and Thompson, S.A. (2001) Integrating Composting and Vermicomposting in the Treatment and Bioconversion of Biosolids. *Bioresource Technology*, **76**, 107-112. http://dx.doi.org/10.1016/S0960-8524(00)00104-8

[7] Tognetti, C., Laos, F., Mazzarino, M.J. and Hernández, M.T. (2005) Composting vs. Vermicomposting: A Comparison of End Product Quality. *Compost Science and Utilization*, **13**, 6-13. http://dx.doi.org/10.1080/1065657X.2005.10702212

[8] Walkley, A. and Black, I.A. (1934) An Examination of Degtja-Reff Method for Determining Soil Organic Matter and a Proposed Modification of the Chromic Acid Titration Method. *Soil Science*, **37**, 29-38. http://dx.doi.org/10.1097/00010694-193401000-00003

[9] Soil Survey Laboratory Staff (1992) Soil Survey Laboratory Methods Manual. Soil Survey Investigation. Report 42, USDA-SCS, Washington DC.

[10] Bouyoucos, G.J. (1962) Hydrometer Method Improved for Making Particle Size Analysis of Soils. *Agronomy Journal*, **54**, 464-465. http://dx.doi.org/10.2134/agronj1962.00021962005400050028x

[11] Akinremi, O.O., Amisen, N., Kashem, M.A. and Janzen, H.H. (2003) Evaluation of Analytical Methods for Total Phosphorus in Organic Amendments. *Communications in Soil Science and Plant Analysis*, **34**, 2981-2991. http://dx.doi.org/10.1081/CSS-120025220

[12] Soil Resources Development Institute (SRDI) (2005) Upazila Land and Soil Resource Utilization Guide. Chittagong, Bangladesh.

[13] Minitab Inc. (1996) Minitab User Guide Release 11. Minitab, State College.

[14] Edwards, C.A. and Burrows, I. (1988) The Potential of Earthworm Composts as Plant Growth Media. In: Edwards, C.A. and Neuhauser, E.F., Eds., *Earthworms in Waste and Environmental Management*, SPB Academic Publishing, The Hague, 211-219.

[15] Wilson, D.P. and Carlile, W.R. (1989) Plant Growth in Potting Media Containing Worm-Worked Duck Waste. *Acta Horticulture*, **238**, 205-220. http://dx.doi.org/10.17660/actahortic.1989.238.24

[16] Shadanpour, F., Mohammadi, T.A. and Hashemi, M.K. (2011) The Effect of Cow Manure Vermicompost as the Planting Medium on the Growth of Marigold. *Annals of Biological Research*, **2**, 109-115.

[17] Goutam, K.C., Goutam Bhunia, B. and Susanta, K. (2011) The Effect of Vermicompost and Other Fertilizers on Cultivation of Tomato Plants. *Journal of Horticulture and Forestry*, **3**, 42-45.

[18] Tomati, U. and Galli, E. (1995) Earthworms, Soil Fertility and Plant Productivity. Proceedings of the International Colloquium on Soil Zoology. *Acta Zoologica Fennica*, **196**, 11-14.

[19] Edwards, C.A. (1988) Breakdown of Animal, Vegetable and Industrial Organic Wastes by Earthworms. In: Edwards, C.A. and Neuhauser, E.F., Eds., *Earthworms in Waste and Environmental Management*, SPB Academic Publishing, The Hague, 21-31.

[20] Tomati, U., Grappelli, A. and Galli, E. (1988) The Hormone-Like Effect of Earthworm Casts on Plant Growth. *Biology and Fertility of Soils*, **5**, 288-294. http://dx.doi.org/10.1007/BF00262133

18

Origin and Distribution of Heavy Minerals in the Surficial and Subsurficial Sediments of the Alluvial Nile River Terraces

Magboul M. Sulieman[1*], Ibrahim S. Ibrahim[1], Jamal T. Elfaki[2], Mohamed S. Dafa-Allah[1]

[1]Department of Soil and Environment Sciences, Faculty of Agriculture, University of Khartoum, Khartoum, Sudan

[2]Department of Seeds Technology, Nile Valley University, Khartoum, Sudan

Email: *magboul@uofk.edu

Abstract

Origin and distribution of the heavy minerals of surficial and subsurficial sediments has been investigated in the alluvial Nile River terraces, Khartoum North, Sudan. Heavy mineral assemblages in the very fine sand fraction (0.063 - 0.125 mm) of 10 sediment samples were identified using petrography microscope. Results of descriptive statistical parameters revealed that most sediments samples belonged within very poorly sorted to extremely poorly sorted, strongly negative skewed to strongly positive skewed and mesokurtic to very leptokurtic. The quartz was the dominant in the opaque minerals in all sediments. The non-opaque heavy minerals were dominant by zircon, tourmaline, rutile, garnet, sillimanite, and andalusite. Results revealed that the ultrastable minerals (zircon, tourmaline and rutile) were found in all sediments with range from (2% - 47.36%, 2.08% - 29% and 3% - 24.99%), respectively. Garnet, sillimanite and andalusite were also found with range from (5% - 67%, 1% - 9.09% and 1% - 50%), respectively. Heavy mineral assemblage indentifies sources that are not bounded to the local origin. The proportion and presence of heavy minerals from outside source rocks indicated relatively strong reworking of zircon sand from the outer-shelf to inner-shelf as well relatively long distance of transport. Fluvial and Aeolian sediments were the dominant environments in the investigated area. We conclude that most heavy minerals in the study area are originally derived from gneisses and schist metamorphic rocks and some igneous rocks of the Ethiopian plateau.

Keywords

Alluvial Nile River Terraces, Khartoum North, Petrography Microscope, Aeolian Sediments,

*Corresponding author.

Ethiopian Plateau

1. Introduction

Heavy minerals are defined as high density minerals, which have specific gravities of 2.9 g·cm^{-3} or higher [1]. About 56 translucent species of heavy minerals are described by [2] and have been regarded as indicators of sediment sources [3].

The heavy mineral assemblage in sediments usually reflects their parent rocks as well their origin. Thus, over the last two decades, specific heavy minerals including zircon, monazite, garnet, tourmaline, apatite, rutile, and Ti-Fe oxide minerals have been widely used to decipher the provenances of marine and Nile river sediments in terms of their unique varietal characteristics [4].

According to [5] the factors which influence the assemblage of the heavy minerals include weathering at different stages between the original source rocks and sedimentary environments, mechanical abrasion during transportation, physical sorting and diagenetic processes during buried. Furthermore, laboratory errors caused by heavy mineral separation during sample preparation could influence the heavy mineral assemblages.

The percentage of zircon, tourmaline and rutile (ZTR index) in the non-opaque heavy mineral is important to assess the mineralogical maturity of the sediments [6]. Sediments which have highest concentrations of ultrastable heavy minerals indicate the most mature mineral composition.

Many researchers mention that the study of opaque minerals provides little information about the origin of the sediments. Therefore, we discuss and propose that this study should be focused on non-opaque heavy minerals assemblages [5] and [6].

The aim of this study was to identify the non-opaque heavy minerals assemblage of the surficial and subsurficial sediments in the Nile terraces at Khartoum North, Sudan in an attempt to refer them to their origin.

2. Materials and Methods

2.1. Study Area

The investigated area is part of the perilagoonal belt located in the northeastern part of Khartoum North, Sudan between the River Nile at Khartoum North, lies between latitudes (451437N to 453478N and longitudes 1747982E to 1750070E, and extending eastwards till the piedmont plain (**Figure 1**). Ideally, the study area received sediments from Ethiopian plateau and is not restricted to local sources. These sediments are interbedded with various metamorphic rocks, mostly gneiss and schist in addition to igneous rocks [7].

The study area belongs within the semi arid climatic zone [8]. The average annual rainfall varies from 100 - 225 mm. Mean maximum temperature of the hottest months (May and June) is 40°C and 42°C, respectively. Mean minimum temperature of the coldest month (January) is 13°C - 16°C. According to the American System of soil classification, the soils of the study area belong within Entisols and Aridisols soil orders [9].

2.2. Soil Sampling

Transect of 3 profiles (total of 10 representative sediment surface and subsurface samples) was collected from three Nile River terraces at Khartoum North, Sudan (**Figure 1**). Samples 1, 2 and 3 represent the very recent (first terrace) Nile terrace soils. Samples 4, 5 and 6 were collected from the second Nile terrace, while samples 7, 8, 9 and 10 were collected from the intermixing of the alluvial plain and the piedmont. Samples 1 (0 - 30 cm), 4 (0 - 18 cm) and 7 (0 - 15 cm) were surficial sediments, while samples 2 (30 - 80 cm), 3 (80 - 130 cm), 5 (18 - 50 cm), 6 (50 - 90 cm), 8 (15 - 40 cm), 9 (40 - 70 cm) and 10 (70 - 120 cm) were subsurficial sediments. The reason for selected 70% of subsurficial sediments to avoid the bias to surficial sediments and to decrease the errors in case if the surficial sediments has been accumulated with Aeolian, particularly at the intermixing of alluvial plain and piedmont.

2.3. Global Positing System and ArcGIS

All samples were located using a global position system (GPS, Model 12 XL Garmin), and map of the study

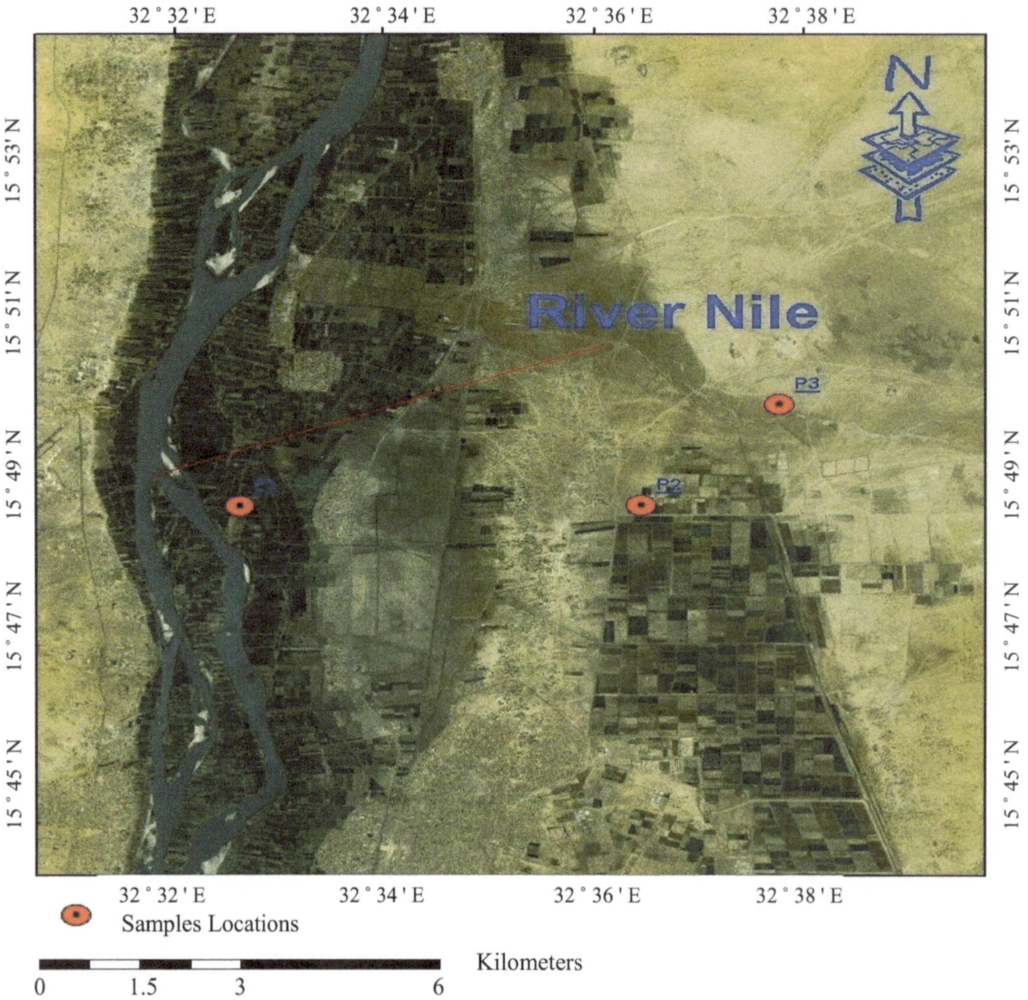

Figure 1. The study area and sites of the soil samples.

area was created using Arc GIS software version 10.2 (Arc GIS, Esri, USA).

2.4. Samples Preparation and Analysis

In the laboratory, soil samples were spread to air dry at room temperature, ground using wood pestles and mortars and sieved to pass 2 mm sieves. Then, samples < 2 mm were oven dried at 50°C to insure the complete dryness.

2.5. Grain Size Analysis

For grain size distribution, samples were treated with 10% HCl to destroy calcium carbonate, using H_2O_2 to remove organic matter, washed with distilled water to remove soluble salts and dispersed chemically with Calgon (Sodium hexameta phosphate), and mechanically with a dispersion machine. The Hydrometer method was used to determine the clay and silt fraction, while sand fraction was obtained by subtraction from 100%. The textural class was obtained using textural triangle [10].

2.6. Determination of Sand Fractionations

The sand fractionation was done using the appropriate set of sieve (0.063 - 1.0 mm). The sediments and suspension were quantitavely transferred from the 1 liter sedimentation cylinder through a 63 micrometer sieve. The sediment onto the 63 micrometer screen was washed using a wash bottle or gentle stream of water. The sand fraction on each sieve was transferred to a tarred beaker or aluminum weigh dish, oven dried at 105°C for 24 hr, and each sand fraction weighed for second times. Individual sand separates were calculated as a percentage of the < 2 mm whole soil sample. Total sand was calculated as the sum of the sand separates [9].

2.7. Heavy Minerals Analysis (Non-Opaque Minerals)

According to [2], about 150 g of each representative sample was treated with 10% HCl to remove $CaCO_3$ and 15% H_2O_2 to remove O.M (60°C - 70°C for 15 minutes). The analysis of heavy minerals were carried out into three steps; 1) separation of very fine sand (VFS) and fine sand (FS) fractions (0.063 - 0.125 mm and 0.125 - 0.25 mm), respectively: The sediments and suspension were quantitatively transferred from the 1 liter sedimentation cylinder through a 0.063 mm sieve, washed with tap water, then sand fraction on 0.063 - 0.125 mm and 0.125 - 0.25 mm were used for magnetic separations; 2) separation of the magnetic from non-magnetic heavy minerals using potassium polytungstate solution with a density of 2.85 gcm^{-3}: 1g subsamples of VFS and 1.5 g of FS fractions were placed individually in centrifuge tubes, then centrifuged at 3000 rpm for 15 minutes [11]. Two groups of funnels with filter papers were prepared to receive the heavy and light minerals, washed 3 times to recover the heavy liquid, and the last wash was performed with alcohol to ensure that all heavy liquid was cleaned from the heavy minerals and 3) identification, characterization and calculation the different heavy minerals using a polarizing microscope: Heavy mineral grains separated from VFS were mounted on a glass slide using Canda balsma (RI = 1.5). Heavy mineral analyses were carried out in the Central petroleum Laboratory, Ministry of Energy, Sudan.

2.8. Calculation of Zircon, Tourmaline, Rutile (ZTR Index):

The three ultrastable minerals (*i.e.* zircon, rutile and tourmaline) were used to calculate the ZTR index of the sediments (*i.e.* their mineralogical maturity). The ZTR for each of the selected samples was calculated using the following equation: $ZTR = \frac{Z + T + R}{\text{Total no. of N.O}} \times 100$

where: N.O = Non-opaque minerals, Z = Zircon, T = Tourmaline and R = Rutile.

This equation is referred to as [6] scheme. From the calculated percentage, ZTR < 75% implies immature to sub-mature sediments; ZTR > 75% indicates mineralogically matured sediments.

2.9. Statistical Analysis

Descriptive statistical analysis includes; mean, standard deviation, skewness and kurtosis were used for grain size, while mean and standard deviation were used for non-opaque heavy minerals. All statistical analyses were performed using SPSS software version 16.0 (SPSS Inc, USA).

3. Results and Discussion

3.1. Grain Size Distribution and Textural Class

Table 1 summarizes the grain size distribution and the textural class of the sediment samples of the investigated area. Results revealed that 50% of the total samples were predominantly of silt and values ranged from 11.9% to 59.52% with an average of 34.64%, sand was dominant in 40% of total samples and values ranged between 14.06% to 52.15% with an average of 32.51%, whereas; silt was found to be dominant only in 10% of samples and values ranged from 24.04% to 52.61%, with an average of 32.85%. Sand sized sediments were obtained to be mostly dominant in the samples from second terrace and the intermixing of the alluvial plain and piedmont. These results could be due to addition of fine sand by Aeolian depositional. Whereas, silt sized were found to be mostly dominant in the first and intermixing terraces. The results for textural class showed that the studied samples belong within 7 classes namely; silty clay loam, clay loam, sandy clay loam, sandy clay, clay, silt loam and loam.

3.2. Descriptive Statistical Parameters of the Grain Size

The descriptive statistical analyses of the grain size are presented in **Table 2**. The mean grain size values of samples ranged from 2.34 to 8.84. The standard deviation values varied from 0.24 to 6.13, which indicate very

Table 1. Percentage of grain size distribution (clay, silt and sand) in the sediment samples.

Location	Sample No.	Depth (cm)	Grain size distribution (%)			Textural class (USDA*)
			Clay	Silt	Sand	
First terrace	1	0 - 30	33.56	52.38	14.06	Silty clay loam
	2	30 - 80	38.32	39.25	22.43	Clay loam
	3	80 - 130	33.56	42.86	23.58	Clay loam
Second terrace	4	0 - 18	33.56	11.91	54.53	Sandy clay loam
	5	18 - 50	35.94	11.91	52.15	Sandy clay
	6	50 - 90	52.61	11.9	35.49	Clay
Alluvial plain and the piedmont	7	0 - 15	26.42	59.52	14.06	Silt loam
	8	15 - 40	26.42	59.52	14.06	Silt loam
	9	40 - 70	24.04	28.57	47.39	Loam
	10	70 - 120	24.04	28.58	47.38	Loam

*United State Department of Agriculture.

Table 2. Descriptive statistical parameters for the grain size in the studied sediment samples.

Location	Sample No.	Descriptive analysis			
		Mean	STD*	Skewness	Kurtosis
First terrace	1	2.34	2.61	0.92	2.04
	2	6.46	6.20	0.43	3.01
	3	6.47	5.13	0.34	1.48
Second terrace	4	8.84	6.13	0.01	0.16
	5	5.99	3.02	0.85	0.22
	6	5.99	4.28	−0.27	1.82
Alluvial plain and the piedmont	7	8.84	2.41	0.38	1.51
	8	5.99	0.24	−0.33	2.87
	9	5.99	2.91	−0.81	0.96
	10	7.89	0.76	0.47	1.96

*Standard deviation.

well sorted to extremely poorly sorted samples. Overall, most samples in the study area falls within very poorly sorted to extremely poorly sorted samples. These results indicate that the energy of the depositing was not sufficient enough to enhance the degree of sorting to separate the sand fractions into different classes. These findings were in agreement with [12]. The skewness results of all studied samples indicate strongly negative skewed to strongly positive skewed and values ranged from −0.81 to 0.92. These results indicate that the sediments samples are dominated by fine and coarse sand fractions. These results are disagreed with [13], who mention that the river sands are usually positively. The kurtosis values varied from 0.16 to 3.01, which indicate very platykurtic to extremely leptokurtic. In fact, 70% of the studied samples varied from mesokurtic to very leptokurtic and valued ranged from 0.96 to 2.87. Similar results were obtained by [2]. According to [14], results of skewness and kurtosis of the studied sediments suggest that fluvial and Aeolian sediments are the dominant environments in the investigated area.

3.3. Non-Opaque Heavy Minerals

The microscopic inspection of the Nile terrace sediments samples showed that the non-opaque heavy minerals included: zircon, tourmaline, rutile, garnet, sillimanite, and andalusite (**Plate 1**).

1) Zircon is the most abundant ultrastabile minerals, its grains are found as rounded and subrounded shape, pale and dark colored, different sizes and inclusions (**Plate 1(b)** and **Plate 1(d)**), and observed in all sediments. Rounded zircon grains were identified in the sediments of the first and intermixing terraces with range from 2% - 15%, while subrounded/angular zircon grains were observed in the soil of the three terraces with range

from 3% - 47.36%. The mean content of zircon in the sediments from first, second and intermixing terraces were; 32.79%, 8.52% and 5.23%, respectively. The distribution of zircon in all sediments shows irregular abundance throughout the entire depth, but decreased far away from the Nile course (**Figures 2-5**). Presents of rounded and sub rounded zircon may indicate a relatively long distance of transport as well relatively strong reworking and probably most coming from igneous or metamorphic rocks derived from the Ethiopian plateau. These findings were in agreement with [12].

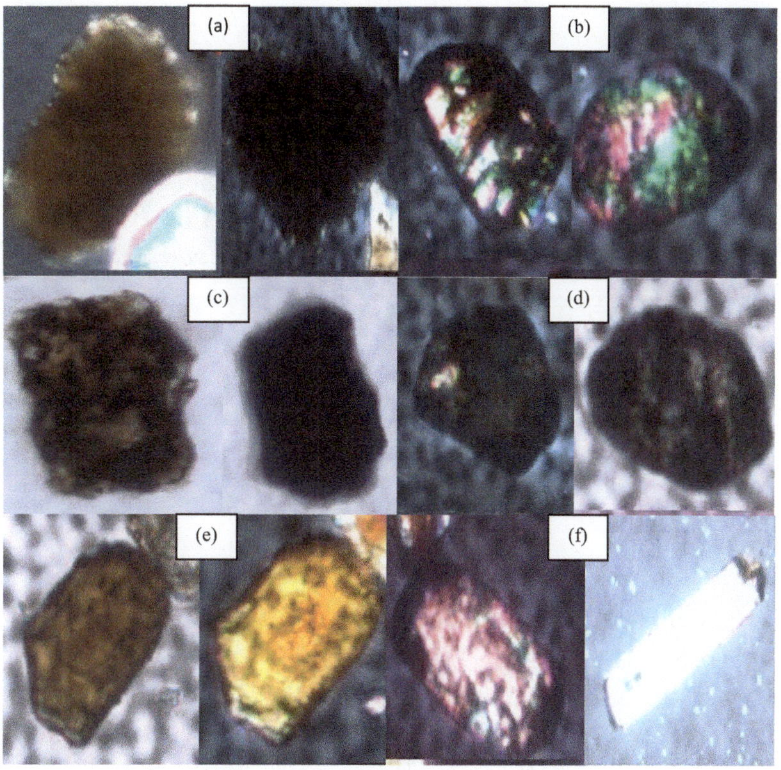

Plate 1. Polarizing microscope photomicrographs (Mag. ×100) of selected heavy minerals identified in the sediments of the study area. (a) Andalusite, (b) sub-rounded zircon, (c) garnet, (d) rounded zircon, (e) types of rutile, (f) types of tourmaline.

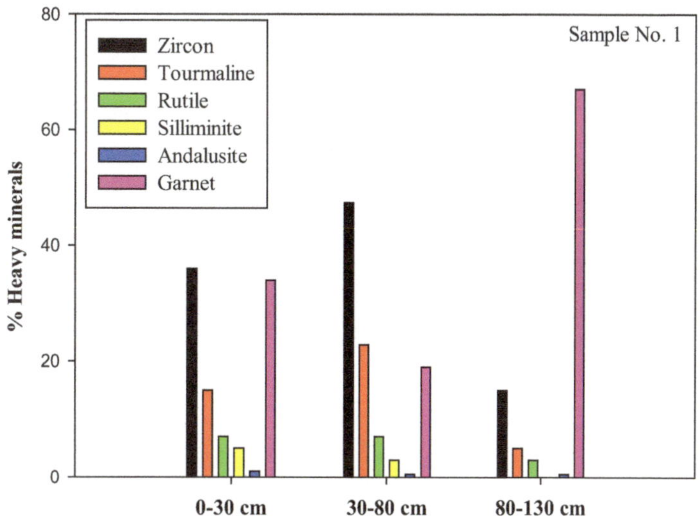

Figure 2. Variation of percent heavy minerals with depth in the sample 1.

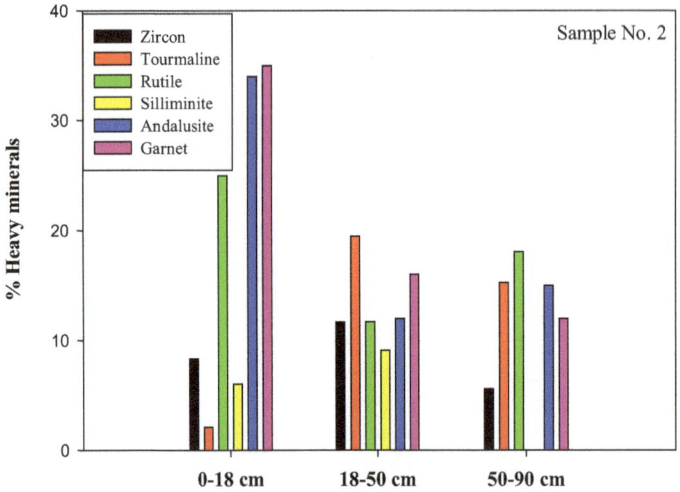

Figure 3. Variation of percent heavy minerals with depth in the sample 2.

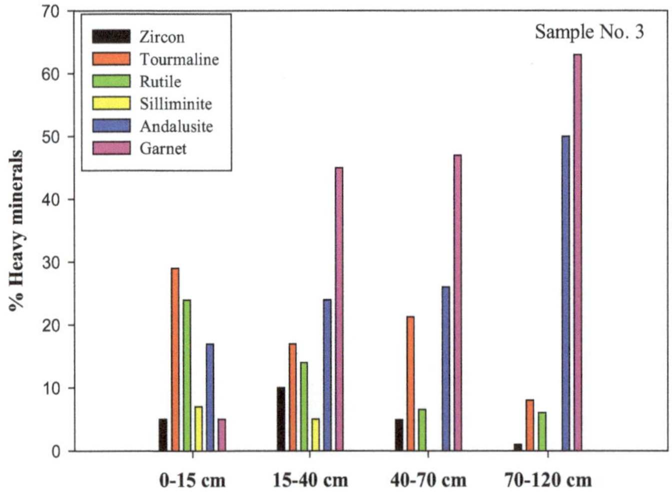

Figure 4. Variation of percent heavy minerals with depth in the sample 3.

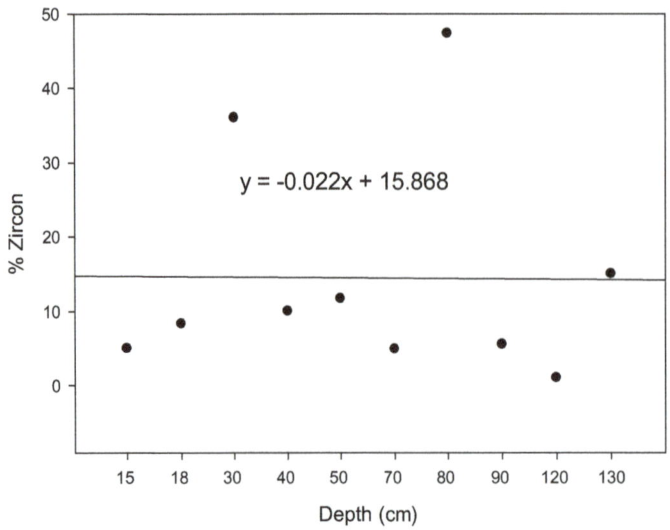

Figure 5. Regression of zircon percentage with depth of all samples.

2) Tourmaline is the second abundant ultrastable mineral in the study area, found as prismatic, elongate, indigo to black grains **Plate 1(f)**, and observed in all sediments, with mean content ranged from 12.27% - 18.83%. The distribution of tourmaline shows irregular pattern throughout the entire depth, but increased far away from the Nile course (**Figures 2-4** and **Figure 6**). This may suggest relatively long distance of transport and most probably igneous or metamorphic (particularly, gneisses and schist) source rocks which could be derived from the Ethiopian plateau. Similar findings were obtained by [15].

3) Rutile was found as very high relief, irregular prismatic, yellowish brown and the dark reddish brown grains **Plate 1(e)**. Rutile was found in the least ultrastable minerals content compare with zircon and tourmaline, with mean ranging from 5.67% to 18.24%. The highest mean abundance (18.24%) was found in the second terrace. The distribution of rutile shows irregular pattern throughout the entire depth and from the Nile stream (**Figures 2-4** and **Figure 7**). The presence of rounded rutile indicated reworked sediments most probably coming from metamorphic source rocks. These findings were in agreement with [16].

4) Garnet mostly relatively resistant mineral, and the grains were found as cubic, pale and dark red **Plate 1(c)**, and observed in the three terraces with mean content of 40, 21% and 40% from first, second and intermixing terraces, respectively. The distribution of garnet shows irregular pattern throughout the entire depths of the first

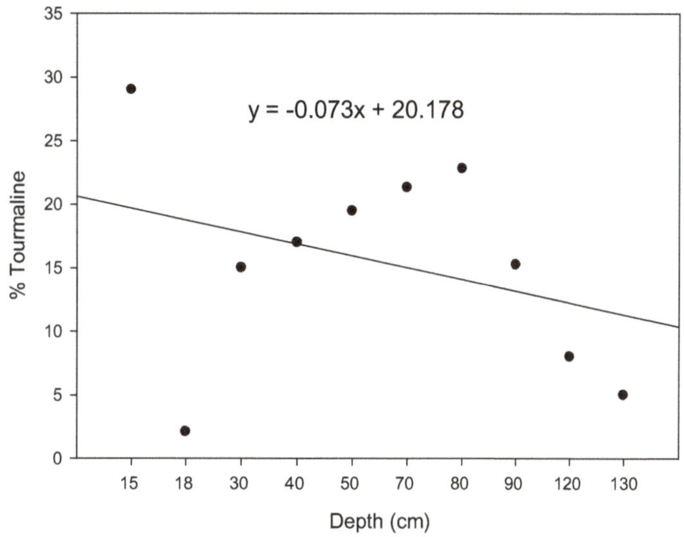

Figure 6. Regression of tourmaline percentage with depth of all samples.

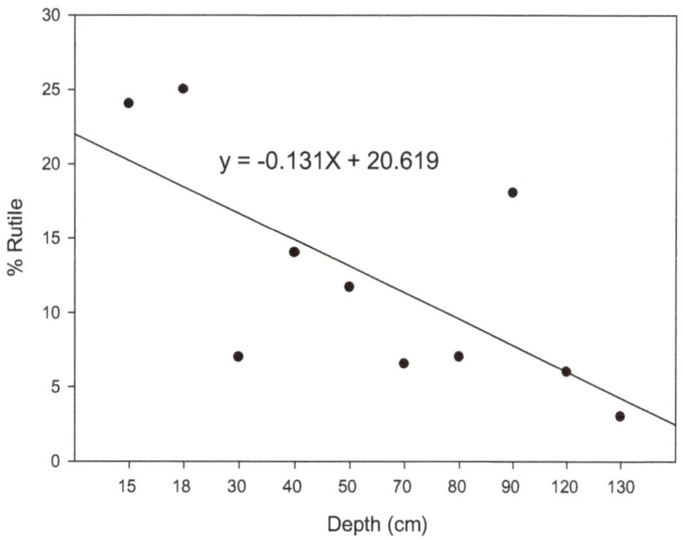

Figure 7. Regression of rutile percentage with depth of all samples.

terrace, while decreased with depth in the second terrace, and increased with depth in the intermixing terrace. The presence of sub rounded to rounded garnet may indicate relatively long distance of transport and most probably derived from metamorphic source rocks. Similar results were found by [17].

5) Sillimanite found as prismatic, colorless, and observed in all sediments with small mean content range from 2.67% - 5%, with irregular distribution throughout the entire depths and from the Nile stream (**Figures 2-4**). According to [2], the presence of sillimanite in small amount might be sourced by metamorphic rocks or granite and genesis complex which contains significant amounts of sillimanite.

6) Andalusite found as square prisms, pink color **Plate 1(a)**, and observed in all sediments with mean range from 0.33% - 29%. The distribution of andalusite shows irregular pattern throughout the entire depths and from the Nile stream (**Figures 2-4**). The presence of the andalusite may indicate reworked sediments and most probably derived from argillaceous rocks subjected to contact metamorphism around igneous intrusions [18] (**Table 3**).

3.4. Proportion of Zircon, Tourmaline and Rutile (ZTR Index)

Table 4 presents data of ZTR values. The results revealed that the average zircon, tourmaline and rutile proportions (ZTR index), was distributed as 54.18%, 48.22% and 36.25%, in the first, second and intermixing terraces, respectively. The heavy minerals assemblages in the sediments of the study area were characterized by low zircon, tourmaline and rutile proportion (ZTR index) which is used as clue for the identification of the source and as measure of the maturity of the sediments. According to [6], these soils mineralogically are immature sediments. The regression of zircon with ZTR, tourmaline with ZTR and rutile with ZTR are shown in (**Figures 8-10**), respectively.

Table 3. Descriptive statistical parameters of heavy minerals of the sediments in the study area.

Vairables	First terrace		Second terrace		Intermixing of 2nd terrace and piedmont	
	Mean	STD	Mean	STD	Mean	STD
Zircon	32.79	16.42	8.52	3.06	5.23	3.69
Tourmaline	14.26	8.92	12.27	9.07	18.83	8.76
Rutile	5.67	2.31	18.24	6.66	12.64	8.41
Silliminite	2.67	2.52	5.03	4.62	3.00	3.56
Andalusite	0.33	0.58	20.33	11.93	29.25	14.36
Garnet	40.00	24.56	21.00	12.29	40.00	24.68

STD—standard deviation.

Table 4. Zircon, tourmaline and rutile index (ZTR index).

Location	Depth (cm)	Zircon	Tourmaline	Rutile	ZTR index	Average ZTR
First terrace	0 - 30	36	15	7	59.18	54.18
	30 - 80	47.36	22.79	7.01	77.81	
	80 - 130	15	5	3	25.56	
Second terrace	0 - 18	8.33	2.08	24.99	32.07	48.22
	18 - 50	11.67	19.47	11.68	53.59	
	50 - 90	5.55	15.26	18.04	59.00	
Alluvial plain and the piedmont	0 - 15	5	29	24	66.67	36.25
	15 - 40	10	17	14	35.65	
	40 - 70	4.9	21.3	6.54	30.96	
	70 - 120	2	8	6	11.72	

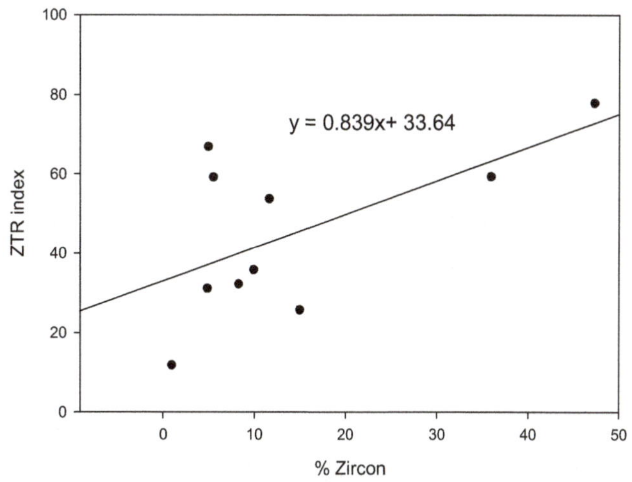

Figure 8. Regression of zircon percentage with ZTR index of all samples.

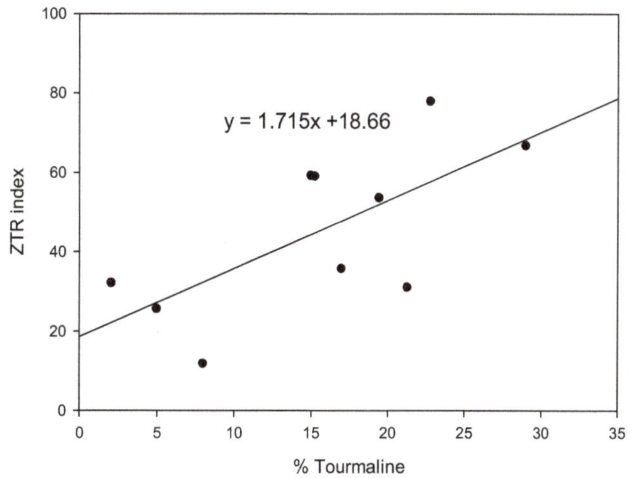

Figure 9. Regression of tourmaline percentage with ZTR index of all samples.

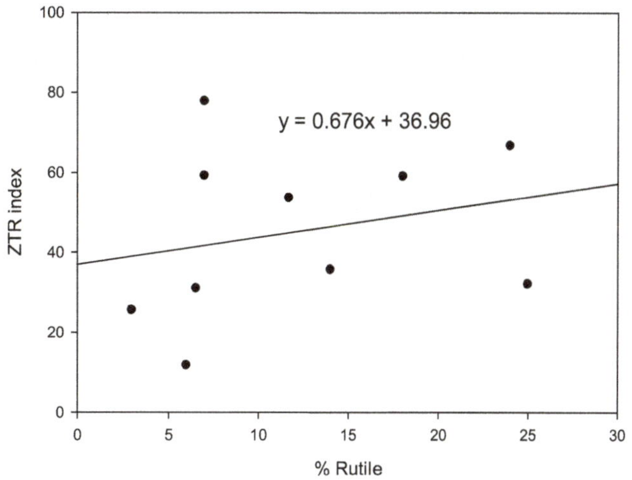

Figure 10. Regression of rutile percentage with ZTR index of all samples.

4. Conclusion

The descriptive statistical parameters of the grain size reveal that most sediments samples belong within very poorly sorted to extremely poorly sorted, strongly negative skewed to strongly positive skewed, mesokurtic to very leptokurtic which indicate that the energy of the depositing is not sufficient enough to enhance the degree of sorting as well as fluvial and Aeolian sediments which are the dominant environments in the investigated area. The quartz is the dominant in the opaque minerals in all sediment and represents about 95% of the total light minerals. In addition, six non-opaque heavy minerals were found in the very fine sand fraction (0.063 - 0.125 mm) in the surficial and subsurficial sediments. These minerals were zircon, tourmaline, rutile, garnet, sillimi-nite and andulsite. Therefore, we conclude that the non-opaque heavy minerals in the Nile river terraces at Khartoum North are originally derived from gneisses and schist metamorphic rocks as well as igneous rocks of the Ethiopian plateau.

Acknowledgements

The authors are indebted their greatly thank to Technical Assistant Khalid El-Enazi, Laboratory of Remote Sensing and GIS, College of Food and Agric. Sciences, King Saud University for his help. Thanks is also extend to all staff of the Central petroleum Laboratory, Ministry of Energy, Sudan, for using all laboratory facilities, their help and valuable suggestions during samples preparation and analyses of heavy minerals.

References

[1] Muller, L.D. (1997) Laboratory Methods of Mineral Separation. In: Zussman, J., Ed., *Physical Methods in Determination Mineralogy*, 2nd Edition, Academic Press, London, 1-34.

[2] Mange, M.A. and Maurer, H.F.W. (1992) Heavy Minerals in Colour. Chapman and Hall, London, 147. http://dx.doi.org/10.1007/978-94-011-2308-2

[3] Yang, Sh., Wang, Z., Guo, Y., Liand, G. and Cai, J. (2009) Heavy Mineral Compositions of the Changjiang (Yangtze River) Sediments and Their Provenance. Tracing Implication. *Journal of Asian Earth Sciences*, **35**, 56-65. http://dx.doi.org/10.1016/j.jseaes.2008.12.002

[4] Raiswell, R. and Anderson, T.F. (2005) Reactive Iron Enrichment in Sediments Deposited Beneath Euxinic Bottom Waters: Constraints on Supply by Shelf Recycling. In: McDonald, I., Boyce, A.J., Butler, I., Herrington, R.J. and Polya, D. (Eds.), *Mineral Deposits and Earth Evolution*, Geological Society London Special Publications No. 218, 179-194. http://dx.doi.org/10.1144/gsl.sp.2005.248.01.10

[5] Morton, A.C. and Hallsworth, C.R. (1999) Processes Controlling the Composition of Detrital Heavy Mineral Assemblages in Sandstones. *Sedimentary Geology*, **124**, 3-29. http://dx.doi.org/10.1016/S0037-0738(98)00118-3

[6] Hubert, J.F. (1962) A Zircon-Tourmaline-Rutile Maturity Index and Independence of Composition of Heavy Mineral Assemblages with Gross Composition and Texture of Sandstone. *Journal of Sedimentary Petrology*, **32**, 440-450.

[7] Suliman, M.M., Ibrahim, I.S. and Elfaki, J. (2015) Land Suitability Characterization for Crop and Fruit Production of Some River Nile Terraces, Khartoum North, Sudan. International Journal of Scientific and Research Publications V.5, Issue 10.

[8] Van der Kevie, W. (1973) Climatic Zones in the Sudan. Soil. Survey Department, Wad Medani.

[9] Soil Survey Staff (2014) Keys to Soil Taxonomy. 12 Edition, United States Department of Agriculture, Natural Resources Conservation Service, Lincoln, 97, 123, 287.

[10] Soil Survey Staff (2014) Kellogg Soil Survey Laboratory Methods Manual. In: Burt, R., Soil Survey Staff (Eds.), *Soil Survey Investigations Report* No. 42, Version 5.0, US Department of Agric. Natural Resources Conservation Service, Lincoln.

[11] Poppe, L.J., Commeau, J.A. and Luepke, G. (1995) Silt Fraction Heavy-Mineral Distributions in a Lateritic Environment: The Rivers and Insular Shelf of North Central Puerto Rico. *Sedimentary Geology*, **95**, 251-268. http://dx.doi.org/10.1016/0037-0738(94)00114-a

[12] Samy, Y and Abou El-Anwar, E. (2013) Textural and Mineralogical Characters of the Sand Fraction of Some Quaternary Sediments on Giza-Fayium District, Western side of the Nile Valley, Egypt. *Australian Journal of Basic and Applied Sciences*, **7**, 770-779.

[13] Friedman, G.M. (1979) Differences in Size Distribution of Populations of Particles among Sand Grains of Various Origins. *Sedimentology*, **26**, 3-20. http://dx.doi.org/10.1111/j.1365-3091.1979.tb00336.x

[14] Mason, C.C. and Folk, R.L. (1958) Differentiation of Beach, Dune and Aeolian Flat Environments by Size Analysis,

Mustang Island, Texas. *Journal of Sedimentary Research*, **28**, 211-226.

[15] Haredy, R.A. (2003) Distribution and Provenance of Heavy Minerals in the Surficial Sediments of the Minnamurra and Bass Point New South Wales, Australia. Research Thesis, School of Geosciences, University of Wollongong, Wollongong.

[16] Krippner, A. and Bahlburg, H. (2012) Provenance of Pleistocene Rhine River Middle Terrace sands between the Swiss-German border and Cologne Based on U-Pb Detrital Zircon Ages. *International Journal of Earth*, **102**, 917-932. http://dx.doi.org/10.1007/s00531-012-0842-8

[17] Kroner, U., Mansy, J.L., Mazur, S., Aleksandrowski, P., Hann, H.P., Huckriede, H., Lacquement, F., Öamarche, J., Ledru, P., Pharaoh, T.C., Zedler, H., Zeh, A. and Zulauf, G. (2008) Variscan Tectonics. In: McCann, T., Ed., *The Geology of Central Europe*, Vol. 1, Precambrian and Palaeozoic, 303-381.

[18] Deer, W.A., Howie, R.A. and Zussman, J. (1992) An Introduction to the Rock-Forming Minerals. 2nd Edition, Prentice Hall, Harlow.

19

Long-Term Effects of Alternative Residue Management Practices on Soil Water Retention in a Wheat-Soybean, Double-Crop System in Eastern Arkansas

Ryan Norman[1], Kristofor R. Brye[1], Edward E. Gbur[2], Pengyin Chen[1], John Rupe[3]

[1]Department of Crop, Soil, and Environmental Sciences, University of Arkansas, Fayetteville, USA
[2]Agricultural Statistics Laboratory, University of Arkansas, Fayetteville, USA
[3]Department of Plant Pathology, University of Arkansas, Fayetteville, USA
Email: kbrye@uark.edu

Abstract

Soil water retention is a critical aspect of agricultural management, especially in areas such as the Lower Mississippi River Alluvial Valley that face potential water shortages in the near future. Previous studies have linked changes in soil water retention characteristics to agricultural management practices, especially as they affect the accumulation of soil organic matter (SOM). Therefore, the objective of this study was to determine the relationship between soil water potential and gravimetric soil water content in the top 7.5 cm as affected by nitrogen (N) fertilization/residue level (high and low), residue burning (burning and non-burning), tillage (conventional and no-tillage), and irrigation (irrigated and non-irrigated) after 12 complete cropping cycles in a wheat (*Triticum aestivum* L.)-soybean [*Glycine max* (L.) Merr.], double-crop production system in the Delta region of eastern Arkansas using soil wetting curves. The soil investigated was a Calloway silt loam (fine silty, mixed, active, thermic Glossaquic Fraglossudalf). The slope characterizing the relationship between the natural logarithm of the soil water potential and the gravimetric soil water content was only affected ($P < 0.05$) by the N-fertilization/residue-level treatment, while the intercept was unaffected by any field treatment. Averaged across tillage, burning, and irrigation, soil water contents under the high- exceeded those under low-N-fertilization/residue-level treatment at the same water potential, with the greatest differences observed at water contents > 0.12 $g \cdot g^{-1}$. Understanding the ways in which alternative residue management practices affect soil water retention characteristics is an important component of conserving irrigation water resources.

Keywords

Water Retention, Tillage, Residue Burning

1. Introduction

Agronomic management practices that promote the formation of soil organic matter (SOM) and soil aggregation, such as reduced tillage and diversifying crop rotations, can increase plant available water in the soil [1] and likely have many more positive, long-term effects on soil water characteristics. For example, significant differences have been observed between soil water retention curves in the top 10 cm for native prairie (SOM = 22 $g \cdot kg^{-1}$) and cultivated agricultural soil (SOM = 10.8 $g \cdot kg^{-1}$) in eastern Arkansas [2]. Specifically, the native prairie soil contained a greater soil water content than the cultivated agricultural soil at the same water potential, indicating a possible correlation between increased SOM and water retention. Similarly, decreased soil water retention under conventional tillage (CT) management was reported compared to increased soil water retention and unsaturated hydraulic conductivity under no-tillage (NT) management in a continuous corn (*Zea mays* L.) study on Mollisols in Iowa [3]. Verkler *et al.* [4] reported slower soil dry down after wetting under non-burned management compared to burned residue management, as well as slower soil dry down under NT compared to CT when examining soil water content dynamics in a wheat (*Triticum aestivum* L.)-soybean [*Glycine max* (L.) Merr.], double-crop system on a silt-loam soil in eastern Arkansas after three years and four complete cropping cycles. Clearly, residue and field management practices influence soil water retention characteristics, which may be related to agricultural management effects on soil aggregation and SOM.

Increases in SOM have been associated with increased infiltration, greater hydraulic conductivity, and increased water retention [5]. Therefore, management practices such as tillage and nitrogen (N) fertilization that may affect the accumulation of SOM may also affect soil water retention characteristics. In a previous study of alternative residue management practice effects on near-surface soil properties in a wheat-soybean, double-crop production system on a silt-loam soil in eastern Arkansas, Amuri *et al.* [6] reported increasing soil carbon (C) and SOM over time in the top 10 cm across all treatment combinations over the course of six years and seven complete wheat-soybean cropping cycles following conversion to alternative management practices, where trends were likely due to the increase in crop residue returned to the soil as a result of converting from monoculture soybean to a wheat-soybean double-crop system. Smith *et al.* [7] reported that the abundance of water-stable aggregates was significantly affected by tillage, irrigation, and N-fertilization treatments. Nitrogen fertilization promotes wheat biomass, which may eventually contribute to an increase in SOM and soil aggregation. Therefore, N fertilization, and other management practices that promote SOM and soil aggregation, may affect the relationship between soil water potential and the soil water content, hence soil water retention. For example, Bowman and Halvorson [8] reported significant increases in soil organic C (SOC), and therefore SOM, in the top 5 cm under increased N-fertilization management. Similarly, SOC and SOM increased at a greater rate under a high (134 $kg \cdot N \cdot ha^{-1} \cdot yr^{-1}$) than under low N-rate (<90 $kg \cdot N \cdot ha^{-1} \cdot yr^{-1}$) treatments in a wheat-containing rotation managed consistently for 10 yr near Akron, Colorado [9].

An understanding of traditional and alternative residue management practices on soil water retention is critical to determining the best management practices in highly productive agricultural regions, especially in areas such as the Delta region of eastern Arkansas that face potential water shortages in the future. Scott *et al.* [10] used a regression equation based on annual water use rate to determine that 75% of the Alluvial Aquifer, the shallowest aquifer underlying most of the Delta region of eastern Arkansas, will be depleted from large irrigation withdrawls by 2041. Therefore, the objective of this study was to examine the effects of N fertilization/residue level (high and low), residue burning (burning and non-burning), tillage (conventional and no-tillage), and irrigation (irrigated and non-irrigated) on the relationship between soil water potential and gravimetric soil water content in the top 7.5 cm after 12 complete cropping cycles in a wheat-soybean, double-crop production system in eastern Arkansas using soil wetting curves. It was hypothesized that water contents under NT will be greater than that under CT at the same water potential due to effects on aggregation and that the history of irrigation or dryland production will have little effect on soil water retention using the wetting curve approach. It was also hypothesized that the cumulative effects of 12 years of residue burning would render soil water retention in the burned treatment lower than that in the non-burned treatment due to the hydrophobic characteristics of the added ash. In addition, it was hypothesized that increased above and belowground biomass inputs would contribute to greater water retention in the high than in the low N-fertilization/residue-level treatment.

2. Materials and Methods

2.1. Site Description

A field study was initiated in Fall 2001 at the Lon Mann Cotton Branch Experiment Station (N34°44'2.26"; W90°45'51.56") [11] in the Lower Mississippi River Alluvial Valley near Marianna, AR. The study site lies within the Southern Mississippi Valley Loess [Major Land Resource Area (MLRA) 134] [12]. This region consists of a series of loess covered hills and alluvial terraces. Despite being prone to erosion, this region has been highly agriculturally productive for decades [12]. The field site is on a Calloway silt loam (fine silty, mixed, active, thermic Glossaquic Fraglossudalf) [13], which consists of 16% sand, 73% silt, and 11% clay in the top 10 cm [14]. The 30-yr mean air annual temperature of the region is 15.6°C and the 30-yr mean annual precipitation is 128 cm [15]. The 30-yr mean maximum and minimum air temperatures of the region are 32.8°C in July and 2.4°C in January, respectively [15]. This study follows a series of several previous studies at this same study site that have evaluated a variety of short- and long-term effects of alternative management practices effects on plant and soil properties [6] [7] [16].

2.2. Experimental Design

Between Fall 2001 and Spring 2005, the study consisted of a three-factor, split-strip-plot, randomized complete block experimental design with six replications of each of eight treatment combinations [11]. The three experimental factors were i) N fertilization/residue level (*i.e.*, a high N fertilization/residue level, achieved with a split application of N fertilizer, and a low N/fertilization/residue level, achieved with minimal to no N additions); ii) residue burning (*i.e.*, burning or non-burning); and iii) tillage (*i.e.*, CT or NT) [11]. However, an irrigation factor was introduced in 2005 by dividing the site into an irrigated half and a non-irrigated half [4]. Since 2005, the experimental area has consisted of 48, 3- × 6-m plots with six replications for every N-fertilization/residue-level-burning-tillage treatment combination and three replications for every N-fertilization/residue-level-burning-tillage-irrigation treatment combination (**Figure 1**) [6].

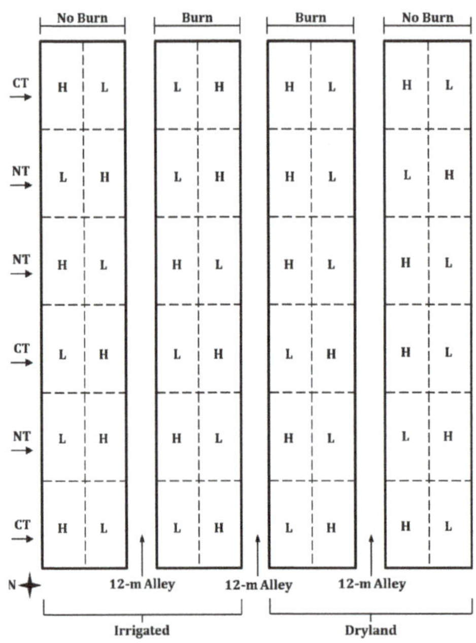

Figure 1. Experimental layout at the Lon Mann Cotton Branch Experiment Station in eastern Arkansas depicting 48, 3- × 6-m plots under residue-level [high (H) and low (L)], burn, tillage [conventional tillage (CT) and no-tillage (NT)], and irrigation treatments.

2.3. Field Management

Prior to the initiation of the study, the site was managed as a continuous, mono-cropped soybean system using CT [11]. Due to the consistent field management prior to beginning this long-term study in 2001, near-surface soil properties throughout entire study area were assumed uniform and any subsequent observed differences in measured soil and/or plant properties were assumed be to the result of imposed field treatments rather than inherent differences among plots [14].

To prepare for this study, initial field preparations in Fall 2001 involved disking twice followed by broadcast applications of N, phosphorous, potassium, and pelletized limestone at rates of 20, 22.5, 56, and 1120 $kg \cdot ha^{-1}$, respectively, prior to wheat planting. Wheat was drill seeded with a 19-cm row spacing each fall thereafter. All plots were manually broadcast fertilized in early March 2002 through 2004 with urea (46% N) at the rate of 101 $kg \cdot N \cdot ha^{-1}$. High-N-fertilization/residue-level plots ($n = 24$) were manually broadcast fertilized in late March at approximately the late-jointing stage with an additional 101 $kg \cdot N \cdot ha^{-1}$ to produce different levels of wheat residue. No N-fertilizer was applied in Spring 2005 due to a failure to establish wheat stands caused by prolonged wet soil conditions in Fall 2004. Since 2006, initial applications of 56 $kg \cdot N \cdot ha^{-1}$ as urea were broadcast on high-N-fertilization/residue-level plots in approximately late February, followed by a split application of an additional 56 $kg \cdot N \cdot ha^{-1}$ at the late-jointing stage in approximately late March. Since 2006, the low-N-fertilization/residue-level plots have received no N fertilization in order to achieve the desired residue-level difference.

Wheat was harvested using a plot combine in approximately early June each year. Wheat residue left behind the plot combine was uniformly spread by hand over each plot immediately following wheat harvest. Any remaining wheat stubble was mowed with a rotary mower to a height of ~3 cm from the soil surface in order to achieve a uniform residue-covered surface for soybean planting. Following mowing, the burning treatment was imposed on half of the plots by propane flaming. The residue-burning treatment was not able to be imposed in 2005, 2007, and 2012 due to the absence of a wheat stand in Spring 2005, prolonged wet soil conditions in Spring 2007, and overly weedy conditions in 2012. Imposition of the burning treatment was followed by impos ing the tillage treatment each year. The CT plots were disked at least twice with a tandem disk to a depth of ~10 cm followed by seedbed smoothing with at least three passes of a soil conditioner. The CT practices used in this study are representative of widely used pre-soybean-planting tillage operations throughout the region.

A glyphosate-resistant soybean cultivar, maturity group 5.3 or 5.4, was drill-seeded with 19-cm row spacing at a rate of approximately 47 kg seed ha^{-1} in early to mid-June each year. Potassium fertilizer was applied according to recommended rates [17] when the previous year's soil test indicated potassium was needed. In 2002 through 2004, all plots were furrow-irrigated as needed, three to four times each soybean-growing season. A levee was created in 2005 to exclude furrow-irrigation water from the non-irrigated (*i.e.*, dryland) treatment, which received only natural rainfall. Weeds and insects were managed consistently throughout the entire study area as necessary based on University of Arkansas Cooperative Extension Service's recommendations, which generally consisted of herbicide and insecticide applications during both the wheat- and soybean-growing seasons [17]. Soybean were harvested with a plot combine from late October to early November each year. Each year within two weeks of soybean harvest, the subsequent wheat crop was sown into the soybean residue, which was left in place without any manipulations.

2.4. Plant Sample Collection and Processing

To verify achievement of the N-fertilization/residue-level treatment, immediately following wheat harvest and rotary mowing and before residue burning each year, aboveground residue was assessed. A residue sample was collected from within a 0.25-m^2 metal frame, oven-dried for 3 to 7 days at 55°C, and weighed to obtain an estimate of aboveground residue mass into which soybean were subsequently planted.

2.5. Soil Sample Collection and Processing

Consistent with prior annual soil sampling conducted at this field site [6] [7] [14] [16], at wheat harvest in May 2014, a single soil sample was collected from the top 10 cm in each plot using a 4.8-cm-diameter stainless steel core chamber beveled to the outside to minimize compaction. Soil samples were oven-dried for 48 hr at 70°C, weighed for bulk density determinations, ground to pass through a 2-mm mesh screen, and analyzed for selected soil chemical properties. Soil pH was determined potentiometrically in a 1:2 (w/v) soil-to-water suspension. Soil OM was determined by weight-loss-on-ignition after 2 hr at 360°C. Total soil C and N were determined by high-temperature combustion with an Elementar VarioMAX Total C and N Analyzer (Elementar Americas Inc., Mt. Laurel, NJ). All measured soil C was assumed to be organic given the lack of effervescence upon treatment with dilute hydrochloric acid [14]. The soil C: N ratio was calculated from measured C and N concentrations. Total C and N and SOM contents ($kg \cdot m^{-2}$) were calculated from measured concentrations ($g \cdot kg^{-1}$) and measured bulk densities in the top 10 cm from the May 2014 soil sampling.

To assess field treatment effects on the relationship between soil water potential and soil water content, 12, ~2-cm-diameter soil samples were collected from the top 7.5 cm of each plot at wheat harvest in May 2014 and combined into one composite soil sample per plot. Each sample was manually homogenized and air-dried for approximately 5 d, ground, and sieved to pass through a 2-mm mesh screen. Subsamples were weighed, oven-dried at 70˚C for 48 hr, and reweighed to obtain the initial moisture content of the air-dried sample. Following the procedures of Brye [2], seven, 5 ± 0.01-g subsamples of air-dried soil from each of the 48 plots were added to small mixing cups. Drops of distilled water (*i.e.*, 2, 4, 6, 10, 12, 15, and 20 drops) were added to each of the seven mixing cups with an eyedropper and manually mixed with a spatula to achieve a range of gravimetric soil water contents. The moist soil in each mixing cup was transferred to small plastic instrument cups, 4 cm in diameter by 1 cm tall, and lightly packed to a uniform bulk density of ~0.7 $g \cdot cm^{-3}$. The plastic instrument cups were capped and allowed to equilibrate overnight to room temperature (*i.e.*, ~20˚C). The next morning after over-night temperature equilibration, the water potential was subsequently measured with a WP4 Dewpoint PotentiaMeter (Decagon Devices, Inc., Pullman, WA), which was calibrated using a standard potassium chloride solution. After the water potential was recorded, each instrument cup was weighed, oven-dried at 70˚C for 48 hr, then reweighed for gravimetric water content determinations. Measured water potentials were natural-logarithm transformed to facilitate statistical analyses.

In addition, to determine actual residue management effects on near-surface bulk density, a third set of soil cores were collected between 8 and 10 weeks after soybean planting in 2014, similar to previous annual bulk densities assessments [6] [7] [14] [16]. A single 4.8-cm-diameter soil core was extracted from the top 10 cm in each plot using a stainless steel core chamber and slidehammer. Samples were oven-dried at 70˚C for 48 hr, and weighed for an additional bulk density measurement.

2.6. Statistical Analyses

An analysis of variance (ANOVA) was conducted using SAS (version 9.3, SAS Institute, Inc., Cary, NC) to evaluate the N-fertilization/residue-level treatment effect on aboveground residue mass for annual measurements conducted between 2007 and 2014. In addition, an ANOVA was also conducted using SAS, based on the strip-split-plot experimental layout of the field treatments (**Figure 1**), to evaluate the effects of N fertilization/residue level, burning, tillage, irrigation, and their interactions on soil pH, SOM, total N, and C contents, C:N ratio, and bulk density in the top 10 cm measured in 2014, which represents the cumulative effects of 12 consecutive years of consistent management. Due to practical limitations of the study area, the addition of the irrigation treatment since 2005 was superimposed on the burning treatment (**Figure 1**). Therefore, irrigation and burning treatments could not be simultaneously analyzed within this experimental design. Therefore, two separate ANOVAs were conducted, each excluding one of the confounding factors (*i.e.*, burning and irrigation). When appropriate, treatment means were also separated by least significant difference (LSD) at the 0.05 level.

To evaluate field treatment effects on soil water retention, an analysis of covariance (ANCOVA) was conducted using SAS to examine the long-term effects of N fertilization/residue level, burning, tillage, irrigation, and their interactions on the linear relationship between the natural logarithm (log) of the measured soil water potential and the gravimetric water content from the soil wetting-curve data. For the purposes of simultaneously evaluating all field treatments and their interactions, the experimental design was assumed to be completely random with three replications of each of 16 treatment combinations. The full ANCOVA model was reduced using a hierarchal principle to remove non-significant terms, and non-significant terms were only included the final model when they participated in higher-order, complex treatment interactions. When appropriate, treatment means for slopes and intercepts from the log-transformed relationships were separated by LSD at the 0.05 level.

3. Results and Discussion

3.1. Aboveground Residue Levels

Following the 13 wheat crops produced in this study between 2002 and 2014, the N-fertilization scheme used annually produced numerically greater aboveground residue amounts, into which the subsequent soybean crop was planted, in the high than in the low N-fertilization/residue-level treatment in 12 of the 13 years, with 2002 being the exception [6] [18]. In addition, significantly ($P < 0.05$) greater aboveground residue amounts were produced in the high than in the low N-fertilization/residue-level treatment in 10 of the 13 years, with 2002, 2004 [6], and 2010 being the exceptions [18]. Therefore, it is clear that the intended residue-level differences were achieved in more than the majority of the 13 years to justify the N-fertilization/residue-level treatment.

3.2. 2014 Soil Properties

After 13 complete wheat-soybean cropping cycles (*i.e.*, 2001 to 2014) and 12 years of consistent management, soil C and N contents, soil C:N ratio, and soil pH in the top 10 cm were affected ($P < 0.05$) by field treatments. When irrigation was excluded from the model, soil C content ($P = 0.038$) and soil C: N ratio ($P = 0.033$) differed between burn treatments in 2014. Averaged across tillage, N fertilization/residue level, and irrigation, soil C content averaged 1.22 and 1.42 $kg{\cdot}m^{-2}$, while the soil C:N ratio averaged 9.2 and 10.2 under burning and non-burning, respectively. However, when burning was excluded from the model, soil N content ($P = 0.032$) and the C:N ratio ($P = 0.021$) differed between the N-fertilizer/residue-level treatments (**Table 1**). Averaged across tillage, burning, and irrigation, soil N content averaged 0.14 and 0.13 $kg{\cdot}m^{-2}$ under the high and low N-fertilization/residue-level treatments, respectively. Similarly, averaged across tillage, burning, and irrigation, the soil C: N ratio averaged 9.4 and 10.0 under the high and low N-fertilization/residue-level treatments, respectively.

In 2014, soil pH in the top 10 cm differed ($P = 0.021$) between irrigation treatments within N-fertilizer/residue-level treatments (**Table 1**) when burning was excluded from the model. Averaged across tillage and burning, soil pH was greater under irrigation regardless of N-fertilization/residue level, where soil pH averaged 7.26 under the high and 7.28 under the low N fertilization/residue level, than that under the dryland treatment, where soil pH averaged 6.67 under the low, which was greater than that under the high N fertilization/residue level (*i.e.*, pH averaged 6.48). However, all soil pH values, regardless of management, exceeded the minimum soil pH threshold of 6.0, below which soybean yield reductions can be expected on silt-loam soils in eastern Arkansas [19]. Therefore, the differences in soil pH among irrigation and N-fertilization/residue-level treatment combinations were agronomically non-significant with regards to soybean production on silt-loam soils in eastern Arkansas.

In contrast to other initial soil properties, after 13 complete wheat-soybean cropping cycles (*i.e.*, 2001 to 2014) and 12 years of consistent management, bulk density and SOM contents were unaffected ($P > 0.05$) by any of the field treatments in 2014 regardless of whether burning (**Table 1**) or irrigation (data not shown) were excluded in the statistical model. Therefore, bulk density averaged 1.21 $g{\cdot}cm^{-3}$ [standard error (SE) = 0.01] and SOM content averaged 2.9 $kg{\cdot}m^{-2}$ (SE = 0.06) across all field treatments in 2014.

Table 1. Analysis of variance summary of the effects of N-fertilization/residue level (residue level), tillage, irrigation, and their interactions on soil bulk density, pH, soil organic matter (SOM), total carbon (TC), total nitrogen (TN), and the soil C: N ratio in the top 10 cm from spring 2014 after 12 complete cropping cycles in a wheat-soybean, double-crop production system in eastern Arkansas.

Source of Variation	Bulk Density	pH	SOM	TC	TN	C:N
			P			
Residue level	ns†	ns	ns	ns	0.032	0.021
Tillage	ns	ns	ns	ns	ns	ns
Irrigation	ns	0.023	ns	ns	ns	ns
Residue level × tillage	ns	ns	ns	ns	ns	ns
Residue level × irrigation	ns	0.021	ns	ns	ns	ns
Tillage × irrigation	ns	ns	ns	ns	ns	ns
Residue level × tillage × irrigation	ns	ns	ns	ns	ns	ns

†Not significant (ns, $P > 0.05$).

3.3. Soil Water Retention

As was expected, the relationship between soil water potential and gravimetric water content across all data and treatment combinations followed a curvilinear pattern, where the water potential increased exponentially as gravimetric soil water content increased (**Figure 2**). Though soil properties, and water retention, were assumed to be uniform at the beginning of the study site in 2001, after 13 complete wheat-soybean cropping cycles (*i.e.*, 2001 to 2014) and 12 years of consistent management, the slope characterizing the linear relationship between the natural logarithm of water potential and gravimetric water content in the top 7.5 cm was affected ($P = 0.007$) by only the N-fertilization/residue-level treatment (**Table 2**; **Figure 3**), and was unaffected ($P > 0.05$) by tillage, burning, irrigation or any of their interactions (**Table 3**). The intercept characterizing the linear relationship between the natural logarithm of water potential and gravimetric water content was unaffected ($P > 0.05$) by any field treatment (**Table 2**; **Figure 3**). Averaged across tillage, burning, and irrigation, the slope for the low was greater (*i.e.*, more negative) than that for the high N-fertilization/residue-level treatment (**Table 3**). This result indicated that, on average, there was a greater change in water content per unit change in water potential associated with the soil from the low compared to the high N-fertilization/residue-level treatment. The greatest differences between high and low N-fertilization/residue treatments were observed at water contents > ~0.12 $g \cdot g^{-1}$ (**Figure 3**). Conversely, as soil water potential decreased, gravimetric soil water contents became increasingly similar under both N-fertilization/residue-levels treatments.

The results of this study on a silt-loam, loessial soil from the Delta region of eastern Arkansas were similar to the soil moisture characteristic curve results reported by Brye [2] using a similar wetting-curve approach. Brye [2] demonstrated that soil water contents in both native prairie and cultivated agricultural silt-loam soils in east-central Arkansas became increasingly similar as soil water potential approached permanent wilting point (*i.e.*, −1.5 MPa), regardless of field treatments imposed. Verkler *et al.* [4] also reported numerically greater maximum soil water contents at the 7.5 cm depth under the high compared with the low N-fertilization/residue-level treatment, although the differences were statistically non-significant. Management practices that increase the amount of crop residue returned to the soil, such as with greater above- and belowground biomass achieved with differential N fertilization, have been shown to increase infiltration, bulk density, and water storage capacity [20].

Though N fertilization/residue level did not significantly affect SOM or C contents in the top 10 cm in 2014 after 13 years of consistent residue management, one possible explanation for the significant effect of N fertili-

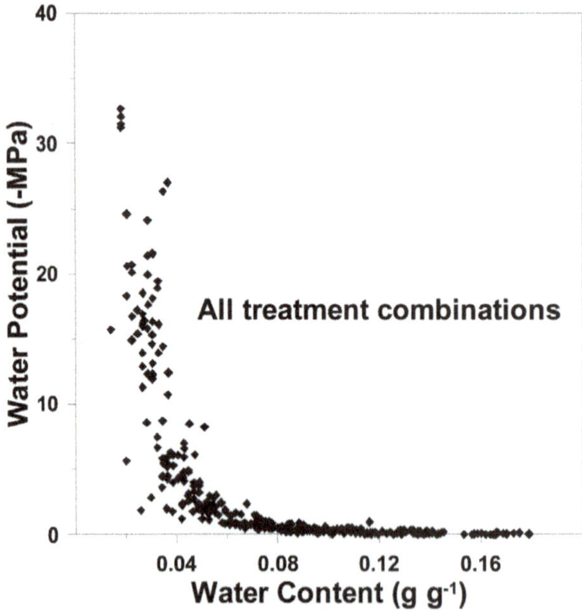

Figure 2. Raw data from all treatment combinations depicting the relationship between soil water potential and gravimetric soil water content from soil wetting curves for the top 7.5 cm in a wheat-soybean, double-crop system in eastern Arkansas after 13 years (*i.e.*, 2001 to 2014) of consistent residue management.

Table 2. Analysis of covariance summary of the effects of N-fertilization/residue level (residue level), burning (burn), tillage, irrigation, and their interactions on the linear relationship between the natural logarithm of soil water potential and gravimetric soil water content using a soil wetting curve approach from the top 7.5 cm after 12 complete cropping cycles in a wheat-soybean, double-crop production system in eastern Arkansas.

Source of variation†	*P*
Intercept term	
Residue level	ns††
Burn	ns
Tillage	ns
Irrigation	ns
Residue level × burn	ns
Residue level × tillage	ns
Residue level × irrigation	ns
Burn × tillage	ns
Burn × irrigation	ns
Tillage × irrigation	ns
Residue level × burn × tillage	ns
Residue level × burn × irrigation	ns
Residue level × tillage × irrigation	ns
Burn × tillage × irrigation	ns
Residue level × burn × tillage × irrigation	ns
Liner term	
Water content	<0.001
Residue level × water content	0.007
Burn × water content	ns
Tillage × water content	ns
Irrigation × water content	ns
Residue level × burn × water content	ns
Residue level × tillage × water content	ns
Residue level × irrigation × water content	ns
Burn × tillage × water content	ns
Burn × irrigation × water content	ns
Tillage × irrigation × water content	ns
Residue level × burn × tillage × water content	ns
Residue level × burn × irrigation × water content	ns
Residue level × tillage × irrigation × water content	ns
Burn × tillage × irrigation × water content	ns
Residue level × burn × tillage × irrigation × water content	ns

†Non-significant interactions ($P > 0.05$) were removed in the final model, except when non-significant terms participated in higher-order, complex treatment combinations. ††Not significant (ns, $P > 0.05$).

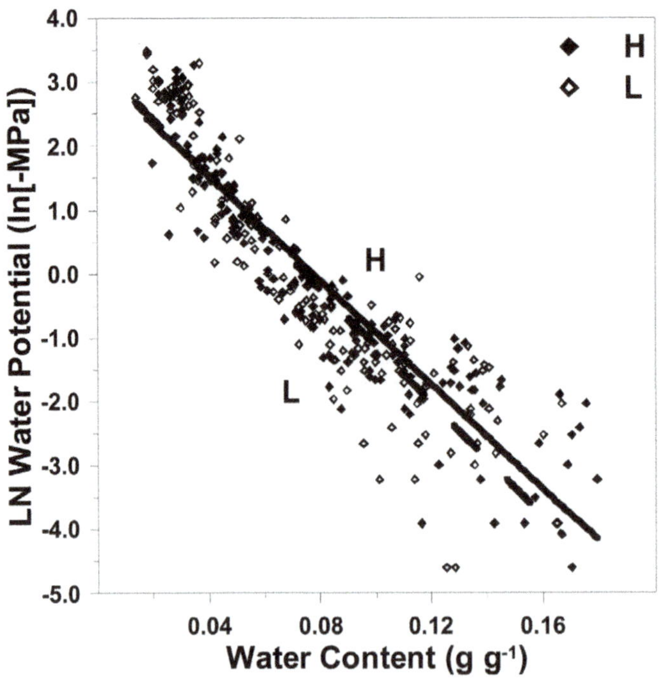

Figure 3. Influence of residue level [high (H, ——) and low (L, -----)] on the relationship between the natural logarithm (LN) of the water potential and the gravimetric soil water content from soil wetting curves for the top 7.5 cm in a wheat-soybean, double-crop system in eastern Arkansas after 13 years (*i.e.*, 2001 to 2014) of consistent residue management.

Table 3. Summary of estimated regression parameters for the N-fertilization/residue-level [High (H) and Low (L)] treatment for the linear relationship between the natural logarithm of soil water potential (−MPa) and gravimetric soil water content ($g \cdot g^{-1}$) in the top 7.5 cm after 12 complete cropping cycles in a wheat-soybean, double-crop production system in eastern Arkansas. Coefficient estimates with the same lower case letter do not differ ($P > 0.05$).

Regression Term	Treatment	Coefficient Estimate	$P^{\dagger}$
Intercept	High	3.108 a	<0.001
	Low	3.357 a	0.002
Linear	High	−39.712 a	<0.001
	Low	−45.207 b	0.001

†$P < 0.05$ indicates coefficient estimate was significantly different from 0.

zation/residue level on the relationship between the natural logarithm of water potential and the gravimetric soil water content was that the high N-fertilization/residue-level treatment promoted increased soil structure development and SOM more than the low N-fertilization/residue-level treatment. While it was concluded in a previous study analyzing soil properties in the same plots used in the current study that N fertilization/residue level alone had no obvious, observable effects on the trend in SOM content in the top 10 cm over time between 2007 and 2014, N fertilization/residue level did affect ($P < 0.05$) the trend in SOM content over time as part of complex treatment combinations [18].

It is also possible that the N-fertilization/residue-level treatment may have impacted SOM content and soil aggregates in the top 7.5 cm differently than in the top 10 cm due to the greater accumulation of both above- and below-ground plant biomass concentrated near the soil surface. A previous study analyzing soil aggregation in the top 10 cm in the same plots used in the current study reported that the concentration of water-stable aggregates was 11% greater in the top 5 cm than in the 5 to 10 cm depth interval after 7 years of consistent management [7] suggesting that SOM and soil aggregates may be more concentrated in the 7.5-cm depth samples used for the current study than in the 10-cm depth samples used for previous studies [6] [18]. Therefore, it is possible that the N-fertilization/residue-level treatment affected the <2-mm-sized soil aggregates, which may have occluded SOM, in the top 7.5 cm, without clearly and obviously affecting SOM contained in the aggregate size classes larger than 2 mm in the top 10 cm. Such an increase in occluded SOM in smaller aggregates might account for an increase in soil water content [5] [21]. For example, Brye [2] reported greater soil water contents in the top 10 cm of a native prairie soil than soil water contents of cultivated agricultural soils at the same water potential and ascribed at least a partial explanation to the greater SOM content in the prairie soil (SOM = 22 $g \cdot kg^{-1}$) compared wo the cultivated agricultural soil (SOM = 10.8 $g \cdot kg^{-1}$) examined.

Somewhat surprisingly, neither burning nor tillage had an observable effect ($P > 0.05$; **Table 2**) on the relationship between the natural logarithm of water potential and the gravimetric soil water content after 13 years of consistent residue management and 12 complete cropping cycles. These results were in contrast with the original hypothesis that tillage would strongly affect the relationship between soil water potential and soil water content, such that soil water content under NT would exceed that under CT at the same water potential. Similarly, these results were in contrast to the original hypothesis that residue burning would render soil retention in the burned treatment lower than that in the non-burned treatment due to cumulative effects of hydrophobic ash additions. It is possible that any potential effects of burning and/or tillage on water retention characteristics were masked as a result of the sample preparation procedure (*i.e.*, grinding and sieving to ≤2 mm), which rendered the soil structure and natural aggregation and porosities highly disrupted from their in-situ, undisturbed state. As hypothesized, the cumulative effects of irrigation or dryland soybean production had no effect on resulting water retention characteristics as determined using soil wetting curves.

In contrast to the results of this study, other soil water retention studies have reported significant correlations between cultivation and near-surface soil water retention characteristics [2] [3] [5], likely due to the increased hydraulic conductivity and infiltration rates associated with relatively undisturbed NT compared to highly disturbed CT soils. Verkler *et al.* [4] reported that residue burning significantly affected maximum soil water contents during irrigation events, and that the mean maximum soil water content was 3% (v/v) greater under residue burning compared with non-burning. However, the water content measurements conducted by Verkler *et al.* [4] were made in-situ (*i.e.*, in undisturbed soil in the field), and are therefore fundamentally different from the oven-dried, ground, sieved, and rewetted soil samples used in the current study.

4. Conclusion

Following conversion to alternative residue and water management practices and after 13 consecutive years (*i.e.*, 2001 to 2014) of management, N fertilization/residue level, but not burning, tillage, or irrigation treatments, significantly affected the linear relationship between the natural logarithm of water potential and the gravimetric soil water content as determined by soil wetting curves. The wetting-curve procedure provided a simple, replicatable, and useful approach to evaluate long-term effects of various field treatment on soil water retention. It can be inferred from the results of this study that differences in N-fertilization/residue-level management can have a cumulative effect on soil water retention characteristics, possibly as a result of the increased soil aggregation and occluded SOM associated with increased crop residue inputs, both above-and belowground, under a high compared to a low residue-level management scheme. Consideration of soil water retention characteristics is vital to planning sustainable use of irrigation water, especially in areas such as the Delta region of eastern Arkansas that will face potential water shortages in the near future.

Acknowledgements

Field assistance provided by Matt Gregory at the Cotton Branch Experiment Station is gratefully acknowledged. This work was funded by the Arkansas Soybean Research and Promotion board.

References

[1] Nielsen, D.C., Vigil, M.F., Anderson, R.L., Bowman, R.A., Benjamin, J.G. and Halvorson, A.D. (2002) Cropping System Influence on Planting Water Content and Yield of Winter Wheat. *Agronomy Journal*, **94**, 962-967. http://dx.doi.org/10.2134/agronj2002.0962

[2] Brye, K.R. (2003) Long-Term Effects of Cultivation on Particle Size and Water-Retention Characteristics Determined using Wetting Curves. *Soil Science*, **168**, 459-468. http://dx.doi.org/10.1097/01.ss.0000080331.10341.36

[3] Hill, R.L., Horton, R. and Cruse, R.M. (1985) Tillage Effects on Soil Water Retention and Pore Size Distribution of Two Mollisols. *Soil Science Society of America Journal*, **49**, 1264-1270.

http://dx.doi.org/10.2136/sssaj1985.03615995004900050039x

[4] Verkler, T.L., Brye, K.R., Gbur, E.E., Popp, J.H. and Amuri, N. (2008) Residue Management and Water Delivery Effects on Season-Long Surface Soil Water Dynamics in Soybean. *Soil Science*, **173**, 444-455. http://dx.doi.org/10.1097/SS.0b013e31817b6687

[5] Azooz, R.H. and Arshad, M.A. (1996) Soil Infiltration and Hydraulic Conductivity under Long-Term No-Tillage and Conventional Tillage Systems. *Canadian Journal of Soil Science*, **76**, 143-152. http://dx.doi.org/10.4141/cjss96-021

[6] Amuri, N., Brye, K.R., Gbur, E.E., Popp, J. and Chen, P. (2008) Soil Property and Soybean Yield Trends in Response to Alternative Wheat Residue Management Practices in a Wheat-Soybean, Double-Crop Production System in Eastern Arkansas. *Journal of Integrative Biosciences*, **6**, 64-86.

[7] Smith, S.F., Brye, K.R., Gbur, E.E., Chen, P. and Korth, K. (2014) Residue and Water Management Effects on Aggregate Stability and Aggregate-Associated Carbon and Nitrogen in a Wheat-Soybean, Double-Crop System. *Soil Science Society of America Journal*, **78**, 1378-1391. http://dx.doi.org/10.2136/sssaj2013.12.0534

[8] Bowman, R.A. and Halvorson, A.D. (1998) Soil Chemical Changes after Nine Years of Differential N Fertilization in a No-Till Dryland Wheat-Corn-Fallow Rotation. *Soil Science*, **163**, 241-247. http://dx.doi.org/10.1097/00010694-199803000-00009

[9] Halvorson, A.D., Reule, C.A. and Follett, R.F. (1999) Nitrogen Fertilization Effects on Soil Carbon and Nitrogen in a Dryland Cropping System. *Soil Science Society of America Journal*, **63**, 912-917. http://dx.doi.org/10.2136/sssaj1999.634912x

[10] Scott, H.D., Ferguson, J.A., Hanson, L., Fugitt, T. and Smith, E. (1998) Agricultural Water Management in the Mississippi Delta Region of Arkansas. Arkansas Agricultural Experiment Station, Division of Agriculture, University of Arkansas, Fayetteville.

[11] Cordell, M.L., Brye, K.R., Longer, D.E. and Gbur, E.E. (2006) Residue Management Practice Effects on Soybean Establishment and Growth in a Young Wheat-Soybean Double-Cropping System. *Journal of Sustainable Agriculture*, **29**, 97-120. http://dx.doi.org/10.1300/J064v29n02_08

[12] Brye, K.R., Mersiovsky, E., Hernandez, L. and Ward, L. (2013) Soils of Arkansas. Arkansas Agricultural Experiment Station, University of Arkansas System Division of Agriculture, Fayetteville, AR. 136 p.

[13] Natural Resources Conservation Service (NRCS), Soil Survey Staff (2012) Web Soil Survey: Soil Data Mart. http://websoilsurvey.nrcs.usda.gov

[14] Brye, K.R., Cordell, M.L., Longer, D.E. and Gbur, E.E. (2006) Residue Management Practice Effects on Soil Surface Properties in a Young Wheat-Soybean Double-Crop System. *Journal of Sustainable Agriculture*, **29**, 121-150. http://dx.doi.org/10.1300/J064v29n02_09

[15] National Oceanic and Atmospheric Administration (NOAA) (2002) Climatography of the United States No. 81, Monthly Station Normals of Temperature, Precipitation, and Heating and Cooling Degree Days 1971-2000: Arkansas. United States Department of Commerce, National Climatic Data Center, Asheville.

[16] Verkler, T.L., Brye, K.R., Popp, J.H., Gbur, E.E., Chen, P. and Amuri, N. (2009) Soil Properties, Soybean Response, and Economic Return as Affected by Residue and Water Management Practices. *Journal of Sustainable Agriculture*, **33**, 716-744. http://dx.doi.org/10.1080/10440040903220724

[17] University of Arkansas Cooperative Extension Service (UACES) (2000) Arkansas Soybean Handbook. Miscellaneous Publication 197. Cooperative Extension Service, University of Arkansas, Little Rock.

[18] Norman, C.R. (2015) Long-Term Effects of Alternative Residue Management Practices on Near-Surface Soil Properties and Soybean Production in a Wheat-Soybean, Double-Crop System in Eastern Arkansas. M.S. Thesis, University of Arkansas, Fayetteville.

[19] Slaton, N., Roberts, T. and Ross, J. (2013) Fertilization and Liming Practices. In: Extension Soybean Commodity Committee, Ed., *Arkansas Soybean Production Handbook*, University of Arkansas Research and Extension, Little Rock. http://www.uaex.edu/publications/pdf/mp197/chapter5.pdf

[20] Shaver, T.M., Peterson, G.A., Ahuja, L.R., Westfall, D.G., Sherrod, L.A. and Dunn, G. (2002) Surface Soil Physical Properties after Twelve Years of Dryland No-Till Management. *Soil Science Society of America Journal*, **66**, 1296-1303. http://dx.doi.org/10.2136/sssaj2002.1296

[21] Dao, T.H. (1993) Tillage and Winter Wheat Residue Management Effects on Water Infiltration and Storage. *Soil Science Society of America Journal*, **57**, 158-159. http://dx.doi.org/10.2136/sssaj1993.03615995005700060032x

20

An Integrated Geochemical and Mineralogical Approach for the Evaluation of Zn Distribution in Long-Term Sludge-Amended Soil

Dominique Proust

LIENSs UMR 7266 CNRS, La Rochelle, France

Email: dominique.proust@univ-lr.fr

Abstract

This research work was designed to compare the Zn distribution in a long-term sludge-amended soil with that in a control soil. Two complementary approaches were performed: 1) a geochemical approach at the metric scale of the bulk soil horizons and 2) a mineralogical approach at the micrometric scale of the primary minerals weathering microsites. The geochemical approach revealed that Zn in the control soil was inherited from the weathering parent-rock. Its concentration was always lower than in the amended soil where Zn was supplied at the surface by the spread sludges and moves downwards. The mineralogical approach showed that the clay minerals, produced by the weathering of the primary minerals (amphiboles and plagioclases), or filling the fissure network were made up of smectites (saponite and montmorillonite) at the bottom and kaolinite at the top of the two soil profiles. Each clay mineral, with its specific sorption capacity, controlled the Zn distribution within the soil: the smectites produced by the amphiboles had high sorption capacity and favored Zn retention in the upper horizons of the soil. Conversely, the kaolinites produced by the plagioclases had lower sorption capacity, did not retain Zn in the surface horizons, and allowed it to migrate to deeper horizons where it was sorbed onto the montmorillonites.

Keywords

Weathering, Soil, Clay Minerals, Heavy Metal, Zinc

1. Introduction

The spreading of sewage sludges onto farmlands can be a beneficial method for soil amendment because the sludges can be a source of plant nutrients (especially N, P) and organic matter. However, much of these sludges results from the treatment of industrial and urban by-products and may contain toxic organic and inorganic compounds. Among the inorganic components, the heavy metals have been critically examined since some of them can be toxic to the biosphere at very low concentrations: for instance, Cd concentration cannot exceed 2 $\mu g \cdot g^{-1}$ for soils to be selected for sludge spreading [1]. The origin of the heavy metals in a soil is either geogenic if they are inherited from the weathering of the parent-rock [2]-[4], or anthropogenic if they come, among other sources, from contamination by repeated application of inorganic fertilizers [5], sludge spreading [6], and/or diffuse atmospheric contamination [7]. One of the most valuable properties of soil as regards sludge-spreading is its ability to absorb and retain the heavy metals ions. The heavy metals adsorption/desorption reactions occur at the soil solid/solution interface and involve especially the clay minerals which are the essential reactive solid components in soils. These clay minerals, with their natural sorptive capacities for heavy metals, are of particular relevance for sludge spreading and have received considerable attention in the last decades [8]-[10].

The adsorption of heavy metals onto clay minerals operates in particular sites with permanent or variable negative charges [11] [12]. The sites with permanent charges are located at the interlayer basal surfaces of the clay minerals and result from the isomorphic substitutions of Al^{3+} for Si^{4+} in the silica tetrahedral sheet and/or Fe^{2+}/Mg^{2+} for Al^{3+} in the alumina octahedral sheet. These interlayer negative charges are neutralized by pH-independent adsorption of cations as outer-sphere complexes (cation exchange reactions). The sites with variable charges are located at the crystal edge surfaces and originate in the protonation or deprotonation of surface hydroxyl (SOH) groups. These sites promote pH-dependent adsorption attributed to surface complexation reactions with silanol and aluminol groups to form inner-sphere complexes.

The major part of the clay minerals found in the soils results from the weathering of the parent-rock in two simultaneous processes: i) the dissolution of the primary rock-forming minerals with the release of the cations and the heavy metals; and ii) the crystallization of clay minerals from the soil solutions with the sorption of the previously released heavy metals. Mineralogical studies have shown that the chemical weathering reactions and their associated metals release are active in specific soil microsystems with their own solid and solution chemical properties [13]-[15]. They are typically micrometric composite aggregates of clay minerals (specific to the primary mineral being weathered), oxihydroxides, and organic matter moistened with soil solutions.

Up to now, studies dealing with polluted soils described heavy metals contaminations at the soil scale, *i.e.* in the fine earth (<2 mm) of each soil horizon [16] [17] and/or in separated soil fractions, e.g. <2 μm clay fraction or 2 - 50 μm silt fraction [18] [19]. These conventional pedological approaches, however, underestimate the influence of the individual soil components, especially the weathering microsystems, upon the adsorption and retention of heavy metals in soils.

Even though the bulk concentrations of heavy metals can be decisive in assessing global soil pollution, they result from the addition of the partial heavy metals concentrations in each of the weathering microsystems. As a consequence, the detailed analysis of a soil contaminated with heavy metals requires both the knowledge of its bulk metals concentration but also the partial metals contents in the microsystems in which they are retained. By taking advantage of a long-term spreading on soil of highly Zn concentrated sewage sludges (498 $mg \cdot kg^{-1}$), this study is undertaken to estimate i) the impact of the sludge application upon Zn distribution and migration in the soil and ii) the influence of the clay mineral species upon Zn retention in the weathering microsystems.

2. Materials and Methods

2.1. Soil Characteristics

The soils investigated in this study were selected and sampled during the geological and pedological field survey of the "Saint-jean-Ligoure" diorite massif in Haute-Vienne (Limousin, France), 25 km in the South of Limoges (**Figure 1**). They are inceptisols where wet sewage sludges were deposited during the 90's, each year for ten years, at an application rate of 2500 $kg \cdot ha^{-1} \cdot year^{-1}$ with mean Zn concentration of 498 $mg \cdot kg^{-1}$. Two soils were sampled on the same dioritic parent-rock: the Amended Soil (AS) that received the sewage sludges for ten years and the Control Soil (CS) that is free of sludge spreading.

The two soils are moderately drained, deep with the weathered rock found at 100 - 150 cm depth and show the typical A, Bw and C horizons sequence of inceptisols (**Figure 2(a)**, **Figure 2(b)**). The R horizon, *i.e.* the coherent unweathered parent rock (**Figure 2(c)**), is a diorite which appears to the naked eye as a "salt and pepper" coarse-grained rock made up of major amphiboles, plagioclases, and minor quartz. The two soils have

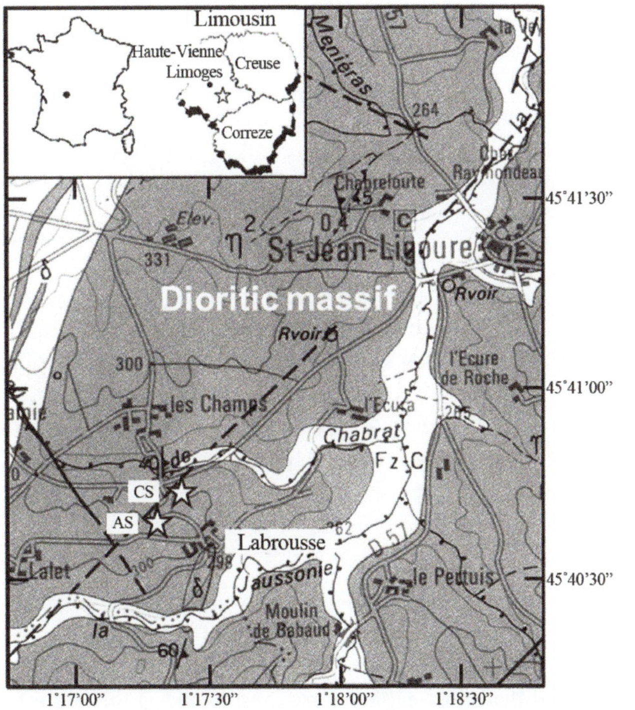

Figure 1. Geological map of the sudied area in the Limousin region, France, and location of the sampled soil profiles; CS:control soil, AS: Amended Soil.

(a) (b)

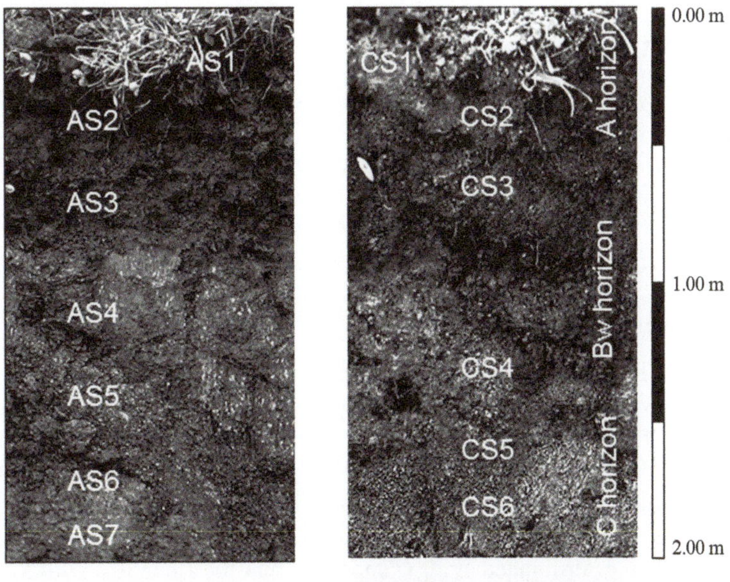

(d)

Soils	CS	AS	CS	AS	CS	AS
Horizons	A	A	Bw	Bw	C	C
pH water	6.10	6.50	6.40	6.80	6.50	6.80
Org. C%	1.92	1.82	0.63	0.85		
C.E.C.*	31.90	26.90	30.80	27.90	27.50	26.80
Clay %	27.90	29.70	28.70	31.80	25.20	24.20

(c)

Figure 2. Sketch of the soil profiles; (a) Amended Soil (AS); (b) Control Soil (CS); (c) Parent-rock (R horizon); (d) Bulk physico-chemical properties of soils.

very similar physicochemical properties (**Figure 2(d)**) with slightly acidic to almost neutral pH in all horizons (6.1 - 6.8). The pH values increase with depth, suggesting some leaching of base cations or probably the effect of organic matter with 1.82% to 1.95 % organic carbon content in the A horizons, decreasing to 0.85% - 0.63 % in the Bw horizon and disappearing in the C horizon. The cation exchange capacities of the two soils are similar with medium values ranging between 26.8 and 31.9 $cmol \cdot kg^{-1}$, whereas the clay contents, typical of the clay loam texture observed in the inceptisols, increase slightly in the Bw horizon (cambic horizon).

2.2. Sampling Methods

Each soil was carefully sampled using plastic cylindrical corers to prevent the soil metal contamination. The cylinder, 10 cm long with a diameter of 10 cm, had cutting edges. It was smoothly and continuously pushed into the soil to collect undisturbed blocks which retained the original soil structure. Each core was divided into two subsamples. The first was disaggregated by gentle shaking in water, sieved to 2 mm, air dried and reserved for the mineral grains separation, the bulk chemical and the X-ray diffraction (XRD) analyses.

The second subsample was preserved undisturbed for thin sections preparation, electron probe microanalyses (EPMA) and scanning electron microscope (SEM) study.

2.3. Mineralogical Separation

The organic matter in the first <2 mm subsample was removed using 33% hydrogen peroxide. 200 grams of this sample were shaken in water to destroy the aggregates, sieved into the sandy fractions 50 - 100 μm, 100 - 150 μm, 150 - 250 μm, 250 - 500 μm, and dried at 50°C for 24 h. The observation of each sieved fraction under optical microscopy (OM) revealed that most euhedral monomineralic grains were found in the 250 - 500 μm fraction. The smaller sized fractions were made up of broken primary minerals. About 500 monomineralic grains, 250 to 500 μm in size, of amphiboles and plagioclases (the two most important minerals in the parent-rock and soils) were then needle-sorted under the stereomicroscope for the study of their specific weathering products. Separation efficiency was checked by XRD.

2.4. Bulk Chemical Analyses

The bulk chemical analyses were obtained from the first subsample, freed of its organic matter. An aliquot quantity of matter (5 g) was collected by quartering with a riffle splitter and crushed down to less than 50 μm size into agate mortar. 300 mg were taken from the total powder, fused with $LiBO_2$ at 1050°C and dissolved in 1N HNO_3 for the bulk chemical analyses. The major and trace elements were analysed using ICP-AES and ICP-MS at the Service d'Analyse des Roches et des Minéraux, (SARM), CRPG-CNRS (Vendoeuvre-lès-Nancy, France). The major and trace elements concentrations were expressed in wt.% and $mg \cdot kg^{-1}$, respectively.

2.5. X-Ray Diffraction Analyses

The bulk clay fractions, together with the clay minerals extracted from the weathered monomineralic grains by sonification, were analysed by XRD on Ca-saturated oriented preparations, air dried and glycolated. The diffractometer was a PHILIPS PW 1730 (40 kV, 40 mA) with a Fe-filtered CoKα radiation and a stepping motor driven with a DACO-MP recorder and the Diffrac-AT software (Socabim, Munich, Germany). The bulk clay fraction <2 μm was extracted from 10 g of the <2 mm first subsample. The 10 grams sampled were first dispersed in 350 ml deionized water, sonicated for 2 min at 300 W/20 kHz, then mechanically shaken for 3 h. The stable suspension was then centrifugated at 20°C and 1000 rpm for 2 mn 30 s with a Jouan JR4.22 centrifuge in order to separate the <2 μm fraction. The effectiveness of this clay separation was then checked using a laser diffraction granulometer IP Malvern Mastersizer.

2.6. Microanalyses of the Soil Samples

The second subsample was devoted to the EPMA and the SEM studies. Thin sections were first prepared according to the procedure of Camuti and McGuire [20]. Samples were hardened under vacuum (100 mbars) with an epoxy resin (ARALDITE 2020) mixed with 20% acetone thinner. After polymerization, samples were cut and thin sections obtained through polishing with silicon carbide (17 μm and 9 μm particle sizes) and diamond calibrated powders (6 μm, 3 μm, 1 μm, and 0.25 μm particle sizes).

Representative pedofeatures (clay coatings, clay cutans, etc.) and weathering microsites were first located on the thin sections, marked with black circle under an optical microscope and then "*in situ*" analysed for major and trace elements. EPMA were obtained using a CAMECA SX 50 electron microprobe (Service CAMPARIS, Université Paris VI) equipped with wavelength-dispersive spectrometers (WDS). The microprobe was calibrated using synthetic and natural oxides. Corrections were made with a ZAF program. A specific trace program with the electron microprobe was developed [21] to analyse heavy metals in small volumes with detection limits closed to 6 - 8 $mg \cdot kg^{-1}$. Major elements analyses were performed at 15 kV and 4 nA prior to trace element analysis performed at higher voltage (30 kV) and higher beam current (100 nA and 500 nA for small and large minerals, respectively) in order to improve detection limits. The spot sizes varied from 1 to 5 - 10 μm to limit beam damage. The counting time was 10 s per major element and 1000s per trace element. Major and trace elements are expressed respectively in wt% and $mg \cdot kg^{-1}$.

For each dominant primary rock forming mineral, *i.e.* amphibole and plagioclase, ten grains were selected from the 250 - 500 μm needle-sorted fraction and affixed to glass plates with double-sided adhesive tape. The two glass plates were then gold-coated for the micromorphological SEM study of the weathering products.

2.7. Geochemical Mass Balances

The geochemical mass balances were obtained from the calculation of enrichment factors (*EF*) using the method of Hernandez *et al.* [7] and Sterckeman *et al.* [4] with the equation:

$$EF = (X/R)_{\text{altered}} / (X/R)_{\text{reference}} \quad (1)$$

where the content of element *X* is normalized to the content of the reference element *R*. A value of 1 for *EF* indicates no element enrichment or depletion. *EF* values >1 indicate element enrichment and when <1 indicate element depletion. The aluminium, with its narrow range of variation in the two soils (17.53% to 20.08% in **Table 1**) was chosen as the reference element. The CS8 sample with the lowest loss on ignition suffered very low weathering and was selected as the reference rock for major elements and Zn bulk analyses.

Table 1. Chemical analyses of major elements (wt%) and Zn ($mg \cdot kg^{-1}$) in control soil (CS) and amended soil (AS) samples.

Sample	CS8	CS7	CS6	CS5	CS4	CS3	CS2	CS1	AS7	AS6	AS5	AS4	AS3	AS2	AS1
Depth (cm)	485	450	180	160	130	65	40	20	190	170	140	110	70	40	20
SiO_2	48.79	49.00	47.99	47.53	47.15	49.12	47.48	46.07	51.2	50.21	51.27	52.34	50.83	51.23	50.95
TiO_2	1.08	1.32	1.14	1.32	1.33	1.30	1.22	1.37	1.28	1.24	1.27	1.31	1.42	1.48	1.51
Al_2O_3	18.78	18.25	19.10	19.37	19.22	19.84	20.08	18.98	18.76	18.47	18.41	17.93	18.67	17.53	18.74
Fe_2O_3[a]	9.58	10.04	10.03	10.08	10.29	9.41	10.13	9.75	8.87	9.53	8.99	9.47	9.26	9.65	9.13
MnO	0.17	0.21	0.18	0.19	0.21	0.20	0.17	0.16	0.14	0.15	0.15	0.12	0.11	0.12	0.11
MgO	4.47	4.55	4.20	4.06	3.89	3.88	4.09	3.52	3.25	3.44	3.19	3.29	3.36	3.36	3.08
CaO	7.52	6.97	6.97	7.05	6.69	6.19	4.15	6.26	5.7	5.83	5.24	2.11	3.55	2.41	4.26
Na_2O	3.54	3.29	3.36	3.13	2.94	3.25	3.88	2.82	3.19	3.08	3.19	3.5	2.96	3.18	3.25
K_2O	1.44	1.40	1.50	1.19	1.37	1.27	1.51	1.47	1.54	1.35	1.69	1.68	1.66	1.64	1.63
P_2O_5	0.59	0.70	0.60	0.66	0.70	0.67	0.47	0.57	0.00	0.00	0.00	0.00	0.00	0.00	0.00
LOI[b]	3.77	3.94	4.68	5.21	5.04	4.62	6.50	8.53	5.91	6.48	6.19	7.77	7.74	8.49	6.72
Total	99.73	99.67	99.75	99.80	98.85	99.73	99.62	99.50	99.84	99.78	99.59	99.52	99.56	99.09	99.38
Zn	111.18	124.46	113.06	119.32	126.08	120.42	108.24	115.56	115.36	103.04	80.64	72.15	79.64	68.27	69.29

[a]Total iron expressed as Fe_2O_3. [b]LOI: Loss on ignition.

3. Results

3.1. Parent-Rock Mineralogy

The two soils are derived from the same dioritic parent rock which outcrops at the bottom of the quarry exposing the control soil profile (R horizon in **Figure 2(c)**). The diorite appears to the naked eye as a "salt and pepper" coarse-grained rock made up of major dark-green amphiboles, white plagioclases and minor vitreous quartz. The modal rock composition was obtained from the sample CS8, *i.e.* the unweathered coherent rock with high apparent density (2.92 $g \cdot cm^{-3}$) and the lowest loss on ignition (3.77% in **Table 1**). The counting of 3000 points on thin section gives the mineralogical composition of 38% amphibole, 32% plagioclase, 12% orthoclase and minor quartz (8%), albite (7%), and titanomagnétite (3%). The EPMA of the major elements in the rock forming minerals show that the amphiboles are calcic hornblendes and plagioclases are andesine with 48% anorthite content. The EPMA of Zn (**Table 2**) reveal that the amphiboles are the major Zn-bearing minerals with high Zn concentrations of 180 $mg \cdot kg^{-1}$ versus lower Zn contents in the plagioclases (17 $mg \cdot kg^{-1}$), orthoclase (20 $mg \cdot kg^{-1}$) and albite (14 $mg \cdot kg^{-1}$).

3.2. Geochemical Mass Balances

The enrichment factors calculated using the Equation (1) and **Table 1** are plotted versus depth in **Figure 3**. The chemical evolution trends from rock to saprolite and soil appear similar for the major elements in the control and amended soils. The first chemical change observed during weathering is the general increase in *EF* calculated for the loss on ignition (*EF-L.O.I.*) from bottom to top of the two soil profiles (**Figure 3(a)**). *EF-L.O.I.* increases

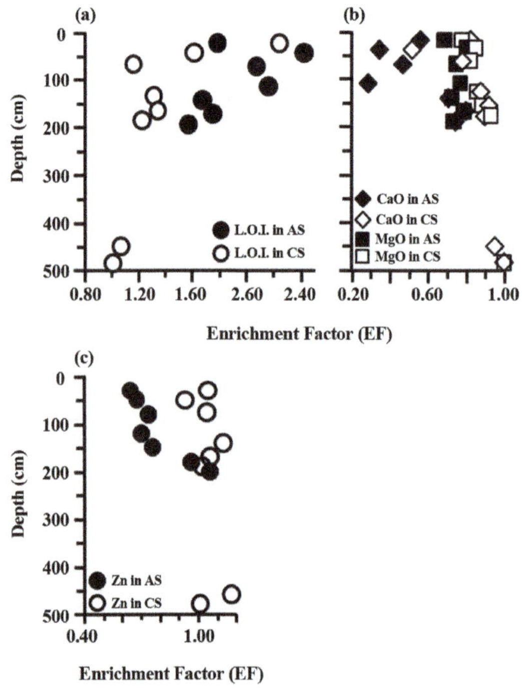

Figure 3. Variations in the enrichment factor (*EF*) calculated for (a) Loss on Ignition (*L.O.I.*); (b) CaO and MgO; (c) Zn in the amended (AS) and control (CS) soils.

Table 2. Mean EPMA of Zn (nanalyses, $mg \cdot kg^{-1}$) in the rock-forming minerals.

	Amphibole		Plagioclase		Orthoclase		Albite	
	$n = 14^a$	SE^b	$n = 10$	SE	$n = 10$	SE	$n = 10$	SE
Zn	180	4	17	2	20	2	14	6

[a]Each analysis is already the mean of ten measurements. [b]Standard error.

more than twice from the sample reference CS8 (*EF-L.O.I.* = 1) to the top of the soils (*EF-L.O.I.* reaches 2.41 in the sample AS2 at the top of the amended soil). This trend reflects the general increase in the bulk water content (loss on ignition at 1050˚C, *i.e.* the hydrate content in the absence of the CO_2 found in carbonates and after organic matter destruction), and is related to the increase in the hydrated clay minerals crystallizations as weathering proceeds. In the same way as aluminium that was chosen for reference element, the measured SiO_2 and Fe_2O_3 concentrations (**Table 1**) remain stable throughout the two soil profiles and give *EF* values close to 1 (0.91 - 1.12 for *EF*-SiO_2 and 0.93 - 1.08 for *EF*-Fe_2O_3). Conversely, the alkaline earth elements Mg and Ca decrease regularly from the bottom to the top of the two soils: The *EF*-MgO and *EF*-CaO reach respectively 0.69 and 0.34 in the sample AS2 at the top of the amended soil (**Figure 3(b)**). This regular and general decrease in the two soils is indicative of the weathering processes which affect the amphiboles (*EF*-MgO and *EF*-CaO) and the plagioclases (*EF*-CaO) in the saprolite and soil.

The chemical analyses of Zn in the bulk soil horizons (**Table 1**) reveal that, although the sewage sludges spread on the amended soil were heavily loaded with Zn (498 $mg\cdot kg^{-1}$ as mean concentration), its concentration in the two soils never exceeds the threshold value for anthropogenic contamination (300 $mg\cdot kg^{-1}$, [22]). The Zn enrichment factors, when calculated from **Table 1** and plotted versus depth in **Figure 3(c)**, reveals two different trends depending on whether the control soil or the amended soil is considered: The distribution of *EF*-Zn in the control soil does not show any significant variation throughout the soil profile, whereas *EF*s in the amended soil indicate pronounced depletion of Zn in the surface horizons that may suggest migration of this metal to depth (down to 190 cm, sample AS7). This Zn migration to depth has been already observed by Scokart *et al.* [23].

3.3. Behaviour of Zn in the Soil Clay Fractions

The clay minerals identified in the <2 μm granulometric fractions are similar in the control and amended soils. The XRD on Ca-oriented preparations reveals the typical 001 reflections of smectite, at 15.10 Å in air-dried state and shifting to 17.05 Å after ethylene-glycol solvation. The smectite is associated with a kaolinite which is typified, in air-dried and ethylene-glycol solvated states, by 001 and 002 reflections at 7.15 Å and 3.55 Å respectively. These reflections disappear after 450˚C heating. No reflection of iron oxide or hydroxide was detected on the XRD patterns of the random powders.

The Zn contents in the soil clay fractions were measured using the electron microprobe and are given with their mean standard errors in **Table 3**. The EPMA are plotted versus depth in **Figure 4(a)** to follow the Zn distribution in the two soil profiles. It appears in **Table 3** that the Zn concentrations in the bulk clay fractions of the amended soil range from 144 to 203 $mg\cdot kg^{-1}$ and are always higher than in the control soil where they range from 10 to 24 $mg\cdot kg^{-1}$. The shape of the Zn distribution profile in the amended soil (**Figure 4(a)**) is typically that of a downward migrating element, with Zn concentrations about ten times that measured in the control soil and a regular decrease from the top to the bottom of the amended soil. The highest Zn concentrations are measured in the surface horizons, from 195 to 203 $mg\cdot kg^{-1}$ at 20 cm (sample AS1) and 40 cm (sample AS2) depths, decreasing to 150 $mg\cdot kg^{-1}$ at the bottom of the soil (190 cm depth, sample (AS7)). For comparison, the Zn distribution profile of the control soil does not show significant variations when standard errors are considered

Table 3. Mean EPMA of Zn (*n* analyses, $mg\cdot kg^{-1}$) in the clay fractions of amended (AS) and control (CS) soils.

	AS7		AS6		AS5		AS4		AS3		AS2		AS1			
Depth (cm)	190		170		140		110		70		40		20			
	$n = 4^a$	SE[b]	$n = 4$	SE	$n = 4$	SE	$n = 4$	SE	$n = 4$	SE	$n = 4$	SE	$n = 4$	SE		
Zn	150	12	158	12	144	11	153	10	177	9	203	13	195	12		
Sample	CS8		CS7		CS6		CS5		CS4		CS3		CS2		CS1	
Depth (cm)	485		450		180		160		130		65		40		20	
	$n = 4$	SE	$n = 4$	SE	$n = 4$	SE	$n = 4$	SE	$n = 4$	SE	$n = 4$	SE	$n = 4$	SE	$n = 4$	SE
Zn	15	2	14	1	14	2	17	2	18	3	19	3	24	5	10	3

[a]Each analysis is already the mean of ten measurements. [b]Standard error.

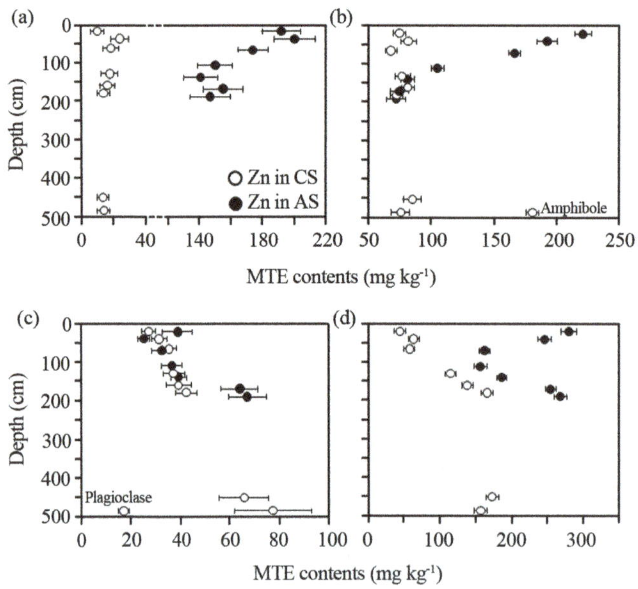

Figure 4. Vertical distribution profile of Zn in the amended (AS) and control (CS) soils; (a) Zn concentrations in the <2 µm clay fraction; (b) Zn concentrations in the amphiboles and their clay minerals weathering products; (c) Zn concentrations in the plagioclases and their clay minerals weathering products; (d) Zn concentrations in the fissural clay minerals.

(**Figure 4(a)**). The Zn concentrations are much lower than those measured in the amended soil, and range from 10 to 24 mg·kg^{-1} at 20 cm (sample CS1) and 40 cm (sample CS2) depths, and from 14 to 15 mg·kg^{-1} at 180 cm (sample CS6) and 485 cm (sample CS8) depths.

3.4. Behaviour of Zn in the Amphibole Weathering Microsystems

The first signs of amphiboles weathering are found in the C horizons of the two soils. The OM observation of the thin sections reveals that euhedral amphibole prismatic crystals begin to weather along the enlarged cleavage planes (**Figure 5(a)**) where SEM shows the crystallization of clay minerals with the typical honeycomb texture of the smectite (**Figure 5(b)**). These characteristic features are commonly observed in the weathered rocks and soils [24] [25]. The weathering intensity increases in the Bw horizons where amphibole crystals appears highly fragmented into 20 - 100 µm organized residues by the opening of the intramineral microcracks, up to 50 µm wide (**Figure 5(c)**). The SEM reveals that the clay minerals filling these microcracks exhibit the same honeycomb morphology of smectite clay minerals, already observed in the C horizons; they are covering the saw-tooth terminations of the dissolving amphibole crystals (**Figure 5(d)**). The most advanced weathering stages are found in the surface A horizons where the amphiboles crystals are almost entirely replaced by a clayey weathering plasma (**Figure 5(e)**) made up of smectites with the characteristic honeycomb texture (**Figure 5(f)**).

The clay minerals, extracted from the amphiboles grains by sonication, were identified by XRD. The XRD patterns are similar in the C, Bw, and A horizons with the same typical reflections of smectite. The Ca-oriented preparations reveals strong 001 reflections ranging from 14.50 Å to 15.49 Å in air-dried state, shifting to 16.87 - 17.14 Å with higher orders at 8.46 Å and 3.40 Å after ethylene-glycol solvation. This expansion to 17 Å with ethylene-glycol is typical of smectite layers.

The EPMA of the amphibole weathering products characterize two chemically distinct smectite populations. The first is detected in the C horizons of soils and is made up of magnesian smectites belonging to the trioctahedral group of saponites. The second population is found in more advanced weathering stages (Bw and A horizons) as aluminous smectites belonging to the dioctahedral smectite group of montmorillonites.

The Zn contents in the clay minerals produced by the weathering of amphiboles were measured using the electron microprobe and are given in **Table 4** with their mean standard errors. The EPMA are plotted versus depth in **Figure 4(b)** to follow the behaviour of Zn in the two soil profiles. The early weathering of amphiboles into saponites (R horizon, samples CS8 and CS7) results into the partial release of Zn, from 180 mg·kg^{-1} in the

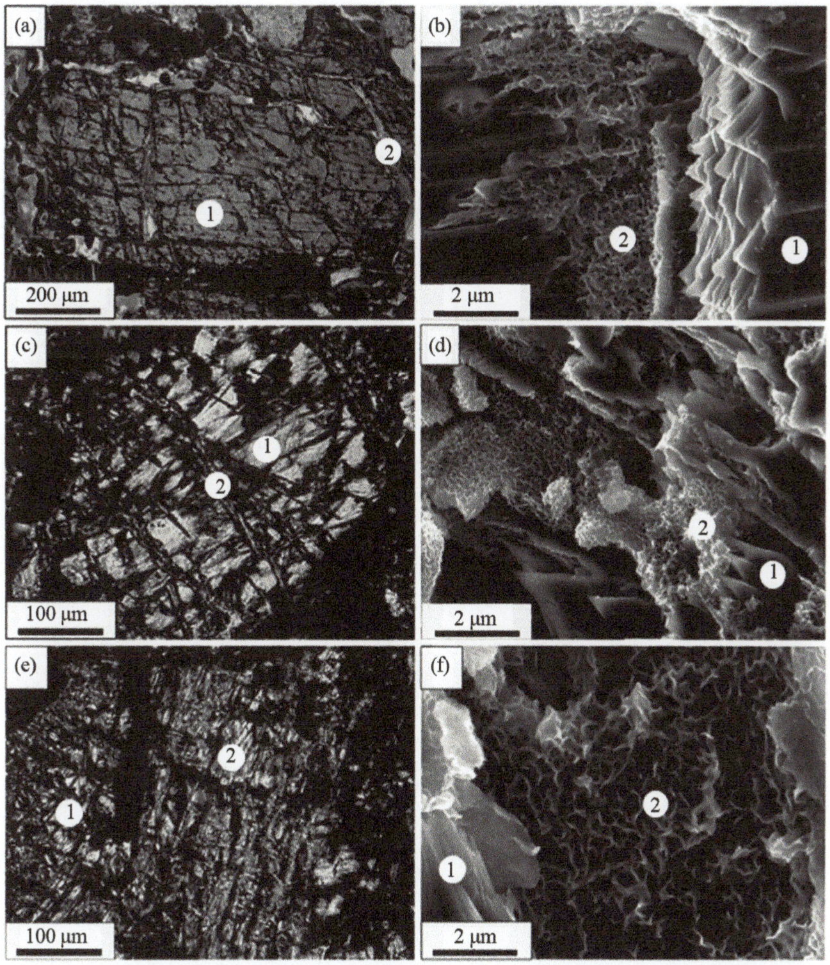

Figure 5. Optical Microscopy (OM) and Scanning Electron Microscopy (SEM) of the amphibole weathering. (a) OM view of an early weathering microsite in the C horizon of the control soil, with 1: elongated amphibole crystal; 2: intramineralmicrocrack with clay minerals infillings. (b) SEM view of the microsite a) with 1: amphibole crystal; 2: honeycomb texture of clay minerals in a microcrack. (c) OM view of a weathering microsite in the Bw horizon of the control soil with 1: amphibole fragments; 2: dense microcrack network with clay minerals infillings. (d) SEM view of the microsite c) with 1: saw-tooth amphibole terminations; 2: honeycomb clay minerals cover. (e) OM view of a weathering microsite in the A horizon of the control soil with 1: increasing amphibole fragmentation with clay minerals crystallizations; 2: complete replacement of amphibole fragments by a clayey plasma. (f) SEM view of the microsite e) with 1: amphibole fragment; 2: honeycomb-textured clay mineral.

fresh amphibole (**Table 2**) to 76 - 85 $mg \cdot kg^{-1}$ in the saponites. Higher weathering levels in the C, Bw, and A horizons produce different Zn distributions depending on whether the control or the amended soil is considered.

When taking into account the mean standard errors, the Zn concentrations in saponites and montmorillonites appear almost constant throughout the control soil profile (**Figure 4(b)**). The behaviour of Zn in the amended soil appears much more complicated. Initially, the Zn concentrations measured in the saponites derived from amphiboles are very similar in the C horizons of the amended and control soil profiles: 72 to 82 $mg \cdot kg^{-1}$ in AS7, AS6, AS5 versus 72 to 81 $mg \cdot kg^{-1}$ in CS6 and CS5 (**Table 4**, **Figure 4(b)**). In a second stage, the Zn concentrations measured in the montmorillonites from the Bw and A horizons of the amended soil differ strongly from that in the control soil. The concentrations increase in the amended soil from 105 to 220 $mg \cdot kg^{-1}$ in AS4 and AS1 versus 68 to 82 $mg \cdot kg^{-1}$ in CS4 and CS1 (**Table 4**, **Figure 4(b)**).

Table 4. Mean EPMA of Zn (n analyses, $mg \cdot kg^{-1}$) in clays from amphiboles and plagioclases weathering, and fissure infillings in control (CS) and amended (AS) soils.

Amphiboles	Saponites								Montmorillonites								Saponites						Montmorillonites							
Sample	CS8		CS7		CS6		CS5		CS4		CS3		CS2		CS1		AS7		AS6		AS5		AS4		AS3		AS2		AS1	
Depth (cm)	485		450		180		160		130		65		40		20		190		170		140		110		70		40		20	
	$n = 4$[a]	SE[b]	$n = 4$	SE	$n = 4$	SE	$n = 4$	SE	$n = 4$	SE	$n = 4$	SE	$n = 4$	SE	$n = 4$	SE	$n = 4$	SE	$n = 4$	SE	$n = 4$	SE	$n = 4$	SE	$n = 4$	SE	$n = 4$	SE	$n = 4$	SE
Zn	76	7	85	7	72	4	81	5	77	4	68	4	82	6	75	4	72	8	75	7	82	2	105	4	165	4	192	8	220	6
Plagioclases	Montmorillonites								K/S mixed layers				Kaolinites				Montmorillonites						K/S mixed layers				Kaolinites			
Sample	CS8		CS7		CS6		CS5		CS4		CS3		CS2		CS1		AS7		AS6		AS5		AS4		AS3		AS2		AS1	
Depth (cm)	485		450		180		160		130		65		40		20		190		170		140		110		70		40		20	
	$n = 4$	SE	$n = 4$	SE	$n = 4$	SE	$n = 4$	SE	$n = 4$	SE	$n = 4$	SE	$n = 4$	SE	$n = 4$	SE	$n = 4$	SE	$n = 4$	SE	$n = 4$	SE	$n = 4$	SE	$n = 4$	SE	$n = 4$	SE	$n = 4$	SE
Zn	77	15	66	10	42	4	39	5	37	4	35	3	31	3	27	2	67	8	64	7	39	3	36	4	32	4	25	2	30	2
Fissures	Montmorillonites								Montmorillonites + Kaolinites								Montmorillonites						Montmorillonites + kaolinites							
Sample	CS8		CS7		CS6		CS5		CS4		CS3		CS2		CS1		AS7		AS6		AS5		AS4		AS3		AS2		AS1	
Depth (cm)	485		450		180		160		130		65		40		20		190		170		140		110		70		40		20	
	$n = 4$	SE	$n = 4$	SE	$n = 4$	SE	$n = 4$	SE	$n = 4$	SE	$n = 4$	SE	$n = 4$	SE	$n = 4$	SE	$n = 4$	SE	$n = 4$	SE	$n = 4$	SE	$n = 4$	SE	$n = 4$	SE	$n = 4$	SE	$n = 4$	SE
Zn	156	8	172	9	165	8	138	7	115	6	58	3	64	4	45	3	268	10	254	6	185	6	156	8	162	7	246	9	280	11

[a]Each analysis is already the mean of ten measurements. [b]Standard error.

3.5. Behaviour of Zn in the Plagioclase Weathering Microsystems

The first weathered plagioclases were observed under OM in the R and C horizons of the two soils. The heart of the plagioclases is dotted with etch pits which are filled by newly-formed clay minerals (**Figure 6(a)**). The SEM gives more details about these clay minerals which display a typical honeycomb habit, previously observed for the smectites in the amphiboles (**Figure 6(b)**). The plagioclases observed under OM in the Bw horizons show more advanced weathering stage. The etch pits, more numerous and connected, lead to the fragmentation of the plagioclase crystals into small pieces, 200 μm as largest size, with clay minerals patches (**Figure 6(c)**). The SEM reveals that these clay minerals have a spheroidal habit (**Figure 6(d)**) very different from the honeycombs observed in the first weathering stage. The most weathered plagioclases are observed in the A horizons. The plagioclase crystals are now almost entirely replaced by a light-grey clayey weathering plasma (**Figure 6(e)**). The SEM reveals that these clay minerals have an exfoliated "booklet-type" morphology (**Figure 6(f)**) which is

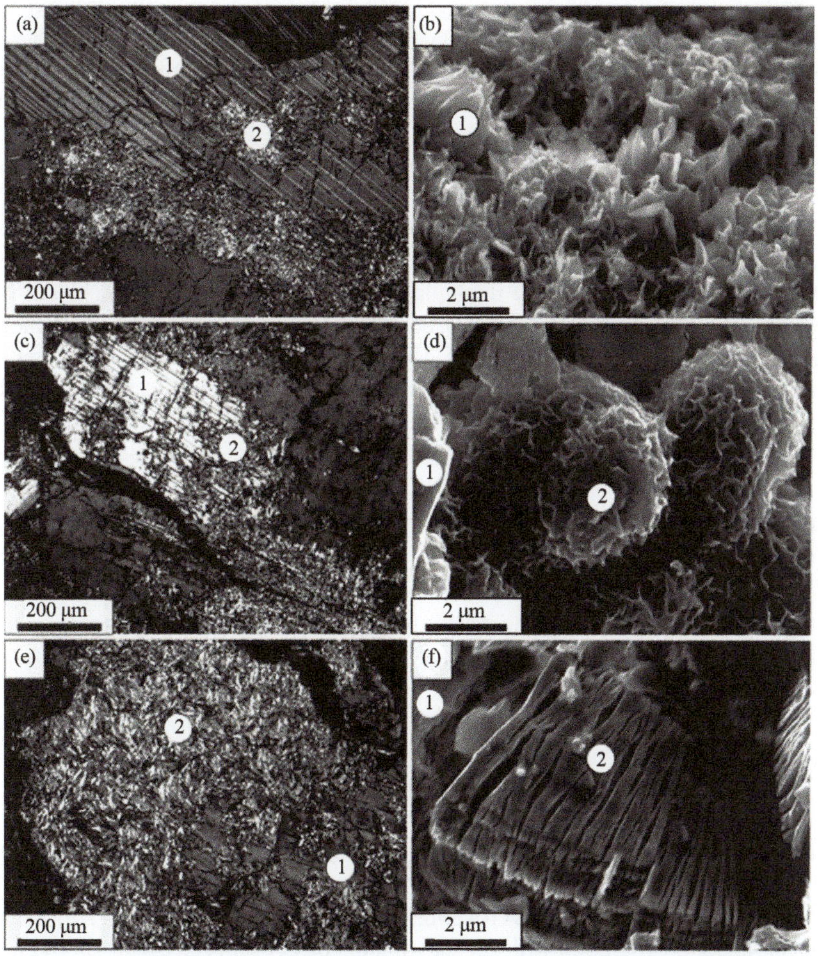

Figure 6. Optical Microscopy (OM) and Scanning Electron Microscopy (SEM) of the plagioclase weathering. (a) OM view of an early weathering microsite in the C horizon of the control soil, with 1: tabular plagioclase crystal; 2: clay minerals infilling dissolution pits. (b) SEM view of the microsite a) with 1: honeycomb texture of clay minerals in a dissolution pit. (c) OM view of a weathering microsite in the Bw horizon of the control soil with 1: plagioclase fragmentation; 2: clay minerals infillings in enlarged dissolution pits. (d) SEM view of the microsite c) with 1: plagioclase remnant; 2: spheroidal clay minerals filling a dissolution void. (e) OM view of a weathering microsite in the A horizon of the control soil with 1: increasing plagioclase fragmentation with clay minerals crystallizations; 2: complete replacement of plagioclase fragments by a clayey plasma. (f) SEM view of the microsite e) with 1: plagioclase fragment; 2: "booklet-type"-textured clay mineral.

commonly observed for the kaolinite crystallizations in the altered rocks [26] [27].

The XRD patterns of the clay minerals extracted from the plagioclase grains differ according to their location in the two soil profiles. The clay minerals in the C horizon of the two soils are similar and made up of a smectite with a 15.56 Å reflection in Ca-oriented, air-dried state, shifting to 17.06 Å after ethylene-glycol solvation, with a second order reflection at 8.48 Å. The XRD patterns of the clay minerals with spheroidal habit (**Figure 6(d)**) in the Bw horizons are more complex, but similar in the two soils. At the bottom of the Bw horizons (140 - 130 cm depth) XRD characterizes a random kaolinite/smectite (K/S) mixed layer with 65% smectite component. This clay mineral is typified by its reflections at 15.03 Å, 7.36 Å, and 3.55 Å in Ca-oriented, air-dried state, shifting to 17.14 Å, 8.28 Å, and 3.42 Å after ethylene-glycol solvation. This K/S is replaced at the top of the Bw horizons (70 - 65 cm depth) by a second random K/S with only 5% smectite component. This K/S differs from the first one by its very low expansion at 7.36 Å (17.14 Å for the first K/S) in Ca-oriented, ethylene-glycol solvated state. The XRD of the clay minerals with the exfoliated "booklet-type" observed in the A horizons characterize a kaolinite clay mineral with a typical 7.17 Å-4.44 Å-3.57 Å reflection group in Ca-oriented, air-dried state, unchanged in the Ca-ethylene-glycol-solvated state.

The EPMA of the clay minerals in the weathered plagioclases are in good agreement with the XRD results. The chemical analyses characterize two weathering sequences producing different clay minerals. The first sequence appears in the parent-rock and the C horizons of soils with the early weathering of plagioclases into aluminous smectites of the montmorillonite group. The second sequence occurs in the Bw and A horizons where montmorillonite weathers first into K/S (Bw horizons) and finally into kaolinite (A horizons).

The Zn concentrations in the clay minerals produced by the weathering of plagioclases are given in **Table 4** and plotted versus depth in **Figure 4(c)**. The early plagioclase weathering into montmorillonites leads to a general Zn enrichment from 17 $mg \cdot kg^{-1}$ in the fresh plagioclase (**Table 2**) to 77 - 66 $mg \cdot kg^{-1}$ (R horizons, samples CS8 and CS7). The Zn distribution in the C, Bw, and A horizons show similar trends in the control and amended soils. The Zn concentrations in the control soil decrease regularly from 42 $mg \cdot kg^{-1}$ at 180 cm depth to 27 $mg \cdot kg^{-1}$ at the soil surface (20 cm depth). A similar evolution is observed in the amended soil, but concentrations in the deepest levels are higher: 67 $mg \cdot kg^{-1}$ at 190 cm and 64 $mg \cdot kg^{-1}$ at 170 cm depths (**Figure 4(c)**). In both soils, the decrease in Zn concentrations from bottom to top follow the decrease in smectite contents observed in the clay minerals: 100% smectite in the C horizons (montmorillonites), followed by 65% and 5% smectite in the K/S from the Bw horizons and 0% smectite in the kaolinite from the A horizons.

3.6. Behaviour of Zn in the Fissural Microsystems

The fissural microsystems are made up of grey to light-green clay minerals that aggregate in the fissural network of the weathered rock and soil. This fissural network ranges from 10 μm to millimeter in width and the clay minerals are observed as infilling clay material. Since clay minerals could not be separated from the bulk soil samples in sufficient pure amount to use XRD identification, they were characterized in situ from thin sections using EPMA. The clay mineralogy deduced from the EPMA of the major elements evolves from pure montmorillonite at the bottom of the two soils to a mixture of montmorillonite and kaolinite at the top. The vertical distribution profile of Zn (**Figure 4(d)**) obtained from the EPMA of the clay minerals (**Table 4**) shows firstly the anthropogenic contamination of the fissural clays by Zn in the amended soil surface, when compared to the control soil. The Zn concentrations measured in the amended soil are always higher than in control soil (156 to 280 $mg \cdot kg^{-1}$ versus 45 to 172 $mg \cdot kg^{-1}$, **Table 4**). The vertical distribution of Zn in the control soil is typical of a downward migrating element, in good correlation with the clay mineralogy (**Figure 4(d)**). The lowest content (45 $mg \cdot kg^{-1}$, **Table 4**) is measured in the fissural clays from the surface sample (sample CS1, 20 cm depth) which are dominated by the kaolinite with very low sorption capacity. Zn is then allowed to migrate downwards, to be retained in the level where fissural infillings are montmorillonites with high sorption capacities. This occurs from 160 cm depth in the fissure of the sample CS5 (138 $mg \cdot kg^{-1}$) down to 485 cm in the sample CS8 (156 $mg \cdot kg^{-1}$). The vertical distribution of Zn in the amended soil is very similar but with higher concentrations and Zn accumulation in the surface horizons, *i.e.* 280 - 246 $mg \cdot kg^{-1}$ in samples AS1 and AS2 at 20 - 40 cm depth (**Table 4**). This surface accumulation, not observed in the control soil, can be directly related to the spreading of the sewage sludges.

4. Discussion and Conclusions

In the first place, two meaningful observations must be discussed at the scale of the soil profiles. The first observation is that the control and the amended soils have the same diorite parent-rock and developed C, Bw and A horizons with similar physiochemical properties, unmodified by sludge spreading (**Figire 2(d)**). This is particularly true for the pH which can affect the variable-charge sites in the clay minerals through protonation or deprotonation of the hydroxyl (SOH) groups. The higher the pH, the higher the adsorption of the heavy metals at the clay-water interfaces [28]-[31] and the lower their mobility [32]. The pH values measured in the amended soil are even closer to neutrality (6.5 - 6.8) than in the control soil (6.1 - 6.5), and thus it will favor adsorption of the heavy metals at the clay surfaces. The second observation is that the geochemical mass balances display the same major elements distribution in the two soil profiles (**Figure 3(a)**, **Figure 3(b)**). This distribution is typical of similar weathering and pedogenetic processes with increasing *LOI* content (crystallization of newly-formed clay minerals), alkaline earth leaching (weathering of primary rock-forming minerals) and stability of SiO_2 and Fe_2O_3. This latter observation indicates that SiO_2 and Fe_2O_3, initially released by the weathering of the rock-forming minerals, are precipitated in the newly-formed clay minerals and do not leave the soil profiles.

These observations show that the two soils are formed from the same parent-rock and have suffered similar weathering and pedogenetic processes which lead to the same horizons development and the same weathering clay mineralogy (smectites, K/S, kaolinite). These converging data validate the choice of these two soils to follow the behaviour of the Zn when it is applied to the amended soil by long-term sludge spreading.

The bulk Zn concentrations in the soil samples (**Table 1**) remain always below the value of 300 $mg \cdot kg^{-1}$, upper threshold value for non-polluted soil in the French regulation. Thus, repeated sludge spreadings on the area do not induce Zn pollution. The distribution of the *EF*-Zn versus depth in the amended soil (**Figure 3(c)**) shows a pronounced Zn depletion from the surface horizons which weakens downward and indicates Zn migration down to 190 cm depth. These data reflect the behaviour of zinc at the scale of the soil sample, but a further study of the soil components indicates that local Zn concentrations may occur in specific soil compartments such as the <2 μm clay fraction, the weathering microsites at the scale of the rock-forming minerals, or the fissural network.

The Zn concentrations in the <2 μm clay fractions of the amended soil (**Table 3**) reveal clearly the anthropogenic contamination of the clay minerals in the surface horizons (20 to 40 cm depth) when compared to the control soil. This surface contamination is associated with the downward Zn migration down to 190 cm depth (**Figure 4(a)**). These clay fractions, however, are the collection of the clay minerals which have formed during the weathering of the rock-forming minerals or have migrated in the fissural network. Thus, the study of Zn behaviour in the two soils will be greatly improved by considering the clay minerals in their crystallization sites, *i.e.* at the micrometric scale of the thin section.

The Zn distribution profiles collected from the amphiboles weathering microsites (**Figure 4(b)**) reveal a general anthropogenic contamination of the amended surface horizons with Zn. This contamination can be related to the crystallization of large amount of montmorillonite as weathering product of amphibole. This montmorillonite, with high sorption capacity of 90 to 127 $cmol \cdot kg^{-1}$ [33] [34] may fix a part of Zn provided by the sludges in the surface horizons. This surface accumulation is followed by a regular Zn decrease in the deeper horizons of the amended soil, in relation with the decrease in amphibole weathering intensity and, accordingly, a decrease in smectite content and sorption capacity with depth.

Conversely, the Zn concentration profiles obtained from the plagioclases weathering microsites (**Figure 4(c)**) show a Zn depletion of the clay minerals in the amended surface horizons and a Zn enrichment deeper in the amended soil. This particular behaviour appears to be controlled by the clay mineral weathering sequence of plagioclases. Kaolinite with low sorption capacity of 16 to 34 $cmol \cdot kg^{-1}$ [35] [36] crystallizes at the top of the amended soil and does not retain Zn. Thus, Zn can migrate downwards to more smectitic assemblages (K/S) and finally to montmorillonite where it is retained at 190 cm depth.

Although Zn concentrations in the fissural network of the amended soil are always higher than in the control soil, the general trend is similar for the C and Bw horizons (**Figure 4(d)**) and typical of a downward migrating element. The behaviour of Zn in the surface horizons appears very different depending on whether one considers the control or the amended soil: Zn in the control soil behaves like a migrating element and follows the same trend as in C and Bw horizons; conversely, Zn in the amended soil shows a surface accumulation that can be directly related to the spreading of the sewage sludges.

Regardless of the sites studied, the present work points out the migration of Zn within the sludge-amended soil profile. The mineralogical multiscale approach demonstrates that such a behaviour is highly controlled by the clay minerals and their associated sorption capacities. Clays study is thus essential to understand the sorption reactions occurring at the solid/solution interface in soils. As demonstrated in this paper, the clayey components in the soils are complex assemblages of several clay minerals which formed in distinct weathering microsystems with their own physicochemical properties. Therefore, there is a need for accurate studies at the scale of these weathering microsystems to identify each clay mineral species in order to infer the sorption capacity of the weathering microsite, and to estimate the potential mobility of the heavy metals in the contaminated soils. This micrometric mineralogical approach could be relevant to describe the location and the behaviour of heavy metals in contaminated soils and improve the modeling of the heavy metals surface complexation in soils [37].

References

[1] Planquart, P., Bonin, G., Prone, A. and Massiani, C. (1999) Distribution, Movement and Plant Availability of Trace Metals in Soils Amended with Sewage Sludge Compost: Application to Low Metal Loadings. *Science of the Total Environment*, **241**, 161-179. http://dx.doi.org/10.1016/S0048-9697(99)00338-1

[2] Baize, D. and Sterckeman, T. (2001) Of the Necessity of Knowledge of the Natural Pedo-Geochemical Background Content in the Evaluation of the Contamination of Soils by Trace Elements. *Science of the Total Environment*, **264**, 127-139. http://dx.doi.org/10.1016/S0048-9697(00)00615-X

[3] Horckmans, L., Swennen, R., Deckers, J. and Maquil, R. (2005) Local Background Concentrations of Trace Elements in Soils: A Case Study in the Grand Duchy of Luxembourg. *Catena*, **59**, 279-304. http://dx.doi.org/10.1016/j.catena.2004.09.004

[4] Sterckeman, T., Douay, F., Baize, D., Fourier, H., Proix, N. and Schvartz, C. (2006) Trace Elements in Soils Developed in Sedimentary Materials from Northern France. *Geoderma*, **136**, 912-929. http://dx.doi.org/10.1016/j.geoderma.2006.06.010

[5] Huang, B., Kuo, S. and Bembenek, R. (2004) Availability of Cadmium in Some Phosphorous Fertilizer s to Field-Grown Lettuce. *Water, Air, and Soil Pollution*, **158**, 37-51. http://dx.doi.org/10.1023/B:WATE.0000044832.04770.41

[6] Mbila, M.O., Thompson, M.L., Mbagwu, J.S.C. and Laird, D.A. (2001) Distribution and Movement of Sludge-Derived Trace Metals in Selected Nigerian Soils. *Journal of Environmental Quality*, **30**, 1667-1674. http://dx.doi.org/10.2134/jeq2001.3051667x

[7] Hernandez, L., Probst, A., Probst, J.L. and Ulrich, E. (2003) Heavy Metal Distribution in Some French Forest Soil: Evidence for Atmospheric Contamination. *Science of the Total Environment*, **312**, 195-219. http://dx.doi.org/10.1016/S0048-9697(03)00223-7

[8] Kraepiel, A.M.L., Keller, K. and Morel, F.M.M. (1999) A Model for Metal Adsorption on Montmorillonite. *Journal of Colloid and Interface Science*, **210**, 43-54. http://dx.doi.org/10.1006/jcis.1998.5947

[9] Srivastava, P., Singh, B. and Angove, M. (2005) Competitive Adsorption Behavior of Heavy Metals on Kaolinite. *Journal of Colloid and Interface Science*, **290**, 28-38. http://dx.doi.org/10.1016/j.jcis.2005.04.036

[10] Zhu, J., Cozzolino, V., Pigna, M., Huang, Q., Caporale, A.G. and Violante, A. (2011) Sorption of Cu, Pb and Cr on Na-Montmorillonite: Competition and Effect of Major Elements. *Chemosphere*, **84**, 484-489. http://dx.doi.org/10.1016/j.chemosphere.2011.03.020

[11] Farrah, H., Hatton, D. and Pickering, W.F. (1980) The Affinity of Metal Ions for Clay Surfaces. *Chemical Geology*, **28**, 55-68. http://dx.doi.org/10.1016/0009-2541(80)90035-2

[12] Wahba, M.M. and Zaghloul, A.M. (2007) Adsorption Characteristics of Some Heavy Metals by Some Soil Minerals. *Journal of Applied Sciences Research*, **3**, 421-426.

[13] Wilson, M.J. (2004) Weathering of the Primary Rock-Forming Minerals: Processes, Products and Rates. *Clay Minerals*, **39**, 233-266. http://dx.doi.org/10.1180/0009855043930133

[14] Caillaud, J., Proust, D. and Righi, D. (2006) Weathering Sequences of Rock-Forming Minerals in a Serpentinite: Influence of Microsystems on Clay Mineralogy. *Clays and Clay Minerals*, **54**, 87-100. http://dx.doi.org/10.1346/CCMN.2006.0540111

[15] Proust, D., Caillaud, J. and Fontaine, C. (2006) Clay Minerals in Early Amphibole Weathering: Tri- to Dioctahedral Sequence as a Function of Crystallization Sites in the Amphibole. *Clays and Clay Minerals*, **54**, 351-362. http://dx.doi.org/10.1346/CCMN.2006.0540306

[16] Keller, C., McGrath, S.P. and Dunham, S.J. (2002) Trace Metal Leaching through a Soil-Grassland System after Sewage Sludge Application. *Journal of Environmental Quality*, **31**, 1550-1560.

[17] Martinez Cortizas, A., Garcia-Rodeja Gayoso, E., Novoa Munoz, J.C., Pontevedra Pombal, X., Burman, P. and Terribile, F. (2003) Distribution of Some Selected Major and Trace Elements in Four Italian Soils Developed from the Deposits of the Gauro and Vico Volcanoes. *Geoderma*, **117**, 215-224. http://dx.doi.org/10.1016/S0016-7061(03)00124-1

[18] Atteia, O., Thélin, P., Pfeifer, H.R., Dubois, J.P. and Hunziker, J.C. (1995) A Search for the Origin of Cadmium in the Soil of the Swiss Jura. *Geoderma*, **68**, 149-172. http://dx.doi.org/10.1016/0016-7061(95)00037-O

[19] Hardy, M. and Cornu, S. (2006) Location of Natural Trace Elements in Silty Soils Using Particle-Size Fractionation. *Geoderma*, **133**, 295-308. http://dx.doi.org/10.1016/j.geoderma.2005.07.015

[20] Camuti, K.S. and McGuire, P.T. (1999) Preparation of Polished Thin Sections from Poorly Consolidated Regolith and Sediment Materials. *Sedimentary Geology*, **128**, 171-178. http://dx.doi.org/10.1016/S0037-0738(99)00073-1

[21] Fialin, M., Rémy, H., Richard, C. and Wagner, C. (1999) Trace Element Analysis with the Electron Microprobe: New Data and Perspectives. *American Mineralogist*, **84**, 70-77.

[22] Baize, D. (1997) Teneurs totales en Eléments Traces métalliques dans les sols (France). INRA, Paris, 408 p.

[23] Scokart, P.O., Meeus-Verdinne, K. and De Borger, R. (1983) Mobility of Heavy Metals in Polluted Soils. *Water, Air, and Soil Pollution*, **20**, 451-463. http://dx.doi.org/10.1007/BF00208519

[24] Arslan, M., Abdioglu, E. and Kadi, S. (2010) Mineralogy, Geochemistry, and Origin of Bentonite in Upper Cretaceous Pyroclastic Units of the Tirebolu Area, Giresun, Northeast Turkey. *Clays and Clay Minerals*, **58**, 120-141. http://dx.doi.org/10.1346/CCMN.2010.0580112

[25] Karakaya, M.C., Karakaya, N. and Kupeli, S. (2011) Mineralogical and Geochemical Properties of the Na- and Ca-Bentonites of Ordu (NE Turkey). *Clays and Clay Minerals*, **59**, 75-94. http://dx.doi.org/10.1346/CCMN.2011.0590109

[26] Jeong, G.Y. (2000) The Dependence of Localized Crystallization of Halloysite and Kaolinite on Primary Minerals in the Weathering Profile of Granite. *Clays and Clay Minerals*, **48**, 196-203. http://dx.doi.org/10.1346/CCMN.2000.0480205

[27] Papoulis, D., Tsolis-Katagas, P. and Katagas, C. (2004) Progressive Stages in the Formation of Kaolin Minerals of Different Morphologies in the Weathering of Plagioclase. *Clays and Clay Minerals*, **52**, 275-286. http://dx.doi.org/10.1346/CCMN.2004.0520303

[28] Coles, C.A. and Yong, R.N. (2002) Aspects of Kaolinite Characterization and Retention of Pb and Cd. *Applied Clay Science*, **22**, 39-45. http://dx.doi.org/10.1016/S0169-1317(02)00110-2

[29] Eloussaief, M. and Benzina, M. (2010) Efficiency of Natural and Acid-Activated Clays in the Removal of Pb(II) from Aqueous Solutions. *Journal of Hazardous Materials*, **178**, 753-757. http://dx.doi.org/10.1016/j.jhazmat.2010.02.004

[30] Farrah, H. and Pickering, W.F. (1977) Influence of Clay-Solute Interactions on Aqueous Heavy Metal Ion Levels. *Water, Air, and Soil Pollution*, **8**, 189-197. http://dx.doi.org/10.1007/BF00294042

[31] Veli, S. and Alyüz, B. (2007) Adsorption of Copper and Zinc from Aqueous Solutions by Using Natural Clay. *Journal of Hazardous Materials*, **149**, 226-233. http://dx.doi.org/10.1016/j.jhazmat.2007.04.109

[32] Kirkham, M.B. (2006) Cadmium in Plants on Polluted Soils: Effects of Soil Factors, Hyperaccumulation, and Amendments. *Geoderma*, **137**, 19-32. http://dx.doi.org/10.1016/j.geoderma.2006.08.024

[33] Ayari, F., Srasra, E. and Trabelsi-Ayadi, M. (2005) Characterization of Bentonitic Clays and Their Use as Adsorbent. *Desalination*, **185**, 391-397. http://dx.doi.org/10.1016/j.desal.2005.04.046

[34] Srodon, J. and McCarty, D.K. (2008) Surface Area and Layer Charge of Smectite from CEC and EGME/H2O-Retention Measurements. *Clays and Clay Minerals*, **56**, 155-174. http://dx.doi.org/10.1346/CCMN.2008.0560203

[35] Ferris, A.P. and Jepson, W.B. (1975) The Exchange Capacities of Kaolinite and the Preparation of Homoionic Clays. *Journal of Colloid and Interface Science*, **51**, 245-259. http://dx.doi.org/10.1016/0021-9797(75)90110-1

[36] Ma, C. and Eggleton, R.A. (1999) Cation Exchange Capacity of Kaolinite. *Clays and Clay Minerals*, **47**, 174-180. http://dx.doi.org/10.1346/CCMN.1999.0470207

[37] Hizal, J. and Apak, R. (2006) Modeling of Copper (II) and Lead (II) Adsorption on Kaolinite-Based Clay Minerals Individually and in the Presence of Humic Acid. *Journal of Colloid and Interface Science*, **295**, 1-13. http://dx.doi.org/10.1016/j.jcis.2005.08.005

21

Tillage and Irrigation Requirements of Sorghum (*Sorghum bicolor* L.) at Hamelmalo, Anseba Region of Eritrea

Ramesh P. Tripathi*, Isaac Kafil, Woldeselassie Ogbazghi

Department of Land Resources and Environment, Hamelmalo Agricultural College, Keren, Eritrea
Email: *rp.tripathi52@gmail.com, nrd70hnk@yahoo.com, wogbazghi@gmail.com

Abstract

Most Eritrean farmers do not adopt soil conservation measures and till even sloppy fields 2 - 4 times for planting sorghum (*Sorghum bicolor* L.) with a view to facilitate rainwater intake. Field experiments were conducted at Hamelmalo to optimize tillage and irrigation requirements of sorghum in loamy sand. Tillage treatments were conventional tillage (4 times) on existing slopes (CTf), conventional tillage on managed plots (terraced) with residue (CTm + R) and without residue (CTm – R), reduced tillage (single tillage 4 days after heavy rainfall) on managed plots with residue (RTm + R) and without residue (RTm – R) and no tillage (direct planting) on managed plots with residue (NTm + R) and without residue (NTm – R) randomized in four replications. Tillage in CTm and CTf was same. Experiment was repeated in year II along with a new experiment in split plot design with same tillage treatments in main plots and 4 irrigation treatments in subplots in 3 replications. Irrigation treatments were rainfed (I_0), 70 mm irrigation at 50% depletion of soil moisture in CTm – R from 1 m profile after end of monsoon (I_1), 70 mm irrigation 7 days after irrigation in I_1 (I_2), and 70 mm irrigation 7 days after irrigation in I_2 (I_3). Bulk density increased and infiltration rates decreased by harvesting due to tillage but changes were lower in residue plots of NT and RT than CT. Optimum soil moisture for emergence of sorghum was within 0.145 ± 0.002 $m^3\ m^{-3}$ at which soil strength was well below critical level for root growth. Soil strength in tilled layer due to intermittent wetting and drying following planting exceeded 2000 k Pa when dried below 0.143 $m^3\ m^{-3}$ moisture. Soil profile in CTf did not recharge by rainfall even by end of the rainy season, whereas it was fully wetted in level and terraced plots. Conservation measures resulted 80 - 150 mm of residual moisture per 2 m of soil profile at sorghum harvesting. Residual moisture was relatively more in residue and irrigated plots than in nonresidue and CTf plots. Soil bunding and levelling alone raised sorghum yields in RT + R to 2887 kg ha^{-1} under rainfed and 3980 kg ha^{-1} under 70 mm irrigation 21 days after last rainfall of the season (I_1). Corresponding yields in CTf were 501 kg ha^{-1} under rainfed and 1161 kg ha^{-1} under irrigation. Single

*Corresponding author.

preplanting tillage 4 days after heavy rainfall (RT) was as good as 2 - 4 tillage (CT) practiced by farmers. Sorghum yields in Hamelmalo could be about 2752 kg ha^{-1} by water use of 344 mm and 4009 kg ha^{-1} by 432 mm. Water use in CTf was lowest (208 mm) under rainfed.

Keywords

Rainwater Conservation, Residue Mulch, Semiarid, Soil Properties, Sorghum Yield, Tillage

1. Introduction

Farmers in Hamelmalo region generally till twice before and once after broadcasting of sorghum (*Sorghum bicolor* L.) and once again about 25 - 30 days from planting on slopes <2% - 35% [1]-[3]. Farmers partly practice contour tillage on existing slopes but do not adopt any other conservation measure to prevent runoff and soil loss. Sorghum yields were relatively higher (0.8 - 1 t ha^{-1}) in the initial years than now (0.2 - 0.6 t ha^{-1}) perhaps due to release of nutrients by rapid oxidation of organic matter by tillage and relatively improved rainwater intake in soils [1] [3]-[5]. High yields in the initial years encouraged farmers to over-till the land without questioning its actual need and consequent deterioration of soil structure, organic matter and soil biota, accelerated soil erosion and overall declining soil quality [6]-[12]. However, it was slowly realised that conventional tillage practice without any conservation measures on the fragile land slopes in the region was unsustainable in terms of production and resources conservation [2] [13]-[16]. The conventional tillage on steep slopes resulted significant loss of top soil (>150 t ha^{-1} y^{-1}) and consequently crop yields in various parts of Africa [1] [3] [13] [15] [17]. Many smallholder farmers' fields in Anseba region are severely affected by sheet and gully erosion, which is greatly accelerated by repeated tillage. Degree of soil degradation can be assessed from the fact that average sorghum yields even in good rainfall years were less than 0.2 - 0.6 t ha^{-1} [1] [18].

Temesgen *et al.* [5] reported that traditional tillage in Ethiopian highlands was precisely to improve rainwater infiltration and reduce runoff and evaporation. However, only 30% - 40% of rainwater infiltrated in the tilled plots on 1% - 6% slopes and about 10% - 25% in non-agricultural lands on slopes > 10% - 30% under the conventional practice of management at Hamelmalo farm [3]. Regardless of the tillage system, almost 100% rainwater could infiltrate only on level and properly bunded fields covered 100% by plant residue mulch [2] [9] [12] [19].

No-till system has been advocated for more than 3 decades as effective option to combat land degradation, increase rainwater conservation and raise yields in semiarid regions like that of Eritrea [12] [20]-[23]. Rockstrom *et al.* [24] observed that minimum tillage increased water productivity and crop yields due to better water harvesting and improved fertilizer use from applications along the ripped and sub-soiled planting lines. Improved field water harvesting through bunding, levelling and mulching under no-till system has also shown to reduce soil erosion, increase biomass production and soil organic matter [2] [25]-[29]. Slower organic matter mineralization in reduced-till than conventional-till systems and consequent nutrient releases was important for resource-constrained farmers of Hamelmalo who suffer most from the consequences of poor soil fertility and soil degradation [3] [30]. Limited field experiments to evaluate tillage requirement of crops to date have been conducted in Eritrea. Ministry of Agriculture initiated research on conservation agriculture in collaboration with FAO [31]. The results were highly encouraging. Another experiment on no-till system with sorghum and groundnut was initiated in 2008 at Hamelmalo Agricultural College in collaboration with Australia. Results substantiated the potential of reduced and no-till systems in resources conservation and raising crop yields. However, as suggested by Giller *et al.* [30], conservation tillage practices must be evaluated to quantify benefits and role of different competing uses of crop residue and other inputs to raise crop yields and arrest soil degradation under local ecological and socio-economic conditions. Reports also indicated decreased yields with conservation tillage, increased labour requirements when herbicides were not used, increased labour burden to women and lack of mulch due to poor productivity and priority given to feeding of livestock with crop residues [30]. Effort to raise crop yields through optimization of tillage and supplementary irrigation requirements have shown potential in resolving many social and soil conservation issues [2] [3].

A preliminary survey of area in 2007 showed that more than 90% of the agricultural land in sub Zoba

Hamelmalo has some form of degradation, including severe soil erosion, sparse vegetation, poor rainwater infiltration, structural degradation and compaction [2] [3]. Rocks and stone pieces outcropping in a large proportion of the land surfaces all around speak of the severity of erosion. More than 90% of the land in Zoba Anseba is uncultivated due to steep slopes, which contributes 70% - 90% of the rainfall as runoff [3]. Whereas much of the runoff goes to neighbouring country and Red sea, it also raises groundwater table along the rivers and their tributaries criss-crossing the area. The groundwater is of good quality and is traditionally being used by farmers for drinking and irrigating orchards and vegetable crops. Farmers generally practice irrigation at every 4 - 6 days interval. Sorghum and pearl millet are common rainfed crops raised during the monsoon season. Although rainfall is sufficient for a good crop, yields are poor due to severe water stress during the grain filling stage (September to October first week) at which supplementary irrigations may increase grain yield prospects significantly [3] [32]. Objective of this research was thus to optimize tillage and supplementary irrigation requirements of sorghum at Hamelmalo to obtain sustainable high yields.

2. Materials and Methods

2.1. Soil

Soils of the Hamelmalo region have developed from fluvial deposits. Experimental soil was loamy sand with 83% sand, 11% silt and 6% clay overlying a layer of sandy loam down to 1.3 m followed by sand (89%) with <20% cobbles and boulders forming a porous bed (**Table 1**). Soil organic matter was 0.65% in the surface layer, which reduced by 50% in lower layers. Soil pH ranged from 7.82 in the surface layer to 8.4 below 1.3 m and electrical conductivity ranged from 0.08 - 0.15 d Sm^{-1}. Available nutrients were low but Ca^{2+}, Mg^{2+} and Na^{+} contents increased with depth. Average bulk density of the surface soil was 1.59 Mg m^{-3}, which increased to 1.69 Mg m^{-3} in 0.2 - 0.5 m layer but reduced underneath to 1.54 Mg m^{-3}. Field capacity moisture was 0.195 m^3 m^{-3} and groundwater table fluctuated from <2.5 m during rainy season to >5 m in dry season.

2.2. Experimental Details

Experiments were conducted during 2007 and 2009 at Hamelmalo Agricultural College farm on predominantly occurring loamy sand in the area. The Hamelmalo Agricultural College is located at 15˚52'20.6"N and 38˚27'57.6"E and 1280 m above msl in the Anseba region of Eritrea. It is about 12 km north of the Keren town on the Keren-Nakfa road adjacent to the river Anseba. Mean annual rainfall in the past 7 years was 488 mm with a minimum of 370 mm and a maximum of 663.1 mm. Total rainfall was 460 mm in 2007 and 390 mm in 2009. Highest mean monthly temperature occurred in May (35.7˚C) and lowest in January (11.1˚C).

In 2007, effect of 7 tillage treatments viz., conventional tillage farmers practice (CTf, slope < 3% - 6%), conventional tillage on managed plots with residue (CTm + R) and without residue (CTm − R), reduced tillage on managed plots with residue (RTm + R) and without residue (RTm − R) and no tillage on managed plots with residue (NTm + R) and without residue (NTm − R), randomized in four replications in 8 m × 6 m plots, were evaluated on rainfed sorghum. Managed plots refer to terraced or level plots with bunds all around. CTf refers to

Table 1. Important properties of the experimental soil.

Depth, m	Soil fractions, %			Texture	pH (1:5)	EC (1:5), dS m^{-1}	OM, %	N, %	P, mg kg^{-1}	Exchangeable cations, cmolc kg^{-1}			
	Sand	Silt	Clay							Ca^{2+}	Mg^{2+}	K^{+}	Na^{+}
0 - 0.2	83	11	6	Loamy sand	7.8	0.08	0.65	0.06	9.32	11.5	3	0.15	0.35
0.2 - 0.5	70	14	16	Sandy loam	8.2	0.08	0.42	0.05	3.71	15.0	5	0.10	0.47
0.5 - 0.3	61	20	19	Sandy loam	8.2	0.14	0.42	0.05	2.91	20.0	5	0.14	0.55
>1.3	89	7	4	Sand	8.4	0.15	0.32	0.04	3.61	29.0	8	0.11	0.51

3 preplanting tillage followed by another 25 days after planting. Tillage in CTm was same as in CTf. RT refers to single preplanting tillage 4 days after heavy rainfall and NT refers to direct planting on managed plots. Except in CTf, bunds of height 0.30 m were made around each plot to prevent runoff. Entire experimental area was surrounded by trench to avoid any run-on. Each plot was separated by 2 m passage, which was sloping towards the main drain. Residue of the previous crop, chopped to 0.1 m, was applied uniformly @ 2.5 t ha^{-1} before tillage operations. The residue cover after planting was 66%, 33% and 63% in NT + R, CTm + R and RT + R plots, respectively.

Sorghum variety PP290 Shambuko was planted on 11 July 2007 at a seed rate of 12 kg ha^{-1} in rows 0.6 m apart at an average depth of 0.03 m using a seeder developed for this purpose. Wherever necessary, plants were thinned to about 0.2 m distance within row in about 20 days after planting. Diammonium phosphate (DAP) was applied @ 100 kg ha^{-1} and urea @ 50 kg ha^{-1} as per recommendations of NARI, Halhale, Eritrea. Entire DAP was applied at the last tillage before sowing and urea was top-dressed around 22 and 45 days from planting following rainfall. Weeds were removed manually on 15 and 35 days from planting. Except smut no major pests and diseases were observed. All infested plants were removed and destroyed after weighing their dry biomass.

Results of the 2007 experiment showed that supplemental irrigations after cessation of monsoon in September may cause significant increase in sorghum yield in the region as dry period coincides with milk to grain development stages. The experiment was thus repeated in 2009 along with a new experiment in split plot design with same 7 tillage treatments in main plots and 4 irrigation levels in subplots in 3 replications. The irrigation treatments were rainfed (I_0), 70 mm irrigation at 50% depletion of soil moisture in CTm – R from 1 m profile (I_1), 70 mm irrigation 7 days after irrigationin I_1 (I_2), and 70 mm irrigation 7 days after irrigation in I_2 (I_3). The crop was planted on July 9, 2009 with similar seed rate and management levels. Plot size of the repeated experiment was same but that of new experiment in split plot design was 5 m × 4 m. The crop was harvested from the central 3 m × 3 m area of the plots. The panicles of mature plants were cut, dried, threshed manually, cleaned and the grain was weighed on a precision electronic balance from each treatment and reported in kg ha^{-1} at 14% moisture content. Biomass yield was reported separately after air drying.

2.3. Measurement of Soil Properties

Soil properties determined were bulk density by core method, infiltration rate using double ring infiltrometer and soil moisture content by gravimetric method at planting, about a month after planting, end of monsoon, harvesting and before irrigations. Sampling for moisture determination was done in 50 mm cylindrical soil cores in duplicate from centre of the 0.2 m layer down to 1 m in one of the replications. High bulk density and low infiltration rates due to intermittent wetting and drying after planting prompted us to determine soil strength in the tilled layer by penetrometer to optimize soil moisture content during germination and emergence. For this reason, series of soil strength measurements were made at different moisture contents following rainfall to establish the relationship between soil moisture content and soil strength.

2.4. Water Use

Crop water use (ET, mm) in n days was calculated from the soil moisture data using water balance equation appropriate for the experimental conditions.

$$ET = RFe + IR - RO - DR \pm \Delta S \quad (1)$$

where RFe, IR, RO, DR and ΔS refer to effective rainfall (mm), irrigation depth (mm), runoff (mm), drainage (mm), and change in soil water storage or loss (mm). Estimation of DR was done from the following considerations [33].

$$DR = 0 \text{ for } S < Sp, \text{ and } DR = S - Sp \text{ for } S > Sp \quad (2)$$

where Sp is potential water storage capacity, considered as soil moisture storage at field capacity. RO was zero from all the plots except CTf in which it was estimated using the rainfall-runoff relationship Equation (3) developed on 3% slope from measurements made in the model watershed [3].

$$Runoff = 0.52 \times rainfall - 2.0, \quad (3)$$

Both rainfall and runoff are in mm.

3. Results and Discussion

3.1. Soil Properties

3.1.1. Bulk Density, Infiltration Rate and Soil Strength

Effect of tillage and crop residue on bulk density and infiltration rates at planting and harvesting were not significant (**Table 2**). But bulk density at harvesting increased by ≈6% and infiltration rate decreased by ≈12% from planting. The effect was lowest in NT and RT plots and maximum in conventional tillage farmers practice (CTf) plots followed by CTm plots. Lower effects in NT and RT plots might be due to lesser degradation of organic matter compared to CTm and CTf plots [6] [27] [34]. Presence of high exchangeable Ca^{2+} and Mg^{2+} in soil (**Table 1**) might have also contributed to formation of denser structure due to series of wetting and drying cycles before harvesting. Tripathi [35] and Tripathi *et al.* [36] reported that both tillage and residue effects may become significant after 4 - 8 years of continuous practice.

Although there was no appreciable swelling and shrinking yet soil strength was rapidly changing with drying (**Figure 1**). Soil strength was lost on wetting close to field capacity and was below 2000 kPa, critical for seedling emergence and root proliferation [2] [37], until soil drying to 0.143 $m^3\ m^{-3}$ moisture but exceeded 2500 kPa with further drying to 0.138 $m^3\ m^{-3}$ moisture at which root growth and water uptake are greatly affected [38]. Farmers also observed that emergence and early establishment of sorghum was better at higher water contents around seed zone [2] [3]. Isaac [2] reported that optimum soil moisture for emergence of sorghum was 0.145 ± 0.002 $m^3\ m^{-3}$ at which soil strength was well below the critical level. Since rainfall was low and erratic, CT was believed to optimize wetness in seed zone in the conventional sloppy fields [5] but crusting in such over till plots due to rapid drying following rainfall after planting cannot be ruled out. A number of reports also indicate that moisture conservation can be better optimized through mulching, terracing, levelling and bunding than tillage

Table 2. Bulk density and infiltration rates at planting and harvesting in different tillage and residue plots.

Treatments	Bulk density ($Mg\ m^{-3}$)		Infiltration rate ($mm\ h^{-1}$)	
	At planting	At harvesting	At planting	At harvesting
CTf	1.61	1.70	7.8	6.6
NT − R	1.60	1.69	8.1	7.1
NT + R	1.59	1.68	8.3	7.3
CTm − R	1.58	1.68	7.9	6.7
CTm + R	1.57	1.67	8.2	7.3
RT − R	1.60	1.68	8.0	7.2
RT + R	1.59	1.66	8.4	7.5
Mean	1.59	1.68	8.1	7.1
LSD (5%)	NS	NS	NS	NS

CTf = Conventional practice on existing slopes, NT = No tillage, CTm = CTf on managed plots, m = managed (terraced or level plots with bunds), −R = Without residue, +R = With residue, NS = Not significant.

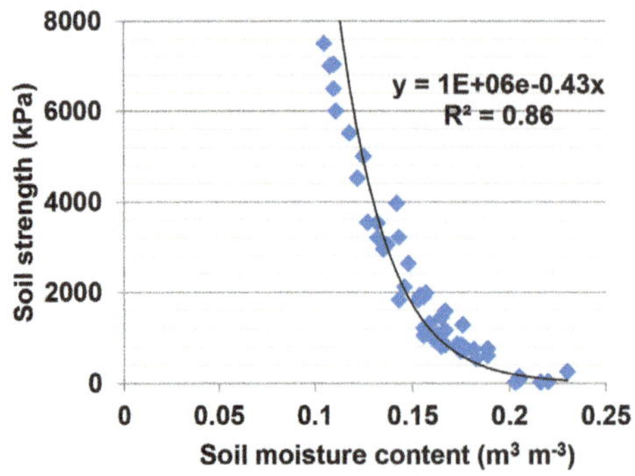

Figure 1. Soil strength as a function of soil moisture.

[3] [6] [34] [35] [39]-[41]. Optimization of tillage would not only reduce cost of cultivation but also facilitate soil and water conservation through improvements in soil structure [6] [35] [40].

3.1.2. Soil Moisture Content

At planting (July 11, 2007), soil moisture in 0 - 0.2 m layer was almost similar in all the plots and, therefore, moisture distribution only in CTf has been presented (**Figure 2**). Soil moisture at planting was about 4% below field capacity (0.195 $m^3 m^{-3}$) in 0 - 0.2 m layer and lower layers were still wetting (**Figure 2**). On 9 August (29 days from planting), all non-residue plots including CTf were still wetting in the lower layers and water content was much below field capacity but all residue plots were wetted beyond field capacity below 0.2 m depth. Total rainfall during June 27 to July 4 was 39.5 mm and that during July 10 to August 9 was 248 mm. Differences in soil moisture contents in CTf and NT-R (**Figure 2**) indicate the effect of levelling and bunding (terracing) and that in the residue plots was due to levelling, bunding and residue additions. Effect of residue on soil moisture was more in NT and RT plots perhaps because of its mulch action due to lesser incorporation in soil than in CT. Wetting in CTf and non-residue plots in lower layers continued until end of the rainy season (8 Sep) but developed uniform upward soil moisture gradient in all the plots before 23 September. The upward gradient in CTf was more because of drying of upper layers than by wetting of lower layers. Soil moisture distribution patterns (**Figure 2**) thus show that a) soil profile in the sloppy fields are not recharged by rainfall even by end of the rainy season, b) rainfall retention as soil water was unaffected by tillage and c) soil water contents in bunded and level plots of NT, RT and CT were always higher in residue than in nonresidue plots. Better capture of rainwater as soil water by crop residue than tillage has been reported by many researchers [25] [27] [34] [39] [41] [42]. Conventional tillage on sloppy fields (CTf) also encourages soil erosion compared to minimum tillage that improves water and fertilizer use [24].

In 2009, rainfall during June 28 to July 8 was 52.6 mm, which penetrated only down to 0.3 m in CTf plots but down to 1 m in the bunded and level plots (**Figure 3**). Except in CTf, wetness was greater and almost to the same depth in residue plots. In nonresidue plots, soil moisture was highest in RT – R and lowest in CTm – R.

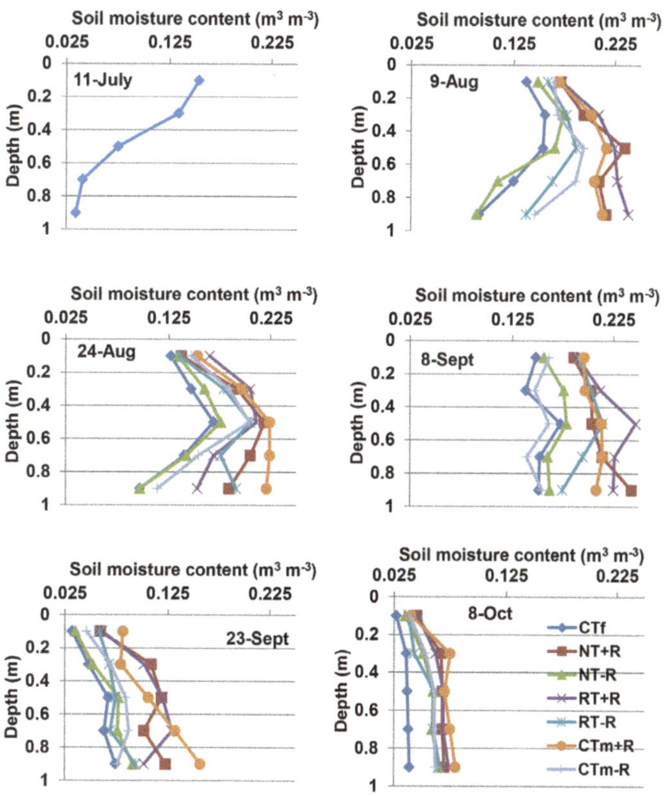

Figure 2. Soil moisture with depth in different tillage and residue plots on various dates during the crop season.

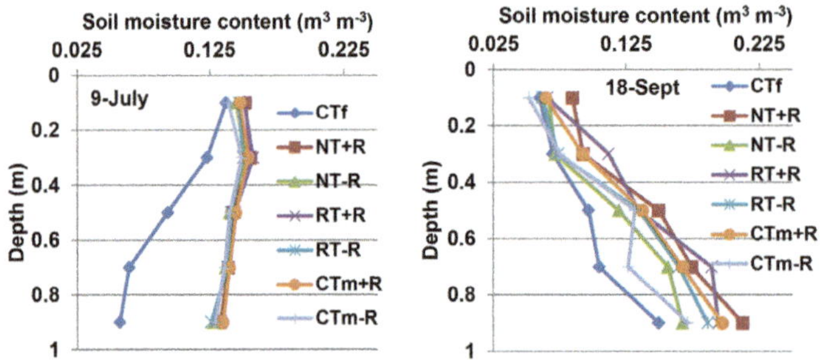

Figure 3. Soil moisture with depth in different tillage and residue plots on July 9, 2009 at planting and Sept 18, 2009 before irrigations.

Relatively more drying in CTm – R might be due to repeated tillage before planting. On September 18, surface 0.4 m soil profile in nonresidue plots dried to less than 7% moisture but wetness in the residue plots was close to field capacity below 0.6 m depth. Irrigations during September 20 to 1st week of October left more residual moisture that can be used by next crop in rotation. At harvesting, surface 0.1m layer dried almost to air dryness but, except in CTf, soil water was 40 - 80 mm m^{-1} of soil profile in lower layers (**Figure 4**). The residual moisture was relatively more in residue plots than in nonresidue plots and in I_3 (Irrigated 14 days after I_1) than in I_0. Results of 2009 confirm the observations of 2007 on tillage and residue effects on rainwater harvesting as soil water and residual moisture left in the soil profile after sorghum harvesting.

3.2. Yield

3.2.1. Tillage-Residue Effects

In 2007, highest yield of grain (2711 kg ha^{-1}) and stover (5812 kg ha^{-1}) under rainfed was observed in CTm + R, which was statistically at par with that in RT + R but significantly greater than in the remaining treatments (**Table 3**). It was lowest (1200 kg ha^{-1} grain and 3000 kg ha^{-1}stover) in CTf. All tillage treatments with residue performed significantly better than those without residue, perhaps due to better rainwater conservation (**Figure 2** and **Figure 3**). Yields were statistically equal in CTm – R and RT – R but significantly greater than in NT – R. This indicates superiority of RT over CT and NT. Similar results were also observed in 2009. Significantly lower yields under NT might be due to poor crop growth in the initial stages because of greater competition for nutrients and water with weeds compared to that under RT in which first flush of weeds that emerged in 4 days after rainfall were turned into the soil by tillage. The crop under NT also showed nitrogen deficiency, which recovered after weeding and urea applications. Two additional weedings were necessary in NT than in other plots. Results show that single preplanting tillage 4 days after heavy rainfall (RT) in well bunded and level plots that ensured rainwater conservation was sufficient to raise sorghum yields in the range of 2405 - 2797 kg ha^{-1} in Hamelmalo region. Crop yield improvements due to minimum tillage were also observed by Mbagwu [39], Steyn *et al.* [34] and Rockstrom *et al.* [24]. Despite potential benefits in crop yields, use of crop residue as mulch has been questioned by Giller *et al.* [30] in semiarid Africa due to its competing uses particularly as animal feed. The argument is valid if farmers are satisfied with current yield levels (1 - 2 t ha^{-1} sorghum stover) but if yields improve to the level obtained (>4 - 5 t ha^{-1}) through better management then at least half of the stover could be diverted back to the soil to which it belongs.

3.2.2. Tillage and Irrigation Effects

Mean grain yield due to tillage was highest in CTm + R (3823 kg ha^{-1}), which was at par with that in RT + R (3756 kg ha^{-1}) but significantly greater than in the remaining treatments (**Table 4**). Similarly mean yield due to irrigations was highest in I_1 (3548 kg ha^{-1}), which was as at par with that in I_2 (3492 kg ha^{-1}) but significantly greater than in I_0 (2416 kg ha^{-1}) and I_3 (3398 kg ha^{-1}). Interaction effects also show that grain yields in CTm + R and RT + R were at par, which further confirmed the superiority of reduced tillage. Except in I_1 and I_2 of CTm, grain yields due to residue were not significant (**Table 4**). Under rainfed (I_0), grain yield was lowest in CTf (501 kg ha^{-1}) and highest in CTm + R (2989 kg ha^{-1}), which was at par with that in RT + R (2887 kg ha^{-1}). Similar effects were observed on stover yields. First irrigation given on 21-day after last rainfall of the season (I_1) raised grain yield by 660 kg ha^{-1} in CTf and 1102 - 1408 kg ha^{-1} in the remaining treatments from that under rainfed.

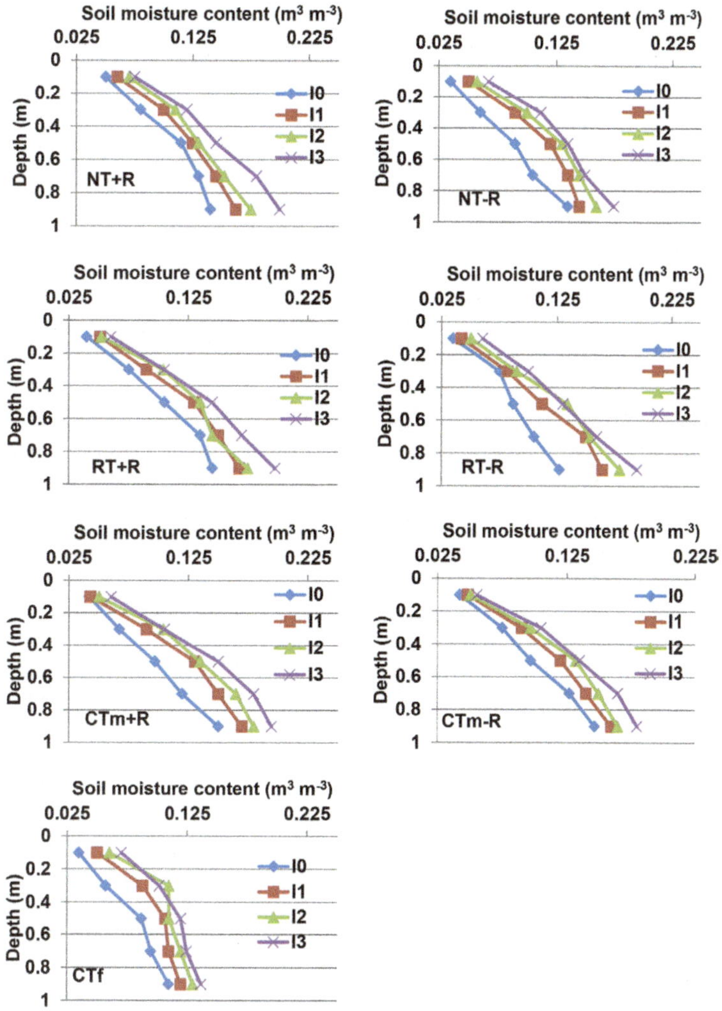

Figure 4. Soil moisture with depth at harvesting in different tillage, residue and irrigation plots.

Table 3. Grain and stover yields of rainfed sorghum in 2007 and 2009.

Treatments	Yield in 2007 (kg ha^{-1})		Yield in 2009 (kg ha^{-1})	
	Grain	Stover	Grain	Stover
CTf	1200	3000	629	1693
NT + R	2447	5474	2640	4975
NT − R	2085	5202	2450	4868
CTm + R	2711	5812	2880	5348
CTm − R	2440	5540	2676	5198
RT+R	2605	5704	2797	5301
RT − R	2405	5439	2590	5158
Mean	2270	5167	2409	4649
LSD (5%)	110	282	126	185

CTf = Conventional practice, NT = No tillage, CTm = CTf on well bunded and level plots, −R = Without residue, +R = With residue.

Table 4. Grain and stover yields of sorghum under different tillage and irrigations.

Tillage treatment, T	Grain yield (kg ha^{-1})					Stover yield (kg ha^{-1})				
	Irrigation levels, I					Irrigation levels, I				
	I_0	I_1	I_2	I_3	Mean	I_0	I_1	I_2	I_3	Mean
CTf	501	1161	983	879	881	1533	2266	2098	1983	1970
NT + R	2530	3880	3803	3731	3486	4973	5912	5872	5743	5625
NT – R	2397	3805	3657	3574	3358	4697	5815	5902	5823	5559
CTm + R	2989	4151	4156	3994	3823	5483	6310	6328	6328	6113
CTm – R	2802	3956	3893	3803	3614	5180	6008	6060.	6155	5851
RT + R	2887	3980	4108	4048	3756	5313	6285	6282	6207	6022
RT – R	2804	3906	3842	3754	3576	5188	5922	5957	5905	5743
Mean	2416	3548	3492	3398	3213	4624	5502	5500	5449	5269
Factors:		T	I	T × I		T	I	T × I		
LSD (P = 0.05)		122	64	185		174	96	271		

CTf = Conventional practice, NT = No tillage, CTm = CTf on well bunded and level plots, -R = Without residue, +R = With residue, I_0 = Rainfed, I_1 = 70 mm irrigation at 50% depletion of soil moisture from 1 m profile, I_2 = 70 mm irrigation 7 days after irrigation in I_1, I_3 = 70 mm irrigation 7 days after irrigation in I_2.

Similarly stover yields increased by 733 - 1118 kg ha^{-1}. The interaction effects justify the significance of reduced tillage (RT: one preplanting tillage 4 days after good rainfall) for optimum yields in well bunded and level plots. Maturity was delayed by 3 to 4 days in I_1 and 2 to 6 days in I_3.

Results thus show that soil bunding and levelling for rainwater conservation with single preplanting tillage can raise sorghum yields to more than 2405 kg ha^{-1} under rainfed and more than 3906 kg ha^{-1} under 70 mm irrigation 21 days after last rainfall of the season (**Table 4**). The three tillage practices (NT, RT, and CT) with residue showed lower weed incidence and produced better crop than without residue. Results also show that irrigations cannot bring significant increases in yield unless fields are level and bunded to arrest runoff and at least part of the residue is returned back to soil to maintain its quality.

3.3. Water Use and Production Function

Water use by rainfed sorghum was minimum (208 mm) in CTf and maximum (353 mm) in NT + R (**Figure 5**). Much of the irrigation water was also lost as runoff in CTf and, therefore, crop water use did not improve as much as in the other treatments. Sorghum crop at first irrigation was at grain filling stage and plants in CTf were almost drying due to inadequate soil water in the root zone but plants were perhaps within recoverable limits in other plots. Greater water use than irrigation amounts in the terraced plots (**Figure 5**) indicates improved upward flow of soil moisture into the root zone from lower layers and efficient root water extraction by mildly stressed plants [43].

A linear relationship between water use and grain yield (**Figure 6**) showed that yield increase due to single irrigation after cessation of monsoon was as important as rainfall during the crop season. Crop response to water availability in the root zone was thus crucial during September 20 to 1st week of October due to rapid drying. Irrigations raised not only the sorghum yield but also left more residual moisture for use by next crop in rotation. Production function (**Figure 6**) shows that farmers can raise rainfed sorghum yields from current <600 kg ha^{-1} [1] [18] to >2752 kg ha^{-1} with a water use of 344 mm through adoption of RT in terraced plots under dryland conditions of Hamelmalo. Grain yields may further increase to 4009 kg ha^{-1} by 70 mm irrigation about 21 days after cessation of the monsoon raising water use to 432 mm.

4. Conclusions

1) Single preplanting tillage 4 days after heavy rainfall (RT) was sufficient for optimum yield of sorghum in well bunded and level fields.

2) Soil bunding and levelling for rainwater conservation can raise rainfed sorghum yields to more than 2400 kg ha^{-1}.

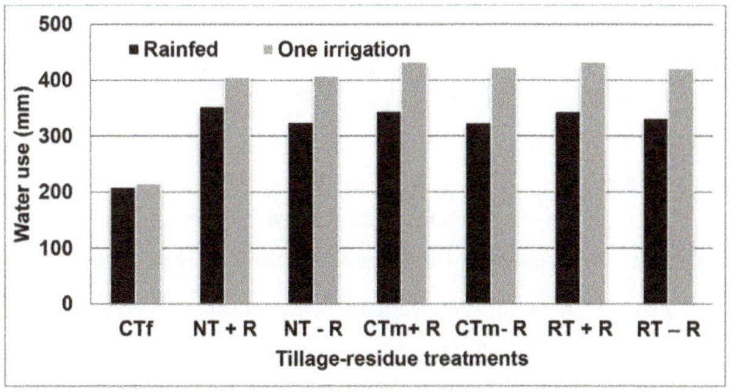

Figure 5. Water use by sorghum indifferent tillage, residue and irrigation plots.

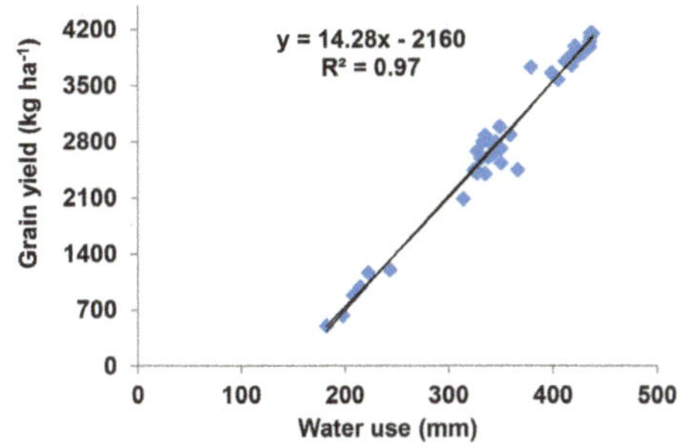

Figure 6. Sorghum yield as a function of water use.

3) Sorghum yields in Hamelmalo region can be raised to more than 3900 kg ha^{-1} by 70 mm irrigation applied 21 days after cessation of rainfall in September.

4) Grain yields would be better under residue than nonresidue conditions.

5) Crop response to water availability in the root zone is crucial for sorghum during September 20 to first week of October.

6) Sorghum root zone in the conventional cultivated sloppy fields is not recharged by rainfall even by end of the rainy season.

7) Irrigations cannot bring significant increases in yield unless fields are level and bunded.

Acknowledgements

Authors are grateful to the Centre for Development and Environment, Institute of Geography, University of Berne, Switzerland for providing financial support through Eastern and Southern Africa Partnership Programme to carry out this research.

References

[1] MOA (Ministry of Agriculture) (2005) Area and Production by Zoba from 1992-2005. Asmara, Eritrea.

[2] Isaac, K. (2008) Effect of Conservation Tillage Practices on Soil Properties, Growth and Yield of Sorghum (*Sorghum bicolour* L.) in the Semiarid Region of Eritrea. M.Sc. Thesis, Department of Land Resources and Environment, Hamelmalo Agricultural College, Eritrea.

[3] Tripathi, R.P. and Ogbazghi, W. (2010) Development and Management of a Hilly Watershed in Hamelmalo Agricultural College Farm, as a Demonstration Site for Farmers and a Study Site for Students. Final Technical Report of the Project Financed by Eastern and Southern Africa Partnership Programme (ESAAP), Hamelmalo Agricultural College, Keren, 59 p.

[4] Haile, A., Araiya, W., Omer, M.K., Ogbazghi, W. and Tewelde, M. (1995) Rehabilitation of Degraded Lands in Eritrea. Ministry of Agriculture, University of Asmara, International Development Center, Technical Paper No. 1.

[5] Temesgen, M., Rockstrom, J., Savenije, H.H.C., Hoogmoed, W.B. and Alemu, D. (2008) Determination of Tillage Frequency among Smallholder Farmers in Two Semiarid Areas in Ethiopia. *Physics and Chemistry of the Earth*, **33**,

183-191. http://dx.doi.org/10.1016/j.pce.2007.04.012

[6] Aina, P.O., Lal, R. and Roose, E.J. (1991) Tillage Methods and Soil and Water Conservation in West Africa. *Soil & Tillage Research*, **20**, 165-186. http://dx.doi.org/10.1016/0167-1987(91)90038-Y

[7] Arnon, I. (1992) Agriculture in Drylands: Principles and Practices. Elsevier Science Publishers, B.V., Amsterdam.

[8] Rockstrom J. and Jonsson, L.O. (1999) Conservation Tillage Systems for Dryland Farming: On-Farm Research and Extension Experiences. *East African Agricultural and Forestry Journal*, **65**, 101-114.

[9] Derpsch, R. (2000) Frontiers in Conservation Tillage and Advances in Conservation Practice. FAO, Rome

[10] Karlen, D.L, Andrews, S.S. and Doran, J.W. (2001) Soil Quality: Current Concepts and Applications. *Advances in Agronomy*, **74**, 1-40. http://dx.doi.org/10.1016/S0065-2113(01)74029-1

[11] Tripathi, R.P., Sharma, P. and Singh, S. (2005) Tilth Index: An Approach to Optimize Tillage in Rice Wheat System. *Soil & Tillage Research*, **80**, 125-137. http://dx.doi.org/10.1016/j.still.2004.03.004

[12] Bollinger, A., Magid, J., Amado, T.J.C., Neto, F.S., Ribeiro, M.D.D., Calegari, A., Ralisch, R. and Neergaard, A.D. (2006) Taking Stock of the Brazilian "Zero-Till Revolution": A Review of Land Mark Research and Farmers' Practice. *Advances in Agronomy*, **91**, 47-110. http://dx.doi.org/10.1016/S0065-2113(06)91002-5

[13] Elwell, H.A. and Stocking, M.A. (1988) Loss of Soil Nutrients by Sheet Erosion Is a Major Hidden Farming Cost. *The Zimbabwe Science News*, **22**, 79-82.

[14] FAO (1994) Agriculture Sector Review for Eritrea. FAO, Rome.

[15] FAO (1998) Press Release 98/42. FAO: Conventional Tillage Severely Erodes Soil; New Concepts for Soil Conservation Required. http://www.fao.org/WAICENT/OIS/PRESS NE/PRESSENG/1998/pren 9842.htm

[16] Rockstrom, J. (2001) Green Water Security for the Food Makers of Tomorrow: Windows of Opportunity in Drought-Prone Savannahs. *Water Science & Technology*, **43**, 71-78.

[17] African Conservation Tillage Network, 2008. www.act-africa.org/

[18] FAO (2005) Global Information and Early Warning System on Food and Agriculture World Food Programme. Special report FAO/WFP Crop and Food Supply Assessment Mission to Eritrea.

[19] Rockstrom, J. and Falkenmark, M. (2000) Semiarid Crop Production from a Hydrological Perspective: Gap between Potential and Actual Yields. *Plant Science*, **19**, 319-346. http://dx.doi.org/10.1080/07352680091139259

[20] Laryea, K.B., Pathak, P. and Klaiji, M.C. (1991) Tillage Systems and Soils in the Semi-Arid Tropics. *Soil & Tillage Research*, **20**, 201-218. http://dx.doi.org/10.1016/0167-1987(91)90040-5

[21] Fowler, R. and Rockstrom, J. (2001) Conservation Tillage for Sustainable Agriculture: An Agrarian Revolution Gathers Momentum in Africa. *Soil & Tillage Research*, **61**, 93-107. http://dx.doi.org/10.1016/S0167-1987(01)00181-7

[22] Hobbs, P.R. (2007) Conservation Agriculture: What Is It and Why Is It Important for Future Sustainable Food Production? *The Journal of Agricultural Science*, **145**, 127-137. http://dx.doi.org/10.1017/S0021859607006892

[23] Hobbs, P.R., Sayre, K. and Gupta, R. (2008) The Role of Conservation Agriculture in Sustainable Agriculture. *Philosophical Transactions of the Royal Society B*, **363**, 543-555. http://dx.doi.org/10.1098/rstb.2007.2169

[24] Rockstrom, J., Kaumbutho, P., Mwalley, J., Nzabi, A.W., Temesgen, M., Mawenya, I., Barron, J. and Damgaard-Larsen, S. (2008) Conservation Farming Strategies in East and South Africa: Yields and Rain Water Productivity from On-Farm Action Research. *Soil & Tillage Research*, **103**, 23-32. http://dx.doi.org/10.1016/j.still.2008.09.013

[25] Lal, R. (1998) Soil Erosion Impact on Agronomic Productivity and Environment Quality. *Critical Reviews in Plant Sciences*, **17**, 319-464. http://dx.doi.org/10.1016/S0735-2689(98)00363-3

[26] Erenstein, O. (2002) Crop Residue Mulching in Tropical and Semi-Tropical Countries: An Evaluation of Residue Availability and Other Technological Implications. *Soil & Tillage Research*, **67**, 115-133. http://dx.doi.org/10.1016/S0167-1987(02)00062-4

[27] Scopel, E., Findeling, A., Guerra, E.C. and Corbeels, M. (2005) Impact of Direct Sowing Mulch-Based Cropping Systems on Soil Carbon, Soil Erosion and Maize Yield. *Agronomy for Sustainable Development*, **25**, 425-432. http://dx.doi.org/10.1051/agro:2005041

[28] Zingore, S., Manyame, C., Nyamugafara, P. and Giller, K.E. (2005) Long-Term Changes in Organic Matter of Woodland Soils Cleared for Arable Cropping in Zimbabwe. *European Journal of Soil Science*, **56**, 727-736. http://dx.doi.org/10.1111/j.1365-2389.2005.00707.x

[29] Farage, P.K., Ardo, J., Olsson, L., Rienzi, E.A., Ball, A.S. and Pretty, J.N. (2007) The Potential for Soil Carbon Sequestration in Three Tropical Dryland Systems of Africa and Latin America: A Modelling Approach. *Soil & Tillage Research*, **94**, 457-472. http://dx.doi.org/10.1016/j.still.2006.09.006

[30] Giller, K.E., Witter, E., Corbeel, M. and Tittonell, P. (2009) Consertvation Agriculture and Smallholder Farming in Africa: The Heretics' View. *Field Crops Research*, **74**, 1-12.

[31] NARI (National Agricultural Research Institute) Eritrea (2001-2003) Reports on Introduction of Conservation Agriculture in Eritrea, and Conservation Agriculture in Eritrea.

[32] Rockstrom, J., Barron, J. and Fox, P. (2002) Rainwater Management for Increased Productivity among Small Holder Farmers in Drought Prone Environments. *Physics and Chemistry of the Earth*, **27**, 949-959. http://dx.doi.org/10.1016/S1474-7065(02)00098-0

[33] Tripathi, R.P. and Mishra, R.K. (1986) Wheat Root Growth and Seasonal Water-Use as Affected by Irrigation under Shallow Water Table Conditions. *Plant and Soil*, **92**, 181-188. http://dx.doi.org/10.1007/BF02372632

[34] Steyn, J.T., Tolmay, J.P.C., Human, J.J. and Kilian, W.H. (1995) The Effects of Tillage Systems on Soil Bulk Density and Penetrometer Resistance of a Sandy Clay Loam Soil. *South African Journal of Plant and Soil*, **12**, 86-90. http://dx.doi.org/10.1080/02571862.1995.10634342

[35] Tripathi, R.P. (1992) Physical Properties and Tillage of Rice Soils in Rice-Wheat System. In: Pandey, R.K., Dwivedi, B.S. and Sharma, A.K., Eds., *Rice-Wheat Cropping System*, Project Directorate for Cropping Systems Research, Modipuram, 53-67.

[36] Tripathi, R.P., Sharma, P. and Singh, S. (2006) Soil Physical Response to Multi-Year Rice-Wheat Production in India. *International Journal of Soil Science*, **1**, 91-107. http://dx.doi.org/10.3923/ijss.2006.91.107

[37] Da Silva, A.P. and Kay, B.D. (1997) Estimating the Least Limiting Water Range of Soil from Properties and Management. *Soil Science Society of America Journal*, **61**, 877-883. http://dx.doi.org/10.2136/sssaj1997.03615995006100030023x

[38] Phillips, R.E. and Kirkham, D. (1962) Mechanical Impedance and Corn Seedling Growth. *Soil Science Society of America Proceedings*, **26**, 319-322. http://dx.doi.org/10.2136/sssaj1962.03615995002600040005x

[39] Mbagwu, J. (1990) Mulch and Tillage Effects on Water Transmission Characteristics of an Ultisol and Maize Grain Yield in SE Nigeria. *Pedologie*, **40**, 155-168.

[40] Lal, R. (1991) Tillage and Agricultural Sustainability. *Soil & Tillage Research*, **20**, 133-146. http://dx.doi.org/10.1016/0167-1987(91)90036-W

[41] Kronen, M. (1994) Water Harvesting and Conservation Techniques for Smallholder Crop Production Systems. *Soil & Tillage Research*, **32**, 71-86. http://dx.doi.org/10.1016/0167-1987(94)90034-5

[42] Bationo, A., Kihara, J., Vanlauwe, B., Waswa, B. and Kimetu, J. (2007) Soil Organic Carbon Dynamics, Functions and Management in West African Agro-Ecosystems. *Agricultural Systems*, **94**, 13-25. http://dx.doi.org/10.1016/j.agsy.2005.08.011

[43] Kramer, P.J. (1978) Plant and Soil Water Relationships: A Modern Synthesis. McGraw-Hill, Inc., New York, 482 p.

22

Estimating Potential Nitrogen Mineralisation Using the Solvita Soil Respiration System

Richard L. Haney[1], Elizabeth B. Haney[2*], Douglas R. Smith[1], Michael J. White[1]

[1]Grassland, Soil and Water Research Laboratory, Agricultural Research Service, United States Department of Agriculture (USDA-ARS), Temple, TX, USA
[2]Texas AgriLife Research & Extension Center, Texas A & M University, Temple, TX, USA
Email: *lhaney@brc.tamus.edu

Abstract

Nitrogen (N) mineralisation contributes considerably to crop growth in fertilized and unfertilized fields. It is useful to be able to assess potential N mineralisation to increase fertilizer application efficiency, prevent excessive N runoff, and improve environmental system models. The microbes present in soil mineralize N based on many factors, including soil temperature and moisture, tillage, and levels of organic C and N. The measurement of soil's ability to mineralize N is considered a good indicator of soil quality. Many methods have been developed to estimate N mineralisation in the laboratory and field. The 7-day anaerobic N mineralisation method developed in the 1960's is considered reliable and is often used to compare new N-mineralisation testing methods. This study examines the use of soil CO_2 evolution as determined using the Solvita Soil Respiration System (Solvita) for estimating N mineralisation by comparing it directly to the anaerobic N mineralisation test. Measured CO_2 using Solvita was strongly correlated with anaerobic N mineralisation (r^2 = 0.82). Results indicate that the Solvita Soil Respiration System can be used to rapidly assess soil respiration and relative N mineralisation potential in any given soil and is considerably faster and easier to perform in a laboratory setting than the anaerobic N mineralisation test.

Keywords

Anaerobic N, Nitrogen Mineralisation, Soil Respiration

1. Introduction

Nitrogen (N) and phosphorus inputs are said to be the primary nonpoint sources of pollution in many parts of the

*Corresponding author.

United States [1]. Plants obtain N from fertilizer applications, atmospheric deposition, precipitation, and biological fixation and mineralisation. Fertilizer recommendations are routinely based on the amount NO_3^- determined during soil testing, negating the contribution of N due to biological activity and resulting in over application of fertilizer. Long-term use of fertilizer has a negative environmental impact by contributing to excessive N and P loading in streams, rivers, and ultimately the ocean. It is therefore important to be able to accurately predict N mineralisation in the environment; however, accurate soil N mineralisation estimates are difficult to obtain rapidly in the laboratory setting [2].

The mineralisation of soil N is controlled by microbial activity as they decompose organic material to obtain energy. Microbial activity is dependent upon soil temperature and moisture, agricultural practices, and physical and chemical soil characteristics. Smith and Humfeld [3] observed that during decomposition studies of green manures, bacteria counts trended with CO_2 evolution, rising rapidly during the first 4 days and then declining to a constant level. Previously in 1919, Gainey [4] had noticed the parallel formation of CO_2, NH_4-N, and NO_3-N in soil. Lebedjantzev [5] indicated that drying soil at a low temperature in the laboratory or the field appeared to increase the fertility of the soil. CO_2 respiration from soil has been studied for over 90 years as an indicator of the rates of C, N, or P mineralisation and ultimately the relative fertility of various soils [4]-[6]. It is now commonly accepted N mineralisation is a measurement of soil quality [7] that is dependent upon the biogeochemical cycling systems of soils including microbial activity.

The Solvita Soil Respiration System (Solvita) is a new tool to evaluate soil microbial respiration rates in an efficient and cost-effective manner, without the need for reagent handling and long incubation periods. Solvita was designed as a complete procedure to quantify the relative differences between varying types of compost by measuring the amount of CO_2 evolved in a short time period [8]. The amount of CO_2 evolved is interpreted as an indication of the completeness of active degradation, also called a maturity index. In composts and manures, Solvita has been used in combination with other indices to determine compost maturity and C and N mineralisation (see www.Solvita.com for complete publications list). The Solvita system has also been suggested as a potential tool for assessing N mineralisation in soils and is currently being utilized by several commercial and government laboratories throughout the United States for this purpose [9]-[11].

There are many published methods for the determination of N mineralisation including long-term aerobic incubation (3 days and 24 days [12], 41 weeks [13]), 7-day anaerobic incubation [14], *in situ* field methods [2], and others [15]. These methods are either too time consuming and laborious for commercial labs to perform at a cost effective price or are not adequate for predicting N mineralisation.

The 7-day anaerobic N mineralisation method is considered the best biological indicator of potentially available N and is commonly used to compare new N-mineralisation testing methods [15]. The objective of this study is to apply the Solvita measurement of CO_2 in soils to determine N mineralisation and compare it to the 7-day anaerobic N mineralisation method developed by Waring and Bremner [14].

2. Materials and Methods

A random sub-sampling (207) of soil samples from the contiguous United States and England, UK that were submitted to the United States Department of Agriculture-Agricultural Research Service lab in Temple, TX, from 2010 to 2014 were utilized in this study. Samples were obtained from Texas, Idaho, Maine, West Virginia, Oklahoma, Georgia, and Louisiana. Some samples obtained were not specifically identified when submitted to the laboratory. In addition, soil-testing results from 50 samples provided by an independent private soil-testing lab were utilized, bringing the total number of samples in the study to 257. The locations for the samples provided from the soil-testing lab were not provided. Methods utilized by both labs were identical.

Soil samples were obtained from the top 15 cm of the soil profile. All soil samples were dried at 50˚C for 24 h and ground to pass a 2-mm sieve. Soil organic matter content was determined using loss on ignition [16]. Ten-gram soil samples were analyzed prior to incubation and extracted with 40 ml of 2 M KCl to determine initial NH_4^+-N content. Nitrogen mineralisation was determined using the anaerobic incubation test described by Waring and Bremner [14], and equals the NH_4^+-N detected on an OI Analytical RFA after incubation minus the NH_4^+-N initially present in the soil.

The flush of CO_2-C following rewetting of dried soil was determined in 40-g subsamples of each soil in 50-ml polypropylene disposable beakers (Fisherbrand Cat. No. 01 - 291 - 10) with four to five 6.35-mm holes drilled in the bottom. A whatman GF/D 4.25-cm glass microfiber filter (Cat No. 1823 - 042) was placed in the bottom of the beaker to prevent soil loss. The beaker and Solvita Soil Respiration paddle were placed in a gas-tight

250-mL glass jar filled with 20 mL of water and sealed with a screw-top lid. The glass jar has a convex bottom to allow for drainage. Capillary action was used to rewet soil according to its water holding capability [17]. Soils were incubated at 25°C and respired CO_2-C was measured after 24 h. The quantity of CO_2-C released was determined using a digital-color reader (DCR) as described by Haney [9]. The Solvita system for estimating the flush of CO_2-C following rewetting of dried soil has been shown to be highly correlated with the commonly used titration method and CO_2-C IRGA method [9]. SigmaStat imbedded in SigmaPlot ver. 12.2 was used for linear regression analysis [18].

3. Results and Discussion

The range in soil organic matter for the soil samples was 0.8% to 7.5% (**Figure 1**). The Solvita results in mg C per kg soil are highly correlated with anaerobic 7-day N mineralisation values (**Figure 2**). These data correspond with findings by [12], [19] [20] that the flush of CO_2 following rewetting dried soil is strongly correlated with potential N mineralisation. While both the anaerobic 7-day mineralisation test and the Solvita test do not require preliminary analysis to determine the amount of water added to each soil, the 7-day N mineralisation test does require preliminary NH_4^+-N determination and Solvita does not. In addition, the Solvita test can be performed without the use of reagents to extract NH_4^+-N, laboratory steam distillation and titration (which introduces considerable human error), and can be completed in just over 24 h versus 7 days. Furthermore, Wienhold *et al.* [2] indicate that anaerobic N mineralisation underestimates field N mineralisation when compared to aerobic N mineralisation. The ability to rapidly assess potential N mineralisation with a simple test should be highly attractive to commercial soil testing labs.

One possible limitation to predicting N mineralisation from the anaerobic test or directly from Solvita is that, regardless of the respiration number, you cannot predict when a given soil is in a state of N immobilization. Only *in situ* field mineralisation tests, which are not practical at a large spatial or commercial laboratory temporal scale, would possibly be able to predict field conditions under any given time. The relationship between Solvita and mineralisation will be sufficient to predict field N mineralisation.

4. Conclusions

Based on the strong correlation between the 7-day N mineralisation test and Solvita test results, across a diverse set of soils, it would be justified to utilize Solvita to predict N mineralisation using the equation below (r^2 = 0.82):

$$\text{N mineralised}(\text{mg N per kg soil}) = 1.23\times \text{ Solvia respiration result} - 9.7$$

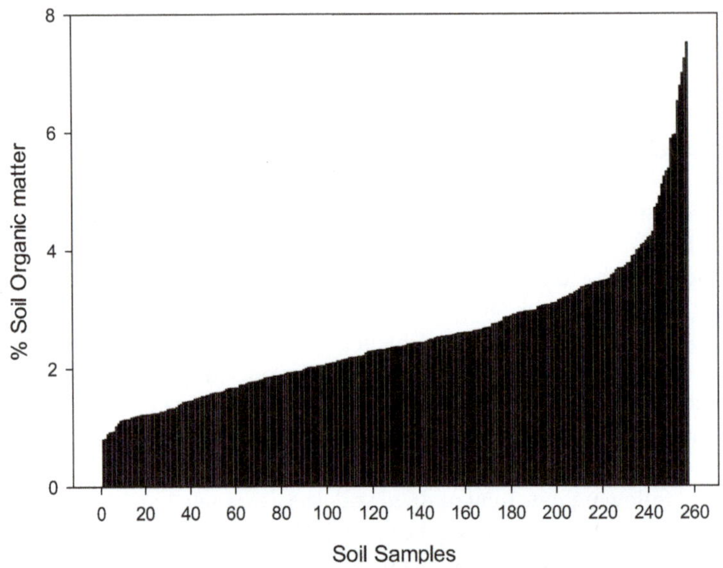

Figure 1. Range in soil organic matter (%) in 257 soil samples analyzed.

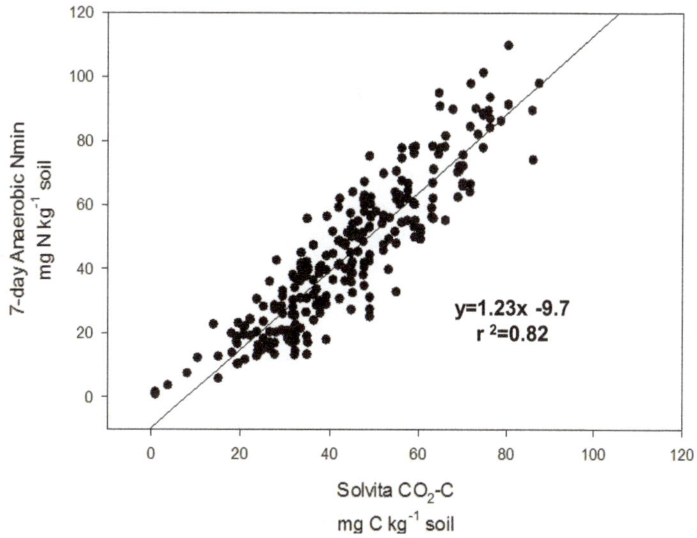

Figure 2. Solvita 1-day C vs. 7-day anaerobic N mineralisation from 257 soil samples.

Solvita can be used to rapidly assess soil respiration and relative N mineralisation potential in any given soil under laboratory conditions. The advantage to utilizing the Solvita test is that: the test only takes 1 day versus 7 days to perform and the analytical equipment is less expensive (Solvita machine versus colorimetric analysis). These results are novel in that Solvita is the first test that uses soil-microbe respiration to estimate N mineralization which is, in fact, microbially driven.

References

[1] Jha, M.K., Gassman, P.W. and Panagopoulos, Y. (2015) Regional Changes in Nitrate Loadings in the Upper Mississippi River Basin under Predicted Mid-Century Climate. *Regional Environmental Change*, **15**, 449-460. http://dx.doi.org/10.1007/s10113-013-0539-y

[2] Wienhold, B.J. (2007) Comparison of Laboratory Methods and an *in Situ* Method for Estimating Nitrogen Mineralisation in an Irrigated Silt-Loam Soil. *Communications in Soil Science and Plant Analysis*, **38**, 1721-1732. http://dx.doi.org/10.1080/00103620701435498

[3] Smith, N.R. and Humfeld, H. (1931) The Decomposition of Green Manures Grown on a Soil and Turned under Compared to the Decomposition of Green Manures Added to a Fallow Soil. *Journal of Agricultural Research*, **43**, 715-731.

[4] Gainey, P.L. (1919) Parallel Formation of Carbon Dioxide, Ammonia, and Nitrate in Soil. *Soil Science*, **7**, 293-311. http://dx.doi.org/10.1097/00010694-191904000-00007

[5] Lebedjantzev, A.N. (1924) Drying of Soil, as One of the Natural Factors in Maintaining Soil Fertility. *Soil Science*, **18**, 419-447. http://dx.doi.org/10.1097/00010694-192412000-00001

[6] Birch, H.F. (1960) Nitrification in Soils after Different Periods of Dryness. *Plant Soil*, **7**, 81-96.

[7] Deenik, J. (2006) Nitrogen Mineralization Potential in Important Agricultural Soils of Hawai'I. University of Hawaii, Honolulu (HI), 5 p. (Soil and Crop Management; SCM-15).

[8] Doran, J., Kettler T. and Tsivou M. (1997) Field and Laboratory Solvita Soil Test Evaluation. Manuscript University of Nebraska USDA-ARS, Lincoln.

[9] Haney, R., Brinton W. and Evans E. (2008) Estimating Soil Carbon, Nitrogen, and Phosphorus Mineralization from Short-Term CO_2 Respiration. *Communications in Soil Science and Plant Analysis*, **39**, 2706-2720. http://dx.doi.org/10.1080/00103620802358862

[10] Jin, V.L., Johnson M-V.V., Haney R.L. and Arnold J.G. (2011) Potential Carbon and Nitrogen Mineralization in Soils from a Perennial Forage Production System Amended with Class B Biosolids. *Agriculture, Ecosystems & Environment*, **141**, 461-465. http://dx.doi.org/10.1016/j.agee.2011.03.016

[11] Harmel, R.D. and Haney, R.L. (2013) Initial Field Evaluation of the Agro-Economic Effects of Determining Nitrogen Fertilizer Rates with a Recently-Developed Soil Test Methodology. *Open Journal of Soil Science*, **3**, 91-99. http://dx.doi.org/10.4236/ojss.2013.32010

[12] Franzluebbers, A.J., Haney, R.L., Honeycutt, C.W., Schomberg, H.H. and Hons, F.M. (2000) Flush of Carbon Dioxide Following Rewetting of Dried Soil Relates to Active Organic Pools. *Soil Science Society of America Journal*, **64**, 613-623. http://dx.doi.org/10.2136/sssaj2000.642613x

[13] Wang, W.J., Smith, C.J. and Chen, D. (2003) Towards a Standardized Procedure for Determining the Potentially Mineralizable Nitrogen of Soil. *Biology and Fertility of Soils*, **37**, 362-374.

[14] Waring, S.A. and Bremner, J.M. (1964) Ammonium Production in Soil under Water-Logged Conditions as an Index of Nitrogen Availability. *Nature*, **201**, 951-952. http://dx.doi.org/10.1038/201951a0

[15] Schomberg, H.H., Weitholter, S., Griffin, T.S., Reeves, D.W., Cabrera, M.L., Fisher, D.S., Endale, D.M., Novak, J.M., Balkcom, K.S., Raper, R.L., Kitchen, N.R., Locke, M.A., *et al.* (2009) Assessing Indices for Predicting Potential Nitrogen Mineralization in Soils under Different Management Systems. *Soil Science Society of America Journal*, **73**, 1575-1586. http://dx.doi.org/10.2136/sssaj2008.0303

[16] Schulte, E.E. and Hopkins, B.G. (1996) Estimation of Organic Matter by Weight Loss-on-Ignition. In: Magdoff, F.R., *et al.*, Eds., *Soil Organic Matter: Analysis and Interpretation*, SSSA Special Publication Number 46, SSSA, Madison, 21-31.

[17] Haney, R.L. and Haney, E.B. (2010) Simple and Rapid Laboratory Method for Rewetting Dry Soil for Incubations. *Communications in Soil and Plant Analysis*, **41**, 1493-1501. http://dx.doi.org/10.1080/00103624.2010.482171

[18] Systat Software, Inc. (2012) SigmaPlot for Windows, Version 12.5, Germany.

[19] Franzluebbers, A.J. and Haney, R.L. (2006) Flush of CO_2 as a Soil Biological Quality Indicator. *Proceedings of the 17th Conference of the International Soil Tillage Research Organization* (*CD-ROM*), Kiel, 3 August-26 September 2006, 736-740.

[20] Haney, R.L., Franzluebbers, A.J., Jin, V.L., Johnson, M.-V., Haney, E.B., White, M.J. and Harmel, R.D. (2012) Soil Organic C:N vs. Water-Extractable Organic C:N. *Open Journal of Soil Science*, **2**, 269-274. http://dx.doi.org/10.4236/ojss.2012.23032

23

In-Season Side-Dressing of Urea and Ammonium Nitrate to Cotton on No-Till Soils with High Residual Nitrogen and Pre-Plant Nitrogen Application

Xinhua Yin

Department of Plant Sciences, The University of Tennessee, Jackson, TN, USA
Email: xyin2@utk.edu

Abstract

It is essential to develop innovative approaches that can apply N more efficiently. The objective of this study was to examine in-season side-dress urea and ammonium nitrate (UAN) applications to cotton on no-till soils with high residual N fertility. A field trial was conducted near Milan, TN in 2011 and 2012 with strip plots in a RCB design with three replicates. The following six in-season side-dress fluid UAN treatments were compared: 1) zero N; 2) low uniform-rate N application of 56 kg·N·ha^{-1}; 3) high uniform-rate N application of 78.4 kg·N·ha^{-1}; 4) ordinary variable-rate N application algorithm for each sub plot based on the average Normalized Difference Vegetation Index (NDVI) value in that sub plot; 5) reversed variable-rate N application algorithm for each sub plot based on average NDVI of that sub plot; and 6) N application rate based on the average NDVI value in each strip plot. All plots received 26 kg·N·ha^{-1} as diammonium phosphate before cotton planting each year. Leaf N concentrations were mostly enhanced with all side-dress N applications ranging from 56 to 78 kg·N·ha^{-1} relative to zero N during early to late bloom although this upland field had high initial soil N fertility and received pre-plant application of 26 kg·N·ha^{-1} across the treatments each year. However, NDVI, plant height, and lint yield were rarely improved with side-dress N application. The three variable-rate N application algorithms consumed 7.8 to 12.3 kg·ha^{-1} more N than the low uniform-rate application of 56 kg·N·ha^{-1}, but 10.1 to 14.6 kg·ha^{-1} less N than the high uniform rate of 78.4 kg·N·ha^{-1}. Our results indicate that the current N recommendations for cotton in Tennessee may be too high on upland soils with high initial N fertility.

Keywords

UAN, Side-Dress, Cotton, Yield, N Consumption

1. Introduction

During the past several decades, the largest increase in the use of agricultural inputs has been fertilizer nitrogen (N) [1]. Nitrogen fertilization is a key production practice in most crops including cotton. Fertilizer N is one of the greatest cost inputs in cotton production. It is also the most difficult nutrient to manage, and has large potential adverse impacts on the environment. Under-application of N on cotton results in poor vegetative and reproductive growth, premature senescence, and reduced yields; while over-application inhibits boll formation and retention, poses serious threats to the environment [2], and reduces grower profitability. Due to substantially increased environmental concern and rising N price during the last decade, there is an urgent need to develop innovative systems and technologies that can apply N more efficiently so as to reduce N losses, increase crop yields and profitability, and improve environmental quality.

Nitrogen fertilizers are recommended to be applied at 67 - 90 kg·N·ha^{-1} (60 - 80 lb N acre^{-1}) on upland soils and 34 - 67 kg·N·ha^{-1} (30 - 60 lb N acre^{-1}) on bottom soils in Tennessee [3]. These recommendations have been used for decades without modifications although cotton production systems have been changed dramatically. For example, new cotton cultivars with high yield potentials have been continuously introduced to the market over the past decades. Due to security concern, urea and ammonium nitrate (UAN) and urea have been increasingly used to replace ammonium nitrate as the N source for cotton production in Tennessee. Presently, UAN is usually injected into the soil, while ammonium nitrate is broadcast on the soil surface. Traditionally, N fertilizer is applied around cotton planting, but split applications (a small portion of N as pre-plant plus a large portion of N as side-dress) have recently gained grower's attention. No-tillage systems are being widely used by producers in cotton production in Tennessee and throughout the southeastern USA. Under no-tillage, N fertilizer is often broadcast on soil surface without incorporation into the soil, while it is generally incorporated after surface application under conventional tillage systems [4]. Overall, information on cotton responses to the application rate, timing, and method of alternate N sources such as UAN and urea is limited under no-till production systems.

Furthermore, current fertilizer N recommendations are developed by individual states on small fields where spatial variances are minimal. Although recommendations may account for soil variations, growers do not have the technology to manage spatial variability within individual fields. Under the current systems, producers use a uniform N fertilizer rate across entire fields, which often results in over- and under-applications of N because crop responses to N fertilization are often variable within individual fields [5], and on some parts of the field more N should be applied or much less to none on other parts of the field [6]. Therefore, the presence of spatial variability within individual fields is a critical issue demanding careful management. In order to solve this problem, it is essential to develop and adopt innovative technologies and management systems that can generate and implement variable-rate fertilizer N recommendations based on spatial variances within individual fields.

Measuring crop N nutrition status during the growing season by optically sensing crop canopy has been developed into a viable precision N management tool during the last several years [7] [8]. Researchers utilize on-vehicle, real-time optical sensing of crop canopy reflectance which can be used to develop indices such as the Normalized Difference Vegetation Index (NDVI) to assess crop growth, health, or N nutritional status and use these indices to adjust the fertilizer N rate accordingly. Integration of optical sensing and variable-rate application technologies enables on-the-go non-destructive diagnoses of crop N deficiency, sufficiency, or excesses without soil or plant tissue diagnostic testing, real-time application of N fertilizer at variable rates to correct those deficiencies or excesses, and precisely treating each part of the field sensed. Since this is real-time sensing, it is done without pre-processing of data or determining location within field beforehand. This innovative precision technology allows variable application of N fertilizers at very high resolutions. So far this system is the most efficient technology for precision N management as it can change uniform-rate N application to variable-rate N applications, minimizing time, labor, and cost of implementing variable-rate N applications [8] [9], which is vastly different from the traditional N management systems of uniform rate N applications across entire fields.

Investigations have shown significant economic and environmental benefits with optical sensor-based variable-rate precision N management technologies and systems. Research on corn and winter wheat have shown 10% - 15% increases in N use efficiency and some significant yield increases with these precision N management systems relative to the traditional uniform-rate N management systems [6] [8]. Biermacher *et al.* (2006) [10] in Oklahoma even reported as high as 59% - 82% reductions in N fertilizer use by winter wheat under the optical sensor-based variable-rate N management systems. Ortiz-Monasterio and Raun (2007) [11] showed that optical sensor-based precision N management systems can save 71 kg·N·ha^{-1} on spring wheat and increase grower profitability by $148 ha^{-1}. However, investigations on cotton responses to variable-rate N applications based on canopy NDVI were limited.

The objective of this study was to examine in-season side-dress UAN application systems (both variable-rate and uniform rate) for cotton on no-till soils with high residual N fertility and pre-plant N application in terms of N fertilizer consumption, lint yield and quality, and residual N in the soil profile.

2. Materials and Methods

A field trial was conducted on the University of Tennessee Research and Education Center at Milan in 2011 and 2012 with the following six in-season side-dress UAN treatments evaluated in strip plots. Each strip plot was divided into 8 sub plots (7.5 m wide × 15 m long).

1) Zero N;
2) Low uniform-rate N application of 56 kg·N·ha^{-1} (50 lb N a^{-1}) over each strip plot;
3) High uniform-rate N application rate of 78.4 kg·N·ha^{-1} (70 lb N a^{-1}) over each strip plot;
4) Ordinary variable-rate N application algorithm in the range of 33.6 to 100.8 kg·ha^{-1} (30 to 90 lb a^{-1}) of N for each sub plot based on the average NDVI value in that sub plot;
5) Reversed variable-rate N application algorithm in the range of 33.6 to 100.8 kg·ha^{-1} (30 to 90 lb a^{-1}) of N for each sub plot based on the average NDVI value in that sub plot;
6) Nitrogen application rate based on the average NDVI value in each strip plot.

No side-dress UAN fertilizer was applied in any strip or sub plot in Treatment 1. Nitrogen fertilizer of 56 kg·N·ha^{-1} was applied to all sub plots in each strip plot in Treatment 2. Nitrogen fertilizer at 78.4 kg·N·ha^{-1} was applied to all sub plots in each strip plot in Treatment 3. The N application rates in Treatments 2 & 3 roughly represented the lower and higher limits of side-dress N applications used by cotton producers after approximately 22 to 34 kg·N·ha^{-1} (20 to 30 lb N a^{-1}) was pre-plant applied as monoammonium phosphate or diammonium phosphate. Nitrogen applications in Treatments 4 to 6 were based on NDVI readings. In Treatment 4, when the NDVI reading was lower, more N fertilizer was applied. However, in Treatment 5, when the NDVI reading was lower, less N fertilizer was applied. In-season fluid N application rate ranged from 33.6 to 100.8 kg·ha^{-1} for both Treatments 4 and 5. In Treatment 6, the N application rate was identical for all sub plots within a strip plot, but it might be different among the three strip plots of the three replicates since it was based on the average NDVI value in each strip plot. Fluid N fertilizer UAN (32-0-0) was injected 3.8 cm deep into the soil and 25.4 cm to one side of the row with a KBH 8-row pull-type coulter injector for all the in-season applied N treatments (Treatments 2 to 6). In addition to the above in-season N treatments, all the plots received 26 kg·N·ha^{-1} and 67 kg P_2O_5 ha^{-1} as diammonium phosphate and 67 kg K_2O ha^{-1} as muriate of potash regardless of treatment in fall 2010. In March 2012, 26 kg·N·ha^{-1}, 67 kg P_2O_5 ha^{-1}, and 101 kg K_2O ha^{-1} were applied to all the plots irrespective of the treatment. The P and K fertilizers were broadcast applied based on the soil testing results each year.

The test field had four different soil types: Calloway, Falaya, Grenada, and Lexington. The initial soil nitrate- and ammonium-N content within the top 60 cm of soil varied substantially ranging from 3 to 57 mg·kg^{-1}. All these suggest that the test field is spatially variable.

The dates of cotton planting, N treatment implementation, and other major field operations for this test are presented in **Table 1**. A composite soil sample was collected at a depth of 60-cm for nitrate and ammonium in the soil profile on a sub plot basis prior to treatment initiation. The following sampling and measurements were taken from each sub plot each year: Canopy NDVI data were recorded at the early square and early, mid, and late bloom growth stages using the GreenSeeker® RT 200 Data Collection and Mapping System (NTech Industries, Inc., CA). A composite leaf sample (10 blades + 10 petioles) was collected four times at about the same dates when NDVI data were taken. All leaf samples were analyzed for N concentrations using a LECO Tru-Spec Analyzer. Cotton harvest was completed timely for lint yield and gin turnout by harvesting the center 4 rows. In addition, fiber quality attributes were determined on a strip plot basis each year. A post-harvest soil sample was taken for soil nitrate and ammonium at a 60-cm depth.

Analysis of variance was conducted for each measurement with a randomized complete block model using SAS statistical software (SAS Institute, Cary, NC). Treatment means were separated with the Fisher's protected

Table 1. Major field operations performed on this trial in 2011 and 2012.

Year	List of Operations Performed	Date Performed
2011	Planted DPL 0912 with a 8-row planter	05/11/11
	Collected pre-treatment soil samples to 60 cm deep (2 cores plot^{-1})	05/26/11
	Recorded canopy NDVI prior to N treatment	06/24/11
	Side-dressed fluid N treatments by plot	07/12/11
	Collected early square leaf samples (10 leaves plot^{-1})	07/12/11
	Recorded early square plant height (10 plants plot^{-1})	07/12/11
	Recorded canopy NDVI at early bloom	07/21/11
	Collected early bloom leaf samples (10 leaves plot^{-1})	07/22/11
	Recorded early bloom plant height (10 plants plot^{-1})	07/22/11
	Recorded canopy NDVI at mid bloom	08/01/11
	Collected mid bloom leaf samples (10 leaves plot^{-1})	08/02/11
	Recorded mid bloom plant height (10 plants plot^{-1})	08/02/11
	Recorded canopy NDVI at late bloom	08/15/11
	Collected late bloom leaf samples (10 leaves plot^{-1})	08/12/11
	Recorded late bloom plant height (10 plants plot^{-1})	08/12/11
	Harvested center 4 rows of each 8 row plot & collected seed cotton samples by strip plot	10/11/11
	Collected post-harvest soil samples to 60 cm deep (2 cores plot^{-1})	11/07/11
2012	Planted DPL 0912 with a 8-row planter	05/10/12
	Recorded canopy NDVI prior to N treatment	07/11/12
	Collected early square leaf samples (10 leaves plot^{-1})	07/11/12
	Recorded early square plant height (10 plants plot^{-1})	07/11/12
	Side-dressed fluid N treatments by plot	07/18/12
	Recorded canopy NDVI at early bloom	07/19/12
	Collected early bloom leaf samples (10 leaves plot^{-1})	07/19/12
	Recorded early bloom plant height (10 plants plot^{-1})	07/19/12
	Recorded canopy NDVI at mid bloom	07/26/12
	Collected mid bloom leaf samples (10 leaves plot^{-1})	07/26/12
	Recorded mid bloom plant height (10 plants plot^{-1})	07/26/12
	Recorded canopy NDVI at late bloom	08/08/12
	Collected late bloom leaf samples (10 leaves plot^{-1})	08/08/12
	Recorded late bloom plant height (10 plants plot^{-1})	08/08/12
	Harvested center 4 rows of each 8 row plot & collected seed cotton samples by strip plot	10/25/12
	Collected post-harvest soil samples to 60 cm deep (2 cores plot^{-1})	11/09/12

LSD method. In addition, nonorthogonal contrasts were conducted in order to compare means of treatment combinations. Probability levels lower than 0.05 were designated as significant.

3. Results and Discussion

3.1. Initial Soil N Fertility

Initial inorganic N (NO_3^--N + NH_4^+-N) content in the top 60 cm of soil varied substantially among the 144 sub plots with a range of the minimum 3 mg·kg^{-1} to the maximum 57 mg·kg^{-1} after this field had received 26 kg·N ha^{-1} as diammonium phosphate in fall 2010 but before the in-season side-dress N treatments were applied in spring 2011 in this study (data not presented). These variations might relate to the soil types in this field. The average initial inorganic N level in the top 60 cm of soil was 6.8, 7.4, 9.8, 8.0, 10.4, and 8.8 mg·kg^{-1}, equivalent to 60.8, 66.4, 88.0, 71.6, 93.2 and 78.8 kg·ha^{-1}, for the plots allocated to Treatments 1, 2, 3, 4, 5, and 6, respectively, before treatment application in spring 2011. Obviously this field had pretty high residual inorganic N level in the soil profile from the previous crop production. High residual inorganic N level in soil profile has been increasingly observed in cotton and corn production in Tennessee and other states.

3.2. Effects of In-Season Side-Dress N Applications on Leaf N Concentrations

Leaf N concentrations were similar among the plots allocated to Treatments 1, 2, 3, 4, 5, and 6, respectively at early square before the in-season side-dress N treatments were applied when the results of 2011 and 2012 were combined (**Table 2**). However, significant differences in leaf N were observed among the six treatments at the early, mid, and late bloom stages (**Table 2**). Leaf N concentrations were mostly significantly higher under the five N applied treatments than those with the zero N treatment (Treatment 1) during early bloom to late bloom. Treatments 3, 4, and 5 had significantly higher leaf N concentrations than Treatment 1, and significantly or numerically higher leaf N levels than Treatments 2 and 6 at the early, mid, and late bloom stages.

Under nonorthogonal contrasts, Applied N, Uniform N, and Variable N all resulted in significantly higher leaf N concentrations than Zero N at the early, mid, and late bloom stages (**Table 2**). However, Variable N had similar leaf N level as Uniform N regardless of growth stage.

Both ANOVA and nonorthogonal contrast analyses of this study indicate that in-season side-dress N application improves cotton N nutrition irrespective of the application method (uniform-rate or variable-rate); but variable-rate N applications do not improve cotton N nutrition compared with uniform-rate N applications on this upland field with high initial soil N fertility.

It was expected that leaf N concentration was pretty high under the zero N treatment regardless of growth stage, which might relate to the fact that the initial inorganic N level was high in the top 60 cm of soil profile

Table 2. Effects of in-season side-dress N applications on leaf N concentrations at different growth stages averaged over 2011 and 2012.

Treatment	Early Square	Early Bloom	Mid Bloom	Late Bloom
	g·kg^{-1}	g·kg^{-1}	g·kg^{-1}	g·kg^{-1}
1	40.5a†	38.3c	35.8c	34.4b
2	40.4a	40.8ab	36.8bc	36.9ab
3	40.7a	41.7ab	38.4a	38.8a
4	41.3a	41.4ab	37.6ab	37.3a
5	41.8a	42.3a	37.7ab	38.4a
6	39.6a	40.2bc	37.0bc	36.5ab
Significance	ns‡	**	**	*
Contrast				
Applied N vs. Zero N§	ns	***	**	**
Uniform N vs. Zero N	ns	**	**	**
Variable N vs. Zero N	ns	**	**	**
Variable N vs. Uniform N	ns	ns	ns	ns

†Values within a column followed by the same letter are not significantly different with Fisher's protected LSD at the 0.05 probability level. ‡*significant at $P < 0.05$; **significant at $P < 0.01$; ***significant at $P < 0.001$; ns, not significant at $P < 0.05$. §Applied N vs. Zero N refers to the comparison of average over the five N applied treatments (Treatments 2, 3, 4, 5, and 6) with the Zero N treatment (Treatment 1); Uniform N vs. Zero N refers to the comparison of average over the two uniform-rate N treatments (Treatments 2 and 3) with the Zero N treatment; Variable N vs. Zero N refers to the comparison of average over the three variable-rate N treatments (Treatments 4, 5, and 6) with the zero N treatment; Variable N vs. Uniform N refers to the comparison of average over the three variable-rate N treatments with the two uniform-rate N treatments.

and 26 kg·N·ha^{-1} was applied across the treatments as diammonium phosphate before cotton planting in both years.

The sensitivity of cotton leaf N to N application is greater at early bloom than at any other growth and developmental stages [12] [13]. This trend indicates monitoring of leaf N nutrition status at early bloom is especially helpful in deciding whether supplemental N application is required. Campbell and Plank (2011) [14] recommended that the range of adequate leaf N concentration be 30 to 45 g·kg^{-1} at early bloom for cotton grown in the southern United States. According to this criterion, leaf N concentration at early bloom was greater than the lower limit of the sufficiency range in the zero N treatment of this study. Similarly, the range of adequate leaf N concentration is recommended to be 30 to 45 g·kg^{-1} at late bloom for cotton [14]. According to this standard, leaf N at late bloom was also above the lower limit of the sufficiency range in zero N. Therefore, cotton yield responses to N applications might not be expected, based on the adequate leaf N nutrition under the zero N treatment, if the recommended sufficiency leaf N ranges during early to late bloom are indicative of final cotton yield, although the in-season N applied treatments mostly resulted in higher N concentrations than the zero N treatment in this study.

3.3. Effects of In-Season Side-Dress N Applications on Canopy NDVI Readings

Canopy NDVI is a good vegetation index that can be used to estimate plant N nutrition status and plant biomass. Canopy NDVI readings were statistically similar among the six treatments at early square before the N treatments were implemented averaged over 2011 and 2012 (**Table 3**). Unlike leaf N concentrations, no significant difference in NDVI was observed among the six treatments at early, mid, or late bloom stage (**Table 3**). Similarly, nonorthogonal contrasts did not show any beneficial effect of Applied N, Uniform N, or Variable N on NDVI over Zero N at any growth stage (**Table 3**).

3.4. Effects of In-Season Side-Dress N Applications on Plant Height

Similar to leaf N and canopy NDVI, plant height did not differ significantly among the six treatments at early square prior to N treatment implementation averaged 2011 and 2012 (**Table 4**). However, plant height was significantly different among the six treatments at the early bloom stage (**Table 4**). Plants at early bloom were

Table 3. Effects of in-season side-dress N applications on canopy NDVI at different growth stages averaged over 2011 and 2012.

Treatment	Early Square	Early Bloom	Mid Bloom	Late Bloom
1	0.55a†	0.69a	0.70a	0.73a
2	0.57a	0.70a	0.71a	0.73a
3	0.58a	0.70a	0.71a	0.73a
4	0.56a	0.69a	0.70a	0.73a
5	0.55a	0.70a	0.72a	0.75a
6	0.58a	0.68a	0.70a	0.72a
Significance	ns‡	ns	ns	ns
Contrast				
Applied N vs. Zero N§	ns	ns	ns	ns
Uniform N vs. Zero N	ns	ns	ns	ns
Variable N vs. Zero N	ns	ns	ns	ns
Variable N vs. Uniform N	ns	ns	ns	ns

†*Values within a column followed by the same letter are not significantly different with Fisher's protected LSD at the 0.05 probability level. ‡ns, not significant at $P < 0.05$. §Applied N vs. Zero N refers to the comparison of average over the five N applied treatments (Treatments 2, 3, 4, 5, and 6) with the zero N treatment (Treatment 1); Uniform N vs. Zero N refers to the comparison of average over the two uniform-rate N treatments (Treatments 2 and 3) with the zero N treatment; Variable N vs. Zero N refers to the comparison of average over the three variable-rate N treatments (Treatments 4, 5, and 6) with the zero N treatment; Variable N vs. Uniform N refers to the comparison of average over the three variable-rate N treatments with the two uniform-rate N treatments.

Table 4. Effects of in-season side-dress N applications on plant height at different growth stages averaged over 2011 and 2012.

Treatment	Early Square	Early Bloom	Mid Bloom	Late Bloom
	cm	cm	cm	cm
1	64.3a†	81.8b	93.0a	98.6a
2	66.5a	81.0b	93.5a	100.6a
3	67.3a	85.9a	97.3a	101.9a
4	66.0a	82.3b	93.7a	99.6a
5	65.5a	81.3b	93.7a	100.6a
6	66.5a	80.5b	92.7a	99.8a
Significance	ns‡	*	ns	ns
Contrast				
Applied N vs. Zero N§	ns	ns	ns	ns
Uniform N vs. Zero N	ns	ns	ns	ns
Variable N vs. Zero N	ns	ns	ns	ns
Variable N vs. Uniform N	ns	*	ns	ns

†*Values within a column followed by the same letter are not significantly different with Fisher's protected LSD at the 0.05 probability level. ‡*significant at $P < 0.05$; ns, not significant at $P < 0.05$. §Applied N vs. Zero N refers to the comparison of average over the five N applied treatments (Treatments 2, 3, 4, 5, and 6) with the zero N treatment (Treatment 1); Uniform N vs. Zero N refers to the comparison of average over the two uniform-rate N treatments (Treatments 2 and 3) with the zero N treatment; Variable N vs. Zero N refers to the comparison of average over the three variable-rate N treatments (Treatments 4, 5, and 6) with the zero N treatment; Variable N vs. Uniform N refers to the comparison of average over the three variable-rate N treatments with the two uniform-rate N treatments.

significantly taller under Treatment 3 which received the highest in-season side-dress N application rate than those with the other treatments.

3.5. Effects of In-Season Side-Dress N Applications on Lint Yield, Gin Turnout, and Fiber Quality

Lint yield responses to in-season side-dress N treatments were not significant on the averages of 2011 and 2012 (**Table 5**). Numerically, lint yields were higher under Treatments 2 and 5 than those with the other treatments. Gin turnout was not significantly affected by the N treatments (data not presented). None of the fiber quality attributes was influenced by the N treatments (**Table 5**).

Similarly, Main *et al.* (2013) [15] reported only 11 of 20 site-years across the Cotton Belt region in which there was a cotton lint yield response to applied N. They observed when 45 kg·N·ha^{-1} was applied, yields were greater than when no N was applied, but were less than yields where 90 to 134 kg·N·ha^{-1} was applied on the N responsive site-years; however, when all the site-years, both N responsive and non-responsive, were considered, 45 kg·N·ha^{-1} increased yields above no applied N, but additional N above 45 kg·N·ha^{-1} did not improve lint yield.

3.6. Effects of In-Season Side-Dress N Applications on In-Season N Fertilizer Consumption

The in-season side-dress N fertilizer consumption was 0, 56.0, 78.4, 68.7, 63.5, and 67.2 kg·N·ha^{-1} for Treatments 1 to 6, respectively, averaged over 2011 and 2012 in this study (**Table 6**). The three variable-rate N application algorithms (Treatments 4, 5, and 6) consumed 7.8 to 12.3 kg·ha^{-1} more N than the low uniform-rate application of 56.0 kg·N·ha^{-1}, but used 10.1 to 14.6 kg·ha^{-1} less of N than the high uniform rate application of 78.4 kg·N·ha^{-1}. Under nonorthogonal contrasts, Applied N, Uniform N, and Variable N all consumed more in-season side-dress N fertilizer than the Zero N treatment, but Variable N used similar in-season side-dress N fertilizer as Uniform N (**Table 6**).

The total annual N application rate (pre-plant N + in-season side-dress N) was 26, 82, 104.4, 94.7, 89.5, and

Table 5. Effects of in-season side-dress N applications on lint yield, gin turnout, and fiber quality averaged over 2011 and 2012.

Treatment	Yield	Micronaire	Strength	Length	Uniformity
	$kg \cdot ha^{-1}$		$g\ tex^{-1}$	cm	%
1	1037.0a†	4.95a	32.9a	2.84a	83.2a
2	1127.7a	4.90a	33.3a	2.82a	83.3a
3	1076.4a	4.95a	32.2a	2.77a	82.3a
4	1089.2a	4.95a	33.0a	2.82a	82.9a
5	1155.5a	4.85a	33.1a	2.82a	83.7a
6	1075.5a	5.00a	32.2a	2.77a	82.7a
Significance	ns‡	ns	ns	ns	ns
Contrast					
Applied N vs. Zero N§	ns	ns	ns	ns	ns
Uniform N vs. Zero N	ns	ns	ns	ns	ns
Variable N vs. Zero N	ns	ns	ns	ns	ns
Variable N vs. Uniform N	ns	ns	ns	ns	ns

†*Values within a column followed by the same letter are not significantly different with Fisher's protected LSD at the 0.05 probability level. ‡ns, not significant at $P < 0.05$. §Applied N vs. Zero N refers to the comparison of average over the five N applied treatments (Treatments 2, 3, 4, 5, and 6) with the zero N treatment (Treatment 1); Uniform N vs. Zero N refers to the comparison of average over the two uniform-rate N treatments (Treatments 2 and 3) with the zero N treatment; Variable N vs. Zero N refers to the comparison of average over the three variable-rate N treatments (Treatments 4, 5, and 6) with the zero N treatment; Variable N vs. Uniform N refers to the comparison of average over the three variable-rate N treatments with the two uniform-rate N treatments.

Table 6. Effects of in-season side-dress N applications on in-season side-dress N consumption and post-harvest soil residual N (0 - 60 cm) averaged over 2011 and 2012.

Treatment	Side-Dress N Consumption	Soil Residual N
	$kg \cdot ha^{-1}$	$mg \cdot kg^{-1}$
1	0c†	7.65a
2	56.0b	7.28a
3	78.4a	8.58a
4	68.7ab	8.95a
5	63.5ab	9.36a
6	67.2ab	7.78a
Significance	***‡	ns
Contrast		
Applied N vs. Zero N§	***	ns
Uniform N vs. Zero N	***	ns
Variable N vs. Zero N	***	ns
Variable N vs. Uniform N	ns	ns

†*Values within a column followed by the same letter are not significantly different with Fisher's protected LSD at the 0.05 probability level. ‡***significant at $P < 0.001$; ns, not significant at $P < 0.05$. §Applied N vs. Zero N refers to the comparison of average over the five N applied treatments (Treatments 2, 3, 4, 5, and 6) with the zero N treatment (Treatment 1); Uniform N vs. Zero N refers to the comparison of average over the two uniform-rate N treatments (Treatments 2 and 3) with the zero N treatment; Variable N vs. Zero N refers to the comparison of average over the three variable-rate N treatments (Treatments 4, 5, and 6) with the Zero N treatment; Variable N vs. Uniform N refers to the comparison of average over the three variable-rate N treatments with the two uniform-rate N treatments.

93.2 kg·N·ha^{-1} for Treatments 1, 2, 3, 4, 5, and 6, respectively, averaged over 2011 and 2012. Since this study was conducted on an upland soil, the results of this study indicate that the current N fertilizer recommendations (67 to 90 kg·N·ha^{-1} for upland soils, and 34 to 67 kg·N·ha^{-1} for bottom soils) made by University of Tennessee [3] may be too high for cotton production on those upland fields with high residual soil inorganic N fertility in Tennessee. Since high residual inorganic N in soil profile has been increasingly observed on corn and cotton fields, it may be beneficial to take the results of pre-plant soil inorganic N test into consideration when the N fertilizer recommendations are made.

3.7. Effects of In-Season Side-Dress N Applications on Post-Harvest Soil Residual N Contents

An important aspect of N management strategies should be the reduction of soil residual N. Reducing N application rates for cotton, when appropriate, will reduce post-harvest residual N levels in the soil profile, and thus decrease the loss of N from soil to ground and surface waters.

No significant differences were observed in post-harvest soil residual N (NO_3^--N + NH_4^+-N) level in the top 60 cm soil profile among the six treatments on the averages of 2011 and 2012 (**Table 6**). Numerically, soil residual N level was higher under Treatments 3, 4, and 5 than those with the other treatments after cotton harvest. Overall, our results suggest that the application of in-season side-dress N ranging from 0 to 78.4 kg·N·ha^{-1} do not significantly affect the post-harvest residual N levels in the soil profile, which may relate to the possibility that applied N was either taken up by cotton plants or lost out of the soil via nitrate leaching to deeper layers or ammonium volatilization to the atmosphere during the growing season.

Boquet and Breitenbeck (2000) [16] reported that N fertilization on cotton for optimal yield was exceptionally efficient and probably did not result in N losses to nearby surface and ground waters. In contrast, soil samples collected after 3 years of continuous N fertilization resulted in a significant increase in NO_3^--N concentration at the 15 - 45 cm depth only in the plots treated with the highest N rate of 224 kg·N·ha^{-1} annually [17]. Results of a 5-yr study showed excessive N buildup in many California cotton soils due to the high N application rates for previous cotton or other rotational crops [18]. Bronson *et al.* (2001) [19] found that residual soil NO_3^--N accumulated to high levels in the low irrigation, high N fertilizer treatments, but remained stable in the high irrigation, low N treatments.

4. Conclusions

Leaf N concentrations were generally improved by in-season side-dress N applications of 56 to 78 kg·N·ha^{-1} compared with those under the zero side-dress N control during early to late bloom even though this upland field held high initial soil inorganic N fertility and 26 kg·N·ha^{-1} was applied across the treatments before cotton planting each year. However, rare significant differences in cotton growth (canopy NDVI and plant height) were observed because of in-season N applications during early or late bloom stages. Lint yield was not significantly improved by either variable-rate or uniform-rate N applications via side-dress during the growing season. The three variable-rate N application algorithms consumed 7.8 to 12.3 kg·ha^{-1} more N than the low uniform-rate application of 56 kg·N·ha^{-1}, but used 10.1 to 14.6 kg·ha^{-1} less of N than the high uniform-rate application of 78.4 kg·N·ha^{-1}. No significant positive effect of in-season side-dress N applications was observed on post-harvest soil residual N levels. Our results suggested that it might be difficult to obtain cotton yield responses to in-season side-dress N applications or compare in-season variable-rate N applications with the traditional uniform-rate N applications in terms of cotton yields on those upland fields that held high initial soil N fertility and had received pre-plant N application.

Since this study was conducted on an upland soil, the results of this study indicated that the current N fertilizer recommendations of 67 to 90 kg·N·ha^{-1} for cotton on upland soils made by University of Tennessee [3] might be too high for cotton production on those upland fields with high initial soil inorganic N fertility in Tennessee. Since high residual inorganic N in soil profile had been increasingly observed on corn and cotton fields, it might be beneficial to take the results of pre-plant soil inorganic N test into consideration when the N fertilizer recommendations were made.

Acknowledgements

This project was supported in part by the Fluid Fertilizer Foundation. Technical assistance was provided by Robert Sharp, Jimmy McClure, Jason Williams, Will Goforth, and others.

References

[1] Johnston, A.E. (2000) Efficient Use of Nutrients in Agricultural Production Systems. *Communications in Soil Science and Plant Analysis*, **31**, 1599-1620. http://dx.doi.org/10.1080/00103620009370527

[2] Prasad, R. and Power, J.F. (1995) Nitrification Inhibitor for Agriculture, Health and Environment. *Advances in Agronomy*, **54**, 233-281. http://dx.doi.org/10.1016/S0065-2113(08)60901-3

[3] Savoy, H.J. and Joines, D. (2009) Lime and Fertilizer Recommendations for the Various Crops of Tennessee. Chapter II. Agronomic Crops. http://soilplantandpest.utk.edu/pdffiles/soiltestandfertrecom/chap2-agronomic_mar2009.pdf

[4] Howard, D.D., Gwathmey, C.O., Essington, M.E., Roberts, R.K. and Mullen, M.D. (2001) Nitrogen Fertilization of No-Till Cotton on Loess-Derived Soils. *Agronomy Journal*, **93**, 157-163. http://dx.doi.org/10.2134/agronj2001.931157x

[5] Vetch, J.A., Malzer, G.L., Robert, P.C. and Huggins, D.R. (1995) Nitrogen Specific Management by Soil Condition: Managing Fertilizer Nitrogen in Corn. In: Robert, P.C., *et al.*, Eds., *Site Specific Management for Agricultural Systems*, ASA, CSSA, and SSSA, Madison, 465-473.

[6] Raun, W.R. and Johnson, G.V. (1999) Improving Nitrogen Use Efficiency for Cereal Production. *Agronomy Journal*, **91**, 357-363. http://dx.doi.org/10.2134/agronj1999.00021962009100030001x

[7] Raun, W.R., Johnson, G.V., Stone, M.L., Solie, J.B., Lukina, E.V. and Thomason, W.E. (2001) In-Season Prediction of Potential Grain Yield in Winter Wheat Using Canopy Reflectance. *Agronomy Journal*, **93**, 131-178. http://dx.doi.org/10.2134/agronj2001.931131x

[8] Raun, W.R., Solie, J.B., Johnson, G.V., Stone, M.L., Mullen, R.W., Freeman, K.W., Thomason, W.E. and Lukina, E.V. (2002) Improving Nitrogen Use Efficiency in Cereal Grain Production with Optical Sensing and Variable Rate Application. *Agronomy Journal*, **94**, 815-820. http://dx.doi.org/10.2134/agronj2002.8150

[9] Tubaña, B.S., Arnall, D.B., Walsh, O., Chung, B., Solie, J.B., Girma, K. and Raun, W.R. (2008) Adjusting Midseason Nitrogen Rate Using a Sensor-Based Optimization Algorithm to Increase Use Efficiency in Corn *(Zea mays* L.). *Journal of Plant Nutrition*, **31**, 1975-1998.

[10] Biermacher, J.T., Epplin, F.M., Brorsen, B.W., Solie, J.B. and Raun, W.R. (2006) Maximum Benefit of a Precise Nitrogen Application System for Wheat. *Precision Agriculture*, **7**, 1-12. http://dx.doi.org/10.1007/s11119-006-9017-6

[11] Ortiz-Monasterio, J.I. and Raun, W.R. (2007) Reduced Nitrogen and Improved Farm Income for Irrigated Spring Wheat in the Yaqui Valley, Mexico, Using Sensor Based Nitrogen Management. *The Journal of Agricultural Science*, **145**, 1-8.

[12] Bell, P.F., Boquet, D.J., Millhollon, E., Moore, S., Ebelhar, W., Mitchell, C.C., Varco, J., Funderburg, E.R., Kennedy, C., Breitenbeck, G.A., Craig, C., Holman, M., Baker, W. and McConnell, J.S. (2003) Relationships between Leaf-Blade Nitrogen and Relative Seedcotton Yields. *Crop Science*, **43**, 1367-1374. http://dx.doi.org/10.2135/cropsci2003.1367

[13] Fritschi, F.B., Roberts, B.A., Travis, R.L., Rains, D.W. and Hutmacher, R.B. (2004) Seasonal Nitrogen Concentration, Uptake, and Partitioning Pattern of Irrigated Acala and Pima Cotton as Influenced by Nitrogen Fertility Level. *Crop Science*, **44**, 516-527. http://dx.doi.org/10.2135/cropsci2004.5160

[14] Campbell, C.R. and Plank, C.O. (2011) Corn. In: Campbell, C.R., Ed., *Reference Sufficiency Ranges for Plant Analysis in the Southern Region of the United States*, Southern Cooperative Series Bulletin 394, Southern Association of Agricultural Experiment Station, Raleigh, 15-18.

[15] Main, C.L., Barber, L.T., Boman, R.K., Chapman, K., Dodds, D.M., Duncan, S., Edmisten, K.L., Horn, P., Jones, M.A., Morgan, G.D., Norton, E.R., Osborne, S., Whitaker, J.R., Nichols, R.L. and Bronson, K.F. (2013) Effects of Nitrogen and Planting Seed Size on Cotton Growth, Development, and Yield. *Agronomy Journal*, **105**, 1853-1859. http://dx.doi.org/10.2134/agronj2013.0154

[16] Boquet, D.J. and Breitenbeck, G.A. (2000) Nitrogen Rate Effect on Partitioning of Nitrogen and Dry Matter by Cotton. *Crop Science*, **40**, 1685-1693. http://dx.doi.org/10.2135/cropsci2000.4061685x

[17] McConnell, J.S., Baker, W.H., Miller, D.M., Frizzell, B.S. and Varvil, J.J. (1993) Nitrogen Fertilization of Cotton Cultivars of Differing Maturity. *Agronomy Journal*, **85**, 1151-1156. http://dx.doi.org/10.2134/agronj1993.00021962008500060011x

[18] Hutmacher, R.B., Travis, R.L., Rains, D.W., Vargas, R.N., Roberts, B.A., Weir, B.L., Wright, S.D., Munk, D.S., Marsh, B.H., Keeley, M.P., Fritschi, F.B., Munier, D.J., Nichols, R.L. and Delgado, R. (2004) Response of Recent Acala Cotton Cultivars to Variable Nitrogen Rates in the San Joaquin Valley of California. *Agronomy Journal*, **96**, 48-62. http://dx.doi.org/10.2134/agronj2004.0048

[19] Bronson, K.F., Onken, A.B., Keeling, J.W., Booker, J.D. and Torbert, H.A. (2001) Nitrogen Response in Cotton as Affected by Tillage System and Irrigation Level. *Soil Science Society of America Journal*, **65**, 1153-1163. http://dx.doi.org/10.2136/sssaj2001.6541153x

24

Assessment of Dithionite and Oxalate Extractable Iron and Aluminium Oxides on a Landscape on Basement Complex Soil in South-Western Nigeria

Olayinka Olakunle Olatunji, Yetunde Oyeyiola, Gbade Oyewunmi Oyediran

Crop Production and Soil Science Department, Ladoke Akintola University of Technology, Ogbomoso, Nigeria
Email: ola_marcus@yahoo.com, ooolatunji@lautech.edu.ng

Abstract

The study investigats the morphological, physical and chemical characteristic of a toposequence on basement complex in Ejioku area, south-western Nigeria on latitude 07˚28.675'N; and longitude 004˚07.219'E; 180 m above sea level at the upper slope. It terminates at the valley bottom on latitude 07˚28.756'N; and longitude 004˚07.229'E; 175 m above sea level. The topography of the site is moderately slope (not greater than 10%). The aim is to examine the two forms of sesquioxides (crystalline-dithionite extractible Fe & Al and amorphous-oxalate Fe & Al) and their distribution on the toposequence. The crystalline form of Fe oxide dominates all the positions and ranges from 7.2 g/kg at the valley bottom to 444.3 g/kg at the crest compared to the amorphous forms that range from 2.7 g/kg to 10.9 g/kg. The crystalline and amorphous aluminium oxide contents of the soils are low (2.9 g/kg - 43.3 g/kg and 1.3 g/kg - 8.7 g/kg respectively). There is significant negative relationship between Fe_d and Fe_d/Fe_o ($r = -0.15$; $P < 0.01$; $n = 16$). There is also a correlation between Fe_o and Al_d ($r = 0.63$; $P < 0.01$; $n = 16$), Fe_o and Fe_o/Fe_d ($r = 0.44$; $P < 0.01$; $n = 16$), signifying a high level of weathering. The relatively high amount of Fe_d virtually at all the horizons indicates that the crystalline and less active forms of the oxides exist more on the landscape and may be responsible for the non-availability of some nutrient like phosphorus that may be sorbed to their crystals lattice. All forms of Al in the soils are low especially, the oxalate extractable forms, when compare with the dithionite extractable forms. However, there is a significant correlation between Al_d and Fe_o/Fe_d ($r = 0.57$; $P < 0.01$). As a result of high accumulation of crystalline form of Fe and Al oxide with increasing depth, most especially, at the crest/upper slope, there is greater amount of concretions, nodules and plinthites which will lead to further deterioration of the soil for agricultural purposes.

Keywords

Plinthites, Sesquioxides, Toposequence, Dithionite Extractable, Oxalate Extractable

1. Introduction

The features observable in a soil profile depend on the activity that goes on in it and how long they have taken place. It is however justifiable to say that characteristics of the soil as observed in a profile at a particular time depend on the interaction of factors and processes of soil formation. These factors have been enumerated as parent material, climate, organisms, relief and time [1]. Topography as a soil forming factor has dominant influence in determining differences among set of soils and their nutrient status. The set of soils are formed as a result of topography acting primarily as soil forming factors are called toposequence. Toposequence is also the succession of soils from a crest/plateau/summit to a valley bottom. The differences between soils of a toposequence are generally related to differences in their positions and drainage patterns, however, slope steepness is about the most important factor that causes variation in moisture condition along a toposequence [2]. According to [2], four groups of pedogenic processes result from slope effects are:

1) Powerful mechanical action (*i.e.* erosion);
2) Less powerful mechanical action *i.e.* lateral movement of clays (in suspension) and other materials in solution;
3) Transport in the form of pseudo and true solution;
4) Local changes in the moisture status and of the redox potential.

The different contribution of these will affect the type of variation observed at different location per time. In the basement complex area, there is a strong relationship between topographic position and soil genesis [2]. The differences in the rate of hydrologic and geomorphic process in various landform position causes differences in the types and intensity of pedogenic processes. This typically results in a non-random distribution of soil classes and properties on the landscape.

In tropical soils, free Fe oxides may be mobilized and deposited in soil profiles as Fe mottles, concretions, and hardpans [4]. Soils of the tropical region are dominantly alfisols, ultisols and oxisols and the dynamics of sesquioxides (iron and aluminum oxides) are progressively the determining factor for release of nutrient to planted crops [5].

Iron (Fe) and Aluminum (Al) found in soils are released during soil weathering and soil development; they are re-precipitated as amorphous or crystalline oxides, hydroxides or oxhydroxides [6]. Crystalline form is the predominant form of Fe in Alfisols which dominate the soils of Savanna region of Nigeria [5].

The crystalline form of Al-oxides is thought to be substituted into crystalline Fe-oxides such as geothites and heamatite [7]. The effect of such substitution is the structural distortion of crystalline Fe oxides with implications for anion retention and surface area [8] [9] studied the relationships between clay content and the acid oxalate extractable Fe and Al in some soils in Nigeria derived from sandstones and finds that the relationships are not significant. He, however, conclude that the amount and nature of the various forms of Fe and A1 oxides with organic complexes can greatly influence the physical and chemical properties of the soil. [10] conclude that the nature, amount and distribution of Iron (Fe) and Aluminum (Al) oxides in soil affect its ionic charge, chemical characteristics, and ion adsorption especially, phosphorus (P) sorption, surface charge and specific surface area. Swelling and aggregate formation may be significantly modified by the presence of amorphous Fe and Al oxides [11].

According to [12], the feature that distinguishes plinthites and accounts for hardness is greater degree of crystallinity and a continuity of the crystalline phase (*i.e.* the ratio of amorphous Iron (Fe_o) to crystalline (Fe_d), low Fe_o/Fe_d ratio indicated high degree of crystallinity).

A large area of the rain forest and savannah zone of south western Nigeria is developed on basement complex, mostly Gneiss [13]-[15] in the classification of soils of south western Nigeria, evaluated some soil association with Fe and Al sesquioxides and pointed out the different soil series in where these oxides are predominant. This should be investigated to provide information on the forms, proportions and relationship among sesquioxides in these soils.

More information is extremely needed for planning, uses and management of the soils. For these reasons Ejioku town, Lagelu Local Government area of Oyo State was selected haven't been an agrarian community producing most of the agricultural products of the state. Thus, this work was carried out to investigate the distribution and forms of sesquioxides on this landscape and determine the interrelationship between Fe and Al oxides with other soil properties.

2. Materials and Methods

2.1. Description of the Study Site (Figure 1)

The study area is Ejioku town; a derived savannah environment in south-western part of Nigeria. The location lies within latitude 07°28.675'N and 07°28.756'N with longitude 004°07.219'E and 004°07.229'E. The elevation was 273.41 m above sea level at the crest and 178.31 m at the valley bottom. The toposequence was about 328 m long. The area is characterised with distinct dry (November to early March) and wet (April to October) seasons. Rainfall pattern is bimodal with peak periods in June and September, mean annual rainfall is about 1312 mm. Temperature is about 27°C, and reach its peak in February and March (29°C to 32°C) respectively. The topography of the site is moderately slope (<10%). The geology of the area is dominated by crystalline rock which forms part of the basement complex of south-western Nigeria and are generally of granitic parent materials.

Over the years, the soil had been put to different uses ranging from arable to permanent crops like maize, cassava, plantain and oil palm while some part had remained fallowed with tall grasses, trees and bush regrowth.

2.2. Field Studies

The whole area was mapped using the rigid grid method of soil survey in which transverses were cut and spaced at 50 m apart. Points of observation were fixed at the intersections of the perpendicularly running transects. After this, soils with similar characteristics were grouped together as same mapping unit. Soils on the landscape were delineated to upper slope, middle slope, lower slope and the valley bottom. Soil profile pits were located and dug (1.5 m × 1.5 m × 2.0 m) to represent each of the identified delineations on the field. Altogether a total of sixteen samples were collected from the four profile pits. Notes on morphological attributes in the soil profile were taken (soil colour, texture, structure, consistence, stoniness, mottles, Cutans and concretions).

2.3. Analytical Procedures

The samples collected were air dried and crushed to pass through 2 mm sieve. Prepared and labelled samples were taken to Soil Physics and Chemo-dynamics Laboratory, Soil Science Department, University of Manitoba,

Figure 1. Map of Oyo State showing Lagelu Local Government area, area of the study site.

Winnipeg, Canada for analysis. Extractable phosphorus was determined in Mehlich-III extractant by equilibrating 2.5 g of air dried soil sample with 25 ml of Mehlich 3 extracting solution for 5 min and filtering through Whatman No. 40 filter paper [16] and the amount of phosphorus was determined using the colorimetric (ascorbic acid) method [17].

Ammonium oxalate extractable Fe, Al and Mn were determined by shaking 0.5 g of soil with 20 ml of ammonium oxalate extracting solution for 4 hours in the dark and then filtered using Whatman No. 40 filter paper [18]. Ammonium oxalate Fe, Al and Mn in the extractant were determined by Inductive coupled plasma technique. The exchangeable cations and the cation exchange capacity (CEC) of the soil were determined using the ammonium acetate method [19]. Particle size distribution was determined using Hydrometer method [20], pH in ratio 1:2 (Soil:H_2O), Organic carbon was by [21].

Dithionite-Citrate Fe, Al and Mn were determined using the [22] soil conservation method. A 0.5 g of soil sample, ground to pass a 35-mesh sieve was weighed into a 50 ml plastic centrifuge tube (0.2 g for clays and 1 g for coarse soils). 25 ml of the sodium citrate ($Na_3C_6H_5O_7 \cdot 2H_2O$), 0.68 M (200 g/L) solution was added. 0.4 g of dithionite (sodium hydrosulphite, $Na_2S_2O_4$) was also added. The centrifuge tube was covered tightly and put in an end-over-end shaker overnight and centrifuged for 20 minutes. The supernatant were filtered and determination of Fe, Al, and Mn was done by ICP-AES (inductive coupled plasma). This treatment is particularly useful for dissolving the "free" (non-silicate) Fe and Al in soils. The difference between the Al_d and Al_o was assumed to be the Al^{3+} substituted for Fe^{3+} in the crystalline [23]. Determination of Total Nitrogen and Total Phosphorus was done by the Kjeldahl Method using a digestion mixture of 350 ml H_2O_2, 0.42 g Se powder, 14 g $Li_2SO_4 \cdot H_2O$ and 420 mL conc. H_2SO_4 [24]. The concentration of P in solution was determined using the colorimetric method: Ascorbic method [25], while nitrogen was by titration method.

2.4. Soil Classification

The soils were classified according to soil taxonomy [22] and [26] soil legend while [15] soil classification was adopted at the series level.

3. Result and Discussion

3.1. Soil Morphological Characteristics

Generally 3 to 5 differentiated horizons were observed in all the profile pits, however; only 3 horizons were observable in the profile located at the valley bottom. This was due to the high water table encountered that prevented deeper examination. The important morphological characteristics of the soil are given in **Table 1**. Designation of the horizons are done according to the USDA soil taxonomy classification: AP represented the ploughed horizon, AB is the transitional point between the ploughed horizon and the B horizon. The B horizon is the subsoil with presence argillic features.

The landscape was delineated to upper slope, mid slope, lower slope and valley bottom. Pedon 1 was located at the upper slope and has a grayish yellow brown (10YR 5/2) topsoil coming down to bright yellowish brown (10YR 6/6) at the subsoil with five identified horizons.

Pedon 2 has a grayish brown (7.5YR 5/2). A horizon coming down to a (7.5YR 5/6) bright brown colour at the subsoil with four observable horizons.

Pedon 3 has grayish brown (7.5YR 4/2) topsoil that gradually changes to orange (7.5YR 6/8) at the subsoil while Pedon 4 was black (7.5YR 2/1) at the top and changes gradually to brown (7.5YR 4/4) at the subsoil.

At the valley bottom, the colour differs from the rest of the slope which was due to high water table content (7.5YR 2/1). The yellowish brown to orange colour observed at the upland subsoil is considered to be due to the mobilization and subsequent immobilization of iron during redox-cycles in the soil [3] [27] advanced this to be the causes of dispersion with progressive oxidation of iron. The bright subsoil colour in some of the profile may be an indication of good internal drainage, except the subsoil of pedon4 which was dull brown (7.5YR 4/4) as a result of high water table and poor internal drainage. Nearly all the soil profiles were mottled, however, mottleness was more prominent at the upland region, and this could be due to lack of mechanical mixing by plant roots and soil [28].

Surface soils were generally coarse sandy loam or sandy clay loam graduating into coarse fine sandy clay at the subsoil. The soil structure ranges from granular coarse at the surface of upland profiles to sub-angular blocky

Table 1. Field morphological description of Pedon studies.

Depth (cm)	Horizon	Colour hue (moist)	Mottles	Stoniness	Soil structure	Consistence (moist)	Texture	Concretion	Drainage
Pedon 1 **IWO series (<u>Arenic Kandiustults</u>)** "(N07°28.675'; E004°07.219'; 180 m asl.)"									
0 - 25	Ap	10 YR5/2	few fine	No stones	granular coarse	dry loose	sandy loam	unidentified few	well drain
26 - 59	AB	10 YR 6/1	few fine	No stones	sub angular blocky medium	loose	sandy clay loam	unidentified few	well drain
60 - 107	B1	10 YR 6/1	common medium	extremely stony	angular blocky v. coarse	extremely Hard	clayey coarse sand	ferruginous v. many	well drain
108 - 142	B2	10 YR6/4	common medium	fairly stony	columnar medium	Slightly Hard	sandy clay	Fe-Mn many	well drain
143 - 198	B3	10 YR 6/6	common medium	fairly stony	columnar medium	slightly Hard	sandy clay	gypsum many	well drain
Pedon 2 **IBADAN series (<u>Kanhaplic Haplustults</u>)** "(N07°28.702'; E004°07.221'; 178 m asl.)"									
0 - 20	Ap	7.5 YR 5/2		no stone	crumb fine or thin	friable	sandy clay loam	unidentified common	well drain
21 - 65	AB	7.5 YR 6/1		stony	granular fine	friable	sandy clay	ferruginous common	well drain
61 - 128	B1	7.5 YR 4/4	common medium	fairly stony	prismatic medium	firm	sandy clay	ferruginous v. many	well drain
129 - 192	B2	7.5 YR 5/6	common medium	fairly stony	platy coarse	firm	coarse sandy clay	ferruginous v. many	well drain
Pedon 3 **APOMU series (<u>Typic Psammaquant</u>)** "(N07°28.747'; E004°07.227'; 177 m asl.)"									
0 - 25	Ap	7.5 YR 4/2	few fine	No stone	granular coarse	friable	slightly sandy clay	unidentified few	well drain
26 - 55	AB	7.5 YR 5/3	few fine	No stone	sub angular blocky medium	friable	sandy clay	unidentified few	well drain
56 - 98	B1	7.5 YR 5/4	few fine	No stone	crumb medium	firm	sandy clay	unidentified few	well drain
99 - 180	B2	7.5 YR 6/8	few fine	No stone	angular blocky	firm	clay sandy	unidentified few	well drain
Pedon 4 **JAGO series (<u>Aquic Haplustults</u>)** "(N07°28.756'; E004°07.229'; 175 m asl.)"									
0 - 12	Ap	7.5 YR 2/1	few fine	fairly stony	sub angular blocky	slightly Sticky	silt clay sand	unidentified few	poor
13 - 59	AB	7.5 YR	few fine	fairly stony	sub angular blocky	sticky fine	sandy clay loam	unidentified few	poor
60 - 85	B	7.5 YR 4/4	few fine	fairly stony	sub angular blocky	very sticky	clayey sand	unidentified few	poor

and fine at the valley bottom but the structure graduated to columnar, platy and sub angular blocky down the profile. This was due to the increase in clay content down the profile. Concretion was more of Fe-Mn component at the upper slope and ferruginous at the middle slope to unidentified and few at the lower slope. Sesquioxides concretion was iron noodles concretion; however, petro-plinthite and quartz materials are dominants in the horizons of Pedon 1 and 2. These are indicative of the granitic parent materials and alternate wetting and drying of the solum. According to [23], the presence of iron stone gravels *in-situ* in residual upland soils suggests that there had been possible segregation of iron in the form of plintholes. This process requires an adequate supply of iron, alternating wet and dry seasons a relatively flat land surface with seasonally wet soils [3] which could be found in the study area. This was also in line with the observations of [23] on some soils of Africa.

3.2. Diagnostic Horizon and Intensity of Weathering

Silt: Clay ratio is an indicator of weathering intensity and a ratio of <0.15 indicates low to moderate (**Table 2**) while a ratio of >0.15 indicates high intensity [2]. Thus all the Pedons showed evidence of high weathering intensity. The high silt content however reflects the probability of the soil being formed from coarse colluvial material (hill wash). This tends to suggest that erosion is an important pedogenic process in the study area (transportation).

3.3. Physico-Chemical Properties of the Soils

The AP horizon of all the pedons has high sand content (78%) with low clay content. Generally, there was a decrease in sand with depth with clay accumulation at the subsoil. The silt/clay ratio is >1 at most horizons, however, in Pedon 4, silt/clay ratio was <1, suggesting a probability of ferallitic pedogenesis [29]. An outstanding feature of these soils across all the location on the topography is the moderate to high silt content (6% - 26%) down the profile. This increase in silt content can be due to the perturbation of soil aggregates during lessiviation and the washing down of the finer soil aggregate down the profile. This feature distinguishes the soils from most sandy soils at south-western Nigeria, which are characterized by low silt content (<10% - 15%) [2]. The high clay content in the deeper horizons of some of the soils (Pedon 1, 2 & 3) with some morphological properties (**Table 1** and **Table 2**) formed the basis for the recognition of argillic horizons in the region. Iron/Manganese (Fe/Mn) concretion ranges from many to common and occurred in most of the upland profiles (**Table 1**), suggesting the tendency of alternating wet and dry cycles, which could infer that plinthization may be a strong pedogenic processes in the study area. The high concentration of concretion may constitute a serious problem to crop production within the study area.

The soils on Pedon 1 was slightly acidic (5.7 - 6.7) to acidic and slightly acid in profile 2 (5.6 - 6.6), slightly acid in Pedon 3 (6.0 - 6.6) while Pedon 4 was strongly acidic (4.9 - 5.3). The factors that could be responsible for the strong acidic nature of Pedon 4 was the high water table condition of the valley bottom and poor drainage. The organic carbon content of the soil was generally low at the subsoil except on Pedon 2 and 3, where the organic carbon was higher at the subsoil. This is suggestive of a lithological discontinuity as a result of pedotransfer. Generally the organic carbon is considered to be low in all the soils which may be due partly to the impact of high temperature and high relative humidity, which favours rapid mineralization of organic matter components [30]. Also the different land use on landscape could be responsible for the lower organic carbon content at the topsoil when comparing the upland and with the valley bottom. However, the amount of organic carbon content cannot conveniently sustain a good crop production programme on a long term basis. Hence, a substantial improvement has to be made through effective crop residue incorporation management strategies.

The effective cation exchange capacity (ECEC) was low ranging from 2.1 to 15.8 Cmol/kg. This could be in part due to the nature of parent material. The value indicate that the soils have low potential for retaining plant nutrients, and therefore call for an adequate soil management strategies such as, split fertilizer application and use of organo-mineral fertilizers.

In addition, these low values of ECEC in combination with low organic carbon and pH are indications of low, inherent soil fertility status, which emphasis the need for improved soil management techniques. This should be incorporated into the traditional methods of bush fallowing if good and long term crop production is to be sustained. Mehlich III extractable phosphorus (P) was generally low and considered inadequate for crop production, when compared with critical values recommended for most tropical crops (8 - 15 ppm) [31].

Table 2. Physical and chemical properties of the soil in the study area.

Depth (cm)	Horizon designation	Sand %	Silt %	Clay %	Texture	Ca^{2+}	K^{+}	Na^{+}	Mg^{2+}	ECEC	TAB	Silt/Clay ratio	pH (H_2O)	Av. P (ppm)	Org. C %
									Cmol/kg						
						Exchangeable cations									
Pedon 1 IWO series (Arenic Kandiustults)															
0 - 25	Ap	92	6	2	sand	1.20	1.27	2.19	2.00	8.69	2.04	3.00	6.17	53.40	6.65
26 - 59	AB	70	26	4	sandy loam	2.18	2.29	3.96	3.63	15.76	3.69	6.50	6.73	7.80	2.09
60 - 107	B1	68	20	12	sandy loam	1.02	1.07	1.85	1.7	7.36	1.72	1.67	6.38	2.60	1.36
108 - 142	B2	66	16	18	sandy loam	1.08	1.14	1.97	1.8	7.83	1.83	0.89	5.72	1.20	0.75
143 - 198	B3	62	18	20	sandy clay loam	0.50	0.53	0.91	0.84	3.64	0.85	0.90	6.2	0.98	0.66
Pedon 2 IBADAN series (Kanhaplic Haplustults)															
0 - 20	Ap	82	12	6	sandy	0.69	0.73	1.26	1.15	5	1.17	2.00	5.8	5.38	2.06
21 - 65	AB	76	6	8	sandy loam	0.54	0.57	0.98	0.9	3.91	0.92	0.75	5.89	4.20	1.41
61 - 128	B1	70	14	16	sandy loam	0.94	0.98	1.7	1.56	6.76	1.58	0.88	5.6	1.70	3.11
129 - 192	B2	66	20	14	sandy loam	1.32	1.39	2.41	2.2	9.57	2.24	1.43	6.56	0.60	3.29
Pedon 3 APOMU series (Typic Psammaquant)															
0 - 25	Ap	88	6	6	sand	0.81	0.86	1.48	1.36	5.89	1.38	1.00	6.2	7.50	3.92
26 - 55	AB	72	14	14	sandy loam	0.87	0.92	1.59	1.45	6.3	1.48	1.00	6.02	2.08	2.35
56 - 98	B1	74	13	13	sandy loam	0.59	0.62	1.07	0.98	4.24	0.99	1.00	6.05	1.00	2.57
99 - 180	B2	70	14	16	sandy loam	1.63	1.72	2.96	2.72	11.79	2.76	0.88	6.56	6.85	7.56
Pedon 4 JAGO series (Aquic Haplustults)															
0 - 12	Ap	80	6	14	sand	1.12	1.18	2.04	1.87	8.1	1.9	0.43	4.93	3.90	8.39
13 - 59	AB	74	10	16	sandy loam	0.29	0.3	0.52	0.48	2.06	0.48	0.63	5.1	7.38	1.80
60 - 85	B	86	16	8	sand	0.77	0.81	1.4	1.29	5.58	1.31	2.00	5.28	2.40	2.78

3.4. Distribution of Dithionite and Oxalate Extractable Sesquioxides

The data on the oxides of Fe and Al are shown in **Table 3**. The distribution of crystalline, non-crystalline and amorphous (poorly crystalline) forms of Fe, Al and Mn oxides in the Pedons and their relationship are presented in **Table 3**. The citrate dithionate extractable only indicate the Fe and Al present in the soil are as free discrete bodies (*i.e.* crystalline and amorphous), and not as part of the structure of silicate minerals [28].

The free discrete bodies collectively referred to as un-combined or pedogenic free forms of the element [28] could either be mobile (amorphous) or immobile (crystalline). There was a wide margin in the distribution of dithionite and oxalate extractable Fe and Al within the horizons. The concentration of dithionite Fe and Al were greater than that of oxalate Fe and Al. However, there was a progressive increase in concentration of both dithionite and oxalate extractable Fe and Al down the profile in pedon1 and Pedon 3, while Pedon 2 did not show any regular pattern in the distribution. Pedon 4 however showed a remarkable decrease in concentration down the profile.

All the Pedons had a lower Fe_o and Al_o at the AB horizon and the highest at B_2. This phenomenon could be associated with topographic, parent material and land use effect on pedogenesis [28]. Among the pedons, B_2 had the highest amount of Fe_d except in Ibadan series where it was highest at B_1. The ratio of free Fe to clay tends to vary regularly in all the Pedons and highest at the AP horizon and decreased down the profile.

Table 3. Distribution of various kinds of Fe and Al oxides in the pedons.

Depth (cm)	Horizon Designation	% $Al_2O_{3(d)}$	% $Fe_2O_{3(d)}$	% $Mn_{(d)}$	% $Al_2O_{3(o)}$	% $Fe_2O_{3(o)}$	% $Mn_{(o)}$	$Fe_2O_{3(d)}$/ Clay	$Fe_2O_{3(O)}$/ $Fe_2O_{3(d)}$	$Al_2O_{3(o)}$/ $Al_2O_{3(d)}$
					Pedon 1					
					IWO Series (Arenic Kandiustults)					
0 - 25	AP	6.66	38.15	1.33	3.04	7.34	1.78	19.08	0.19	0.46
26 - 59	AB	5.96	27.14	1.27	2.33	3.91	1.93	6.78	0.14	0.39
60 - 107	B1	15.77	128.9	4.42	3.49	6.01	4.59	10.74	0.05	0.22
108 - 142	B_2	10.31	42.68	0.51	5.06	2.90	1.05	2.37	0.07	0.49
143 - 198	B_3	43.31	444.3	14.42	8.66	10.86	12.82	22.22	0.02	0.20
					Pedon 2					
					IBADAN Series (Kanhaplic Haplustults)					
0 - 20	AP	22.14	222.75	3.84	6.10	6.75	4.13	37.13	0.03	0.28
21 - 65	AB	4.69	17.15	0.00	1.69	3.28	0.06	2.14	0.19	0.36
61 - 128	B_1	13.06	83.65	0.00	5.55	7.22	0.15	5.23	0.09	0.42
129 - 192	B_2	9.56	61.65	0.00	6.46	6.06	0.22	4.40	0.10	0.68
					Pedon 3					
					APOMU Series (Typic Psammaquant)					
0 - 25	AP	5.77	31.17	3.01	2.07	3.32	4.07	5.19	0.11	0.36
26 - 55	AB	7.76	44.94	2.89	2.40	3.55	3.53	3.21	0.08	0.31
56 - 98	B_1	8.56	57.25	2.22	2.80	3.64	2.74	4.40	0.06	0.33
99 - 180	B_2	9.37	59.70	3.19	3.22	4.23	4.68	3.73	0.07	0.34
					Pedon 4					
					JAGO Series (Aquic Haplustults)					
0 - 12	AP	15.82	104.50	0.25	3.43	12.76	0.51	7.46	0.12	0.22
13 - 59	AB	2.88	7.18	0.00	1.30	2.74	0.08	0.45	0.38	0.45
60 - 85	B	3.57	22.74	0.00	1.95	7.54	0.20	2.84	0.33	0.55

$_d$ = sodium citrate-bicarbonate-dithionite extractable; $_o$ = ammonium oxalate extractable; Fe = iron oxide; Al = aluminium oxide.

The ratio of Fe_o/Fe_d which serves as the index of the proportion of amorphous iron content and degree of crystalline of the soil was highest in the first two horizons at upper and lower slope. It increases down the profile at the valley bottom (Jago) except at the mid-slope where the ratio at the second horizon was almost double those of the other horizon in the Pedon.

Generally, there was significant negative relationship between Fe_d and Fe_d/Fe_o ($r = -0.15$; $P < 0.01$). There are also correlation between Fe_o and Al_d ($r = 0.63$; $P < 0.01$) and Fe_O and Fe_O/Fe_d ($r = 0.44$; $P < 0.01$). The relatively high amount of Fe_d virtually at all the horizons indicates that the crystalline and less active forms of the oxides exist more on the landscape and may be responsible for the non-availability of some nutrient like phosphorus that may be sorbed to their crystals lattice. Less emphasis is given to the distribution of various forms of Al in the soils as the quantities are much lowers especially the oxalate extractable forms. However, there are significant correlation between Al_d and Fe_o ($r = 0.63$; $P < 0.01$) and Al_d and Fe_o/Fe_d ($r = 0.57$; $P < 0.01$).

4. Conclusion

The results of this study have shown that the soils are relatively high in sesquioxides and their toxicity will definitely have an impact on the productive capacity of the soil. Crystalline forms as compared to the amorphous form dominate the free sesquioxides in the soils throughout the landscape. This shows that they have been highly weathered and the high proportion of crystalline form of Fe and Al will lead to structural distortion with implication for anion retention. The formation of hard pans and concretions can also lead to compaction of soil which can inhibit the penetration of plant root. Thus, very good and dynamic management technique and land uses are essential to ensure the preservation of the soil.

Acknowledgements

I acknowledged the contributions of The Canadian Bureau for International Education, through the Canadian Commonwealth Graduate Student's Exchange Program (GSEP) and Department of Soil Science, Faculty of Agricultural and Food Sciences, University of Manitoba, Canada.

References

[1] Juo, A.S.R., Moormann, F.R. and Maduakor, H.O. (1974) Forms and Pedogenetic Distribution of Extractable Iron and Aluminum of P, Fe_d, and Al_d in Selected Soils of Nigeria. *Geoderma*, **11**, 167-179. http://dx.doi.org/10.1016/0016-7061(74)90015-9

[2] Ogunkunle, A.O. (1987) Influence of Micro Topography and Sampling on the Variability of Some Chemical Properties in Southern Nigeria. *Tropical Agriculture* (*Trinidad*), **64**, 111-114.

[3] Fasina, A.S., Omolayo, F.O., Falodun, A.A. and Ajayi, O.S. (2007) Granitic Derived Soils in Humid Forest of South-Western Nigeria—Genesis Classification and Sustainable Management. *American-Eurasian journal of Agiculture and Environmental Science*, **2**, 189-195.

[4] Ojanuga, A.G. (1978) Genesis of Soils in the Metamorphic Forest Region of South-Western Nigeria. *Pedolomorphic Pedologie xxviii*, **1**, 105-117.

[5] Onweremadu, E.U., Omeke, J., Onyia, V.N., Aguand, C.M. and Onwubiko, N.C. (2007) Inter-Horizon Variability in Phosphorus-Sorption Capability of Sesquioxide-Rich Soils Southeastern Nigeria. *Journal of American Science*, **3**, 43-48.

[6] Guertal, W.R. (1994) The Pedologic Nature of Weathered Rock. In: Cremeens, D.L., *et al.*, Eds., *Whole Regolith Pedology*, SSSA, Madison, 21-40.

[7] Ainsworth, C.C. and Sumner, M.E. (1985) Effect of Aluminum Substitution in Goethite on Phosphorus Adsorption. II. Rate of Adsorption. *Soil Science Society of American Journal*, **49**, 1149-1153. http://dx.doi.org/10.2136/sssaj1985.03615995004900050015x

[8] Agbenin, J.O. (2003) Extractable Iron and Aluminum Effects on Phosphate Sorption in a Savanna Alfisol. *Soil Science Society of American Journal*, **67**, 589-595. http://dx.doi.org/10.2136/sssaj2003.5890

[9] Udo, E.J. (1980) Profile Distribution of Iron Sesquioxides Contents in Selected Nigeria Soils. *Journal of Agricultural Science*, **95**, 191-198. http://dx.doi.org/10.1017/S0021859600029439

[10] Torrent, J., Barron, V. and Schwertmann, U. (1990) Phosphate Adsorption and Desorption by Goethites Differing in Crystal Morphology. *Soil Science Society of American Journal*, **54**, 1007-1012. http://dx.doi.org/10.2136/sssaj1990.03615995005400040012x

[11] Angers, D.A. and Chenu, C. (1998) Dynamics of Soil Aggregation and Carbon Sequestration. In: Lal, R., Kimble, J.M., Follett, R.F. and Stewart, B.A., Eds., *Soil Processes and the Carbon Cycle*, Advances in Soil Science, CRC Press, 199-206.

[12] Yaro, D.T. (2006) The Position of Plinthites in a Landscape and Its Effects on Soil Properties. PhD Thesis, Ahmadu Bello University, Zaria, 241.

[13] Okusami, T.A. and Oyediran, G.O. (1985) Slope-Soil Relationship on an Aberrant Toposequence in Ife Area of South-Western Nigeria: Variability in Soil Properties. *Ife Journal of Agriculture*, **7**, 1-15.

[14] Amusan, A.A. and Ashaye, T.I. (1991) Granitic-Gneisis Derived Soils in Humid Forest Tropical South Western Nigeria I: Genesis and Classification. *Ife Journal of Agriculture*, **13**, 1-10.

[15] Smyth, A.J. and Montgomery, R.F. (1962) Soil and Land Use in Central Western Nigeria. Govt. Press, Ibadan.

[16] Mehlich, A. (1984) Mehlich-III Soil Test Extractant: A Modification of Mehlich II Extractant. *Communication in Soil Science and Plant Analysis*, **15**, 1409-1416. http://dx.doi.org/10.1080/00103628409367568

[17] Murphy, J. and Riley, J.P. (1962) A Modified Single Solution Method for the Determination of Phosphate in Natural Waters. *Analytica Chimica Acta*, **27**, 31-36. http://dx.doi.org/10.1016/S0003-2670(00)88444-5

[18] Jackson, M.L., Lim, C.H. and Zelazny, L.W. (1986) Oxides, Hydroxides and Aluminosilicates. In: Klute, A., Ed., *Methods of Soil Analysis*, *Part* 1, *Physical and Mineralogical Methods*, 2nd Edition, Agronomy Monogram 9, ASA and SSSA, Madison, 101-150.

[19] McKeague, J.A. (1978) Manual on Soil Sampling and Methods of Analysis. 2nd Edition, Canadian Society of Soil Science, Ottawa, 212 p.

[20] Gee, G.W. and Or, D. (2002) Particle Size Analysis. In: Dane, J.H. and Topp, G.C., Eds., *Methods of Soil Analysis*, *Part* 4, *Physical Methods*, Soils Science Society of America, Book Series No. 5, Madison, 255-293.

[21] Nelson, D.W. and Sommers, L.E. (1982) Total Carbon, Organic Carbon and Organic Matter. In: Page, A.L., Miller, R.H. and Kenney, D.R., Eds., *Methods of Soil Analysis*, *Part* 2, American Society of Agronomy, Madison, 539-579.

[22] Soil Survey Staff (2006) Keys to Soil Taxonomy. 9th Edition, US Department of Agriculture, NRCS, Washington DC, 281.

[23] Fitzpatrick, R.W. and Schwertmann, U. (1982) Al-Substituted Goethite: An Indicator of Pedogenic and Other Weathering Environments in South Africa. *Geoderma*, **27**, 335-347. http://dx.doi.org/10.1016/0016-7061(82)90022-2

[24] Akinremi, O.O., Armisen, N., Kashemand, A. and Janzen, H.H. (2003) Evaluation of Analytical Methods for Total P in Organic Amendments. *Communications in Soil Science and Plant Analysis*, **34**, 2987-2997. http://dx.doi.org/10.1081/CSS-120025220

[25] Simard, R.R., Beauchemin, S. and Haygarth, P.M. (2000) Potential for Preferential Pathways of Phosphorus Transport. *Journal of Environmental Quality*, **29**, 97-105. http://dx.doi.org/10.2134/jeq2000.00472425002900010012x

[26] FAO-ISRIC-ISSS (Food and Agriculture Organization of the United Nations, International Soil Reference and Information Centre, International Society for Soil Science) (2006) World Reference Base for Soil Resources. Report No. 103, FAO, Rome.

[27] Buol, S.W., Hole, F.D. and McCracken, R.J. (1980) Soil Genesis and Classification. Iowa State University Press, Ames, 404 p.

[28] Obi, J.C., Akinbola, G.E. and Anozie, H.I. (2009) Distribution of Dithionite and Oxalate Extractable Iron Oxides of a Catena in the Basement Complex of South-Western Nigeria. *Nigeria Journal of Soil Science*, **19**, 100-119.

[29] Ashaye, T.I. (1969) Sesquioxide Status and Particle Size Distribution in Twelve Nigerian Soils Derived from Sandstones. *African Soils Journal*, **14**, 85-96.

[30] Ibia, T.O. (2001) Forms of Fe and Al in Soil Particles of Inland Flood Plains of South-Eastern Nigeria. *Proceeding of the 27th Annual Conference of the Soil Science Society of Nigeria*, Calabar, 5-9 November 2001, 165-168.

[31] Obigbesan, G.O. (2001) Plant Mineral Nutrition in Nigeria: My Experience. In: Akorodon, M.O., Ed., *Agronomy in Nigeria*, Department of Agronomy, University of Ibadan, Ibadan, 188-194.

25

Planting Cotton in a Crop Residue in a Semiarid Climate: Water Balance and Lint Yield

Robert J. Lascano[1], Dan R. Krieg[2], Jeffrey T. Baker[3], Timothy S. Goebel[1], Dennis C. Gitz III[1]

[1]Wind Erosion and Water Conservation, USDA-ARS*, Lubbock, TX, USA
[2]Department of Plant and Soil Science, Texas Tech University, Lubbock, TX, USA
[3]Wind Erosion and Water Conservation, USDA-ARS, Big Spring, TX, USA
Email: Robert.Lascano@ars.usda.gov

Abstract

Cotton (*Gossypium hirsutum* L.) is planted on more land area than any other crop on the Texas High Plains. Much of this area is considered highly erodible and requires a conservation compliance program to participate in government farm programs. Because this region is semiarid and because irrigation water is increasingly limited, water conservation and efficient use of water are necessary to maximize cotton lint yields. One popular conservation compliance practice used is to plant cotton into a chemically terminated small grain crop, *i.e.*, residue that provides wind protection to the cotton seedlings. Our hypothesis was that in a semiarid region the use of a small grain cover crop under irrigated conditions would use more water than it conserves compared to conventional tilled cotton, thus reducing cotton lint yields. To test the hypothesis separate field studies over two growing seasons and on two soil textures, a loamy fine sand and a clay loam, were conducted. The main treatments were tillage systems (conventional and conservation using terminated wheat residue). The two split plot treatments were water supply based on replacement of calculated grass reference evapotranspiration (ET_0). Tillage did not affect the amount of water used by the cotton crop at either location (< 7% difference, $P > 0.05$) except for an 80% ET_0 irrigation treatment at a single location where the bare soil treatment used 10% more water than the residue treatments for both years. The residue treatment decreased ($P < 0.05$) cotton lint yields at both locations by 12% except for the 50% ET_0 single irrigation treatment in which the residue treatment yielded 14% more lint than the bare soil treatment. The use of terminated wheat residue had no impact on soil water storage during any part of the year. During a 5-month period associated with wheat growth, the wheat evapotranspiration was 20 to 40 mm more water ($P < 0.05$) than that lost through soil water evaporation from the conventional treatments. The use of termi-

*The US Department of Agriculture (USDA) prohibits discrimination in all its programs and activities on the basis of race, color, national origin, age, disability, and where applicable, sex, marital status, familial status, parental status, religion, sexual orientation, genetic information, political beliefs, reprisal, or because all or part of an individual's income is derived from any public assistance program.

nated wheat residue did not benefit the water balance of the cotton crop, and was associated with decreased cotton lint yields. The results were consistent with our working hypothesis, and disproved the idea that planting cotton into wheat stubble cover increases water use efficiency.

Keywords

Soil Water, Evapotranspiration, Cropping Systems, Semi-Arid, Wheat Residue

1. Introduction

In 2013 and 2014, about 1.2 million ha were planted with upland cotton on the Texas South Plains [1]. The annual mean precipitation in this region is 467 ± 159 mm and about 50% of this rainfall occurs during the 4-month (May to August) growing season. About half of the planted area is irrigated with a water supply ranging from 2.5 to 7.5 mm·d^{-1} and the potential evaporative demand is on average 9.8 ± 0.8 mm·d^{-1} during the four-month growing season [2]. The source of the irrigation water is the Ogallala aquifer, where withdrawals exceed natural recharge, resulting in a declining water table [3]. Much of the water applied to the cotton is lost to bare soil water evaporation (E_{soil}) in conventional tillage systems [4] [5]. In this semiarid region, where cotton is planted in a wide row spacing (1-m) the crop seldom reaches a leaf area index of 3 m^2·m^{-2} that is necessary to achieve canopy closure [6] [7]. Furthermore this exposed and bare soil is conducive to increase the amount of E_{soil} [8].

Given the decline of irrigation water from the Ogallala aquifer [3] [9] with a decreasing pumping capacity, particularly during periods of low rainfall which are frequent in this region, led to develop agronomic practices to maximize cotton's water use efficiency (WUE), *i.e.*, production of lint yield per unit of water taken up by the plant [10]. An example of this conservation practice is to plant cotton into a winter wheat (*Triticum aestivum L.*) crop that is chemically terminated in the spring before the cotton crop is planted [11]. Residues may protect cotton seedlings by providing shelter to wind damage and sand abrasion [12]. Also, the residue may reduce E_{soil}; however, in some cases it may use more water than it conserves [5] [8] via wicking [13]. This is the process whereby the wheat residues may contribute to additional water loss via capillary action, *i.e.*, wicking effect [14]. In this situation, the wheat residue may increase the ability to transport water from the seedbed to the atmosphere via evaporation and thus resulting in a reduction of soil water and negatively affecting cotton emergence when planted in the wheat residue.

In the cotton cropping systems of the Texas High Plains, the effect of crop residues on water conservation and lint yield, are a function of the water supply, from irrigation and rainfall, soil type, and tillage practices [15] [16]. Planting cotton into a wheat residue on a sandy loam soil only increased cotton lint yield when the crop was irrigated at the grass reference evapotranspiration demand (ET_o). However, this practice is not suited to the availability of irrigation water from the Ogallala aquifer. In field trialsand over a five-year period, in a loamy fine sandy soil the residue did not impact lint yield under irrigated conditions, but significantly and adversely affected dryland lint yields. In a clay loam soil, the lint yield of cotton planted in a wheat residue increased on average by 13% under irrigation and reduced dryland cotton lint yield by 8% [17] [18]. From these results, it can be concluded that the effects of wheat residues on cotton lint yield depend primarily on the water supply. The texture effect is related to the amount of water that can be stored in the soil, particularly from the time the wheat is terminated to when the cotton is planted. For example, in the surface 0.20 m of the soil profile, a clay soil can store about twice as much water as a sandy soil and for dryland conditions the additional stored water would be available during cotton planting and favor seed germination and emergence of the cotton plant.

In first analysis for the Texas High Plains, the amount of water required to establish the wheat residue is often ignored in the water balance of dryland and limited irrigated cotton cropping systems [4] [5] [8]. To establish a wheat residue that would not impact the germination and emergence of the subsequent cotton crop planted in the residue, and of sufficient growth to be beneficial to cotton seedlings requires a net gain of water. In the Texas High Plains, winter wheat is planted in months (November to February) where the average monthly rainfall is 16 ± 2 mm [2]. Clearly, in a dryland system the establishment of the residue comes with the risk of using stored water that otherwise would be available at the time needed when the cotton crop is planted [19]. Furthermore, the winter wheat crop requires water to grow and be of benefit to the cotton crop. The measured daily ET rate of irrigated winter wheat varied between 1 and 2 $mm·d^{-1}$ with a maximum of 13 $mm·d^{-1}$ during the late spring [20]. Therefore, it was concluded that for the Texas High Plains under dryland conditions the establishment of a wheat residue is perhaps not practical and will require irrigation to be of value [19].

The short- and long-term benefits of residues on soil physical and chemical properties, and crop production has been widely investigated, e.g., [15] [21]-[23]. Residues increase infiltration of water, from either rain or irrigation, by increasing the surface roughness and flow path tortuosity and thus increasing the time for more water to infiltrate by surface ponding [19] [24] [25]. Residues, also modify the kinetic energy from raindrops impacting the soil surface [26] [27], which leads to less erosion [26] and increasing rain interception [28] by as much as 10% of the annual rainfall [29]. A long-term benefit of residues is to increase water storage and thus providing more water for crop production [30] [31].

The impact of residues on E_{soil} is well documented [4] [5] [8] [32]-[34]. Residues can reduce the rate of E_{soil} in the so-called first stage [25] [30] [34]-[37]. The type of residue also affects the E_{soil} rate and wheat, for example, is more effective than cotton [35]. The presence of a residue reduces E_{soil}; however, the total evaporation equilibrates between residue covered soil and bare soil after a long period of time between water inputs [35]. For example, the rate of E_{soil} from a bare soil and a residue-covered soil will be different; however, the total amount of water lost from the bare and residue-covered soil will be very similar after a long period of time [5] [8]. The rate of E_{soil} is defined by exchanges of energy and water between the soil surface and the surrounding environment [32] and determined by the combined water and energy balance of the soil surface. The aerodynamic resistance to vapor and latent heat flux from the soil surface to the atmosphere is increased by the presence of a residue, which decreases the rate of E_{soil} [5] [8] [38] [39]. Another effect of a residue is to reduce the amount of irradiance that reaches the surface, decreasing soil temperature and E_{soil}.

The presence of a surface residue modifies the E_{soil} and affects the WUE [8]. The daily E_{soil} and crop transpiration (T) for a 100-day growing season of cotton planted into terminated wheat and conventionally tilled soil where compared by Lascano *et al.* [8]. In this experiment daily E_{soil} was measured for a seven-day period using microlysimeters [32] and calculated using the mechanistic ENWATBAL model [4] [40]. Both cotton cropping systems, residue and conventional, had the same seasonal ET. However, the conventional tilled cotton evaporated more E_{soil} and had a lower amount of plant T than the cotton planted into the terminated wheat. The wheat residue increased the seasonal crop T, *i.e.*, more water used for plant growth, and reduced the seasonal E_{soil}, resulting in an increase of lint yield of 35% compared to the cotton planted in the conventional tillage with the same amount of water. The WUE of the cotton planted in the wheat residue was 2.6 g lint per kg of water used in ET, which was 27% more lint than the conventionally planted cotton [8]. In these field experiments the amount of water used to grow and establish the wheat residue were not measured and thus the calculated value of WUE is biased towards the cotton lint yield from the residue.

Our hypothesis is that despite the many benefits that residue offer, in a semiarid environment establishing the residue would use more water than conserved when used along a cotton crop. In our hypothesis the assumption was made that the cover crop that provides the residue and cotton were supplemented as needed with irrigation and each tillage treatment received the same amount of water annually. Further, it was postulated that the use of chemically terminated winter wheat residue would not significantly increase lint yields because less water would available for cotton growth and lint production. The main objective of this study was to examine and compare the water balance of two cotton cropping systems for a two-year period for conventionally tilled cotton, and for cotton planted into terminated wheat. Specifically, our objectives were to 1) compare the water used by wheat vs. soil water lost to evaporation from the conventionally tilled bare soil; 2) determine if wicking occurred after the wheat was terminated and before the cotton was planted; and 3) determine the effects that bare soils and residue-covered soils have on the water available for cotton growth and lint yield.

2. Materials and Methods

This study was done during two growing seasons, 1994 and 1995, at experimental fields within two research farms managed by Texas Tech University, Lubbock, TX. One farm was in central Terry County, TX near Brownfield (33°10'47"N - 102°16'15"W) where the site has an Amarillo loamy fine sand soil [fine-loamy, mixed, thermic, Torrertic Paleustalf]. The other farm was in northern Lubbock County, TX near New Deal (33°44'13.76"N, 101°43'58.04"W) with a Pullman clay loam soil [fine, mixed, thermic, Torrertic Paleustoll]. These two research farms were part of a long-term (1991-2001) study to evaluate dryland cropping systems, cotton and sorghum

(*Sorghum bicolor L.*) in the semiarid climate of the Texas High Plains. Results presented here are partially based on work by Vorheis [41] and Ralston [42].

2.1. Irrigation

Irrigation-water at each location was applied using sprinkler irrigation methods. At Brownfield, TX a center pivot, 400 m long, irrigation system provided 50, 75, and 100% of the weekly crop water use with Low Energy Precision Application (LEPA) spray nozzles [43]. The linear (300 m in length) irrigation system used at New Deal provided 40% or 80% of weekly crop water use with LEPA spray nozzles. Half the field at New Deal was irrigated at 40% ET_o and the other half irrigated at 80% ET_o.

At each location, the amount of water applied was based on calculated daily grass reference ET_o using as weather input net irradiance, air temperature and humidity, and wind speed measured at each site, and using the Penman-Monteith method as given by Allen *et al.* [44]. The daily water use was determined as ET_o multiplied by a locally developed crop coefficient (K_c) that depended on the ground cover [45]. The weekly crop water use was calculated using the appropriate ET_o and K_c for that site. Date and amount of water applications for Brownfield are given in **Table 1** and for New Deal in **Table 2**. Rainfall was measured at each site using a tipping bucket. Irrigation treatments at both sites did not begin until cotton emergence. The irrigation levels were different between the two locations because the two studies were independent of each other, with different soil textures and environmental conditions.

2.2. Experimental Design

The experimental design at both locations was a randomized block, split plot setup. The experimental field at each site was about 46 ha. The major plots were tillage effects and the irrigation treatments were the split plots. All treatments were replicated four times and the plot size of each replicate was 156 × 183 m. The irrigation treatments (*i.e.*, 50% and 100% ET_o at Brownfield, and 40% and 80% ET_o at New Deal) were started after the cotton emerged (late May through early June). Therefore, all plots for each location received identical irrigation amounts between cotton harvest from the previous season until cotton emergence the next season. One half of each irrigation treatment at each location was planted conventionally with cotton (left fallow in winter) and the other half was planted with cotton in terminated winter wheat. Locations were previously used for cotton production prior to this study with half of each field in terminated wheat residue and half of each field under conventional tillage. Treatments were replicated, n = 4, and mean separation was done using the PROC MIXED procedure and with t-type confidence interval of 0.05 using SAS on an Apple® computer (WMware and SAS version 9.2).

2.3. Cultural Practices

Winter wheat at both locations was drilled at a rate of 30 $kg \cdot ha^{-1}$ at 0.2-m row spacing into listed soil resulting in two rows of wheat in each furrow on 0.75-m centers. The tops of the beds were left bare in preparation for the planting of cotton. The wheat (TAM-200, Texas A&M University Foundation Seed, College Station, TX)[#] grew until it was tall, about 0.3 m, enough to provide wind protection, which was usually in mid- to late-April and was then terminated with the herbicide glysophate (Roundup®, Monsanto Company, St. Louis, MO) at a rate of 2 $L \cdot ha^{-1}$. Cotton (HS-26, Paymaster Technology Corp., Scott, MS) was planted into the terminated wheat and on the fallow plots during early to mid-May. A complete fertilizer blend consisting of 50-50-0-10 was applied to the entire area after each cotton harvest and prior to planting the wheat in the fall each year. Supplemental N was provided through the irrigation water at a rate of 0.2 kg N $mm^{-1} \cdot ha^{-1}$ with an irrigation volume of 25 mm for the 100% and 80% irrigation treatments through the third week of flowering. Important dates for different cultural operations at both locations are given in **Table 3**.

2.4. Measurements

Soil water contents were measured gravimetrically at various times throughout the year, usually one to two

[#]Mention of this or other proprietary products is for the convenience of the readers only and does not constitute endorsement or preferential treatment of these products by the USDA-ARS.

Table 1. Irrigation applied and rain for two irrigation treatments and two growing seasons in Brownfield, TX.

Date	Irrigation Treatment [mm]		Rain [mm]	Date	Irrigation Treatment [mm]		Rain [mm]
	50% ET_o	100% ET_o			50% ET_o	100% ET_o	
28-Apr-94			7.4	04-Jan-95			4.6
29-Apr-94	25.4	25.4		20-Jan-95			20.1
29-Apr-94			8.4	27-Jan-95			3.3
30-Apr-94			18.0	18-Feb-95			13.0
3-May-94			1.0	01-Mar-95			1.0
10-May-94			15.5	10-Mar-95			5.6
11-May-94			29.2	07-Apr-95			8.1
12-May-94			54.1	09-Apr-95	25.4	25.4	
13-May-94			32.3	15-Apr-95	25.4	25.4	
18-May-94	6.4	6.4		02-May-95			1.3
25-May-94			3.3	05-May-95			10.9
26-May-94			1.5	06-May-95			4.1
27-May-94			5.6	11-May-95			11.4
31-May-94	25.4	25.4		15-May-95			1.5
3-Jun-94	13.2	13.2		24-May-95			1.1
10-Jun-94	25.4	25.4		30-May-95			41.0
12-Jun-94			1.5	04-Jun-95			6.4
21-Jun-94	25.4	25.4		10-Jun-95			13.5
12-Jul-94	10.8	21.6		23-Jun-95			18.0
13-Jul-94			1.8	24-Jun-95			7.0
14-Jul-94			3.8	27-Jun-95			21.0
22-Jul-94	25.4	50.8		05-Jul-95	14.0	28.0	
28-Jul-94			14.5	15-Jul-95			1.3
30-Jul-94			14.5	18-Jul-95	10.0	20.3	
3-Aug-94			3.0	20-Jul-95			5.3
8-Aug-94	25.4	50.8		21-Jul-95			2.8
27-Aug-94	12.5	25.4		25-Jul-95	10.0	20.3	
1-Sep-94			11.7	28-Jul-95	6.3	12.7	
9-Sep-94			2.0	31-Jul-95			1.3
15-Sep-94			3.8	01-Aug-95			5.1
16-Sep-94			7.9	14-Aug-95			58.0
7-Oct-94			10.7	21-Aug-95			6.0
8-Oct-94			5.3	09-Sep-95			19.3
15-Oct-94			3.6	10-Sep-95			13.7
18-Oct-94			2.3	13-Sep-95			3.0
5-Nov-94			5.8	16-Sep-95			134.0
24-Nov-94			4.3	22-Sep-95			11.0
				28-Sep-95			5.0
Total [mm]	**195.3**	**269.8**	**272.8**		**91.1**	**132.1**	**458.7**

Table 2. Irrigation applied and rain for two irrigation treatments and two growing seasons, in New Deal, TX.

Date	Irrigation Treatment [mm]		Rain [mm]	Date	Irrigation Treatment [mm]		Rain [mm]
	40% ET_o	80% ET_o			40% ET_o	80% ET_o	
26-May-94			14.0	22-Jan-95			6.9
13-Jun-94	25.4	25.4		29-Jan-95			3.0
16-Jun-94			3.0	10-Feb-95	38.1	38.1	
29-Jun-94			19.0	18-Feb-95			3.3
13-Jul-94			51.0	5-Mar-95			1.5
14-Jul-94			10.0	12-Mar-95			6.9
19-Jul-94	25.4	50.8		18-Mar-95			1.3
21-Jul-94			5.0	25-Mar-95			1.3
2-Aug-94			1.0	1-Apr-95			2.3
11-Aug-94	30.8	63.5		8-Apr-95			18.3
14-Aug-94			3.6	15-Apr-95	50.8	50.8	
8-Sep-94			7.4	16-Apr-95			8.1
14-Sep-94			11.4	18-Apr-95			1.0
15-Sep-94			8.4	5-May-95			18.5
7-Oct-94			4.3	6-May-95			5.1
14-Oct-94			1.5	15-May-95			14.5
15-Oct-94			4.6	23-May-95			2.3
18-Oct-94			1.3	24-May-95			1.8
19-Oct-94			3.3	25-May-95			2.0
20-Oct-94			1.8	26-May-95			5.1
4-Nov-94			12.7	29-May-95			16.3
19-Nov-94			2.0	30-May-95			7.9
6-Dec-94			1.0	2-Jun-95			1.0
27-Dec-94			1.3	3-Jun-95			6.4
30-Dec-94			1.0	10-Jun-95			13.5
				2-Jul-95			14.5
				14-Jul-95	16.5	33.0	
				18-Jul-95			3.3
				24-Jul-95	10.1	20.3	
				26-Jul-95	10.1	20.3	
				31-Jul-95			4.8
				1-Aug-95			2.7
				2-Aug-95			1.3
				19-Aug-95	25.4	50.8	
				9-Sep-95			2.8
				10-Sep-95			13.2
				12-Sep-95			13.5
				15-Sep-95			90.0
				19-Sep-95			55.0
				29-Sep-95			100.0
Total [mm]	**81.6**	**139.7**	**168.6**		**151.0**	**213.3**	**449.4**

Table 3. Cultural operations at two sites and growing seasons.

Cultural Practice	New Deal	Brownfield
Wheat Planted	3-Dec-93	15-Dec-93
	1-Dec-94	4-Dec-94
Wheat Terminated	4-May-94	28-Apr-94
	26-Apr-95	21-Apr-95
Cotton Planted	20-May-94	5-May-94
	15-May-95	12-May-95
Cotton Harvested	6-Nov-94	7-Nov-94
	10-Nov-95	2-Nov-94

times per month [46], and these were converted to volumetric water content using appropriate bulk density values provided by Baumhardt *et al.* [47]. On each sampling date, 6 - 10 soil samples per plot were taken in 0.30-m increments to a 1.2-m depth. Water used through E_{soil} and/or T during different crop stages was based on the water balance, inputs and outputs and included precipitation, irrigation and net change in soil water content. When no crops were on the ground, all the water loss was assumed to be through E_{soil}, and if a crop did exist, the water loss was assumed to be a combination of E_{soil} and T [48]. Specifically, the following quantities were determined: 1) water required to grow an adequate amount of residue; 2) water used by the wheat after being chemically terminated; 3) water lost through E_{soil}; 4) water used by each cotton crop; and 5) the effect of each cotton cropping system on cotton lint yields. Runoff was assumed to be negligible because of slopes < 0.1% and soil water content below the 1.2-m depth was not measured because of the existence of a caliche layer around the 1.4-m depth. Cotton lint yield was estimated by hand harvesting 10 m^2 of plot area in each replication of each treatment. Cotton yield components were determined and their relative contribution to total lint yield evaluated.

2.5. Additional Experiment

An additional experiment was done in New Deal, TX in 1996 to measure evaporative losses of cotton planted in a bare soil and wheat terminated residue, without any irrigation. To measure the loss of water by evaporation, gravimetric soil samples were taken in the top 0.60 m of the soil profile in about five-day increments starting on the day the wheat was terminated by applying glysophate, 15 May 1996, and ending on the 3 June 1996, when more than 75% of the cotton planted had emerged. Gravimetric soil water content was converted to a volume basis using the measured soil bulk density and these measurements were replicated six times. The bare soil and wheat residue treatment was replicated three times and each plot was the same size as previously used, *i.e.*, 156 × 183 m.

3. Results and Discussion

3.1. Wheat Water Use

The first research objective was to measure and compare the water used by the wheat to the soil water lost to evaporation from conventionally tilled bare soil. The amount of water used to establish a wheat residue and the amount of water lost to bare E_{soil} for Brownfield and New Deal are given in **Table 4**. During the wheat growth stage, all plots received the same irrigation amount. These results showed that in 1995, the wheat used more ($P \leq 0.05$) water than the bare soil lost E_{soil} in both treatments and locations. At Brownfield, the wheat used 40% more water ($P \leq 0.05$) than the bare soil lost to water evaporation in the loamy fine sand textured soil. At New Deal, the wheat used 36% more water ($P \leq 0.05$) than the bare soil lost to evaporation in the clay loam soil. During the spring of 1995, establishing a wheat residue required more water compared to leaving the soil bare at both locations and on both soil textures.

The water depth in the upper 0.6 m of the soil profile at wheat planting and at wheat termination at both sites,

Table 4. Water balance (inputs and outputs) at two locations for the 1995-growing season.

Year	Brownfield				New Deal			
	Inputs [mm]		Outputs# [mm]		Inputs [mm]		Outputs# [mm]	
			Residue Treatment				Residue Treatment	
	Irrigation*	Rain	Bare Soil	Residue	Irrigation*	Rain	Bare Soil	Residue
1995	53.0	62.0	94.5 (17)	132.5 (16)	89.0	55.0	105 (12)	142.5 (13)

*Irrigation treatments were not active during the wheat growth; #Calculated values of outputs are the mean of all measurements and the standard deviations are given in parenthesis.

Brownfield and New Deal, TX are given in **Table 5**. Even though all the plots within each location received the same amount of irrigation during the wheat growth period, our measurements demonstrated the variability in soil water content at each location. The coefficient of variation of the measured mean soil water content was > 25% (data not shown). This variability could be explained by the irrigation treatments applied during the previous season because overall soil water content could increase over the growing season in the highest irrigation treatments compared to the lower ones.

At both locations, the bare soil treatment resulted in an increase of the soil water content in the top 0.6 m by > 30% ($P \leq 0.05$) between the time when the wheat was terminated and planted (**Table 5**). However, the wheat residue treatments showed no gain in stored water at wheat termination compared to the amount of soil water at wheat planting. In the loamy fine sand, the wheat residue caused a reduction ($P \leq 0.05$) of 14% in soil water content. At Brownfield, the bare soil gained 71% ($P \leq 0.05$) soil water and at New Deal the gain was 37% ($P \leq 0.05$) from wheat planting to wheat termination. The statistical difference ($P \leq 0.05$) between the amount of soil water at wheat termination at Brownfield was 36 mm between the bare soil and wheat residue treatments. The statistical difference ($P \leq 0.05$) between soil water at wheat termination at New Deal was 40 mm of water in the top 0.6 m of soil. Therefore, there was no gain, 36 vs. 40 mm, in stored water by the wheat residue plots over the spring of 1995 at either location, while the bare soil was able to increase the water content in the top 0.6 m of soil by more than 30% ($P \leq 0.05$) over the spring months.

3.2. Wicking

The second objective was to determine if upward movement of water through the wheat residue, *i.e.*, wicking, occurred after the wheat was terminated and before the cotton was planted. The amount of water lost to evaporation from both the bare soil and terminated wheat plots between wheat termination and cotton emergence for 1995 at Brownfield and New Deal are shown as a bar graph in **Figure 1**. These results suggested, and as expected the evaporation of water was greater in the finer textured soil when compared to the coarser textured soil. In the fine loamy sand soil, the bare soil evaporated 44% more ($P \leq 0.05$) soil water than the wheat residue. However, in the clay loam soil the wheat residue evaporated 66% more ($P \leq 0.05$) soil water than the bare soil. This result suggested that there might be a soil textural effect on the amount of water that can evaporate from a bare soil and from a terminated wheat residue [19].

It was speculated that differences in water evaporation from the bare and from the wheat-covered soil were due to the different textures and the process of wicking. The soil at New Deal is a clay loam, *i.e.*, more conducive for upward capillary flow of water; while Brownfield has a loamy fine sand, whose large pores would decrease capillary flow [13]. Under wicking, wheat residue could lose more water to evaporation than bare soil because many roots act as capillaries and the surface area provided by the leaves of the terminated wheat contribute to the evaporation. A possible solution to this problem could be to mechanically sever the roots of the wheat plant after termination at a 0.2 - 0.3 m depth and in this way disrupt the capillary movement of water from the soil to the atmosphere.

The cumulative evaporation of soil water from the bare soil and terminated wheat, between wheat termination to cotton emergence is shown in **Figure 2**. The bare soil had an 18% larger ($P \leq 0.05$) amount of water evaporation loss compared to the wheat residue. This is a result that in part may be explained by the frequent irrigation and rain events that occurred during the study. These wetting events resulted in a bare soil that was wetter than the top of the wheat residue, leading to near free-water evaporation to occur at faster rate than wicking. However,

Table 5. Amount of water in the top 0.6 m of the soil for wheat residue at planting and termination at two locations.

Year	Brownfield Residue Treatment [mm]				New Deal Residue Treatment [mm]			
	Bare Soil		Residue		Bare Soil		Residue	
	Planting	Termination	Planting	Termination	Planting	Termination	Planting	Termination
1995	47.0	80.5	51.5	44.5	104.0	142.5	102.0	103.0

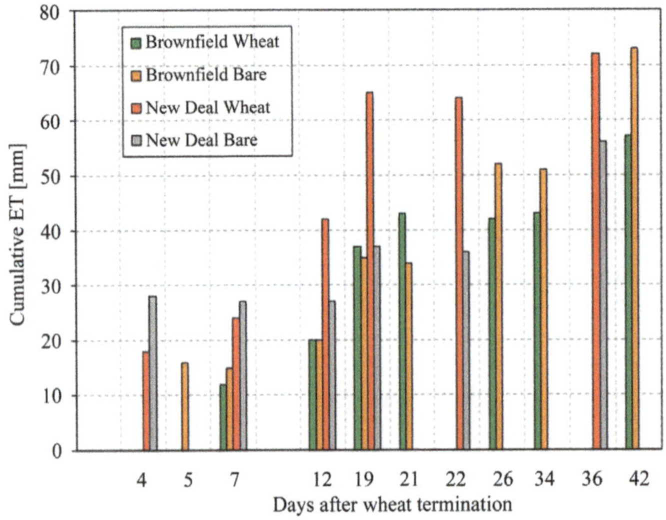

Figure 1. Cumulative measured evapotranspiration (ET, mm) as a function of days after wheat termination at Brownfield (BF) and New Deal (ND) from the wheat and bare soil residue treatment for the 1995-growing season.

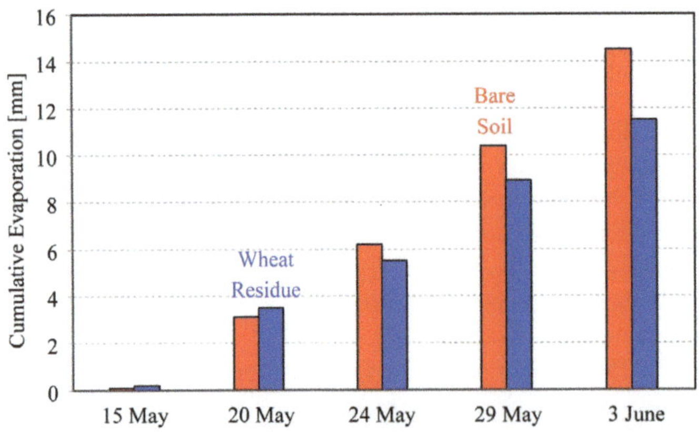

Figure 2. Cumulative measured evaporation in mm from a bare soiland wheat residue treatment for the time period between wheat termination (15 May 1996) to cotton emergence (3 June 1996) in New Deal TX.

the results from this study were not conclusive to determine the effects of wicking in a clay loam soil and the measurement of E_{soil}.

Part of the difficulty in determining differences of E_{soil} under field conditions is that the amount of water that may be lost due to wicking is of the order of 1 mm·d^{-1} and this loss takes place from mainly the surface 0.1 m layer of the soil profile. This rate of water evaporation represents a change of 0.01-m^3·m^{-3} volumetric water content on a daily basis, which is beyond the sensitivity of current instrumentation to measure volumetric soil water content [46]. Nevertheless, a wicking loss of 1 mm·d^{-1} for 20 days represents 20 mm of water, and this amount of water could determine the difference of having adequate water in the seedbed to germinate and to emerge the cotton crop planted in the residue [13].

The water used by the wheat and lost by E_{soil} affects the amount of water available for cotton growth. The amount of water content in the 1.2 m soil profile at cotton planting for both sites and growing seasons season is shown in **Table 6**. The bare soil treatments had more water at cotton planting than the wheat residue treatments at both locations except for the 40% ET_o plots at New Deal and the 50% ET_o plots at Brownfield for 1994. All the treatments in New Deal excluding the 40% ET_o irrigation treatment in 1994 showed a significant increase ($P \leq 0.05$) in soil water for the bare soil treatments compared to the wheat residue treatments, while Brownfield did not have a significant increase ($P \leq 0.05$) in soil water for the bare soil treatments at cotton planting. This stored water is important, particularly at the time when the cotton is planted as it necessary for seed germination and emergence impacting the vitality and vigor of the cotton seedling, which can later affect lint yields [49] [50].

Even though the wheat used more water in Brownfield in 1995 than the bare soil (**Table 5**), it appears that the wheat only extracted water from the top 0.6 m of soil because the water depth in the top 1.2 m of soil at cotton planting (**Table 6**) showed that the soil profile for each residue treatment contained about the same amount of water in the top 1.2 m of soil. However, at New Deal, capillary movement of water by a hydraulic gradient and by wheat roots may have caused a reduction in the amount of water in the top 1.2 m of soil for the wheat residue plots, allowing the bare soil to store more water at the time that cotton was planted (**Table 6**).

3.3. Cotton Water Use

The last and third objective of this study was to determine the effects bare and residue-covered soils have on the water available for cotton growth and development. Water inputs and water used by the cotton in both seasons and at both sites are given in **Table 7**. There were no significant differences ($P \leq 0.05$) in the amount of water used by the cotton for any of the residue treatments at Brownfield (**Table 7**). However, the 80% ET_o bare soil treatments used at least 10% more ($P \leq 0.05$) water to grow the cotton than the residue treatments for both years at New Deal (**Table 7**) and as previously discussed, could be accounted for sampling error (**Table 5**).

Cotton lint yields were used to evaluate the effects residues have on cotton growth and development. Cotton lint yields for both locations and growing seasons are given in **Table 8**. These cotton lint yield values were similar to cotton yields reported for surrounding areas [51] [52]. In 1994, both locations had significantly higher ($P \leq 0.05$) lint yields (at least 8% higher) from the conventionally tilled treatments than the wheat residue plots for all irrigation treatments (**Table 8**). The only treatment with a loss in lint yield from the conventionally tilled treatments was in the in the low irrigation treatment at Brownfield in 1995 with a 14% decrease ($P \leq 0.05$) in lint yields. The greatest gain in lint yield from the conventionally tilled treatments over the terminated wheat treatments occurred in the 40% irrigation treatment at New Deal in 1994 with a 42% gain ($P \leq 0.05$). Therefore, our results suggested that the presence of residues had a significant ($P \leq 0.05$) impact with a loss of lint yield for both years of the study. This result confirms our hypothesis that in a semiarid region the use of a small grain cover crop under irrigated conditions uses more water than it conserves compared to conventional tilled cotton, reducing cotton lint yields.

As expected and in climates where water is a limiting factor in crop production, the irrigation treatments at each location that received the greatest amount of water resulted in higher lint yields compared to the lower irrigation amounts (**Table 8**). For example, Bordovsky *et al.* [45] showed that higher irrigation amounts also gave higher yields, but also reported that at a 50% ET_o irrigation level, conventionally tilled and wheat residue cotton systems yielded about the same ($P \leq 0.05$), while at a 100% ET_o irrigation level, the conventionally tilled cotton cropping system yielded 17% less ($P \leq 0.05$) as compared to the wheat residue treatment at Lamesa, TX, which has a sandy loam soil.

Table 6. Amount of water in the top 1.2 m of the soil when cotton was planted at two locations and two years.

Year	Brownfield Residue Treatment [mm]		New Deal Residue Treatment [mm]	
	Bare Soil	Residue	Bare Soil	Residue
1994	207.0	201.5	296.0	299.0
1995	177.5	169.5	238.0	192.0

Table 7. Water balance (inputs and outputs), for two irrigation treatments at two locations and two years.

Brownfield							
Year	Water Inputs [mm]		Rain [mm]	Water Outputs [mm]			
	Irrigation Treatment			Irrigation Treatment			
	50% ET_o	100% ET_o		50% ET_o		100% ET_o	
				Residue Treatment		Residue Treatment	
				Bare Soil	Residue	Bare Soil	Residue
1994	125	200	93	276	295	395	403
1995	41	81	375	315	314	406	403

New Deal							
Year	Water Inputs [mm]		Rain [mm]	Water Outputs [mm]			
	Irrigation Treatment			Irrigation Treatment			
	40% ET_o	80% ET_o		40% ET_o		80% ET_o	
				Residue Treatment		Residue Treatment	
				Bare Soil	Residue	Bare Soil	Residue
1994	83	140	172	283	304	442	387
1995	93	130	382	541	539	620	558

Table 8. Average cotton lint yields for two locations, and two irrigation and residue treatments, for two years.

Brownfield – Cotton Lint Yield [kg·ha^{-1}]				New Deal Cotton Lint Yield [kg·ha^{-1}]			
Irrigation Treatment				Irrigation Treatment			
50% ET_o		100% ET_o		40% ET_o		80% ET_o	
Residue Treatment		Residue Treatment		Residue Treatment		Residue Treatment	
Bare Soil	Residue	Bare Soil	Residue	Bare Soil	Residue	Bare Soil	Residue
440 (20)*	392 (16)	933 (31)	816 (11)	561 (19)	394 (23)	674 (11)	599 (21)
380 (17)	444 (21)	594 (26)	546 (24)	436 (11)	401 (31)	603 (19)	559 (11)

*The standard error of the mean is in parenthesis. Mean separation was calculated using PROC MIXED with a t-type confidence interval of 0.05.

4. Conclusions

In this study, the cotton water use and lint yield of two cotton-cropping systems in the semiarid climate of the Texas High Plains was compared. These cropping systems were conventionally tilled and cotton planted into terminated wheat, on two sites and a low and high irrigation rate over two growing seasons. Our objective was to evaluate the effectiveness of planting cotton into a crop residue obtained by terminating a winter wheat crop planted in December. The purpose of the residue is mainly to provide a mechanical barrier to wind and thus protect cotton seedlings at a time when they are most vulnerable to sandblast damage due to high wind speeds. The selected sites, allowed us to compare the impact of soil texture, *i.e.*, coarse (Brownfield, TX) and fine (New Deal, TX). The residue treatments did not affect the total amount of water evaporated and transpired during the two seasons. However, at both locations the bare residue treatment had a higher lint yield than the cotton grown under the wheat residue. Wicking, the capillary movement of water through the wheat residue was only statistically significant at the site with the coarser texture (Brownfield, TX) and only in one (1994) of the two years.

These results suggest that the use of terminated wheat residue did not have a positive impact on increasing water storage for the subsequent cotton crop. Furthermore, terminated wheat residue adversely affected cotton lint yield on both years and sites and thus no advantages found in planting cotton into a terminated wheat residue in the semiarid Texas High Plains. Nevertheless, over a longer period of time, the use of a wheat residue may improve soil structure, which may lead to higher infiltration rate reducing runoff and storing more rainfall.

It was postulated that in a semiarid climate growing a small grain cover crop might use more water than it conserves when combined with a cotton crop. During the 1995 wheat growth period, the conservation tillage treatment used 20% more water than the conventional tillage treatment for all treatments at both locations supporting our hypothesis. It was further hypothesized that the use of residue would not increase cotton lint yields, and all the conventionally tilled treatments for two seasons had higher lint yields except for one plot in 1994 at Brownfield over the wheat residue plots. Therefore, for these two locations for the 1994 and 1995 cotton crop seasons, the use of terminated wheat residue did not improve the growing conditions and environment for the cotton crops compared to conventional tillage cropping systems. The significance of this work is that growers using this "conservation compliance" tillage system may not be using water optimally and may even be experiencing decreased cotton lint yields.

Acknowledgements

This work is a tribute to the many research contributions of Dr. Dan Krieg, Professor emeritus, Plant and Soil Science, Texas Tech University, to crop physiology and agronomy of dryland cropping systems. On the 18 June 2015 he lost his battle with cancer.

References

[1] USDA-NASS (2014) Texas Crop Production, USDA—National Agricultural Statistics Service, Issue No. PR-163-14, 10 December 2014. http://www.nass.usda.gov/Statistics_by_State/Texas/Publications/Current_News_Release/2014_Rls/pr16314.pdf

[2] Lascano, R.J. (2000) A General System to Measure and to Calculate Daily Crop Water Use. *Agronomy Journal*, **92**, 821-832. http://dx.doi.org/10.2134/agronj2000.925821x

[3] Colaizzi, P.D., Gowda, P.H., Marek, T.H. and Porter, D.O. (2009) Irrigation in the Texas High Plains: A Brief History and Potential Reductions in Demand. *Irrigation & Drainage*, **58**, 257-274. http://dx.doi.org/10.1002/ird.418

[4] Lascano, R.J., Van Bavel, C.H.M., Hatfield, J.L. and Upchurch, D.R. (1987) Energy and Water Balance of a Sparse Crop: Simulated and Measured Soil and Crop Evaporation. *Soil Science Society of America Journal*, **51**, 1113-1121. http://dx.doi.org/10.2136/sssaj1987.03615995005100050004x

[5] Lascano, R.J. and Baumhardt, R.L. (1996) Effects of Crop Residue on Soil and Plant Water Evaporation in a Dryland Cotton System. *Theoretical Applied Climatology*, **54**, 69-84. http://dx.doi.org/10.1007/BF00863560

[6] Krieg, D.R. (1996) Physiological Aspects of Ultra Narrow Row Cotton Production. Proceedings of the Beltwide Cotton Conference, Nashville, TN. National Cotton Council, Memphis, 9-12 January 1996, 66.

[7] Darawsheh, M.K., Khah, E.M., Aivalakis, G., Chachalis, D. and Sallaku, F. (2009) Cotton Row Spacing and Plant Density Cropping Systems. I. Effects on Accumulation and Partitioning of Dry Mass and LAI. *Journal of Food, Agriculture & Environment*, **7**, 258-261.

[8] Lascano, R.J., Baumhardt R.L., Hicks, S.K. and Heilman, J.L. (1994) Soil and Plant Water Evaporation from Strip-Tilled Cotton: Measurement and Simulation. *Agronomy Journal*, **86**, 987-994. http://dx.doi.org/10.2134/agronj1994.00021962008600060011x

[9] Musick, J.T., Pringle, F.B., Harman, W.L. and Stewart B.A. (1990) Long-Term Irrigation Trends-Texas High Plains. *Applied Engineering Agriculture*, **6**, 717-724. http://dx.doi.org/10.13031/2013.26454

[10] Lascano, R.J. and Nelson, J.R. (2014) Circular Planting to Enhance Rainfall Capture in Dryland Cropping Systems at a Landscape Scale: Measurement and Simulation. In: Ahuja, L.R., Ma, L. and Lascano, R.J., Eds., *Practical Applications of Agricultural System Models to Optimize the Use of Limited Water*, *Advances in Agricultural Systems Modeling*, ASA, CSSA, SSSA, Madison, Volume 5, 85-111. http://dx.do.org/10.2134/advagricsystmodel5.c4

[11] Bordovsky, J.P., Lyle, W.M. and Keeling, J.W. (1994) Crop Rotation and Tillage Effects on Soil Water and Cotton Yield. *Agronomy Journal*, **86**, 1-6. http://dx.doi.org/10.2134/agronj1994.00021962008600010001x

[12] Baker, J.T., McMichael, B., Burke, J.J., Gitz, D.C., Lascano, R.J. and Eprath, J.E. (2009) Sand Abrasion Injury and Biomass Partitioning in Cotton Seedlings. *Agronomy Journal*, **101**, 1297-1303. http://dx.doi.org/10.2134/agronj2009.0052

[13] Koekkoek, E.J.W., Lascano, R.J., Hicks, S.K., Krieg, D.R. and Stroosnijder, L. (1995) Loss of Water through Terminated Wheat Plants: A Wick Effect. *Proceedings of the* 1995 *Annual Meetings*, St. Louis, 29 October-3 November 1995, 15.

[14] Van Rensburg, L.D. (2010) Advances in Soil Physics: Application in Irrigation and Dryland Crop Production, South African. *Journal of Plant and Soil*, **27**, 9-18. http://dx.doi.org/10.1080/02571862.2010.10639966

[15] Unger, P.W. (1994) Residue Management for Winter Wheat and Grain Sorghum Production with Limited Irrigation. *Soil Science Society American Journal*, **58**, 537-542. http://dx.doi.org/10.2136/sssaj1994.03615995005800020041x

[16] Nielsen, D.C., Unger, P.W. and Miller, P.R. (2005) Efficient Water Use in Dryland Cropping Systems in the Great Plains. *Agronomy Journal*, **97**, 364-372. http://dx.doi.org/10.2134/agronj2005.0364

[17] AG-CARES (2004-2008) AG-CARES Annual Report. Texas A & M AgriLife Research & Extension Center, Lubbock. http://lubbock.tamu.edu/ag-cares/

[18] Helms Farm Research Reports (2001-2005) Helms Farm Annual Reports. Texas A & M AgriLife Research & Extension Center, Lubbock. http://lubbock.tamu.edu/programs/disciplines/irrigation-water/helms-farm-research-reports/

[19] Jones, O.R., Hauser, V.L. and Popham, T.W. (1994) No-Tillage Effects on Infiltration, Runoff, and Water Conservation on Dryland. *Transactions of the ASAE*, **37**, 473-479. http://dx.doi.org/10.13031/2013.28099

[20] Howell, T.A., Steiner, J.L., Schneider, A.D. and Evett, S.R. (1995) Evapotranspiration of Irrigated Winter Wheat—Southern High Plains. *Transactions of the ASAE*, **38**, 745-759. http://dx.doi.org/10.13031/2013.27888

[21] Guerif, J., Richard, G., Durr, C., Machet, J.M., Recous, S. and Roger-Estrade, J. (2001) A Review of Tillage Effects on Crop Residue Management, Seedbed Conditions and Seedling Establishment. *Soil & Tillage Research*, **61**, 13-22. http://dx.doi.org/10.1016/S0167-1987(01)00187-8

[22] Hobbs, P.R., Sayre, K. and Gupta, R. (2008) The Role of Conservation Agriculture in Sustainable Agriculture. *Philosophical Transactions of the Royal Society B: Biological Sciences*, **363**, 543-555. http://dx.doi.org/10.1098/rstb.2007.2169

[23] Alvarez, R. and Steinbach, H.S. (2009) A Review of the Effects of Tillage Systems on Some Soil Physical Properties, Water Content, Nitrate Availability and Crops Yield in the Argentine Pampas. *Soil & Tillage Research*, **104**, 1-15. http://dx.doi.org/10.1016/j.still.2009.02.005

[24] Baumhardt, R.L., Keeling, J.W. and Wendt, C.W. (1993) Tillage and Residue Effects on Infiltration into Soils Cropped to Cotton. *Agronomy Journal*, **85**, 379-383. http://dx.doi.org/10.2134/agronj1993.00021962008500020038x

[25] Dao, T.H. (1993) Tillage and Winter Wheat Residue Management Effects on Water Infiltration and Storage. *Soil Science Society of America Journal*, **57**, 1586-1595. http://dx.doi.org/10.2136/sssaj1993.03615995005700060032x

[26] Hoogmoed, W.B. and Stroosnijder, L. (1984) Crust Formation on Sandy Soils in the Sahel. I. Rainfall and Infiltration. *Soil & Tillage Research*, **4**, 5-23. http://dx.doi.org/10.1016/0167-1987(84)90013-8

[27] Lascano, R.J., Vorheis, J.T., Baumhardt, R.L. and Salisbury, D.R. (1997) Computer-Controlled Variable Intensity Rain Simulator. *Soil Science Society of America Journal*, **61**, 1182-1189. http://dx.doi.org/10.2136/sssaj1997.03615995006100040025x

[28] Alberts, E.E. and Neibling, W.H. (1994) Influence of Crop Residues on Water Erosion. In: Unger, P.W., Ed., *Managing Agricultural Residues*, Lewis Publ., Chelsea, MI, 19-39.

[29] Savabi, M.R. and Stott, D.E. (1994) Plant Residue Impact on Rain Interception. *Transactions of the ASAE*, **37**, 1093-1098. http://dx.doi.org/10.13031/2013.28180

[30] Steiner, J.L. (1994) Crop Residue Effects on Water Conservation. In: Unger, P.W., Ed., *Managing Agricultural Residues*, Lewis Publ., Chelsea, MI, 41-76.

[31] Baumhardt, R.L., Schwartz, R., Howell, T.A., Evett, S.R. and Colaizzi, P. (2013) Residue Management Effects on Water Use and Yield of Deficit Irrigated Cotton. *Agronomy Journal*, **105**, 1026-1034. http://dx.doi.org/10.2134/agronj2012.0361

[32] Lascano, R.J. and Van Bavel, C.H.M. (1986) Simulation and Measurement of Evaporation from a Bare Soil. *Soil Science Society of America Journal*, **50**, 1127-1132. http://dx.doi.org/10.2136/sssaj1986.03615995005000050007x

[33] Farahani, H.J. and Ahuja, L.R. (1996) Evapotranspiration Modeling of Partial Canopy/Residue-Covered Fields. *Transactions of the ASAE*, **39**, 2051-2064. http://dx.doi.org/10.13031/2013.27708

[34] Klocke, N.L., Currie, R.S. and Aiken, R.M. (2009) Soil Water Evaporation and Crop Residues. *Transactions of the ASABE*, **52**, 103-110. http://dx.doi.org/10.13031/2013.25951

[35] Steiner, J.L. (1989) Tillage and Surface Residue Effects on Evaporation from Soils. *Soil Science Society American Journal*, **53**, 911-916. http://dx.doi.org/10.2136/sssaj1989.03615995005300030046x

[36] Enz, J.W., Brun, L.J. and Larsen, J.K. (1988) Evaporation and Energy Balance for Bare and Stubble Covered Soil. *Agricultural and Forest Meteorology*, **43**, 59-70. http://dx.doi.org/10.1016/0168-1923(88)90006-8

[37] Horton, R., Kluitenberg, G.J. and Bristow, K.L. (1994) Surface Crop Residue Effects on the Soil Surface Energy Balance. In: Unger, P.W., Ed., *Managing Agricultural Residues*, Lewis Publ., Chelsea, MI, 143-162.

[38] Heilman, J.L., McInnes, K.J., Gesch, R.W. and Lascano, R.J. (1992) Evaporation from Ridge-Tilled Soil Covered with Herbicide-Killed Winter Wheat. *Soil Science Society American Journal*, **56**, 1278-1286. http://dx.doi.org/10.2136/sssaj1992.03615995005600040045x

[39] Fryrear, D.W. and Bilbro, J.D. (1994) Wind Erosion Control with Residues and Related Practices. In: Unger, P.W., Ed., *Managing Agricultural Residues*, Lewis Publ., Chelsea, MI, 7-17.

[40] Evett, S.R. and Lascano, R.J. (1993) ENWATBAL.BAS: A Mechanistic Evapotranspiration Model Written in Compiled Basic. *Agronomy Journal*, **85**, 763-772. http://dx.doi.org/10.2134/agronj1993.00021962008500030044x

[41] Vorheis, J.T. (1997) Water Balance of Cotton Cropping Systems. Master's Thesis, Plant and Soil Science Department, Texas Tech University, Lubbock, 38 p.

[42] Ralston, J.T. (1997) Management Strategies for Dryland Cotton Production in West Texas. Master's Thesis, Plant and Soil Science Department, Texas Tech University, Lubbock, 47 p.

[43] Lyle, W.M. and Bordovsky, J.P. (1981) Low Energy Precision Application (LEPA) Irrigation System. *Transactions of the ASAE*, **24**, 1241-1245. http://dx.doi.org/10.13031/2013.34427

[44] Allen, R.G., Walter, I.A., Elliot, R., Howell, T.A., Itenfisu, D. and Jensen, M.E. (2005) The ASCE Standardized Reference Evapotranspiration Equation. ASCE-EWRI Task Committee Report, January 2005, 70 p. http://www.kimberly.uidaho.edu/water/asceewri/ascestzdetmain2005.pdf

[45] Bordovsky, J.P., Lyle, W.M., Lascano, R.J. and Upchurch, D.R. (1992) Cotton Irrigation Management with LEPA Systems. *Transactions of the ASAE*, **35**, 879-884. http://dx.doi.org/10.13031/2013.28673

[46] Evett, S.R. (2007) Soil Water and Monitoring Technology. In: Lascano, R.J. and Sojka, R.E., Eds., *Irrigation of Agricultural Crops*, 2nd Edition, ASA, CSSA, SSSA, Madison, 25-84.

[47] Baumhardt, R.L., Lascano, R.J. and Krieg, D.R. (1995) Physical and Hydraulic Properties of a Pullman and Amarillo Soil on the Texas South Plains. Technical Report D.R. No. 95-1, Texas A & M University Agricultural Research and Extension Center, Lubbock/Halfway.

[48] Bertuzzi, P., Bruckler, L., Bay, D. and Chanzy, A. (1994) Sampling Strategies for Soil Water Content to Estimate evapotranspiration. *Irrigation Science*, **14**, 105-115. http://dx.doi.org/10.1007/BF00193132

[49] Peng, S., Krieg, D.R. and Hicks, S.K. (1989) Cotton Lint Yield Response to Accumulated Heat Units and Soil Water Supply. *Field Crop Research*, **19**, 253-262. http://dx.doi.org/10.1016/0378-4290(89)90097-X

[50] Morrow, M.R. and Krieg, D.R. (1990) Cotton Management Strategies for a Short-Growing Season Environment: Water-Nitrogen Considerations. *Agronomy Journal*, **92**, 52-56. http://dx.doi.org/10.2134/agronj1990.00021962008200010011x

[51] USDA-NASS (1994) Agricultural Statistics-1994. http://www.nass.usda.gov/Publications/Ag_Statistics/agr4all.pdf

[52] USDA-NASS (1995) Agricultural Statistics-1995. Chapter II—Statistics of Cotton, Tobacco, Sugar Crops, and Honey. http://www.nass.usda.gov/Publications/Ag_Statistics/1995-1996/agr95_2.pdf

26

Chamber Size Effects on Methane Emissions from Rice Production

Alden D. Smartt[1], Kristofor R. Brye[1], Christopher W. Rogers[2], Richard J. Norman[1], Edward E. Gbur[3], Jarrod T. Hardke[4], Trent L. Roberts[1]

[1]Department of Crop, Soil, and Environmental Sciences, University of Arkansas, Fayetteville, USA
[2]Department of Plant, Soil, and Entomological Sciences, University of Idaho, Aberdeen, USA
[3]Agricultural Statistics Laboratory, University of Arkansas, Fayetteville, USA
[4]Department of Crop, Soil, and Environmental Sciences, University of Arkansas, Stuttgart, USA
Email: kbrye@uark.edu

Abstract

Quantifying methane (CH_4) emissions from cultivated rice (*Oryza sativa* L.) production in the field has received increased attention recently due to methane's importance as a greenhouse gas. The enclosed-headspace chamber technique is the standard methodology for field assessments of trace gas emissions. However, to our knowledge, no direct comparisons of measured CH_4 fluxes and emissions from field-grown rice among differing chamber sizes have been reported. Therefore, the objective of this study was to evaluate the effect of chamber size [15.2- and 30.4-cm inside diameter (id)] on CH_4 fluxes and season-long emissions from rice grown on a clay soil in Arkansas. Chamber size did not affect ($P > 0.05$) CH_4 fluxes on 10 sampling dates during the flooded portion of the rice growing season and only affected fluxes on one of four sampling dates after flood release. Total season-long CH_4 emissions from optimally N-fertilized rice were 32.6 and 35.6 kg CH_4-C $ha^{-1} \cdot season^{-1}$, which did not differ, and from bare clay soil were 0.74 and 1.75 kg CH_4-C $ha^{-1} \cdot season^{-1}$, which also did not differ, from the 15.2- and 30.4-cm chambers, respectively. Chamber size (*i.e.*, 15.2- or 30.4-cm id) did not result in differences in cumulative CH_4 emissions from this flooded-rice study that was conducted on a Sharkey clay soil in northeast Arkansas. Results indicate that both 15.2- and 30.4-cm diameter chambers were similarly adequate for measuring CH_4 fluxes and emissions from the clay soil investigated. The similarity in emissions results between chamber sizes also indicates that the 15.2-cm diameter chambers adequately facilitated the quantification of CH_4 emissions in this study.

Keywords

Methane Emissions, Rice Production, Clay Soil, Arkansas

1. Introduction

Rice (*Oryza sativa* L.) is one of the main staple food crops for much of the world's population, with direct human consumption accounting for 85% of rice production compared to 72% and 19% for wheat (*Triticum aestivum* L.) and maize (*Zea mays* L.), respectively [1]. However, globally, agriculture accounts for 47% of total anthropogenic methane (CH_4) emissions, with 64% and 22% of agricultural CH_4 emissions resulting from enteric fermentation and rice cultivation, respectively [2] [3]. Agricultural activities account for 33% of CH_4 emissions in the United States (US), of which enteric fermentation is responsible for approximately 70% and rice cultivation is responsible for 3.7% [4]. Methane emissions from cultivated agricultural activities are significant since CH_4 is a greenhouse gas and has a global warming potential that is 25 times greater than that of CO_2 [5].

For field assessments of trace gas emissions, the enclosed-headspace chamber technique is the standard methodology [6] [7]. Livingston and Hutchinson [6] suggested that chamber areas typically range from 175 cm^2 [*i.e.*, 15.2-cm inside diameter (id)] to 1 m^2, and that chamber areas of 500 to 900 cm^2 (*i.e.*, 25- to 34-cm id, respectively) are most common. The 15.2-cm id has also been suggested as the minimum chamber size to use in assessing trace gas emissions [7]. Thus, there is clearly no uniform chamber size associated with the enclosed-headspace chamber technique.

Recently, several field studies have used multiple chamber diameters to quantify CH_4 emissions. Denier van der Gon *et al.* [8] used 20-cm diameters chambers to assess CH_4 emissions from a clay soil in a pot study at the International Rice Research Institute in the Philippines. Rogers *et al.* [9] used 15.2-cm diameter chambers in a field study on a silt-loam soil in central Arkansas. Rogers *et al.* [10] used 30.4-cm diameter chambers in a field study on a similar silt-loam soil in central Arkansas. Simmonds *et al.* [11] used 30-cm diameter chambers on a silt-loam soil in central Arkansas and a silty-clay in California.

For a variety reasons, there is potential for variation in chamber diameters used in the field for quantifying trace gas emissions. Resource limitations may require smaller chambers to be constructed. Logistics of the crop to be studied and field installations may necessitate the use of smaller-diameter chambers. In particular, if studies are interested in differentiating between in-row and between-row emissions, then chamber diameter may be limited by plant spacing [9]. Statistical considerations may dictate larger chambers as larger chambers tend to mask spatial variability by integrating measurements over a larger soil area [6] [7]. Smaller chambers may be desirable in studies examining environmental gradients [6] [7].

Though there are a number of reasons for potential chamber-diameter variations, to our knowledge, no direct comparisons of measured CH_4 fluxes and emissions from field-grown rice among differing chamber sizes have been reported. Therefore, the objective of this study was to evaluate the effect of chamber size (15.2- and 30.4-cm id) on CH_4 fluxes and season-long emissions from rice grown on a clay soil in Arkansas. It was hypothesized that CH_4 fluxes would be more variable from the 15.2-cm chamber due to the smaller footprint, thus differences in fluxes between chamber sizes would likely not be measured, but that season-long emissions would be greater from the 30.4-cm chamber due to the greater potential for microsite inclusion in the larger area.

2. Materials and Methods

2.1. Site Description

A field research study was conducted throughout the 2012 rice growing season (*i.e.*, April to September) at the University of Arkansas System Division of Agriculture Northeast Research and Extension Center in Keiser, Arkansas (35˚40'N 90˚05'W). The soil throughout the study area was mapped as a Sharkey clay (very-fine, smectitic, thermic Chromic Epiaquerts) [12], which was derived from alluvial sediments in the lower Mississippi River flood plain, and has been in a rice-soybean [*Glycine max* (L.) Merr.] rotation for >25 years. The mean annual air temperature throughout the region is 15.5˚C and the mean annual precipitation is 126 cm [13].

2.2. Treatments and Experimental Design

Small (15.2-cm id) and large (30.4-cm id) chambers were installed in replicated plots (n = 4 per treatment combination) containing optimally nitrogen (N)-fertilized rice and no rice (*i.e.*, bare soil). A randomized completed block (RCB) split-plot design was used, with the whole-plot factor being vegetation (*i.e.*, rice or no rice) and the split-plot factor being chamber size (*i.e.*, 15.2- or 30.4-cm). Sample date (*i.e.*, time) was also treated as a repeated measure since flux measurements were made periodically throughout the growing season.

2.3. Field and Plot Management

Crop residues were incorporated by disking to a depth of ~15 cm followed by land planing the previous fall to prepare the seed bed. Plots (1.6 m wide by 5.0 m long) were seeded at an 18-cm row spacing with the standard stature, pure line, long-grain rice cultivar 'Taggart' at a rate of 112 kg·ha^{-1}, which resulted in nine rice rows per plot. This cultivar was used due to its frequent use throughout the region at the time and its high yield potential [14]. Based on 18-cm row spacing and the surface area enclosed by each chamber, 10 and 40 cm of total row length of plants were included in the small (15.2-cm id) and large chambers (30.4-cm id), respectively, in order to duplicate the plant density of the field plots in both chamber sizes.

After planting, levees were constructed to create a single bay of rice when the permanent flood was established at the 4 - 6 leaf stage. Nitrogen, in the form of urea (46% N), was hand applied to individual vegetated plots only at 150 kg N ha^{-1} one day prior to flood establishment [15]. A typical mid-season split application of 50 kg N ha^{-1} as urea was made at the beginning of internode elongation to the vegetated plots only. A 5 to 10 cm flood depth was maintained throughout the growing season. **Table 1** summarizes the dates of major agronomic activities that occurred in the field.

2.4. Methane Sampling and Analysis

A methodology using enclosed-headspace gas sampling chambers [6] [7] was used in this study. Similar field methods have been used previously in Arkansas [9] [10].

Small (15.2-cm id) and large (30.4-cm id) base collars (30 cm tall), extensions (40 and 60 cm long), and caps (10 cm tall) were constructed out of Schedule 40 polyvinyl chloride (PVC) pipe (**Figure 1**). Base collars were beveled to a 45° angle on the bottom to facilitate installation into the soil and had four 12.5-mm holes drilled just above the soil surface to allow water movement into and out of the base collar. After flood release, the holes were plugged with gray butyl-rubber septa (Voigt Global, part # 73828A-RB, Lawrence, KS) during sampling. To accommodate the growth of the rice plants over the course of the growing season, chamber extensions of varying lengths were used to adjust the chamber height. To minimize temperature fluctuations inside the chambers, chamber extensions and caps were covered with reflective aluminum tape (CS Hyde, Mylar metalized tape, Lake Villa, IL). Chamber extensions were secured to base collars using sections of tire inner tube.

Chamber caps had a 5-mm thick flat sheet of PVC glued to the top with a 15-cm section of 4.5-mm id copper tubing installed in the side of each cap as a vent. Gray butyl-rubber septa were installed into 12.5-mm holes drilled in the top of each cap to serve as a sampling port and a port for measuring the temperature inside the chamber. A 2.5-cm diameter, 9-V battery-powered fan was also installed on the underside of each cap to mix the headspace inside the chamber. Sections of tire inner tube were also used to secure the caps to the extensions. To minimize soil disturbance after the flood was established, wood planking was installed on concrete blocks for access to all chambers in the field plots. The base collars were installed one week prior to sampling, just after initiating flood establishment.

Table 1. Summary of dates for major agronomic field activities involved in the management of methane emissions plots for the 2012 rice growing season at the University of Arkansas System Division of Agriculture Northeast Research and Extension Center in Keiser, Arkansas.

Field Activity	Date
Planting	2 April
Emergence	30 April
Pre-flood Fertilizer Application	7 June
Flood Establishment	8 June
Mid-season Fertilizer Application	10 July
Flood Release	23 August
Harvest	11 September

(a) (b) (c) (d) (e) (f)

Figure 1. Selected images of the different chamber sizes in the field. Both chamber sizes win a plot with no rice plants (a). Both chamber sizes in a plot with rice plants (b). Syringe sampling the 30.4-cm diameter chamber through the cap (c). Syringe sampling the 15.2-cm diameter chamber through the cap (d). Both chambers in multiple field plots (e). Both chambers in multiple field plots while measuring the flood depth (f).

Methane fluxes were measured weekly between flood establishment and flood release (*i.e.*, 10 times total) and every other day after flood release (*i.e.*, four times total). Approximately 16 hours prior to sampling, rice plants were carefully bundled together with tie wire prior to installing the chamber extensions. Once the chamber extensions had been set in place, the tie wire was removed to minimize plant disturbance. The following morning, gas sampling occurred between 0800 and 1000 hours. Gas samples were collected from the chamber headspace at 0, 20, 40, and 60 minutes after capping the chamber. A 20-mL B-D syringe (Beckton Dickson and Co., Franklin Lakes, NJ) was inserted into each chamber through the gray butyl-rubber septa at the top of the cap (**Figure**

1). After collection of a 20-mL gas sample from the mixed chamber headspace, samples were transferred into pre-evacuated 10-mL, crimp-top glass vials (Agilent Technologies, part # 5182-0838, Santa Clara, CA). Methane gas standards were also collected in the field following chamber sampling to evaluate potential leakage prior to actual analysis.

Within 48 hours of collection, all gas samples were analyzed on an Agilent 6890-N gas chromatograph (Agilent Technologies) equipped with a flame-ionization detector and a 30-m-long HP-Plot-Q capillary column (Agilent Technologies). Methane concentrations in $\mu L \cdot L^{-1}$ were determined by peak-area response from the resulting chromatograms. As described by Parkin and Venterea [7] and according to the Ideal Gas Law, fluxes in mg CH_4 $m^{-2} \cdot hr^{-1}$ were calculated by linear regression using the four concentration measurements per chamber and the chamber volume, temperature, and atmospheric pressure measured during sampling. The resulting r^2 values from the linear regressions were always greater than 0.9. Linear interpolation between measurement dates was used to calculate season-long CH_4 emissions from each chamber.

2.5. Plant Sampling

At physiological maturity, all aboveground dry matter (DM) from inside the chambers of both sizes was collected from the vegetated plots. Similarly, plants were cut at the soil surface and collected from one, 1-m row length in each vegetated plot to assess aboveground DM in the bulk plot. All plant samples were dried at 60˚C for 96 hr and weighed. About three weeks after flood release, panicles from three, 1-m row lengths of rice were hand-harvested from each vegetated plot, dried at 60˚C for 96 hr, manually threshed, and weighed for yield determinations. Yields are reported at a grain moisture content of approximately 120 $g \cdot kg^{-1}$.

2.6. Statistical Analyses

The effects of vegetation, chamber size, and their interaction on CH_4 fluxes throughout the growing season were evaluated by analysis of variance (ANOVA) in SAS (version 9.2, SAS Institute, Inc., Cary, NC) using PROC Mixed based on a RCB split-plot repeated-measures design, where the whole-plot factor was vegetation, the split-plot factor was chamber size, and time was treated as a repeated measure. Due to differences in mechanisms of CH_4 release and sampling intervals, flux data were analyzed separately from flooding to flood release and after flood release (*i.e.*, post-flood release). A separate ANOVA was conducted to evaluate the effects of vegetation, chamber size, and their interaction on season-long, area-scaled emissions, post-flood-release emissions, and post-flood-release emissions expressed as a percentage of season-long emissions. In addition, aboveground dry matter was analyzed by ANOVA based on a RCB design with three sampling locations (*i.e.*, inside the 15.2-cm chambers, inside the 30.4-cm chambers, and in the bulk plot). Fisher's protected least significant difference, at the $\alpha = 0.05$ level, was used when appropriate for means separations.

3. Results and Discussion

3.1. Methane Fluxes

Similar to that hypothesized, chamber size did not affect ($P > 0.05$) measured CH_4 fluxes between flooding and flood release (**Table 2**). However, CH_4 fluxes differed ($P < 0.001$) between vegetation treatments over time (**Table 2**). In the first few weeks after flood establishment, averaged across chamber size, CH_4 fluxes were low and near zero in both treatments as the soil oxidation-reduction potential was decreasing due to the persistence of the flood. This result is similar to previous observations from a silt-loam soil [9] [10]. In contrast, after about three weeks from flood establishment, averaged across chamber size, the optimally N-fertilized rice consistently maintained greater CH_4 fluxes throughout the remainder of the growing season until flood release. This result is also similar to previous observations from a silt-loam soil [9] [10].

Similar to during the flooded period of the rice growing season, after flood release, CH_4 fluxes differed ($P = 0.007$) between vegetation treatments over time (**Table 2**). Averaged across chamber size, a significant post-flood-release CH_4 pulse (*i.e.*, increased flux) occurred from the bare soil at 5 days after flooding (DAFR), while no pulse occurred from the vegetated treatment (**Table 2**). However, in contrast to during the flooded period and similar to that hypothesized, CH_4 fluxes also differed ($P = 0.007$) between chamber sizes over time (**Table 2**). Averaged across vegetation treatments, CH_4 fluxes from the 30.4-cm chambers were greater than those from the 15.2-cm chambers at 5 DAFR, while fluxes on the other three post-flood-release sample dates were unaffected

Table 2. Analysis of variance summary of the effects of rice vegetation, chamber size, time, and their interactions on methane fluxes from flooding to flood release and following flood release from a clay soil during the 2012 growing season at the Northeast Research and Extension Center in Keiser, Arkansas.

Source of Variation	Measurement Period	
	Flooding to Flood Release	Post-Flood Release
	P	
Vegetation	<0.001	0.897
Chamber Size	0.788	0.320
Vegetation × Chamber Size	0.795	0.147
Time	<0.001	0.001
Vegetation × Time	<0.001	0.007
Chamber size × Time	0.773	0.007
Vegetation × Chamber Size × Time	0.768	0.183

by chamber size. The chamber-size effect may have been due to differential soil drying that occurred within the base collars of different dimensions. The post-flood-release pulse measured in this study at 5 DAFR was similar to that reported previously from a silt-loam soil [9] [10]. Despite a chamber-size effect on one sample date after flood release, all post-flood-release fluxes were small, <2.75 mg CH_4-C $m^{-2}\cdot hr^{-1}$. Furthermore, post-flood-release emissions have been reported to only account for approximately 10% of the total season-long emissions from a silt-loam soil in eastern Arkansas [9].

Though chamber size did not affect measured fluxes prior to flood release, it is possible that the differing dimensions could have affected the gradient inside the chambers over the 1-hr measurement cycle. If the gradient was affected such that the concentrations leveled off after some time, then the linear regression approach to calculate the chamber flux using the four concentrations measurements may be erroneous. However, in contrast to Rogers *et al.* [16] where non-linear concentrations were measured at times from a silt-loam soil, none of the measurements from either chamber size in this study exhibited any indication of a non-linear CH_4 concentration increase over time (*i.e.*, increasing CH_4 concentrations within the chambers did not substantially affect the molecular diffusion gradient). This result indicated that both 15.2- and 30.4-cm diameter chambers appear suitable for measuring CH_4 fluxes from flooded rice grown on a clay soil with relatively low resulting season-long CH_4 emissions.

3.2. Seasonal Methane Emissions

In contrast to that hypothesized, chamber size did not affect ($P > 0.05$) season-long area-scaled emissions, post-flood-release emissions, or the post-flood-release emissions as a proportion of total emissions, while vegetation affected both season-long, area-scaled emissions ($P < 0.001$) and post-flood-release emissions as a proportion of total emissions ($P = 0.005$; **Table 3**). Total season-long, area-scaled CH_4 emissions from the optimally N-fertilized rice were 32.6 and 35.6 kg CH_4-C $ha^{-1}\cdot season^{-1}$ from the 15.2- and 30.4-cm chambers, respectively, which did not differ and averaged 34.1 kg CH_4-C $ha^{-1}\cdot season^{-1}$ (**Table 4**). Furthermore, chamber size did not impact ($P = 0.951$) aboveground DM as 2.71 kg DM m^{-2} accumulated inside the 15.2-cm chambers compared to 2.77 kg DM m^{-2} inside the 30.4-cm chambers, while the bulk plot accumulated 2.83 kg DM m^{-2}. Total season-long, area-scaled CH_4 emissions from the bare soil were 0.74 and 1.75 kg CH_4-C $ha^{-1}\cdot season^{-1}$ from the 15.2- and 30.4-cm chambers, respectively, which did not differ, and averaged 1.25 kg CH_4-C $ha^{-1}\cdot season^{-1}$, which was significantly lower than that from the optimally N-fertilized rice (**Table 4**).

The differences in emissions observed between the bare soil and optimally N-fertilized rice were consistent with results previously reported from a silt-loam soil in Arkansas [9] [10]. The similarity in emissions results between chamber sizes indicates that the 15.2-cm diameter chambers adequately facilitated the quantification of CH_4 emissions in this study. The post-flood-release emissions as a proportion of total emissions were significantly greater from the bare soil (51%) than from the optimally N-fertilized rice (2%; **Table 4**). This result sup-

Table 3. Summary of the effects of rice vegetation, chamber size, and their interaction on methane (CH_4) emissions from a clay soil during the 2012 growing season at the Northeast Research and Extension Center in Keiser, Arkansas.

Emissions Property	Vegetation	Chamber Size	Vegetation × Chamber Size
		P	
Total Area-Scaled Emissions (kg CH_4-C ha^{-1}·$season^{-1}$)	< 0.001	0.737	0.869
Post-Flood-Release Emissions (kg CH_4-C ha^{-1})	0.723	0.249	0.164
Post-Flood Emissions (% of Total Emissions)	0.005	0.096	0.083

Table 4. Seasonal methane (CH_4) emissions for unvegetated and high-vegetation rice treatments measured with 15.2- and 30.4-cm inside-diameter chambers from a clay soil during the 2012 growing season at the Northeast Research and Extension Center in Keiser, Arkansas.

Property/Chamber Size	Bare Soil	Vegetated	Mean
	Area-Scaled Emissions (kg CH_4-C ha^{-1}·$season^{-1}$)		
15.2-cm	0.74	32.6	16.7 A[‡]
30.4-cm	1.75	35.6	18.7 A
Mean	1.25 b[†]	34.1 a	
	Post-Flood Emissions (kg CH_4-C ha^{-1})		
15.2-cm	0.32	0.73	0.53 A[‡]
30.4-cm	1.27	0.62	0.95 A
Mean	0.80 a[†]	0.68 a	
	Post-Flood Emissions (% of Total Emissions)		
15.2-cm	38.6	2.3	20.5 A[‡]
30.4-cm	63.5	1.6	32.6 A
Mean	51.0 a[†]	2.0 b	

[†]Different lower-case letters within a row for a measured property indicate differences among vegetation treatments ($P < 0.05$); [‡]Different upper-case letters within a column for a measured property indicate differences between chamber-size treatments ($P < 0.05$).

ports previous reports that the presence of the rice plants greatly facilitate passive transport of CH_4 from the soil to the atmosphere throughout the flooded portion of the rice growing season [9] [10]. In contrast, CH_4 builds up in the soil where rice plants are not present (*i.e.*, the bare-soil treatment in this study) until the overlying floodwater is removed, thus allowing for rapid escape of the entrapped CH_4 and the post-flood-release pulse of CH_4 observed in this and previously in other studies.

Regardless of chamber size, the magnitude of season-long CH_4 emissions measured in this study from a clay soil in northeast Arkansas is low compared to similar studies conducted on silt-loam soils in Arkansas [9] [11]. Furthermore, the average season-long CH_4 emissions measured in this study from optimally N-fertilized rice is substantially lower than the US Environmental Protection Agency's reported emissions factor of 178 kg CH_4-C ha^{-1}·$season^{-1}$ for non-California-grown, primary-crop rice production [4].

4. Conclusion

Chamber size did not affect CH_4 fluxes during the flooded portion of the rice growing season and only affected fluxes on one of four samples dates after flood release. Results indicate that both 15.2- and 30.4-cm diameter chambers were similarly adequate for measuring CH_4 fluxes and emissions from the clay soil investigated in this study. Therefore, chamber size does not appear to be an important feature of CH_4 emissions studies that would compromise results from relatively homogeneous study areas and/or from soils that have relatively low season-long CH_4 emissions. If material supplies and/or costs are limiting factors or other logistical constraints require diameter-limited chambers, such as for in-row and between-row comparisons, the smaller, 15.2-cm-diameter chamber size may be the more logical choice. When material supplies and/or costs or other constraints are not limiting factors, then the larger, 30.4-cm diameter chambers would likely be preferred.

Acknowledgements

This research was supported by the Arkansas Rice Research and Promotion Board.

References

[1] Maclean, J.L., Dawe, D.C., Hardy, B. and Hettel, G.P., Eds. (2002) Rice Almanac: Source Book for the Most Important Economic Activity on Earth. 3rd Edition, CABI Publishing, Wallingford, UK.

[2] Smith, P., Martino, D., Cai, Z., Gwary, D., Janzen, H., Kumar, P., McCarl, B., Ogle, S., O'Mara, F., Rice, C., Scholes, B. and Sirotenko, O. (2007) Agriculture. In: Solomon, S., Qin, D., Manning, M., Chen, Z., Marquis, M., Averyt, K.B., Tignor, M. and Miller, H.L., Eds., *Climate Change* 2007: *The Physical Science Basis. Contribution of Working Group I to the Fourth Assessment Report of the Intergovernmental Panel on Climate Change*, Cambridge University Press, Cambridge, United Kingdom and New York, USA.

[3] United States Environmental Protection Agency (USEPA) (2012) Global Anthropogenic Non-CO_2 Greenhouse Gas Emissions: 1990-2030.
http://www.epa.gov/climatechange/Downloads/EPAactivities/EPA_Global_NonCO2_Projections_Dec2012.pdf

[4] United States Environmental Protection Agency (USEPA) (2014) Inventory of US Greenhouse Gas Emissions and Sinks: 1990-2012.
http://www.epa.gov/climatechange/Downloads/ghgemissions/US-GHG-Inventory-2014-Main-Text.pdf

[5] Forster, P., Ramaswamy, V., Artaxo, P., Berntsen, T., Betts, R., Fahey, D.W., Haywood, J., Lean, J., Lowe, D.C., Myhre, G., Nganga, J., Prinn, R., Raga, G., Schulz, M. and Van Dorland, R. (2007) Changes in Atmospheric Constituents and in Radiative Forcing. In: Solomon, S., Qin, D., Manning, M., Chen, Z., Marquis, M., Averyt, K.B., Tignor, M. and Miller, H.L., Eds, *Climate Change* 2007: *The Physical Science Basis. Contribution of Working Group I to the Fourth Assessment Report of the Intergovernmental Panel on Climate Change*, Cambridge University Press, Cambridge, United Kingdom and New York, USA.

[6] Livingston, G. and Hutchinson, G. (1995) Enclosure-Based Measurement of Trace Gas Exchange: Applications and Sources of Error. In: Matson, P.A. and Harris, R.C., Eds., *Biogenic Trace Gases*: *Measuring Emissions from Soil and Water*, Blackwell Sciences Ltd., Osney Mead, Oxford, 14-51.

[7] Parkin, T. and Venterea, R. (2010) Chamber-based Trace Gas Flux Measurements. In: Follett, R., Ed., *Sampling Protocols*, United States Department of Agriculture, Agricultural Research Service, Fort Collins.
http://www.ars.usda.gov/SP2UserFiles/Program/212/Chapter%203.%20GRACEnet%20Trace%20Gas%20Sampling%20Protocols.pdf

[8] Denier van der Gon, H.A.C., Kropff, M.J., van Breemen, N., Wassmann, R., Lantin, R.S., Aduna, E., Corton, T.M. and van Laar, H.H. (2002) Optimizing Grain Yields Reduces CH_4 Emissions from Rice Paddy Fields. *Proceedings of the National Academy of Sciences*, **99**, 12021-12024. http://dx.doi.org/10.1073/pnas.192276599

[9] Rogers, C.W., Brye, K.R., Norman, R.J., Gbur, E.E., Mattice, J.D., Parkin, T.B. and Roberts, T.L. (2013) Methane Emissions from Drill-Seeded, Delayed-Flood Rice Production on a Silt-Loam Soil in Arkansas. *Journal of Environmental Quality*, **42**, 1059-1069. http://dx.doi.org/10.2134/jeq2012.0502

[10] Rogers, C.W., Brye, K.R., Smartt, A.D., Norman, R.J., Gbur, E.E. and Evans-White, M.A. (2014) Cultivar and Previous Crop Effects on Methane Emissions from Drill-Seeded, Delayed-Flood Rice Production on a Silt-Loam Soil. *Soil Science*, **179**, 28-36. http://dx.doi.org/10.1097/SS.0000000000000039

[11] Simmonds, M.B., Anders, M., Adviento-Borbe, M.A, van Kessel, C., McClung, A. and Linquist, B.A. (2015) Seasonal Methane and Nitrous Exide Emissions of Several Rice Cultivars in Direct-Seeded Systems. *Journal of Environmental Quality*, **44**, 103-114. http://dx.doi.org/10.2134/jeq2014.07.0286

[12] Soil Survey Staff, Natural Resources Conservation Service, United States Department of Agriculture (2015) Web Soil Survey. http://websoilsurvey.nrcs.usda.gov/app/WebSoilSurvey.aspx

[13] National Oceanic and Atmospheric Administration (NOAA) (2002) Climatography of the United States No. 81: Monthly Station Normals of Temperature, Precipitation, and Heating and Cooling Degree Days 1971-2000.
http://www.ncdc.noaa.gov/climatenormals/clim81/ARnorm.pdf

[14] Moldenhauer, K.A.K., Gibbons, J.W., Lee, F.N., Bernhardt, J.L., Wilson Jr., C.E., Cartwright, R.D., Norman, R.J., Blocker, M.M., Ahrent, D.K., Boyett, V.A., Bullock, J.M. and Castaneda, E. (2008) Taggart, High Yielding Large Kernel Long-grain Rice Variety. In: Norman, R.J. and Moldenhauer, K.A.K., Eds., *B.R. Wells Rice Research Studies*,

2008, *Arkansas Agricultural Experiment Station Research Series*, **571**, Division of Agriculture, The University of Arkansas System, Fayetteville, 68-73.

[15] Hardke, J.T., Ed. (2013) Arkansas Rice Production Handbook. Misc. Publ. 192. University of Arkansas, Cooperative Extension Service, Little Rock.

[16] Rogers, C.W., Brye, K.R., Norman, R.J., Gasnier, T., Frizzell, D. and Branson, J. (2012) Methane Emissions from a Silt-Loam Soil Under Direct-Seeded, Delayed-Flood Rice Management. In: Norman, R.J. and Moldenhauer, K.A.K., Eds., *B.R. Wells Rice Research Studies*, 2011, *Arkansas Agricultural Experiment Station Research Series*, **600**, Division of Agriculture, The University of Arkansas System, Fayetteville, 240-247.

27

Comparison and Evaluation of Two Analytical Methods for Cation Exchange Capacity and Exchangeable Sodium Percentage of Five Soil Types in Central Sudan

Jamal Elfaki[1], Mohamed Gafer[2], Magboul Sulieman[3*], Mushtaha Ali[4]

[1]Department of seeds technology, Nile Valley University, Atbara, Sudan
[2]College of Agricultural Studies, Sudan University of Science and Technology, Shambat, Sudan
[3]Department of Soil and Environment Sciences, University of Khartoum, Shambat, Sudan
[4]General Administration of Natural Resource and Sustainable Development, Khartoum, Sudan
Email: *magboul@uofk.edu

Abstract

Twenty-six soil samples were collected from five soil profiles at different climatological and ecological regions in central Sudan. Soil profile was dug in each studied area and morphological profile description was carried out for different horizons. All samples were analyzed using two different methods to determine Cation Exchange Capacity (CEC) and exchangeable sodium percentage (ESP). Statistical analysis (T-test) was used in order to investigate the differences between soil samples for the studied locations. Significant differences appeared when compared the two methods for CEC determination at Gedaref area, Wad Medani and Nile flood plain and that appeared in evaluation of ESP at Nile flood plain and Shambat area. The results also revealed that, the developed method used in this study was more practical, simple and reliable for determination of CEC and ESP as the currently used in most soil laboratories. In addition, it will be safer than the other methods in some problematic soils. The adoption of this developed method is advisable because it is less time consuming as it omits the washing step. In contrast, the old method cannot be a good substitute in laboratories which have no possibility to determine sodium by using flame photometer. We conclude that when the developed method is used to determine CEC and ESP time will be saved, that fewer amounts of chemicals will be used and that accurate results will be achieved.

*Corresponding author.

Keywords

Ecological Regions, Central Sudan, Cation Exchange Capacity, Exchangeable Sodium Percentage

1. Introduction

The major soils in Sudan can be divided geographically into three categories: the sandy soils of the northern and west central regions, the clay soils of the central and eastern regions, and the laterite soils of the southern regions.

Most agricultural and environmental planning requires soil analysis, or at least should require analysis for better implementation for any change. Furthermore, better practical analysis methods can rapidly estimate soil properties needed to improve quantitative assessments of land management problems [1].

The basic problem of this research is that the soil laboratories in the Sudan, do not enter the new modern means of analyzing soil, and most of the methods used are very old. Studies have shown the lack of modern efficiency to meet required needs. There are new methods of analysis, with less expenses and time-consuming, and they have not been used in the Sudan.

CEC is important for maintaining adequate quantities of plant available calcium (Ca^{2+}), magnesium (Mg^{2+}) and potassium (K^{+}) in soils. Under acid conditions (pH < 5.5), aluminum (Al^{3+}) may also be present as an exchangeable cation. While a soil with a higher CEC may not necessarily be more fertile, when combined with other measures of soil fertility, CEC is a good indicator of soil quality and productivity.

The Cation Exchange Capacity is an important property of clay minerals. CEC results are frequently used for characterization and quantification of sorbents in clays and soils. Determination of the CEC and exchangeable cations of soils and clays have been performed since the early work [2].

Numerous publications about various methods reflect the limitation of analytical validity of result obtained by CEC procedures for the wide variety of natural materials. However, common CEC methods like ammonium acetate [3] or barium chloride [4], are time consuming, and results for natural materials are often poor. Recent methodological approaches use the higher selectivity of metal-organic complexes, e.g. silver-thiourea method as a one-step procedure compared to common methods to reduce the time factor [5].

Methods have been proposed for calcareous and gypsiferous soils, [4] [6]-[12]. They are not generally applicable to soils containing both $CaCO_3$ and gypsum or they are too cumbersome and demanding for routine determinations.

In general, analytical problems in CEC methods developed through specific interactions between components of soils and clays with the exchange solutions used. The main sources of errors are soluble (Ca^{2+}) phases and in the faster modern methods, hydrophobic interaction can cause unrealistic CEC values. The objective of this research was to compare and evaluate the results of the [6] and [13] methods for determination of Cation Exchange Capacity (CEC) and Exchangeable Sodium Percentage (ESP) of five soil types in central Sudan.

2. Materials and Methods

2.1. Soil Sampling and Characterization

Twenty-six soil samples were collected from different five soil profiles were dug in central Sudan based on different geographical, climatological and ecological regions, which included; Gedaref area (14°10'64"N 35°38'26"E), Soba area (15°52'66"N 32°60'79"E), Wad Madani area (14°39'19"N 33°49'30"E), College farm (15°64'74"N 32°51'76"E), and recent Nile terrace at Khartoum North area (15°65'16"N 32°51'25"E).

Each soil profile was studied in the field, and described following the format of the FAO [14], guidelines of soil profile description and sampled according to genetic horizons and classified on the bases of its diagnostic characteristic used at different categories levels of the American system for soil classification [15].

Each sample was kept in a cloth bag, labeled with; collected data, area, soil profile number, sample depth, then, subjected to physical and chemical analyses at the soil laboratories in Khartoum University. Soil pH was determined on the saturated paste and the electrical conductivity of the saturation extraction was used as a measure of soil salinity. The organic matter (OM) was determined used Walkley-Black method [16]. Texture classes were determined using [17] method. P was analyzed using colorimetric method by spectrophotometer [18].

Flame photometer was used in sodium measurement, after adjusting the flame photometer by the standard solution (100 ppm Na) at 100 reading and distilled water at Zero reading. Extractable Na was obtained by using a known volume of 1 N ammonium acetate and the exchangeable contents of these elements were obtained by difference between the extractable and soluble quantities. While exchangeable sodium percentage (ESP) was calculated, using results of two CEC methods.

$$\text{ESP} = \frac{\text{Exch. Na}}{\text{CEC}} \times 100$$

where: Exch. Na = Exchangeable sodium in meq/100g.
CEC = Cation Exchange Capacity meq/100g.

2.2. Determination of Cation Exchange Capacity (CEC)

According to [6] method, 4 g of each soil sample was placed in centrifuge tube, 33 ml of IM sodium acetate was added, and pH adjusted to 8.2. The soil suspension was shaken for 5 minutes and then centrifuged until the supernatant liquid was clear. The supernatant was then discarded completely. The sample was extracted in this manner a total of tour times. Excess salt of sodium acetate washed four times by adding 33 ml 95% ethanol to the tube. The adsorbed sodium replaced by three extractions with 33 ml 1 M ammonium acetate, shacked, centrifuged and supernatant liquid collected in 100 ml volumetric flask. Finally, Sodium concentration was measured used flame photometer.

According to [13] method, 4 g of each soil sample from different profiles was saturated with sodium by four successive equilibrations with 33ml aliquots of a 60% ethanol solution. During each equilibration the soil suspension was shaken for 5 minutes, and then centrifuged for 5 minutes (2000 rpm/min) until the supernatant was clear after which the supernatant liquid was discarded. The Na-saturated samples plus the occluded saturating solution were extracted three times with 33 ml aliquots of 0.5 M pH 7 solution of $Mg(NO_3)_2$. The exchangeable Na was calculated as the total Na minus the occluded soluble Na that was 5 times the occluded Cl.

2.3. Statistical Analysis

Statistical differences between samples were determined using statistical analysis [19], using T-test with multiple samples where differences were calculated from various measurements. The means of these differences were obtained ($\bar{D}$), the deviation from each measurement was used to get the standard deviation (s_d). Then, the T value was calculated from the equation below:

$$T = \frac{\bar{D}}{s_d}\sqrt{N}$$

where: $T \equiv$ Calculated T value.
$\bar{D} \equiv$ Means of differences.
$s_d \equiv$ Standard deviation.
$N \equiv$ Number of samples.

3. Results and Discussion

3.1. Morphological Properties

The description of the study sites and selected morphological properties of representative soil profiles are presented in (**Table 1** and **Table 2**), respectively. The parent material of Profiles I and II was alluvium/colluvium and old alluvium of the Bule Nile, respectively. While the parent material of profiles III, V, and VI were alluvium. Soil texture of all profiles belong to five textural classes; loam, sandy clay loam, clay loam silty clay and clay. All profiles showed angular/sub-angular blocky structure in the surface horizon and the lower horizons were massive. The quantity of roots in the soil profiles decreased with depth and the boundary between horizons was generally diffused and smooth.

Table 1. Selected site properties of the studied profiles.

Profile No	Location	Coordinates		Parent material	Slope (%)	Land use
		Latitude	Longitude			
P1	Gedaref	14°10'64"	35°38'26"	Alluvium/Colluvium	Flat	Agric.
P2	Soba	15°52'66"	32°60'79"	Old alluvium of the Blue Nile	Flat	Forest
P3	Wad Madani	14°39'19"	33°49'30"	Alluvium	Flat	Agric.
P4	Recent Nile terrace	15°65'16"	32°51'25"	Alluvium	Flat	Agric
P5	College Farm	15°64'74"	32°51'76"	Alluvium	Flat	Agric.

Table 2. Selected morphological properties of the representative profiles.

Profile No.	Depth (cm)	Color (moist)	Texture[a] (field)	Structure[b]	Roots[c]	Boundary[d]	Diagnostic characteristics
P1	0 - 18	2.5Y 3/2	C	2 fabk	1f	cw	Cracks up to 4 cm
	18 - 48	2.5Y 3/2	C	1 fabk	1f	cw	Cracks at the base of the horizon
	48 - 80	2.5Y 3/2	C	1 fsbk	1f	cw	Cracks up to 3 cm
	80 - 105	10YR 6/3	C	ma	1f	cw	Slicken sides not clearly visible
	105 - 150	10YR 6/3	C	2 csbk	1f	cw	-
P2	0 - 30	7.5YR 4/4	L	ma	2f	cs	-
	30 - 45	10YR 4/4	Cl	1 msbk	2f	is	-
	45 - 107	10YR 4/3	C	2 abk	2f	ds	-
	107 - 150	10YR 5/6	C	2 sbk	-	ds	Slightly cemented
P3	0 - 27	2.5YR 3/2	Cl	2 fabk	1f	cw	Cracks up to 4 cm
	27 - 56	2.5YR 3/2	C	1 fabk	2f	cw	Cracks at the base of horizon
	56 - 84	2.5YR 3/2	C	1 cabk	1f	iw	Cracks up to 2cm
	84 - 130	10YR 6/3	C	ma	1f	cw	-
	130 - 150	10YR 6/3	C	2 csbk	2f	cw	-
P4	0 - 12	10YR 3/3	C	3 fg	4f	cs	-
	12 - 48	10YR 3/3	Scl	1 csbk	4vf	cw	-
	48 - 68	10YR 3/2	Scl	2 abk	3vf	cw	Common krotovina
	68 - 86	10YR 3/3	Scl	ma	3vf	aw	-
	86 - 111	10YR 3/2	Scl	ma	3vf	cw	-
	111 - 150	10YR 3/3	Scl	ma	3vf	cw	-
P5	0 - 12	10YR 3/3	C	2 sbk	3f	cs	Few cracks
	12 - 36	10YR 4/3	Sic	1 msbk	1vf	cs	Few soft $CaCO_3$ aggregates
	36 - 57	10YR 3/3	C	1 mabk	1vf	cs	Soft lime aggregates, common termites
	57 - 83	10YR 3/2	C	1 abk	1f	cs	Soft $CaCO_3$, common krotovina
	83 - 111	10YR 3/2	C	ma	-	ws	Hard $CaCO_3$ concretion and $CaSO_4$
	111 - 150	10YR 3/2	C	ma	-	ws	Hard $CaCO_3$s concretion and $CaSO_4$

Texture[a]; C: clay; Scl: Sandy clay loam; Sic: silty clay; Cl: clay loamy; L: loam. Structure[b]; 1: weak; 2: moderate; 3: strong; f: fine; m: medium; c: coarse; sbk: subangular blocky; abk, angular blocky; ma: massive. Roots[c]; 1: very few; 2: few; 3; moderate; 4: common; f: fine; m: medium; c: coarse. Boundary[d]; a: abrupt; c: clear; d: diffuse; i: irregular; s: smooth; w: wavy.

3.2. Physical and Chemical Properties

Some of the physical and chemical properties of the representative profiles were presented in (**Table 3**). Gadarif area soil was non-saline, non-sodic and calcareous, while Soba area soil at Khartoum state was saline-Sodic and calcareous. The soil of Agricultural Research Corporation Farm, Gezira State is non-saline at the depth of 0 - 84

Table 3. Some physical and chemical properties of the representative profiles.

Profile No	Depth cm	Clay %	Silt %	Sand %	Texture Class	pH Paste	E C dSm^{-1}	P ppm	O.M %	CEC Cmol (+)/kg	ESP	$CaCO_3$ %
Gedaref Area	0 - 18	65.9	12	22.1	Clay	7.9	0.75	4.7	0.073	58.5	1.67	9.59
	18 - 48	63.1	11.5	25.4	Clay	7.97	0.7	5.7	0.056	56.1	8.05	8.82
	48 - 80	63.1	11.2	25.7	Clay	7.85	2.13	5.8	0.036	56.2	2.66	8.42
	80 - 105	62.5	11	26.5	Clay	7.8	2.9	3.8	0.033	55.9	6.26	7.47
	105 - 150	65.1	10.8	24.1	Clay	7.9	1.75	4.6	0.031	57.8	10.17	13.46
Khartoum Area	0 - 30	39.5	13.6	47	Sandy Clay	8.17	3.2	5.7	0.057	35.1	32.23	7.46
	30 - 45	49.1	12.8	38.1	Clay	8.27	22	5	0.038	43.7	35.05	6.42
	45 - 107	57.2	12.6	30.2	Clay	8.17	17	4.3	0.036	49.2	34.55	7.88
	107 - 150	57.2	12.1	30.7	Clay	8.4	8.5	5.6	0.034	49.3	38.42	8.31
Gezira State Area	0 - 27	49.8	13.7	36.5	Clay	8.15	1.45	4.8	0.055	44.1	15.19	8.13
	27 - 56	51.5	13.5	35	Clay	8.3	1	6.3	0.043	45.5	20.1	7.93
	56 - 84	58.2	13.1	28.7	Clay	8.35	3.3	4.8	0.041	50.1	25.84	9.32
	84 - 130	63.3	12.9	23.8	Clay	8.11	6	4.7	0.036	55.2	24.76	5.47
	130 - 150	63.3	12.5	24.2	Clay	8.12	8.7	5.6	0.034	55.1	34.52	5.19
Khartoum North Area	0 - 12	54.4	14.6	31	Clay	7.51	1.8	7.6	0.076	47.8	3.49	4.41
	12-48	54.4	14.1	31.5	Clay	7.8	0.7	5.5	0.097	47.6	3.09	4.31
	48 - 68	26.3	14.1	59.6	Sandy Clay	8.12	0.4	4.7	0.095	23.1	3.2	5.2
	68 - 86	26.3	14	59.7	Sandy Clay Loam	8.14	0.35	5.7	0.088	22.9	3.1	4.51
	86 - 111	26.3	14	59.7	Sandy Clay Loam	8.1	0.35	5	0.069	23.1	3.03	3.99
	111 - 150	26.3	14	59.7	Sandy Clay Loam	8.1	0.3	5.6	0.069	23.4	3.03	4.12
College of Agricultural Studies Farm	0 - 12	51.1	11.9	37	Clay	8.1	1.4	4.1	0.074	6.21	45.8	29.96
	12-36	51.5	11.7	36.8	Clay	8.17	1.85	4.9	0.067	6.46	45.5	20.29
	36 - 57	51.2	11.5	37.3	Clay	7.97	3.4	5	0.059	7.05	45.3	16.69
	57 - 83	52.1	11.2	36.7	Clay	8.17	3	5.1	0.06	6.3	45.6	20.17
	83 - 111	54.4	11.1	34.5	Clay	8.07	4.9	5.1	0.053	5.89	45.1	28.76
	111 - 150	50.4	11.4	38.2	Clay	7.94	6.1	5	0.052	5.97	45.2	34.58

cm, slightly saline at the bottom depth 84 - 150 cm, sodic and slightly calcareous. The Nile flood plain soil was non-saline, non-calcareous and non-sodic; it was very suitable for agriculture. The soil of Shambat area—College of Agricultural Studies Farm is non-saline at the top surface 0 - 83cm, and slightly saline at the bottom 83 - 150 cm, sodic and slightly calcareous soil.

3.3. Determination of Cation Exchange Capacity by (Bower *et al.*, 1952) and (Mario and Rhoades, 1977) Methods

The only method used for determination of Cation Exchange Capacity (CEC) in Sudan is [6], which is not replaced or changed although errors can result during measurement of CEC. Errors observed during saturation step with adsorb cation, in the washing step of the excess cation with alcohol, and in the decantation step. [13] mentioned that; during saturation step in Bower *et al.*, (1952) the adsorb Na^+ could not be complete saturated in the soil, because of the competence of other cation at the exchange sites, e.g. Ca^+ which is abundant in calcareous soils. (The study areas of Gedaref, Gezira contained high amount of $CaCO_3$), these areas the most important agricultural areas in Sudan, so accurate results for CEC should be obtained.

During washing step, adsorb Na^+ may be lost by hydrolyses and replaced by other cations (e.g. Ca^+ and gypsum) [10] [13] [20]. Probably, the significant differences at Gedaref and Wad Medani areas when used two methods are due to the above-mentioned reasons.

[13] also mentioned that the loss of organic matter from the soil at the washing step using alcohol, or at the decantation step where some clay minerals could lost, which have a significant role in CEC, especially two layers clay minerals. In this study, the soil samples are poor in organic matter, so what have been mentioned of these lost not affect in the results obtained. Two layers clay minerals (e.g. montmorillonitic, Na^+ ions may trapped between the two layers could not be replaced or washed by alcohol, leads to wrong CEC reading). Soil rich in zeolits and feldspars, and vermiculites assists Na^+ to be trapped at the center extraction [13]. This can affect the CEC readings and spontaneously affect exchangeable sodium percentage (ESP).

[13] method alleviate many errors that could occur in applying Bower method, plies its simplicity and more or less accuracy in calcareous and gypsiferous soils. In addition, [13] omitted the washing step to remove excess sodium. **Figure 1** showed the differences in CEC and ESP of the two methods.

Significant differences appeared in Gedaref area, Wad Medani and Nile flood plain when used these two methods for determination CEC (**Table 4**). Significant differences also appeared in ESP at Nile flood plain and Shambat area (**Table 5**), that may be due to; effect of several washing solution lead to significant dissolution of O.M, which lowered the CEC systematically, agreed by [21]. The permanently added Ca^{2+} ions compete successfully with NH_4^+ during the washing step lead to underestimation of CEC [13]. High layers charge density allows slow cation exchange when NH_4^+ was used.

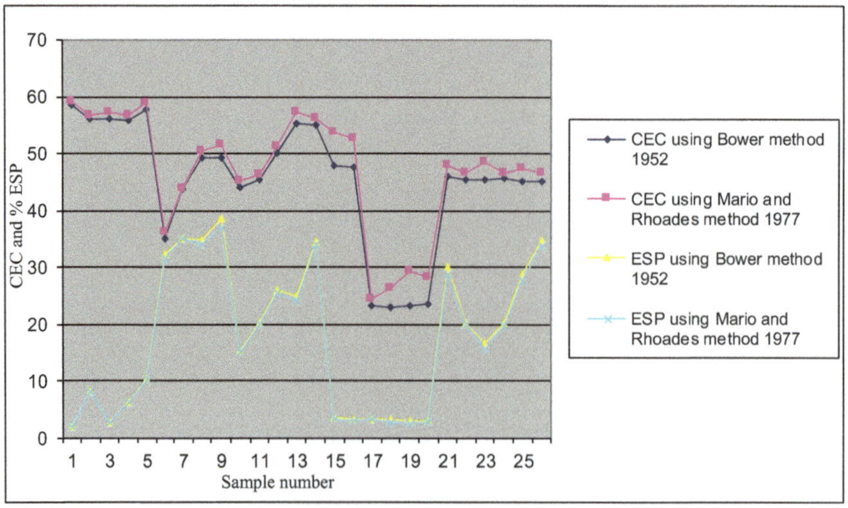

Figure 1. The difference between results by using (Bower *et al.*, 1952) and (Mario and Rhoades, 1977) methods for determination of CEC and ESP.

Table 4. Statistical comparison between CEC results using (Bower *et al.*, 1952) and (Mario and Rhoades, 1977) methods.

Location	Tabulated T-value	Calculated T-value
Gedaref Area	5.598	7.100*
Khartoum Area	7.453	2.730$^{N.S}$
Gezira State Area	5.598	5.813*
Khartoum North Area	4.773	5.244*
College of Agricultural Studies Farm	4.773	4.730$^{N.S}$

$^{N.S}$Non significant. *Significant at ($P < 0.05$).

Table 5. Statistical comparison between ESP results using (Bower *et al.*, 1952) and (Mario and Rhoades, 1977) methods.

Location	Tabulated T-value	Calculated T-value
Gedaref Area	5.598	3.011$^{N.S}$
Khartoum Area	7.453	2.798$^{N.S}$
Gezira State Area	5.598	5.565$^{N.S}$
Khartoum North Area	4.773	5.3245*
College of Agricultural Studies Farm	4.773	5.1412*

$^{N.S}$Non significant. *Significant at ($P < 0.05$).

Yaalon *et al.*, (1962) reported that, the use of NH_4^+ salts should be avoided because of the possibility of NH_4 fixation and release of fixed potassium resulting in underestimation of CEC.

It can be stated that, the newly developed method of Mario and Rhoades used in this study is more practical, simple and reliable for determination of CEC as the currently used Bower method, but will be safer than Bower method in some problematic soils. The adoption of Mario and Rhoades method is advisable because it is less time consuming as it omitted the washing step. It cannot be a good substitute in laboratories which no possibility to determination sodium by flame photometer.

4. Conclusion

After compared and evaluated of Bower *et al.*, (1952) and Mario and Rhoades (1977) methods for determination of Cation Exchange Capacity (CEC) and Exchangeable Sodium Percentage (ESP), we conclude that when Mario and Rhoades (1977) method is used to determine CEC and ESP time will be saved, that less amount of chemicals will be used in laboratories and that accurate results will be achieved.

Acknowledgements

Special thanks to the staff of the laboratories of the Land and Water Research Centre at the Agricultural Research Corporation, Wad Medani, and the Staff of Soil Science and Water Department at College of Agricultural Studies, Sudan University of Science and Technology, for availing their laboratories facilities and for their technical advice.

References

[1] Shepherd, K.D. and Walsh. M.G. (2002) Development of Reflectance Spectral Libraries for Characterization of Soil Properties. *Soil Science Society of America Journal*, **66**, 988-998. http://dx.doi.org/10.2136/sssaj2002.9880

[2] Way, J.T. (1852) On the Power of Soils to Absorb Manure. *Journal of the Royal Agricultural Society of England*, **13**, 123-143.

[3] Lewis, D.R. (1949) Analytical Data on Reference Clay Materials. Sect. 3, Base-Exchange Data, Reference Clay Minerals. A.P.I. Research Project 49, Preliminary Report No. 7, Columbia University, New York, 91.

[4] Mehlich, A. (1942) Rapid Estimation of Base-Exchange Properties of Soils. *Soil Science*, **53**, 1-14. http://dx.doi.org/10.1097/00010694-194201000-00001

[5] Chhabra, R., Pleysier, J. and Cremers, A. (1975) The Measurement of the Cation Exchange Capacity and Exchangeable Cations in Soils: A New Method. Proceedings of the International Clay Conference, 16-23 July 1975, Mexico City, 439-449.

[6] Bower, C.A., Reitemeier, R.F. and Fireman, M. (1952) Exchangeable Cation Analysis of Saline and Alkali Soils. *Soil Science*, **73**, 251-261. http://dx.doi.org/10.1097/00010694-195204000-00001

[7] Yaalon, D.H., Schuylenborgh, J. and Slager, S. (1962) The Determination of Cation Exchange Characteristics of Saline and Calcareous Soils. *Netherlands Journal of Agricultural Science*, **113**, 194-199.

[8] Bascomb, C.L. (1964) Rapid Method for the Determination of Cation Exchange Capacity of Calcareous and Non-Calcareous Soil. *Journal of the Science of Food and Agriculture*, **15**, 821-823. http://dx.doi.org/10.1002/jsfa.2740151201

[9] Carpena, O., Law, A. and Vahtras, K. (1972) Determination of Exchangeable Cations in Calcareous Soils. *Soil Science*, **113**, 194-199. http://dx.doi.org/10.1097/00010694-197203000-00008

[10] Tucker, B.M. (1974) Laboratory Procedures for Cation Exchange Measurement on Soils. Technical Paper No. 23, CSIRO Division of Soils, Brisbane.

[11] Garman, M. and Hesse, P.R. (1975) Cation Exchange Capacity of Gypsic Soils. *Plant and Soil*, **42**, 477-480. http://dx.doi.org/10.1007/BF00010022

[12] Papanicolaou, E.P. (1976) Determination of Cation Exchange Capacity of Calcareous Soils and Their Percent Base Saturation. *Soil Science*, **121**, 65-71. http://dx.doi.org/10.1097/00010694-197602000-00001

[13] Mario, P. and Rhoades, J.D. (1977) Determining Cation Exchange Capacity: A New Procedure for Calcareous and Gypsiferous Soils. *Soil Science Society of America Journal*, **41**, 524-528. http://dx.doi.org/10.2136/sssaj1977.03615995004100030018x

[14] FAO (2006) Report on the Ad Hoc Expert Consultation on Land Evaluation. Rome, Italy, 6-8.

[15] Soil Survey Staff (2014) Key to Soil Taxonomy. A Basic System of Soil Classification for Making and Interpreting Soil Surveys. 2nd Edition, Agriculture Hand Book No. 436, USDA, Washington DC.

[16] Chapman, H.D. and Partt, P.F. (1961) Methods of Analysis of Soils, Plants and Waters. University or California. Berkeley.

[17] Day, P.R. (1965) Particle Fractionation and Particle Size Analysis. Methods of Soil Analysis. Part 1. Agronomy, 9, 545-566. American Society of Agronomy, Madison.

[18] Olsen, S.R. and Sommers, L.E. (1982) Phosphorus. In: Page, A.L. and Miller, R.H., Eds., *Methods of Soil Analysis*, Part 2, 2nd Edition, Agronomy Monograph 9, *ASA and SSSA*, Madison, 403-430.

[19] Snedecor, G.W. (1965) Statistical Methods. 5. Aufl. The Iowa State University Press, Ames.

[20] Okazaki, R., Smith, H.W. and Moodie, C.D. (1962) Development of a Cation Exchange Capacity Procedure with Few Inherent Errors. *Soil Science*, **93**, 343-349. http://dx.doi.org/10.1097/00010694-196205000-00009

[21] Frink, C.R. (1964) The Effects of Wash Solvents on Cation Exchange Capacity Measurements. *Soil Science Society of America Journal*, **28**, 506-511. http://dx.doi.org/10.2136/sssaj1964.03615995002800040018x

Permissions

All chapters in this book were first published in OJSS, by Scientific Research Publishing; hereby published with permission under the Creative Commons Attribution License or equivalent. Every chapter published in this book has been scrutinized by our experts. Their significance has been extensively debated. The topics covered herein carry significant findings which will fuel the growth of the discipline. They may even be implemented as practical applications or may be referred to as a beginning point for another development.

The contributors of this book come from diverse backgrounds, making this book a truly international effort. This book will bring forth new frontiers with its revolutionizing research information and detailed analysis of the nascent developments around the world.

We would like to thank all the contributing authors for lending their expertise to make the book truly unique. They have played a crucial role in the development of this book. Without their invaluable contributions this book wouldn't have been possible. They have made vital efforts to compile up to date information on the varied aspects of this subject to make this book a valuable addition to the collection of many professionals and students.

This book was conceptualized with the vision of imparting up-to-date information and advanced data in this field. To ensure the same, a matchless editorial board was set up. Every individual on the board went through rigorous rounds of assessment to prove their worth. After which they invested a large part of their time researching and compiling the most relevant data for our readers.

The editorial board has been involved in producing this book since its inception. They have spent rigorous hours researching and exploring the diverse topics which have resulted in the successful publishing of this book. They have passed on their knowledge of decades through this book. To expedite this challenging task, the publisher supported the team at every step. A small team of assistant editors was also appointed to further simplify the editing procedure and attain best results for the readers.

Apart from the editorial board, the designing team has also invested a significant amount of their time in understanding the subject and creating the most relevant covers. They scrutinized every image to scout for the most suitable representation of the subject and create an appropriate cover for the book.

The publishing team has been an ardent support to the editorial, designing and production team. Their endless efforts to recruit the best for this project, has resulted in the accomplishment of this book. They are a veteran in the field of academics and their pool of knowledge is as vast as their experience in printing. Their expertise and guidance has proved useful at every step. Their uncompromising quality standards have made this book an exceptional effort. Their encouragement from time to time has been an inspiration for everyone.

The publisher and the editorial board hope that this book will prove to be a valuable piece of knowledge for researchers, students, practitioners and scholars across the globe.

List of Contributors

Katy E. Brantley, Kristofor R. Brye, Mary C. Savin and David E. Longer
Department of Crop, Soil, and Environmental Sciences, University of Arkansas, Fayetteville, USA

Takamitsu Kai, Masaki Mukai, Kiwako S. Araki, Dinesh Adhikari and Motoki Kubo
Department of Biotechnology, Faculty of Life Sciences, Ritsumeikan University, Biwako-Kusatsu Campus, Kusatsu, Shiga 525-8577, Japan

Richard L. Haney, R. Daren Harmel, Jeffrey G. Arnold and Michael J. White
United States Department of Agriculture-Agriculture Research Service, Grassland, Soil and Water Research Laboratory, Temple, TX, USA

Virginia L. Jin
United States Department of Agriculture-Agriculture Research Service, Agroecosystem Management Research Laboratory, University of Nebraska, East Campus, Lincoln, NE, USA

Mari-Vaughn V. Johnson
United States Department of Agriculture-Natural Resource Conservation Service, Temple, TX, USA

Elizabeth B. Haney
Texas A&M University, Texas AgriLife Research & Extension Center, Temple, TX, USA

Valdinar Ferreira Melo and Sandra Cátia Pereira Uchôa
Department of Soil and Agricultural Engineering, Federal University of Roraima, Boa Vista, Brazil

Zachary N. Senwo
Department of Biological & Environmental Sciences, Alabama A&M University, Huntsville, USA

Ronilson José Pedroso Amorim
Agronomy, Federal University of Roraima, Boa Vista, Brazil

Huixia Chai, Sheng Rao, Qi Huang and Xuejie Mou
Chinese Academy for Environmental Planning, MEP, Beijing, China

Reibo Wang
2State Key Laboratory of Resources and Environmental Information System, Institute of Geographical Sciences and Natural Resources Research, CAS, Beijing, China

Jing Liu
Department of Geography, University of Wisconsin-Madison, Madison, USA

Bente Foereid
Bioforsk—Norwegian Institute for Agricultural and Environmental Research, Ås, Norway

Jesper K. V. Nielsen
Carrington Research Station, North Dakota State University, Carrington, ND, USA

Howard J. Woodard
Plant Science Department, South Dakota State University, Brookings, SD, USA

Manbir K. Rakkar, David W. Franzen and Amitava Chatterjee
Soil Science Department, North Dakota State University, Fargo, USA

Revocatus Petro, Mohammed Mijai and Geofrey Nyaradani
Tanzania Forestry Research Institute, Moshi, Tanzania

Francis Laswai and Chelestino Balama
Tanzania Forestry Research Institute, Morogoro, Tanzania

Alayna Jacobs, Kristofor R. Brye, Lisa S. Wood and Larry C. Purcell
Department of Crop, Soil, and Environmental Sciences, University of Arkansas, Fayetteville, USA

Randy King
USDA-NRCS Plant Materials Center, Booneville, USA

Joel Douglas
USDA-NRCS Central National Technology Support Center, Fort Worth, USA

Michael Looper
Department of Animal Science, University of Arkansas, Fayetteville, USA

Deborah S. Page-Dumroese and Joanne M. Tirocke
USDA Forest Service, Rocky Mountain Research Station, Moscow, USA

Matt D. Busse
USDA Forest Service, Pacific Southwest Research Station, Davis, USA

Steven T. Overby
USDA Forest Service, Rocky Mountain Research Station, Flagstaff, USA

Brian D. Gardner
USDA Natural Resource Conservation Service, Moscow, USA

Zofia Fischer
Department of Biotechnology and Environment, John Paul II, Catholic University of Lublin, Lublin, Poland

Pavel Blažka
Department of Science, University of South Bohemia, České Budějovice, Czech

Sutapa Paul, Prasanta Kumar Patra, Niladri Paul and Debatosh Mukherjee
Department of Agricultural Chemistry and Soil Science, Bidhan Chandra Krishi Viswavidyalaya, Mohanpur, India

Arunabha Chakravarty and Premasis Sukul
Department of Agricultural Bio-Chemistry, Bidhan Chandra Krishi Viswavidyalaya, Mohanpur, India

Qingxiang Deng and Defang Zeng
School of Resource and Environmental Engineering, Wuhan University of Technology, Wuhan, China

Serge-Étienne Parent and Léon E. Parent
Department of Soils and Agrifood Engineering, Université Laval, Québec, Canada

Md. Abul Kashem, Ashoka Sarker, Imam Hossain and Md. Shoffikul Islam
Department of Soil Science, University of Chittagong, Chittagong, Bangladesh

Shahin Sultana and Abul Kalam Mohammad Moniruzzaman Mollah
Asian University for Women, Chittagong, Bangladesh

Md. Abul Kashem
Department of Soil Science, Chittagong University, Chittagong, Bangladesh
Asian University for Women, Chittagong, Bangladesh

Magboul M. Sulieman, Ibrahim S. Ibrahim and Mohamed S. Dafa-Allah
Department of Soil and Environment Sciences, Faculty of Agriculture, University of Khartoum, Khartoum, Sudan

Jamal T. Elfaki
Department of Seeds Technology, Nile Valley University, Khartoum, Sudan

Ryan Norman, Kristofor R. Brye and Pengyin Chen
Department of Crop, Soil, and Environmental Sciences, University of Arkansas, Fayetteville, USA

Edward E. Gbur
Agricultural Statistics Laboratory, University of Arkansas, Fayetteville, USA

John Rupe
Department of Plant Pathology, University of Arkansas, Fayetteville, USA

Dominique Proust
LIENSs UMR 7266 CNRS, La Rochelle, France

Ramesh P. Tripathi, Isaac Kafil and Woldeselassie Ogbazghi
Department of Land Resources and Environment, Hamelmalo Agricultural College, Keren, Eritrea

Richard L. Haney, Douglas R. Smith and Michael J. White
Grassland, Soil and Water Research Laboratory, Agricultural Research Service, United States Department of Agriculture (USDA-ARS), Temple, TX, USA

Elizabeth B. Haney
Texas AgriLife Research & Extension Center, Texas A & M University, Temple, TX, USA

Xinhua Yin
Department of Plant Sciences, The University of Tennessee, Jackson, TN, USA

Olayinka Olakunle Olatunji, Yetunde Oyeyiola and Gbade Oyewunmi Oyediran
Crop Production and Soil Science Department, Ladoke Akintola University of Technology, Ogbomoso, Nigeria

Robert J. Lascano
Wind Erosion and Water Conservation, USDA-ARS*, Lubbock, TX, USA

Dan R. Krieg
Department of Plant and Soil Science, Texas Tech University, Lubbock, TX, USA

Jeffrey T. Baker
Wind Erosion and Water Conservation, USDA-ARS, Big Spring, TX, USA

Alden D. Smartt, Kristofor R. Brye, Richard J. Norman and Trent L. Roberts
Department of Crop, Soil, and Environmental Sciences, University of Arkansas, Fayetteville, USA

Christopher W. Rogers
Department of Plant, Soil, and Entomological Sciences, University of Idaho, Aberdeen, USA

Edward E. Gbur
Agricultural Statistics Laboratory, University of Arkansas, Fayetteville, USA

Jarrod T. Hardke
Department of Crop, Soil, and Environmental Sciences, University of Arkansas, Stuttgart, USA

Jamal Elfaki
Department of seeds technology, Nile Valley University, Atbara, Sudan

Mohamed Gafer
College of Agricultural Studies, Sudan University of Science and Technology, Shambat, Sudan

Magboul Sulieman
Department of Soil and Environment Sciences, University of Khartoum, Shambat, Sudan

Mushtaha Ali
General Administration of Natural Resource and Sustainable Development, Khartoum, Sudan